Heinemann

IGCSE Core Mathematics

Colin Nye

Holt, Payne, Rayment, Robinson

www.pearsonIS.com

Free online support
Useful weblinks
24 hour online ordering

Part of Pearson

Heinemann is an imprint of Pearson Education Limited, a company incorporated in England and Wales, having its registered office at Edinburgh Gate, Harlow, Essex, CM20 2JE. Registered company number: 872828

www.heinemann.co.uk
Heinemann is a registered trademark of Pearson Education Limited

First published 2009

13 12 11 10 09
10 9 8 7 6 5 4 3 2 1

British Library Cataloguing in Publication Data is available from the British Library on request.

ISBN 978 0 4359 6685 0

Designed by Tony Richardson
Typeset by HL Studios
Original illustrations © Pearson Education Limited 2009
Illustrated by Adrian Barclay and Mark Raffe
Additional illustration by HL Studios
Cover design by Creative Monkey

Printed in Italy by Rotolito

Acknowledgements
The author and publisher would like to thank the following individuals and organisations for permission to reproduce photographs: Alamy Images p 176; Brand X Picture, Burke Trio Productions p 67; Comstock Images p 211; Corbis pp 8 (Mt Kilimanjaro, Table Mountain), 294, 403; Digital Stock pp 15, 196; Digital Vision pp 8 (Mt Fuji), 226, 329, 382, 402; Getty Images p 421; Guillaume Dargaud p 543; Image Source Ltd pp 8 (Mt Everest), 19, 49, 76; Image Source Ltd Alamy p 478; iStockPhoto p 333; iStockPhoto/Galina Barskaya p 293; iStockPhoto/ Matjaz Slanic p 1; iStockPhoto/Robert St Coeur p383; John Foxx. Alamy p 306; Jonathan Kirn. Illinois Bureau of Tourism p 31; Momentum Creative Group. Alamy p 193; Nimpuno p 72; Pearson Education Ltd. Ikat Design. Ann Cromack pp 45 (Glasgow), 69; Pearson Education Ltd. Devon Olugbena Shaw p 528; Pearson Education Ltd. Jules Selmes pp 205, 224, 238, 295, 339, 376, 379, 479; Pearson Education Ltd. Lord and Leverett pp 392, 480; Pearson Education Ltd. Malcolm Harris p 307; Pearson Education Ltd. Martin Sookias p 194; Pearson Education Ltd. Rob Judges pp 98, 122, 231; Pearson Education Ltd. Studio 8, Clark Wiseman p 384; Photos. com. Jupiterimages p 174; Photodisc pp 12, 380; Photodisc, Bruce W. Heinemann p 272; Photodisc. C Squared Studios pp 328, 336; Photodisc, Doug Menuez p 227; Photodisc, Getty Images p 197; Photodisc, Jules Frazier p 38; Photodisc, Karl Weatherly pp 267, 532; Photodisc, Photolink pp 3, 8 (Matterhorn), 13, 26, 192, 223, 282, 388; Photodisc, Photolink, E. Pollard pp 55; Photodisc, Photolink, T. O'Keefe p441; Photodisc, Sami Sarkis p 45 (Italy); Photodisc, Steve Cole p 198; Ryan McVay, Photodisc p 36; SNCF p 74 and Stockbyte p 315.

Every effort has been made to contact copyright holders of material reproduced in this book. Any omissions will be rectified in subsequent printings if notice is given to the publishers.

Past paper questions are reproduced by permission of the University of Cambridge Local Examinations Syndicate.

Contents

Introduction

This book has many features that will help you during the course. These features are described below.

Colour coding

The chapters have colour-coded page numbers to help you find related chapters quickly.

92 Number

124 Algebra

176 Space and shape

318 Statistics and Probability

Hint boxes

Helpful comments and hints feature throughout the book and key words are explained.

A common mistake is to think that 1 m^3 = 100 cm^3. The correct answer gives one million cm^3 in 1 m^3.

> means 'is greater than'

Example boxes

The book contains worked examples with explanatory notes throughout.

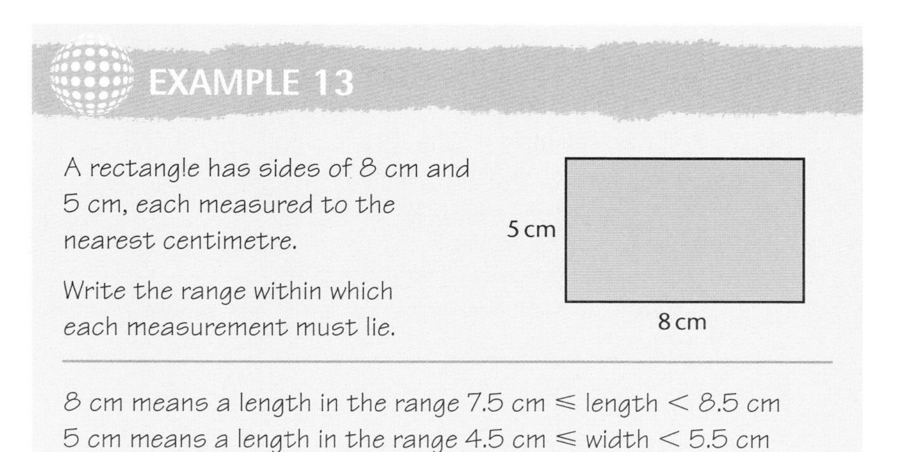

Exercises

Exercises give students plenty of practice and are structured to provide a clear progression path. Icons indicate when use of a calculator is not allowed.

The answers to these exercises are available on the Teacher CD (ISBN 978 0435 966935).

EXERCISE 10B

1 The frequency table shows the type and number of pets treated one week.

Copy and complete the tally marks and the frequency table.

Pet	Tally	Frequency
Dog	卌 ǀǀ	
Cat		9
Bird	卌 ǀ	
Other		
	Total	32

Past paper exam questions

At the end of each chapter, examination questions from past Cambridge IGCSE Mathematics papers are included so that you can test yourself using real past papers.

The answers to these questions are available on the Teacher CD (ISBN 978 0435 966935).

EXAMINATION QUESTIONS

1 Cement, sand, aggregate and water are used to make concrete, in the ratio

Cement : Sand : Aggregate : Water = 2 : 5 : 8 : 1.

(a) Bobbie wants to make 1.2m^3 of concrete.
How much aggregate will he need? [1]

(b) Eddie wants to make concrete.
He uses $0.25\ \text{m}^3$ of cement.

(i) How much sand does he need? [1]

(ii) When water and aggregate have been added, how much concrete will he have? [1]

(CIE Paper 1, Jun 2001)

Chapter 1

Basic rule of number

This chapter will show you how to

- ✔ understand place value for whole numbers and decimal numbers
- ✔ order whole numbers and decimals
- ✔ understand the four rules for positive and negative whole numbers and decimals
- ✔ use the correct order of operations
- ✔ multiply and divide by 10, 100, 1000

If you think you can answer questions on the topics in this chapter, try the revision exercise at the end of the chapter. If you need some help with any of the questions, go to the relevant section of the chapter and look at the examples.

1.1 Whole numbers

Place value for whole numbers

There were 60 479 people at a music festival.

60 479 is a number in the **denary** system, the number system you use every day.

In the denary system, the value of each digit is 10 times the value of the digit on its right-hand side.

The position of a digit in a number tells you its **place value**.

You can write numbers in a place value table.

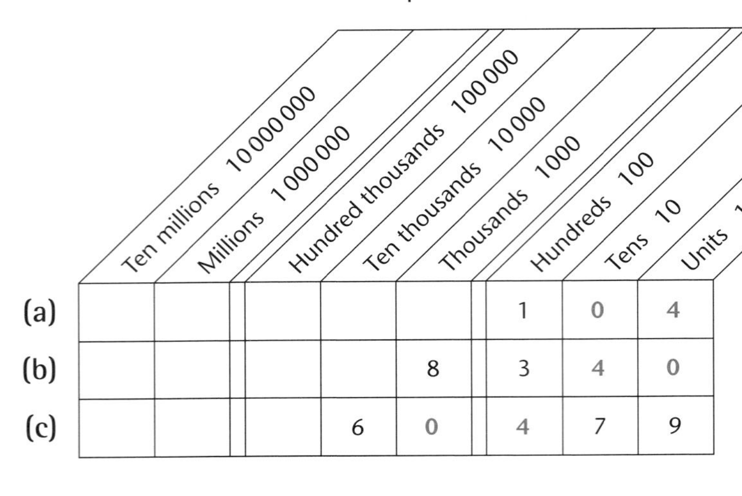

	Ten millions 10 000 000	Millions 1 000 000	Hundred thousands 100 000	Ten thousands 10 000	Thousands 1000	Hundreds 100	Tens 10	Units 1
(a)						1	0	4
(b)					8	3	4	0
(c)				6	0	4	7	9

(a) is a three-digit or three-figure number.
(b) is a four-digit or four-figure number.
(c) is a five-digit or five-figure number.

In (a) the 4 represents 4 units and has value 4.
In (b) the 4 represents 4 tens and has value 40.
In (c) the 4 represents 4 hundreds and has value 400.
Notice that 0 is a **place holder** in all of these numbers.

In **(a)** there are no 'tens', in **(b)** there are no 'units' and in **(c)** there are no 'thousands'. You need to write the zero in each of these places to show this.

EXAMPLE 1

Write down the value of the 6 in each of these numbers.

(a) 46 003 **(b)** 211 068 **(c)** 6 978 123 **(d)** 5006

	TM	M	HT	TT	Th	H	T	U
(a)				4	6	0	0	3
(b)			2	1	1	0	6	8
(c)		6	9	7	8	1	2	3
(d)					5	0	0	6

A place value table makes it easy to see the value of the 6.

(a) The 6 represents 6 thousands, value 6000.
(b) 6 tens, value 60.
(c) 6 millions, value 6 000 000.
(d) 6 units, value 6.

EXERCISE 1A

1 Write down the value of the 4 in these numbers.
   (a) 14 (b) 423
   (c) 64 128 (d) 745 000
   (e) 400 555

2 Johan's new motorbike costs $1376.
   What is the value of the 3 in this number?

3 The 2004 Olympic Games in Athens had 10 625 competitors.
   What is the value of the 1 in this number?

4 Write down the value of the red digit in these numbers.
   (a) 15632 (b) 729
   (c) 11 854 (d) 162 759
   (e) 500 403 (f) 159 077
   (g) 4259 (h) 741 963
   (i) 789 002 (j) 40 000
   (k) 29 876 (l) 103 301

Writing numbers in words and in figures

You read a number from left to right.

	TM	M	HT	TT	Th	H	T	U
(a)						1	0	4
(b)					8	3	4	0
(c)				6	0	4	7	9

(a) 104 is 1 hundred, no tens and 4 units. You say 'one hundred and four'.

(b) 8340 is 8 thousands, 3 hundreds, 4 tens and no units. You say 'eight thousand, three hundred and forty'.

(c) 60 479 is 6 ten-thousands, no thousands, 4 hundreds, 7 tens and 9 units. You say 'sixty thousand, four hundred and seventy nine'.

You read larger numbers in the same way.

TM	M	HT	TT	Th	H	T	U
	4	3	7	6	5	8	1
2	3	7	0	5	0	0	9

3 hundred thousands 7 ten thousands and 6 thousands makes 376 thousand.

2 ten millions and 3 millions makes 23 million.

For 4376 581 you say 'four million, three hundred and seventy six thousand, five hundred and eighty one'.

For 23 705 009 you say 'twenty three million, seven hundred and five thousand, and nine'.

You can use the same ideas in reverse when you are given a number in words to write in figures.

EXAMPLE 2

Write these numbers in figures.

(a) five thousand and forty two

(b) two million, six hundred and fifty thousand and three

(c) eighteen million, one hundred and ninety seven thousand, six hundred and sixty

(d) seventy thousand, nine hundred and twenty nine

	TM	M	HT	TT	Th	H	T	U
(a)					5	0	4	2
(b)		2	6	5	0	0	0	3
(c)	1	8	1	9	7	6	6	0
(d)				7	0	9	2	9

Put them into the correct columns in a place value table.

Remember the 0 as place holder.

EXERCISE 1B

1 Write these numbers in words.

(a) 1625 (b) 21 800 (c) 4004

(d) 3 000 303 (e) 101 010

2 Write these numbers in figures.

(a) one thousand, one hundred and forty five

(b) seven million, seven hundred thousand

(c) fifteen million, five hundred and five

(d) sixty thousand and six

(e) two hundred and two thousand, two hundred and two

3 Pritesh writes a cheque for $8450 for a new car. Write this number in words.

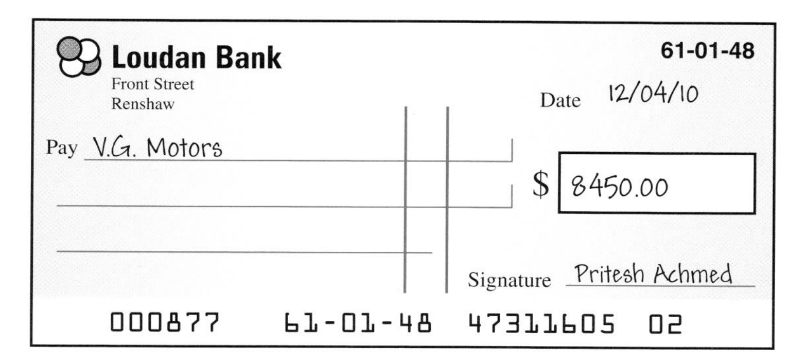

4 According to the local newspaper, there were thirty four thousand, six hundred and three people at the football match this week. Write this number in figures.

5 The average price for different types of homes in a town is shown in the table. Write each value in words.

5 bedroom house	$465 000
4 bedroom house	$280 500
3 bedroom house	$165 250
2 bedroom apartment	$79 995
1 bedroom flat	$53 750

6 A factory made these numbers of tyres in one month.

1st November	Twenty three thousand, four hundred and eight
8th November	Thirty nine thousand and sixty two
15th November	Eighty thousand, three hundred and four
22nd November	One hundred thousand and nine

Write each number in figures.

Ordering whole numbers

The size of a whole number depends on how many digits it has; the more digits, the bigger the number. For example 1234 is bigger than 567.

Sometimes you can easily write numbers in order.

For example 1, 9, 6, 8, 3, 0, 4

written in ascending order is 0, 1, 3, 4, 6, 8, 9

written in descending order is 9, 8, 6, 4, 3, 1, 0

ascending going up
descending going down

To order large numbers, it helps to write them in a place value table. For two numbers with the same number of digits, look at the digit in the highest place value column. The largest digit belongs to the biggest number.

TM	M	HT	TT	Th	H	T	U
				5	2	3	4
				1	2	3	4

5 is bigger than 1.
So 5234 is bigger than 1234.

If the digits in the highest place value column are equal, compare the digits in the next highest place value column, and so on.

TM	M	HT	TT	Th	H	T	U
			1	3	8	2	0
			1	3	1	7	5
		6	0	2	4	5	1
		6	0	2	4	2	8

8 is bigger than 1.
So 13 820 is bigger than 13 175.
5 is bigger than 2.
So 602 451 is bigger than 602 428.

EXAMPLE 3

The crowds at four English Premiership football matches were

Arsenal	44 059	Chelsea	45 904
Everton	44 095	Tottenham	45 094

Write them in order of size, largest first.

Continued ▼

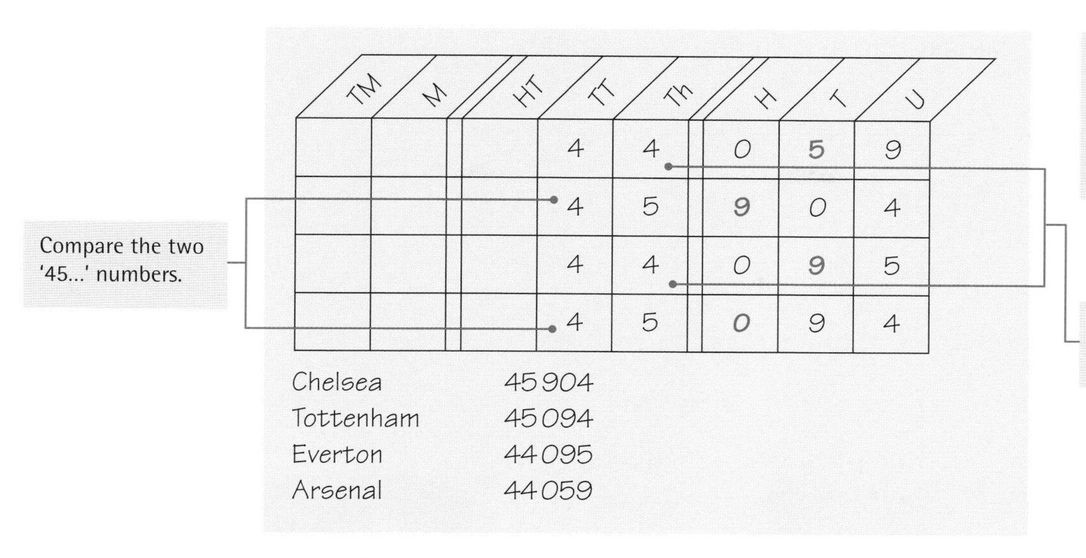

The numbers that start 45... are bigger than the numbers that start 44...

Compare the two '44...' numbers.

EXERCISE 1C

1 Which number in each pair is the biggest?

(a) 1456 1546 **(b)** 125 462 123 654

(c) 5642 5604 **(d)** 19 305 19 350

(e) 723 7203 **(f)** 1 203 000 1 202 999

2 Write these numbers in order from smallest to largest.

(a) 79 93 112 87 108 205 145

(b) 230 203 231 213

(c) 16 324 17 939 16 432 17 940

(d) 236 2345 2349 2136

(e) 1 567 324 1 634 324 563 454

3 The crowds at five concerts were

32 486 32 598 25 654 27 921 25 645

Write these in order, starting with the largest.

4 The prices of different holidays for a family of four are shown in the table. Write them in order, cheapest first.

Hotel in Tenerife	$2130
Holiday camp in France	$1325
Touring Canada by coach	$2360
Cruise in the Caribbean	$2340
Diving in Australia	$3280
Self-catering in Turkey	$1327

5 The number of grains found in different soil samples were

Sample	A	B	C	D
Number	156 300 000	98 700 000	160 983 000	159 999 999

Which sample has the largest number of grains?

6 List these five mountains in order, starting with the lowest.

Matterhorn
4478 m

Mount Everest
8848 m

Kilimanjaro
4559 m

Mount Fuji
3776 m

Table Mountain
1086 m

Using the signs < and >

When you order numbers, you can use special signs for 'less than' and 'greater than'.

8 < 11	means	'8 is less than 11'
56 < 61	means	'56 is less than 61'
40 > 35	means	'40 is greater than 35'
102 > 86	means	'102 is greater than 86'

This also means '11 is greater than 8'.

The **open** (bigger) end of the sign is next to the bigger number.

The **pointed** (smaller) end of the sign is next to the smaller number.

EXAMPLE 4

Use < or > to describe the crowds at the football matches in Example 3.

45 904 > 45 094 > 44 095 > 44 059

1 Copy the pairs of numbers, placing between them the correct sign, $<$ or $>$.

(a) 6 9 (b) 34 41

(c) 53 19 (d) 159 143

2 Write 'true' or 'false' for each statement.

(a) $7 > 3$ (b) $15 < 9$

(c) $168 > 98$ (d) $3452 < 3544$

$486 < n < 488$ means n is a number *between* 486 and 488.
n must be larger than 486 and smaller than 488.

3 Write down a value for n (any whole number) in the following. There may be more than one possible answer.

(a) $486 < n < 488$ (b) $159 > n > 157$

(c) $27\,368 < n < 27\,386$ (d) $6001 > n > 5990$

4 True or false?

(a) $4 < 7 < 11$ (b) $28 > 16 > 9$

(c) $5632 > 3321 > 4892$ (d) $62 < 195 < 950$

(e) $15\,821 > 5811 > 11\,263$

1.2 The four rules for positive whole numbers

Positive whole numbers can also be called
- positive integers
- natural numbers
- counting numbers.

Addition

You can use different words for addition.

Find the sum of 21 and 14. Work out 21 plus 14.
What is the total of 21 and 14?

Ali has \$21 and Benito has \$14. How much do they have altogether?

These all mean $21 + 14$.

- sum
- plus
- find the total of
- how much altogether?

} all mean *add*

EXAMPLE 5

Three buses are taking children on a school trip.

One bus takes 43 children, one takes 39 and the other takes 32.

How many children go on the trip altogether?

```
    4 3
+   3 9
    3 2
  -----
  1 1 4
  -----
  1 1
```

114 children go on the trip altogether.

EXERCISE 1E

1 Add these pairs of numbers.

(a) 16 and 23 (b) 75 and 86 (c) 183 and 138

2 What is 2349 plus 86?

3 Find the sum of 164, 79 and 288.

4 Work out these sums.

(a) $162 + 73 + 439 + 5$ (b) $16\,325 + 438 + 7 + 2821$

(c) $1687 + 8 + 63 + 125$ (d) $17 + 259 + 2 + 1237$

5 A shop sold 23 magazines on Monday, 103 on Tuesday, 257 on Wednesday and 87 on Thursday. How many magazines did it sell altogether?

6 A class are collecting vouchers for school equipment. These are the numbers of vouchers the students brought in.

Table A – 6, 3, 5, 9, 17
Table B – 11, 15, 9, 2, 12
Table C – 4, 19, 14, 5
Table D – 12, 23, 7, 9, 18
Table E – 26, 11, 15, 4, 8

(a) Work out the number of vouchers for each table.

(b) What is the total for the whole class?

7 In this addition pyramid, you add the numbers in two bricks together and write the answer in the brick above. The first one has been done for you. Copy and complete the pyramid.

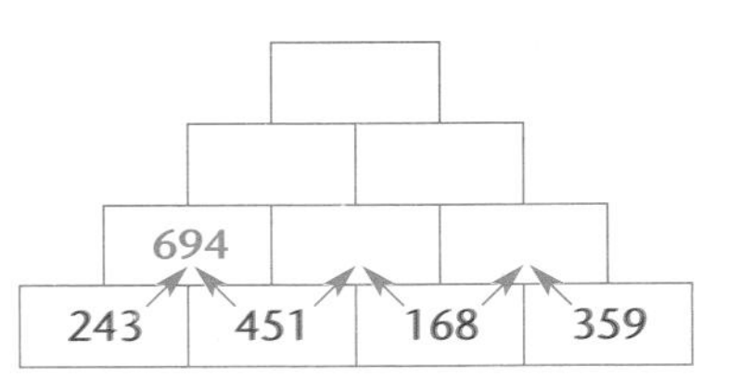

8 In a company, there are 187 people in Sales and 204 in Accounts. How many people is this altogether?

9 On a journey to Mumbai, Angus walked 2 kilometres to the bus station, travelled 27 kilometres by bus to the railway station and then 434 kilometres by train.
How far did he travel altogether?

Multiplication

You can use different words for multiplication.

Multiply 36 by 12. Work out 36 times 12.
Find the product of 36 and 12.

These all mean 36×12.

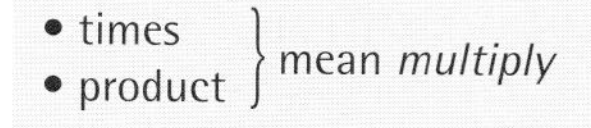

EXAMPLE 6

A shopkeeper buys 12 boxes of crisps each containing 36 packets.

How many packets of crisps are there altogether?

Method A

```
    36
×   12
    72   ← 36×2
+  360   ← 36×10
   432
   1
```

Method B

	10	2
30	300	60
6	60	12

```
  3 6 0
    7 2
  4 3 2
```

You could work out $36 + 36 + 36 + \ldots$ (12 times), but it is much quicker to multiply.

Method C

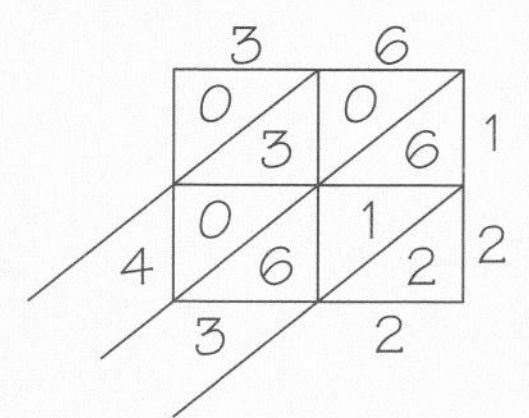

There are 432 packets of crisps.

For method C, add along the diagonals.
Start from the bottom right.

1/2 $6 \times 2 = 12$

$6 + 1 + 6 = 13$, so carry 1.

EXERCISE 1F

1 What is 7 times 94?

2 Multiply 231 by 15.

3 Copy and complete the multiplication square on the right.

×	8	14	28	123
5				
9				
11				

4 Find the product of 24 and 352.

5 Work these out.

(a) 24×13 (b) 62×38

(c) 167×41 (d) 327×54

6 A railway carriage holds 48 passengers. A train has 9 full carriages. How many passengers are there altogether?

7 A box contains 180 tins of beans. How many tins are there in 17 boxes?

8 Find the product of 6, 5 and 11.

Multiply 6 and 5 first.

9 The space shuttle burns 3785 litres of fuel every second. How much fuel is burnt in the first minute of its launch?

10 It costs \$540 per person for a holiday in Florida and \$107 per person for a 3-day pass into the theme parks.

(a) How much would a holiday in Florida cost for 7 people?

(b) How much would it cost 7 people to go to the theme parks?

(c) What is the total cost for the holiday and the 3-day pass for 7 people?

Subtraction

You can use different words for subtraction.

Work out 46 minus 32. Find the difference between 46 and 32. Take 32 from 46.

How much more than 32 is 46? How much less than 46 is 32? 46 take away 32.

These all mean $46 - 32$.

- minus
- difference
- take away
- take ... from ...
- how much more than?
- how much less than?

all mean *subtract*

EXAMPLE 7

A train travels from London to Edinburgh via Newcastle.

London to Edinburgh is 658 kilometres.
London to Newcastle is 456 kilometres.

How far is it from Newcastle to Edinburgh?

Continued ▼

```
  658
- 456
-----
  202
-----
```

It is 202 kilometres from Newcastle to Edinburgh.

You need to think of this as how much more than 456 is 658?

To check your answer of 202, work out 202 + 456. You should get 658.

```
  456
+ 202
-----
  658
```

EXERCISE 1G

1 What is 79 minus 24?

2 What is the difference between 83 and 27?

3 What is left if I take 207 from 532?

4 Work these out.

(a) 783 − 126

(b) 1825 − 532

(c) 402 − 137

(d) 3000 − 1562

5 John has 103 CDs and Ali has 59. How many more CDs does John have than Ali?

6 A sports car costs $23 500 and an off-road car costs $18 700. What is the difference in price between the two cars?

7 To get a free DVD Susie needs to collect 500 vouchers. She has collected 392 so far. How many more does she need?

8 Work out 162 minus 43 minus 29.

Work out 162 minus 43 first.

9 To qualify for the school sports team you need 250 points. Lloyd has 187, Fatima has 208 and Laura has 196. How many more points does each one need to reach 250?

10 It is 3423 kilometres to my holiday destination. I have travelled 987 kilometres so far. How many more kilometres do I have to travel?

Division

You can use different words and phrases for division.

How many times does 7 go into 28? Share 28 by 7.

Work out $\frac{28}{7}$. How many sevens are there in 28?

These all mean $28 \div 7$.

share / how many times does ... go into ...? / how many ... are there in ...? } all mean *divide*

EXAMPLE 8

Eight friends share the sum of \$5136 equally between them.

How much do they each receive?

$$\begin{array}{r} 8\overline{)\,5\,1^{3}3\,6} \\ 6\,4\,2 \end{array}$$

8 goes into 51 6 times, remainder 3.

They each receive \$642.

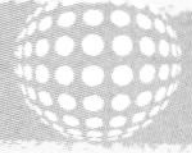

EXAMPLE 9

Work out $3358 \div 23$.

This method is called 'long division'.

146	
23)3358	
23	⟵ 23 goes into 33 once 23×1
105	⟵ Remainder 10, bring down 5
92	⟵ 23 goes into 105 four times 23×4
138	⟵ Remainder 13, bring down 8
138	⟵ 23×6
0	⟵ No remainder

The answer is 146.

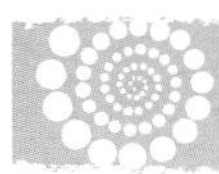

EXERCISE 1H

1 Work these out.

(a) $18 \div 3$ (b) $256 \div 4$ (c) $665 \div 5$

2 Share 56 sweets equally between 4 people.

3 Work these out.

(a) $105 \div 15$ (b) $168 \div 21$ (c) $154 \div 11$

4 Eight friends earned $40 planting seeds. How much will each receive?

5 A box contains 12 eggs. How many boxes can I fill with 156 eggs?

6 Share 1860 bolts equally between 5 containers. How many are in each container?

7 A pack of writing paper contains equal numbers of sheets of 6 colours. There are 186 sheets. How many sheets are there of each colour?

8 In a supermarket 938 tins of fruit are stacked on 7 identical shelves. How many tins are on each shelf?

9 A parrot has 161 nuts to last her 23 days. If she eats the same number each day, how many nuts can she have each day?

10 230 people are going on a trip. A bus holds 41 people. How many buses do they need?

Mixed problems

To solve a problem, you need to decide whether to add, subtract, multiply or divide to find the answer. In some of the questions you might need more than one **operation**.

EXAMPLE 10

A holiday brochure gives the cost of a holiday in Majorca.

	Number of days	
	7	14
Adult	$227	$425
Child aged 3–16	$148	$263
Child under 3	Free	Free

Mr and Mrs Lee and their four children, Kenny (11 yrs), Alice (8 yrs), Liew (4 yrs) and Salma (2 yrs), decide to go for two weeks.

How much will it cost them?

Continued ▼

add + subtract −
multiply × divide ÷
are *operations*

Cost for Mr & Mrs Lee
$= \$425 \times 2 \quad = \$850$

Cost for Kenny, Alice and Liew
$= \$263 \times 3 \quad = \$789$

Cost for Salma $= \$0$ (children under 3 go free)

Total cost $= \$1639$

14 days = 2 weeks.

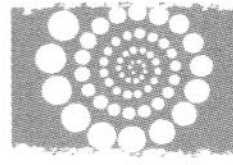

EXERCISE 1I

1 A shop selling newspapers has 347 copies of the *Daily News*, 659 copies of the *Free Print Paper* and 68 copies of *Life and Times*. How many newspapers is this altogether?

2 Johann has 103 CDs in his collection and Tomasina has 87. How many more CDs does Johann have than Tomasina?

3 How many weeks are there in 91 days?

4 I get $50 for my birthday. I buy a jumper for $17 and some shoes for $24.
How much do I have left?

5 Ester shares $168 equally between her three nieces. How much will each get?

6 How many hours are there in 4 days?

7 580 pies are to be packed into packets of 12.
(a) How many full packets will I get?
(b) How many pies will be left over?

8 At a stable there are 15 horses. The owner is inspecting their ears, hooves and tails.
(a) How many ears does the owner inspect?
(b) How many hooves does the owner inspect?
(c) How many hooves, ears and tails does the owner inspect altogether?

9 A box of raisins holds 12 packets and a box of sultanas holds 20 packets. If I have 6 boxes of raisins and 5 boxes of sultanas, how many packets of fruit do I have altogether?

10 'Easy Seat' has sold 1160 chairs in 5 weeks. These are the numbers sold each week.

Week	Week 1	Week 2	Week 3	Week 4	Week 5
Chairs sold	204	198	211	187	

How many chairs did they sell in Week 5?

11 A company sells chocolate eggs in packs of 3 and packs of 6. They produce 531 eggs one day. They decide to make 60 packs containing 6 eggs.

(a) How many eggs will this use?

(b) How many packs containing 3 eggs can they make with the eggs left over?

12 At a sale Li had 84 books, 129 CDs and 75 DVDs to sell. He sold 68 books, 98 CDs and 43 DVDs.

(a) How many items altogether did he have to take home?

(b) He sold books for 20 cents, CDs for 75 cents and DVDs for 50 cents. How much money did he make?

1.3 Order of operations

Brackets — You must work out the value of any brackets **first**.

Index or **Indices** — You must work out any indices **second**.

Divide
Multiply — You must work out any divide and multiply calculations **third**.

Add
Subtract — You must work out any add and subtract calculations **last**.

Indices are 'powers' and you will meet these later. For now we will use the index ^{2} for 'squared'. $5^2 = 5 \times 5 = 25$.

Divide and Multiply can be done in either order. You usually work from left to right.

Add and Subtract can be done in either order. You usually work from left to right.

EXAMPLE 11

Work out these sums.

(a) (i) $2 \times 8 - 4 \times 3$ (ii) $2 \times (8 - 4) \times 3$

(b) (i) $18 + 4 \div 2 - 11$ (ii) $(18 + 4) \div 2 - 11$

(c) (i) $2 + 3^2 - 1 \times 5$ (ii) $(2 + 3)^2 - 1 \times 5$

(a) (i) $2 \times 8 - 4 \times 3 = 16 - 12 = 4$
(Mult, Mult, Sub)

(ii) $2 \times (8 - 4) \times 3 = 2 \times 4 \times 3 = 24$
(Br, Mult, Mult)

(b) (i) $18 + 4 \div 2 - 11 = 18 + 2 - 11 = 9$
(Div, Add, Sub)

(ii) $(18 + 4) \div 2 - 11 = 22 \div 2 - 11 = 11 - 11 = 0$
(Br, Div, Sub)

(c) (i) $2 + 3^2 - 1 \times 5 = 2 + 9 - 1 \times 5 = 2 + 9 - 5 = 6$
(Ind, Mult, Add, Sub)

(ii) $(2 + 3)^2 - 1 \times 5 = 5^2 - 1 \times 5 = 25 - 1 \times 5 = 25 - 5 = 20$
(Br, Ind, Mult, Sub)

After each line of calculation make a note of the operation you used. This helps you to keep the operations in the correct order.

Use short forms of the operations e.g. 'mult' for multiplication.

EXERCISE 1J

1 Work these out.

(a) $3 + 2 \times 7$ (b) $5 \times 8 - 3$ (c) $9 + 4 \times 5 - 2$

(d) $9 \times 5 + 3 \times 6$ (e) $4 - 3 + 7 \times 2$ (f) $3 + 3 \times 5$

2 Find the value of

(a) $4 \times (3 + 4)$ (b) $16 - (2 \times 4)$

(c) $5(13 - 9)$ (d) $(2 + 8) \times (7 - 4)$

(e) $7 + (6 - 4) - 3$ (f) $(13 + 5) \div 3 + 2$

$5(13 - 9)$ is another way of writing $5 \times (13 - 9)$.

3 Work these out.

(a) $6 \times 3 + 5^2$ (b) $(4 + 5) \times 2^2$

(c) $4 + (7 - 3) \times 3$ (d) $110 - (3 + 2)^2$

(e) $3 + 6^2 \div 2$ (f) $4(15 - 13) \div 2^2$

Remember
$2^2 = 2 \times 2$
$5^2 = 5 \times 5$ etc.

4 Copy these and put in the correct signs to make the statements true.

(a) 3 ☐ 2 ☐ 4 = 10
(b) 3 ☐ 2 ☐ 4 = 18
(c) 5 ☐ 1 ☐ 7 ☐ 3 = 60
(d) 10 ☐ 6 ☐ 2 ☐ 3 = 43
(e) 2 ☐ 7 ☐ 6 ☐ 2 = 11
(f) 9 ☐ 2 ☐ 3 ☐ 20 = 25

5 Work out each calculation and match the answer to a letter. Now rearrange your letters to make a word. The starting letter has been done for you. Which letter has not been used?

Answer	7	8	1	4	14	9	5	6
Letter	A	C	I	F	N	S	T	E

(a) $25 - 3 \times (4 + 3) = 25 - 3 \times 7 = 25 - 21 = 4 \quad 4 \rightarrow F$
(b) $7 \times 8 \div (19 - 15)$
(c) $19 - 4 \times 3$
(d) $(18 - 6) \div (24 - 12)$
(e) $81 - (8 \times 6) - (4 \times 7)$
(f) $(8 \times 7 - 36) \div 2^2$
(g) $(15 \times 4) - (17 \times 3)$
(h) $6 \times 4 \div 3 - 1$
(i) $5 \times 4 - 6 \times 2$

1.4 The four rules for positive and negative numbers

Negative numbers

The temperature in Moscow yesterday was −25 °C.

This means that the temperature was 25° **below** 0 °C.

The Himalayas (the mountain range which includes Mount Everest) rise almost 9 km above sea level but have their base about 60 km **below** sea level. This can be written as −60 km.

Numbers can be shown on a number line.

The number line can be horizontal or vertical.

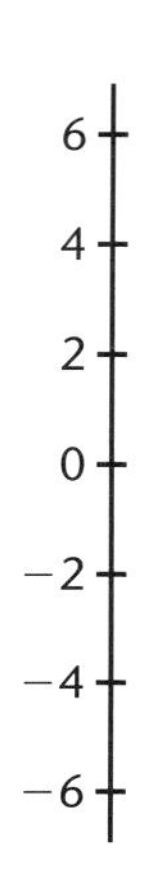

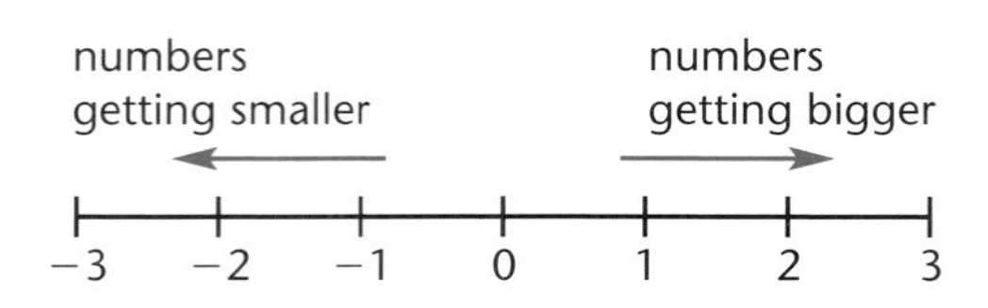

EXAMPLE 12

Use the thermometer to answer the questions.

(a) The temperature in Moscow at 0600 is −2 °C and at noon it is 7 °C.
What is the rise in temperature between these times?

(b) The temperature in my garden this morning was −4 °C.
By the afternoon it had risen by 10 °C.
By the evening it had fallen to 7 °C lower than it was in the afternoon.

(i) What was the temperature in the afternoon?

(ii) What was the temperature in the evening?

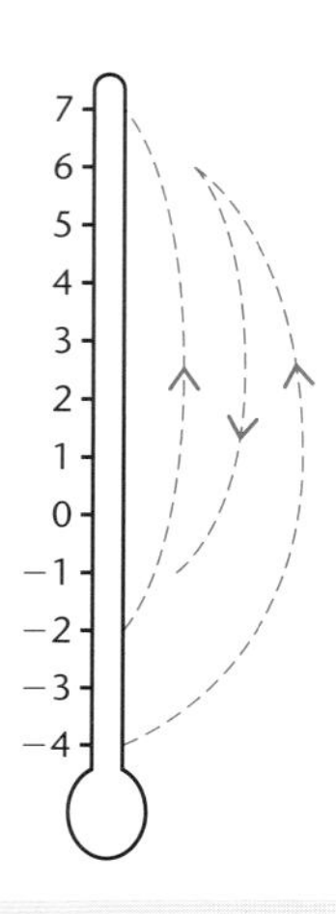

Remember to count 0° as one of the steps because 0 is a place holder.

(a) The rise in temperature is 9 °C.

(b) **(i)** The temperature in the afternoon was 6 °C.

(ii) The temperature in the evening was −1 °C.

Start at −2 °C and count the steps up to 7 °C.

Start at −4 °C and count up 10 steps.

From 6 °C, count down 7 steps

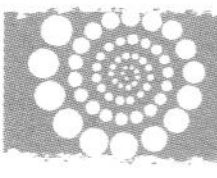

EXERCISE 1K

1 In Helsinki, the temperature at midnight is −6 °C. At noon the next day the temperature is 4 °C.
What is the rise in temperature between these times?

2 On a spring day the temperature at 9 am was −1 °C. By 2 pm the temperature had risen by 12 °C. By 9 pm the temperature had dropped 14 °C from what it was at 2 pm.

(a) What was the temperature at 2 pm?

(b) What was the temperature at 9 pm?

3 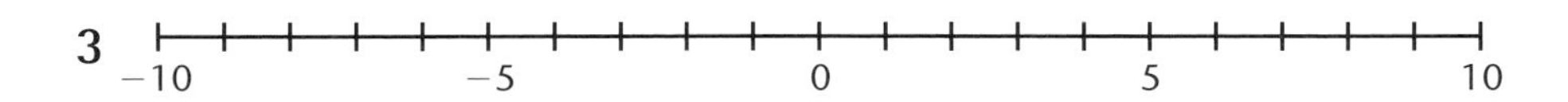

(a) Copy the number line and mark on it

A = 0 B = 7 C = −6 D = 1 E = −2

(b) Write the numbers in part (a) in order from largest to smallest.

Use the number line to help.

4 The temperatures in five cities one December day were

London −2 °C Rome 3 °C New York −5 °C
Moscow −23 °C Madrid −14 °C

Write these temperatures in order, starting with the coldest.

5 (a) Use the scale to find the height of each item (bird, flag, man's head, shark, diver, fish, starfish) compared to sea level.

(b) At what height will the bird be if it dives 20 m?

(c) The diver swims to the bottom of the sea. How far has he dropped?

(d) The fish swims up 15 m and then down 10 m. How far below the surface is the fish now?

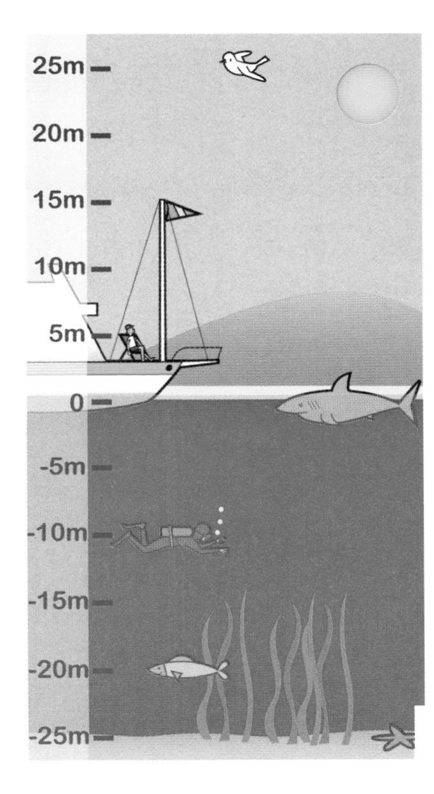

6 The freezer units in a laboratory are kept at different temperatures. They read

−8 °C −2 °C −8 °C −6 °C −4 °C −22 °C

−25 °C −18 °C

Write these temperatures in order from coldest to hottest.

Adding and subtracting positive and negative numbers

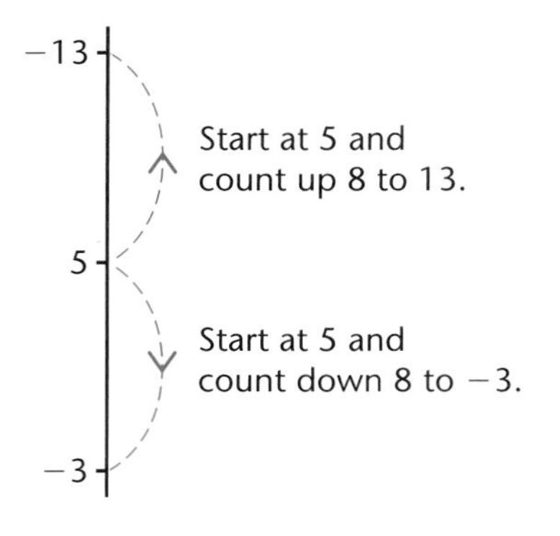

You can add and subtract using a number line.

To add a positive number, count up the number line. $5 + 8 = 13$

You could write $5 + (+8) = 1$

$5 - 8 = -3$ can also be written as
$5 - (+8) = -3$

To subtract a positive number, count down the number line.

To add a negative number, count down.

Putting ice in a drink makes it colder.

So $5 + (-8) = -3$ is the same as $5 - 8 = -3$.

To subtract a negative number, count up.

Taking ice out of a drink makes it warmer.

So $5 - (-8) = 13$ is the same as $5 + 8 = 13$.

In this example, 5 is the starting value.

You can use any number for the starting value, the rules are the same.

For example

$$-4 + (+6) = -4 + 6 = 2$$
$$-4 - (-6) = -4 + 6 = 2$$
$$-4 - (+6) = -4 - 6 = -10$$
$$-4 + (-6) = -4 - 6 = -10$$

The 'rules of signs' are

+ + is the same as +
− − is the same as +
− + is the same as −
+ − is the same as −

EXAMPLE 13

Work out these sums.

(a) $2 + (-5)$ (b) $-2 + (+8)$
(c) $-3 - (+10)$ (d) $-11 - (-7)$

(a) $2 + (-5) = 12 - 5 = 7$ (b) $-2 + (+8) = -2 + 8 = 6$
(c) $-3 - (+10) = -3 - 10 = -13$ (d) $-11 - (-7) = -11 + 7 = -4$

EXERCISE 1L

1 Work these out.

(a) $+7 + (-2)$ (b) $-5 + (+3)$

(c) $-8 + (-3)$ (d) $3 + (-6) + (-8)$

(e) $-5 + (+9) + (-8)$ (f) $-4 + (-3)$

2 Complete these subtractions.

(a) $+8 - (+5)$ (b) $-7 - (+4)$

(c) $-9 - (-3)$ (d) $-6 - (+12)$

(e) $-2 - (+3) - (-4)$ (f) $3 - (-2)$

(g) $4 - 5 + 2$ (h) $10 - 3 + 4$

3 The recorded evening temperatures in 5 cities were $-3\,°C$, $1\,°C$, $-4\,°C$, $2\,°C$ and $-1\,°C$.

(a) Write these in order from coldest to hottest.

(b) What was the difference in temperature between the coldest and the warmest city?

4 (a) The temperature at noon was 8 °C. By midnight it was −3 °C. By how many degrees had the temperature fallen?

(b) In July the temperature in Auckland was 15 °C. In Cairo it was 11° higher. What was the temperature in Cairo?

5 Copy and complete these addition and subtraction tables.

+	4	1	−2	−3
5	9			
2				
−1				
−3				

$5 + 4 = 9$

−	3	1	−2	−3
4	1			
2				
−3				
−5				

$4 - 3 = 1$

6 Copy these calculations and write in the missing numbers.

(a) $-4 + \square = 3$ (b) $8 - \square = 10$

(c) $7 + \square = -2$ (d) $\square - (-3) = 1$

Multiplying and dividing positive and negative numbers

There are similar 'rules of signs' for multiplying and dividing positive and negative numbers.

Muliplying

+	×	+	=	+
+	×	−	=	−
−	×	+	=	−
−	×	−	=	+

Dividing

+	÷	+	=	+
+	÷	−	=	−
−	÷	+	=	−
−	÷	−	=	+

All these rules come from following patterns in multiplication tables.

$2 \times 2 = 4$ (−2)
$1 \times 2 = 2$ (−2)
$0 \times 2 = 0$ (−2)
$-1 \times 2 = -2$
$-2 \times 2 = -4 \rightarrow 2 \times -2 = -4$ (+2)
$1 \times -2 = -2$ (+2)
$0 \times -2 = 0$ (+2)
$-1 \times -2 = 2$
$-2 \times -2 = 4 \rightarrow 4 \div -2 = -2$ etc.
$-3 \times -2 = 6 \rightarrow 6 \div -3 = -2$

EXAMPLE 14

Work these out.

(a) $(-8) \times (+4)$ (b) $(-5) \times (-3)$
(c) $(+20) \div (-2)$ (d) $(+36) \div (+9)$

Work out the sign of the answer first using the 'rules of signs' tables. Then work out the number.

(a) $(-8) \times (+4) = -32$ (b) $(-5) \times (-3) = 15$
(c) $(+20) \div (-2) = -10$ (d) $(+36) \div (+9) = 4$

When the answer is positive you need not write the + sign.

EXERCISE 1M

1 Complete these multiplications.

(a) $(-7) \times (+3)$ (b) $(-5) \times (-4)$
(c) $4 \times (-3)$ (d) $(-2) \times 5$
(e) $(-3) \times (-3)$ (f) $(-4) \times (-2) \times (-3)$

A number without a sign is always positive.

2 Work out these divisions.

(a) $(+15) \div (-3)$ (b) $(-27) \div (-3)$

(c) $(+16) \div (+4)$ (d) $\frac{-16}{-4}$

(e) $\frac{-15}{5}$ (f) $(-36) \div (-6)$

$\frac{-16}{-4}$ means $-16 \div -4$

3 Copy these calculations and write in the missing numbers.

(a) $(-9) \div \square = +3$ (b) $21 \div \square = -3$

(c) $(-40) \div \square = -10$ (d) $\square \div (-6) = -6$

(e) $\square \div 2 = 17$ (f) $\frac{-120}{\square} = 10$

4 Find the missing value for each calculation.

(a) $3 \times \square = -21$ (b) $\square \times (-6) = 24$

(c) $(-3) \times \square = -9$ (d) $6 \times \square = -48$

(e) $(-5) \times \square = 25$ (f) $3 \times \square = 120$

5 In this multiplication pyramid you multiply the numbers in two bricks and write the answer in the brick above. The first one has been done for you.

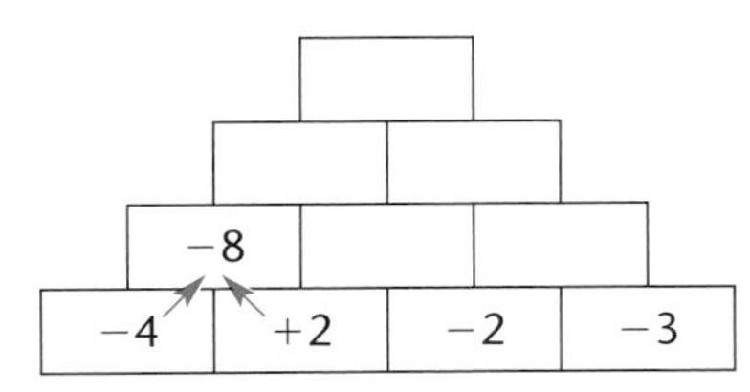

Copy and complete the multiplication pyramid.

6 In these division pyramids you divide the number on the left by the number on its right and write the answer in the brick above.

Copy and complete the division pyramids.

(a)

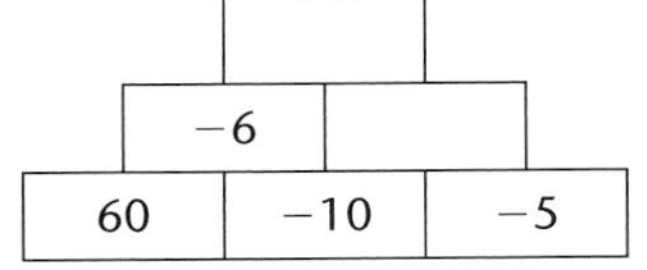

(b)

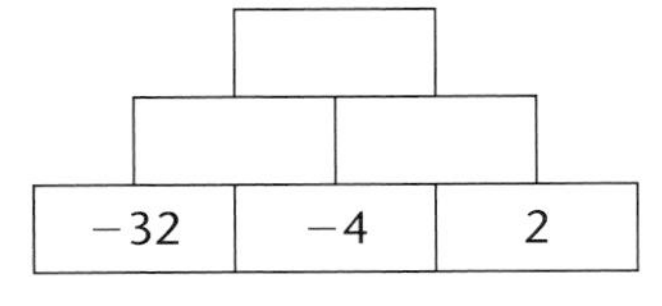

EXERCISE 1N

1 Work out these sums.

(a) $2 - 4$ (b) $-1 - 3$
(c) $-8 + 3$ (d) $4 + (-2)$
(e) $(-10) + 12$ (f) $31 - 45$
(g) $(-2) + (-4)$ (h) $2 + 5$
(i) $5 \times (-4)$ (j) $(-16) \div 4$
(k) $(-3) \times 7$ (l) $18 \div (-3)$
(m) $(-5) \times (-6)$ (n) $0 \div (-9)$
(o) $(-24) \div (-8)$ (p) $110 \div 10$

2 The temperature at 9 p.m. was 5 °C. Find the temperature if it dropped by

(a) 1 °C (b) 5 °C (c) 9 °C
(d) 12 °C (e) 20 °C.

3 The temperature in London is 4 °C. The temperature in Moscow is −8 °C. What is the difference between these temperatures?

4 Copy and complete these calculations.

(a) $-6 + \square = -9$ (b) $5 + \square = -1$
(c) $4 - \square = 1$ (d) $8 - \square = 10$
(e) $5 \times \square = -30$ (f) $\square \times -3 = -12$
(g) $\square \div (-4) = -3$ (h) $\square - (-8) = 6$

5 The depth of the sea can be written as a negative number. For example, 4 m below sea level is −4 m.
A cornetfish swims at −4 m. A bream swims at seven times this depth. How deep is this?

6 One morning the temperature went up from −4 °C to 7 °C. By how many degrees did the temperature rise?

7 The temperature on a rock was 15 °C. At night the temperature fell by 18 °C. What was the temperature on the rock at night?

1.5 Decimals

Place value for decimal numbers

In a place value table the **decimal point** is to the right of the 'units' column.

The columns to the right of the decimal point represent parts of a whole number.

M	HT	TT	Th	H	T	U	.	tenths $\frac{1}{10}$	hundredths $\frac{1}{100}$	thousandths $\frac{1}{1000}$
							.			

The value of each digit is 10 times the value of the digit on its right-hand side.

	HT	TT	Th	H	T	U	.	t	h	th
(a)						2	.	3	6	4
(b)				1	5	7	.	0	3	
(c)					1	8	.	9	0	3

In **(a)** the 3 represents 3 **tenths** and has value $\frac{3}{10}$

the 6 represents 6 **hundredths** and has value $\frac{6}{100}$

the 4 represents 4 **thousandths** and has value $\frac{4}{1000}$

In **(b)** the 3 represents 3 hundredths and has value $\frac{3}{100}$

In **(c)** the 3 represents 3 thousandths and has value $\frac{3}{1000}$

0 is a place holder in **(b)** and **(c)**, there are no 'tenths' in **(b)** and no 'hundredths' in **(c)**.

You put a zero in each of these places to show this.

(a) 2.364 you say 'two point three six four'

(b) 157.03 you say 'one hundred and fifty seven point zero three'

(c) 18.903 you say 'eighteen point nine zero three'

2.364 has 3 **decimal places**

EXAMPLE 15

Write down the value of the 9 in each of these numbers.

(a) 34.097 (b) 108.94 (c) 92.506 (d) 0.129

Write the numbers in a place value table.

	HT	TT	Th	H	T	U	.	t	h	th
(a)					3	4	.	0	9	7
(b)				1	0	8	.	9	4	
(c)					9	2	.	5	0	6
(d)						0	.	1	2	9

You do not need zero here.

(a) 9 hundredths, value $\frac{9}{100}$ (b) 9 tenths, value $\frac{9}{10}$

(c) 9 tens, value 90 (d) 9 thousandths, value $\frac{9}{1000}$

EXERCISE 10

1 What is the value of the 5 in each of these numbers?

(a) 0.005 (b) 0.05 (c) 0.5

2 Write in words.

(a) 0.8 (b) 0.407 (c) 1.256

3 Write these decimal numbers in figures.

(a) four tenths (b) one point four

(c) two thousandths (d) six point one zero three

Write them in a place value table.

4 Write down the value of the 4 in these numbers.

(a) 132.4 (b) 5.24 (c) 7.426

(d) 0.154 (e) 4.23

5 Write these as a decimal number in figures.

(a) $\frac{1}{10}$ (b) $\frac{3}{100}$ (c) $\frac{4}{10} + \frac{3}{100}$

(d) $\frac{27}{100}$ (e) $\frac{3}{10} + \frac{4}{1000}$

6 Statistics show that the families in a country have an average of 2.4 children.
Write this number in words.

7 The makers of a timing device claim that it is accurate to within one thousandth of a second. Write this number as a decimal.

Multiplication and division by 10, 100, 1000 ...

Look at these numbers written in a place value table.

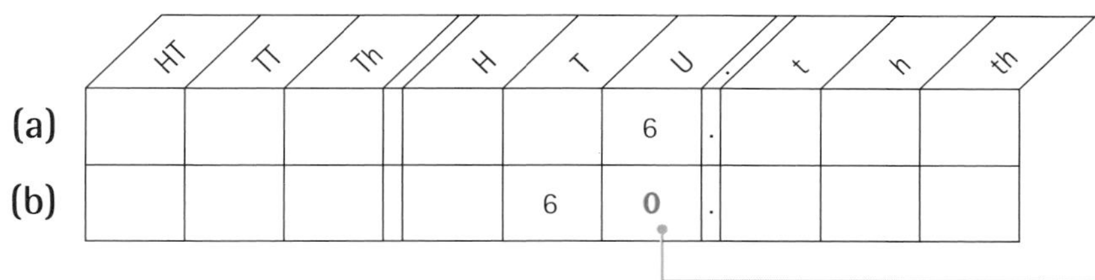

You need 0 as a place holder here.

In **(a)** there is a 6 in the units column so **(a)** has value 6.

In **(b)** the 6 has moved one place to the left and now represents 6 tens, so **(b)** has value 60.

Moving digits one place to the left is the same as multiplying by 10.

This idea can be extended.

	HT	TT	Th	H	T	U	.	t	h	th	
(a)					1	9	.	0	7		
(a) × 10				1	9	0	.	7			19.07 × 10 = 190.7
(a) × 100			1	9	0	7	.				19.07 × 100 = 1907
(a) × 1000		1	9	0	7	**0**	.				19.07 × 1000 = 19 070

You need 0 as a place holder here.

Moving digits one place to the left is the same as *multiplying by 10.*

Moving digits two places to the left is the same as *multiplying by 100.*

Moving digits three places to the left is the same as *multiplying by 1000.*

If you move digits to the right, you divide.

Right is the 'opposite' of left. Divide is the 'opposite' of multiply.

Moving digits one place to the right is the same as *dividing by 10.*

Moving digits two places to the right is the same as *dividing by 100.*

Moving digits three places to the right is the same as *dividing by 1000.*

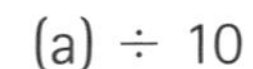

	HT	TT	Th	H	T	U	.	t	h	th	
(a)				4	0	8	.				
(a) ÷ 10					4	0	.	8			408 ÷ 10 = 40.8
(a) ÷ 100						4	.	0	8		408 ÷ 100 = 4.08
(a) ÷ 1000						**0**	.	4	0	8	408 ÷ 1000 = 0.408

You need 0 as a place holder here to show there are no units.

Place holders are needed to fill spaces at the *right* of whole numbers, or at the *left* of decimal numbers.

EXAMPLE 16

Work out **(a)** 3.4 × 100 **(b)** 763 ÷ 10

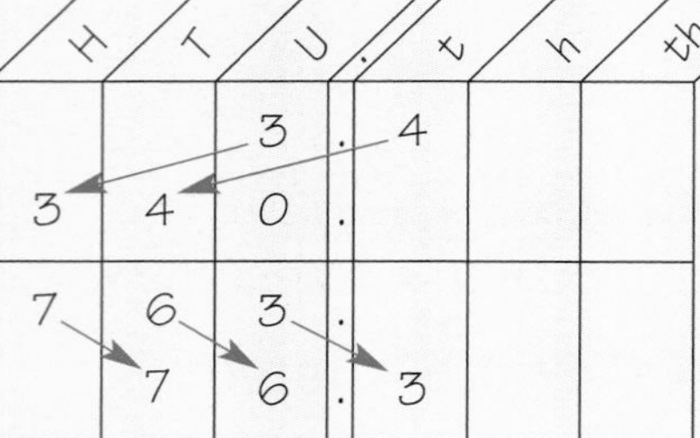

3.4 × 100 = 340

763 ÷ 10 = 76.3

EXERCISE 1P

1 Work out the following. Use a place value table to help you.

(a) 5.2 × 10 **(b)** 5.2 × 100 **(c)** 5.2 × 1000
(d) 0.48 × 10 **(e)** 64 × 100 **(f)** 3.85 × 10
(g) 0.3 × 1000 **(h)** 0.006 × 100 **(i)** 0.066 × 100

2 The cost of one shirt is $9.99. What is the cost of

(a) 10 shirts **(b)** 100 shirts?

3 Work out the following. Use a place value table to help you.

(a) 6.4 ÷ 10 **(b)** 6.4 ÷ 100 **(c)** 9 ÷ 10
(d) 8.5 ÷ 100 **(e)** 13.2 ÷ 10 **(f)** 0.2 ÷ 100
(g) 14 ÷ 1000 **(h)** 0.04 ÷ 10 **(i)** 215 ÷ 100

4 The mass of 10 identical robots is 53.4 kg.
What is the mass of 1 robot?

5 Find the value of

(a) 3.6×100 (b) $4.9 \div 10$ (c) 7.32×10
(d) $12.8 \div 100$ (e) 0.005×100 (f) $7 \div 10$
(g) $15.3 \div 10$ (h) 6.02×100 (i) $5 \div 100$

6 A paperclip has a mass of 0.8 grams.
What is the mass of 1000 paperclips?

7 100 children each pay \$6.75 to visit an animal park. How much is this altogether?

8 Ten tickets for the cinema cost a total of \$23.50.
How much will 1 ticket cost?

9 (a) 170 000 apples are put into bags of 10.
How many bags are there?
(b) 100 of these bags are put into a box.
How many boxes are there?
(c) A van carries 10 boxes each journey. How many journeys does the van need to deliver all the boxes?

10 True or false? If the statement is false, say why.
(a) 50 is ten times as big as 5.
(b) 4000 is ten times bigger than 40.
(c) 70 000 is one hundred times bigger than 700.
(d) 0.9 is the same as $9 \div 10$.
(e) 60 is equal to $6000 \div 10$.

Ordering decimals

Use the same method for ordering decimal numbers as for whole numbers.

Put the numbers in a place value table.

Examine the numbers from the left. The one with the largest digit in the highest place value column is the biggest.

If two or more numbers have equal digits in the highest place value column, compare the digits in the next highest place value column, and so on.

EXAMPLE 17

Write these in order, starting with the largest.

6.504   6.800   7.290   6.510

	H	T	U	.	t	h	th
(a)			6	.	5	0	4
(b)			6	.	8	0	0
(c)			7	.	2	9	0
(d)			6	.	5	1	0

7.290, 6.8, 6.51, 6.504

(c) has the greatest units digit, so it must be the largest.

(b) has a larger 'tenths' digit than either (a) or (d), so (b) is second biggest.

(d) has a larger 'hundredths' digit than (a), so (d) is the third biggest.

You can write $7.29 > 6.8 > 6.51 > 6.504$ using the signs for 'greater than'.

See Section 1.1.

In **(b)** and **(c)** there are some zeros as place holders (in red). These have been included so that you can clearly see the value of the digits in the decimal places.

A common mistake is to look at the decimal parts of the three numbers 6.504, 6.8 and 6.51 and think that because 504 takes up more space than 51 and they are both longer than 8, then the order must be $6.504 > 6.51 > 6.8$. This takes no notice of the place value of the digits and gives the *wrong* order for these three numbers.

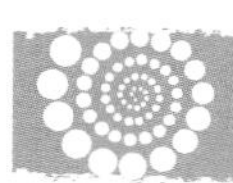

EXERCISE 1Q

1 Which number in each pair is the smaller?

(a) 0.5 and 0.05   **(b)** 1.23 and 1.3

(c) 6.509 and 6.5   **(d)** 4.023 and 4.032

2 Write these numbers in order from smallest to largest.

(a) 0.4   0.6   0.41

(b) 2.36   2.345   2.349

(c) 0.23   0.203   0.231   0.2

(d) 9.503   9.305   9.5   9.53

3 Copy the number line and place the following numbers on it.

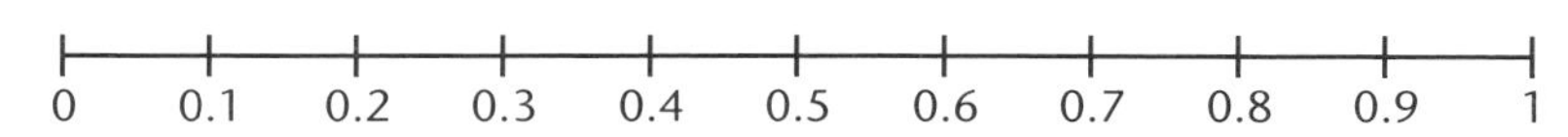

0.75 0.25 0.05 0.65 0.95 0.85

4 Copy and complete each statement using the $<$ or $>$ sign.

(a) 6.4 ☐ 6.04 (b) 0.305 ☐ 0.3

(c) 7.23 ☐ 7.203 (d) 11.206 ☐ 11.026

5 Write down a value for n in the following. (There is more than one possible answer.)

$6 > n > 5$ means a number between 6 and 5. n is smaller than 6 and more than 5.

(a) $6 > n > 5$ (b) $6.1 < n < 6.2$

(c) $115.2 > n > 115$ (d) $3.13 < n < 3.14$

6 The times in seconds for a heat of the 100 m race were

The quickest is the one with the lowest time.

10.32 10.302 10.4 10.325 10.3

Write these in order from the quickest to the slowest.

7 The table shows the average height in centimetres of some flowers. Write them in order from the shortest to the tallest.

Flower	Height in cm
Bluebell	23.4
Primula	33.04
Daffodil	33.4
Tulip	33.35
Iris	33.38

8 What value can n have in the following? Give all possible values.

$\leq$ is less than or equal to
$\geq$ is greater than or equal to

(a) $25.3 \leqslant n \leqslant 25.5$ if n has 1 decimal place

(b) $3.69 \geqslant n \geqslant 3.67$ if n has 2 decimal places

(c) $4.5 \leqslant n \leqslant 4.6$ if n has 2 decimal places

1 decimal place means 1 digit after the decimal point.

Adding decimals

To add decimal numbers, line up the numbers so that digits of equal place value are under each other and the decimal points are in line.

You can then use the same method as for whole numbers, 'carrying' where necessary.

EXAMPLE 18

Find the total of 26.08 and 174.369.
Do not use a calculator and show all your working.

```
   26.080
+ 174.369
  200.449
  1 1  1
```

Decimal points in line.

The 0 place holder helps to line up the numbers correctly.

Digits written in columns as in a place value table.

Decimal point in line in the answer.

EXAMPLE 19

Alison spent $14.83 in the supermarket, 76 cents in the newsagents and $3.05 on her bus fare. How much did she spend altogether?

```
   14.83
    0.76
+   3.05
   18.64
     1 1
```

Write 76 cents in $ as 0.76 so that digits of equal place value are under each other.

Alison spent $18.64 altogether.

EXERCISE 1R

1 Find the total of each of these.
   (a) $0.5 + 0.36$ (b) $5.29 + 7.3 + 8.245$
   (c) $3.5 + 2.46 + 0.723 + 0.28$
   (d) $4.56 + 6 + 7.2 + 0.384$

2 At the cinema, Ranjit spent $2.45 on a ticket, $1.15 on sweets, $1.10 on popcorn and $1.29 on a drink.
   How much did Ranjit spend altogether?

3 Mr Kalik has some fencing wire 2.3 m long, another piece 0.87 m long and a piece 1.125 m long. What is the total length of wire?

4 Sita is comparing costs of a bottle of water at different shops. The cost, in dollars, is given for 8 shops.

Out-of-town	1.59	1.46	1.72	1.62
Town centre	1.67	1.42	1.80	1.58

(a) Find the total cost for the out-of-town shops.

(b) Find the total cost for the town centre shops.

(c) Sita bought all 8 bottles.
How much did she spend in total?

5 The mass of 3 different dogs was 11.2 kg, 15.65 kg and 8.2 kg. What is the total mass of the dogs?

6 On a school trip Sara spent $2.38 on drinks, $0.98 on an ice cream, $1.24 on a pack of stickers for her sister and $14.99 on a DVD. How much did she spend altogether?

Subtracting decimals

To subtract decimal numbers line up the digits, keeping the decimal points in line.

Make sure the number you are subtracting *from* goes on the top.

Always put in any zeros as place holders.

You can use the same method as for whole numbers, 'borrowing' where necessary.

EXAMPLE 20

Work out 11.8 − 2.647. Do not use a calculator and show all your working.

11.8 must go on the top.

```
  1 7 9 1
  1 1 . 8 0 0
−   2 . 6 4 7
    9 . 1 5 3
```

You *must* put in the 0 place holders.

EXAMPLE 21

Grace earns $183.65 a week. $46.09 is deducted for tax and insurance.

How much pay does she take home?

$$\begin{array}{r} {}^{7}1\!\!\!/8\,{}^{1}3.{}^{5}6\!\!\!/\,{}^{1}5 \\ -\ 46.09 \\ \hline 137.56 \\ \hline \end{array}$$

Grace takes home $137.56 per week.

You can check your subtraction by adding the bottom two lines to see if you get the number on the top line. See Example 7.

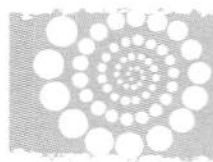

EXERCISE 1S

1 Work out these sums.

(a) 6.2 − 1.7 (b) 3.27 − 1.53

(c) 9.01 − 4.68 (d) 9 − 1.42

2 Ria is 42.5 kg and Miguel is 48.3 kg. How much heavier is Miguel than Ria?

3 What is the difference between 1.634 and 2.58?

4 If I cut 35.6 cm from a piece of string 90.2 cm long, how much will I have left?

5 The times, in seconds, of the five fastest goals scored are given in this table.

Goal	A	B	C	D	E
Time	12.58	34.6	59.84	64.07	72.29

What is the difference between the quickest and the slowest time?

6 If I buy a pack of pens for $1.39 and a pair of compasses for $1.25, how much change will I get from $5?

Multiplying decimals

You multiply decimals in the same way as whole numbers. You need to be able to put the decimal point in the correct place in your answer.

To multiply decimals

1. Ignore the decimal points and multiply the numbers as if they were whole numbers.
2. Count the total number of decimal places in the numbers you are multiplying together.
3. Put the decimal point in your answer so that it has the same total number of decimal places.

These are the digits to the *right* of the decimal point.

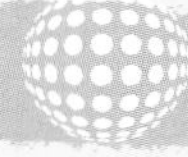

EXAMPLE 22

Work out 4.16×0.2.

```
   416
×    2
   832
```

The answer is 0.832.

Notice, *no* decimal points.

4.16 has 2 decimal places and 0.2 has 1 decimal place.
The answer must have $2 + 1 = 3$ decimal places.

You can check your answer by using your common sense!
4.16×0.2 is approximately the same as 4×0.2, so the answer should be about 0.8 (which it is!).

EXAMPLE 23

Anita needs some floor covering for her bedroom.

The bedroom is a rectangle, 3.8 metres long and 2.75 metres wide.

How many square metres of floor covering does she need?

You need to multiply 3.8 by 2.75

```
     275
×     38
    2200   ← 275 × 8
+   8250   ← 275 × 30
   10450
```

The answer is 10.45 square metres.

Area of a rectangle = length × width.

Ignore the decimal points and multiply 38 by 275.
Put the larger whole number on the top.

2.75 has 2 decimal places and 3.8 has 1 decimal place. The answer must have $2 + 1 = 3$ decimal places.

Ann will probably need to buy 12 square metres because floor covering is usually in a given width, often 3 or 4 metres.

You do not need a zero place holder at the right of decimal numbers.

EXERCISE 1T

1 Work these out.

(a) 32×0.3 (b) 67×0.4

(c) 142×0.05 (d) 53×0.12

2 Work these out.

(a) 4.3×0.7 (b) 6.2×1.3

(c) 15.4×0.08 (d) 16.4×3.5

3 I want to make 21 shelves, each 0.6 m long. What length of wood do I need?

4 A rectangular picture is 6.4 cm wide and 9.5 cm long. What is the area of the picture?

Area of rectangle = length × width.

5 A square garden has side lengths of 4.7 m. What is the area of the garden?

6 Fishing line costs $1.15 per metre. If I buy 4.5 m, how much will it cost?

Dividing decimals

The rules for dividing whole numbers can be used for decimal division.

You need to take care placing the decimal point in your answer.

When you divide by a whole number, the decimal point in the answer is in line with the one in the original number.

EXAMPLE 24

Share $343.80 equally between six people.

```
    4 1
 6)343.80
   57.30
```

Each person receives $57.30.

This is approximately $300 ÷ 6 = $50 so $57.30 is likely to be the correct answer.

To divide by a number which is *not* a whole number, first multiply top and bottom by the same number so that you get a whole number divisor.

The **divisor** is the number you are dividing by.

For example, in the calculation

$\frac{2.832}{0.3} = 9.44$

You can avoid dividing by a decimal if you multiply 0.3 by 10. Remember to multiply 2.832 by 10 as well.

$0.3 \times 10 = 3$.

Then use the method shown in Example 24.

EXAMPLE 25

Work these out.

(a) $2.832 \div 0.3$ **(b)** $14.215 \div 0.05$ **(c)** $0.000764 \div 0.004$

(a) $\frac{2.832}{0.3} = \frac{28.32}{3}$

Multiply top *and* bottom by 10 (the answer will still be the same). This makes the divisor a *whole number.*

$3\overline{)28.32}$ (carries: 1 1)

9.44 Answer = 9.44

Making the number on the *bottom* a *whole number* is the secret to doing correct division.

(b) $\frac{14.215}{0.05} = \frac{1421.5}{5}$

$0.05 \times 100 = 5$.
Multiply top *and* bottom by 100.

$5\overline{)1421.5}$ (carries: 4 2 1)

284.3 Answer = 284.3

(c) $\frac{0.000764}{0.004} = \frac{0.764}{4}$

$0.004 \times 1000 = 4$.
Multiply top *and* bottom by 1000.

$4\overline{)0.764}$ (carry: 3)

0.191 Answer = 0.191

EXERCISE 1U

1 Work these out.

(a) $16.35 \div 3$ (b) $31.4 \div 5$ (c) $43.5 \div 3$

(d) $1.284 \div 4$ (e) $0.34 \div 8$

2 Convert these calculations to give a whole number divisor.

(a) $\frac{1.64}{0.5}$ (b) $\frac{10.432}{0.0001}$ (c) $\frac{7.35}{0.07}$ (d) $\frac{13.36}{0.15}$

3 Work these out.

(a) $6.3 \div 0.3$ (b) $13.4 \div 0.02$ (c) $72.5 \div 0.5$

(d) $39 \div 1.3$ (e) $25 \div 0.1$

4 Work these out.

(a) $6.8 \div 0.2$ (b) $3.9 \div 0.03$ (c) $1.32 \div 0.4$

(d) $12.5 \div 0.25$ (e) $0.03 \div 0.2$

5 Theatre tickets for three friends cost $28.20. If all three seats cost the same amount, what was the cost of one ticket?

6 The bill at a café comes to $11.60. Four people share it equally. How much will each person pay?

7 I have a 4 m length of rope. I want to cut it into 0.6 m lengths. How many whole lengths will I get? What length of rope will be left over?

8 The area of a rectangular garden is 51.66 m^2. If the length is 6.3 m, what is the width?

Area of rectangle = length × width.
Area ÷ length = width.

EXERCISE 1V MIXED QUESTIONS

1 How much change will I get if I pay for a shirt costing $17.75 with a $20 note?

2 The total cost of four tickets for the cinema is $15.40. What is the cost of 1 ticket?

3 A teacher buys 74 text books each costing $6.45. What is the total cost of the books?

4 An apple costs 35 cents.

(a) How many apples can I buy for $5?

(b) How much change will I have?

5 Harry buys a jumper for \$15.99, a DVD for \$11.75, a pair of trainers for \$35.44 and a baseball cap for \$5.79.
  (a) How much did he spend?
  (b) How much change did he have from \$80?

6 A coach company is taking 400 people on a trip. Each coach seats 46 people. The cost is \$120 for each coach.
  (a) How many coaches do they need for everyone?
  (b) What is the total cost for the coaches?
  (c) The total cost is shared equally between the people on the trip. How much does each person pay?

7 Flora earns \$5.25 per hour. One week she worked 15 hours and had \$18.75 deducted for tax. How much did she take home?

8 The cost of electricity is 9.768 cents per unit for the first 100 units, then 5.9 cents per unit for the rest. If you use 235 units, how much will you have to pay?

9 A garden centre buys 375 trays of plants. There are 44 plants in a tray.
  (a) How many plants is this altogether?
  (b) They paid \$1500 for all the plants. They sell small trays of 10 plants for \$1.24. How much profit will they make?

10 Copy the cross-number grid below. Calculate the answers and write them in the squares on the grid. Put in the decimal points on the lines where necessary. The answer to 1 across has been done as an example.

1 5	• 2	2 5	■	3		■	4
	■	5				■	
6			■		■	7	
	■	8	9		■		■
■	10			■	11		
12		■	13	14		■	■
	■	15	■	16		17	18
19			■		■	20	

Clues

Across

1	$3.29 + 1.96$
3	$24.5 \div 0.7$
5	$31.54 - 7.37$
6	$2.52 - 0.8$
7	1.6×4
8	$8.5 \div 0.05$
10	$101\,000 \div 1000$
11	$3843 \div 7$
12	$104.6 - 95.9$
13	11.1×0.5
16	3.046×1000
19	$9 \times 0.9 \times 0.9$
20	27.5×2.4

Down

1	$32.80 + 19.38$
2	$64\,720 - 12\,510$
3	3.1×1000
4	$12.84 \div 6$
7	$8.12 - 1.28$
9	6.5×1.1
10	$1.36 \div 0.8$
11	0.25×2200
12	$0.7 + 2.8 + 23.3 + 60.9$
14	$7.82 - 2.47$
15	0.07×700
17	$4.14 \div 0.9$
18	165×0.04

EXAMINATION QUESTIONS

1 The temperature inside an aeroplane is 18.7 °C and the temperature outside is 51.3 °C lower.
What is the temperature outside? [1]

(CIE, Paper 1, Jun 2000)

---

2 To make some biscuits the following ingredients are needed.

176g of plain flour
60g of rice flour
50g of sugar
110g of butter
55g of walnuts

(a) Calculate the total mass of the ingredients. [1]
(b) The ingredients will make 20 biscuits.
What mass of butter will be needed to make 120 biscuits? [2]

(CIE Paper 3, Jun 2000)

---

3 Work out $\frac{1.9 + 2.3}{3}$. [1]

(CIE Paper 1, Nov 2000)

---

4 The Dead Sea is 393 m below the level of the Mediterranean Sea.
Jerusalem is 747 m above the level of the Mediterranean Sea.
How many metres is Jerusalem above the Dead Sea? [1]

(CIE Paper 1, Nov 2000)

---

5 Dorcas buys 24 cans of orange drink for $12 to sell in her shop.
She sells all the cans for $0.65 each.
Work out her total profit. [1]

(CIE Paper 1, Nov 2000)

---

6 In the village of a ski resort the temperature was 4°C.
At a nearby mountain top the temperature was 12°C lower.
What was the temperature at the mountain top? [1]

(CIE Paper 1, Nov 2001)

---

7 Work out $3.2 \times 5 - 2(4.1 - 2.9)$. [1]

(CIE Paper 2, Nov 2001)

---

8 Work out $7 - 2 \times 4$. [1]

(CIE Paper 1, Jun 2002)

---

9 Work out $50 − $23.46. [1]

(CIE Paper 1, Nov 2002)

---

10 The diagram shows a flood warning post in a river.

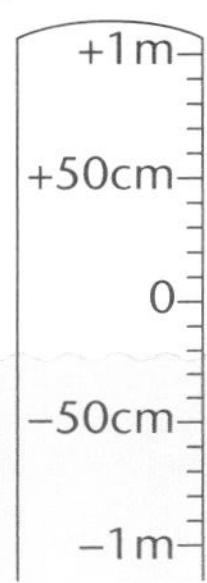

(a) Write down the water level shown in the diagram. [1]
(b) The water level rises by 1 metre.
What is the new level? [1]

(CIE Paper 1, Nov 2002) [1]

---

11 The diagram shows how the water level of a river went down during a drought.

The measurements are in metres.
(a) By how many metres did the water level go down? [1]
(b) A heavy rainfall followed the drought and the water level went up by 1.6 metres.
What was the water level after the rainfall? [1]

(CIE Paper 1, Jun 2003)

---

12 Work out $\frac{2 + 12}{4 + 3 \times 8}$. [1]

(CIE Paper 2, Nov 2003)

---

13 The altitude of Death Valley is -86 metres.
The altitude of Mount Whitney is 4418 metres.
Calculate the difference between these two altitudes. [1]

(CIE Paper 2, Nov 2003)

---

14 The Dead Sea shore is 395 metres **below** sea level.
Hebron is 447 metres **above** sea level.
Find the difference in height. [1]

(CIE Paper 1, Jun 2004)

---

15 At a weather centre the temperature at midnight was $-21°C$.
By noon the next day it had risen to $-4°C$.
By how many degrees had the temperature risen? [1]

(CIE Paper 1, Nov 2004)

---

16 Place brackets in the following calculation to make it a correct statement.
$10 - 5 \times 9 + 3 = 60$ [1]

(CIE Paper 1, Nov 2004)

Chapter 2

Approximations and estimation

This chapter will show you how to

- ✔ understand why you round numbers and use approximations
- ✔ round numbers to the nearest 10, 100, 1000
- ✔ round numbers to a given number of significant figures or decimal places
- ✔ estimate answers to calculations by using approximations

2.1 Rounding

Sometimes you don't need to know an exact answer.

How can the land area of the UK be measured so accurately?

The land area of the UK is 244 110 square kilometres.

As people are born or die, the population of Italy will be changing all the time.

The population of Italy is 58 751 711.

You don't need to know the amount down to the last dollar.

You could use **rounded** or **approximate** values.

The land area of the UK is *about* 245 000 square kilometres.	The population of Italy is *approximately* 59 million.	New York lady gives *nearly* $8 million away to charity.

All of these statements give enough information.

An approximation should be easy to read and understand.

Rounding to the nearest 10, 100, 1000

You can use a place value table to round numbers.

The land area of a small country is 94 251 square kilometres.

	HT	TT	Th	H	T	U	.	t	h	th
		9	4	2	5	1	.			
(a)		9	4	2	5	0	.			
(b)		9	4	3	0	0	.			
(c)		9	4	0	0	0	.			
(d)		9	0	0	0	0	.			

When you round to the nearest 10, the last non-zero digit is in the Tens column.

(a) shows 94 251 rounded to the nearest 10 since 51 is nearer 50 than 60

(b) shows 94 251 rounded to the nearest 100 since 251 is nearer to 300 than 200 (just!)

(c) shows 94 251 rounded to the nearest 1000 since 4251 is nearer to 4000 than 5000

(d) shows 94 251 rounded to the nearest 10 000 since it is nearer to 90 000 than 100 000

A number rounds to different answers, depending on the accuracy required.

In everyday life, the most important thing is to give an answer which is easy to understand and makes sense.

In the examination you are expected to round all answers to 3 significant figures unless the question asks for a different accuracy.

The only exceptions to this are

Money – give the answer to the nearest cent

Exact – where the answer is a whole number give all the figures

Angles – give the angle to 1 decimal place

Estimate – when the question asks you to estimate the value of a calculation then use 1 significant figure for each number

EXAMPLE 1

Round these numbers to the degree of accuracy indicated.

(a) 273 to the nearest 10

(b) 27350 to the nearest 100

(c) 273589 to the nearest 1000

(a)

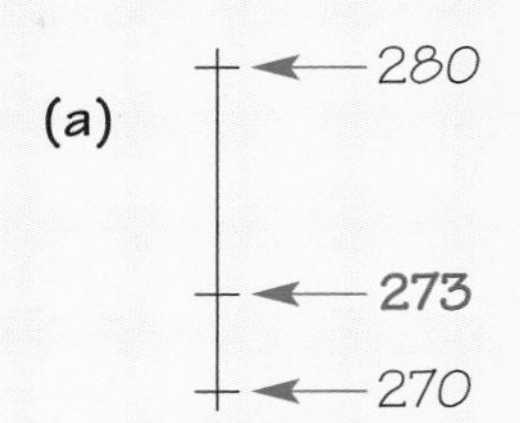

A number line will help you round.

73 is nearer to 70 than 80

273 is 270, to the nearest 10

(b)

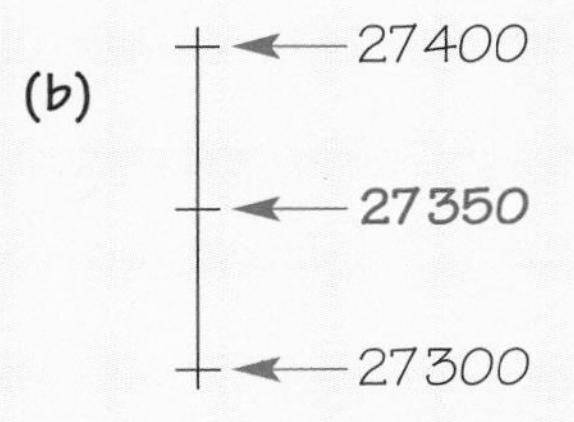

350 is exactly halfway between 300 and 400

27 350 is 27 400, to the nearest 100

When the number falls *exactly* halfway between two limits you always round *upwards*.

(c)

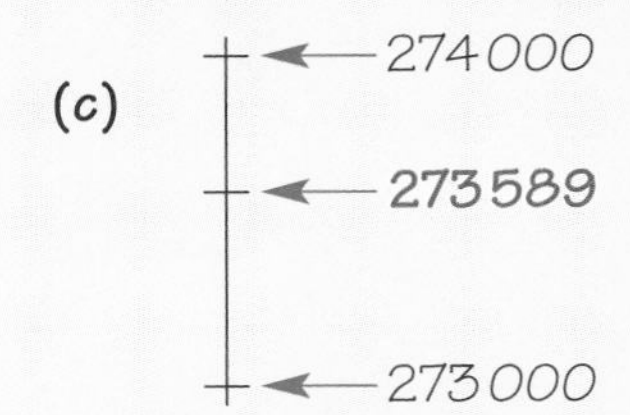

3589 is nearer to 4000 than 3000

273589 is 274000, to the nearest 1000

EXAMPLE 2

Round 8550

(a) to the nearest 100

(b) to the nearest 1000

(a)

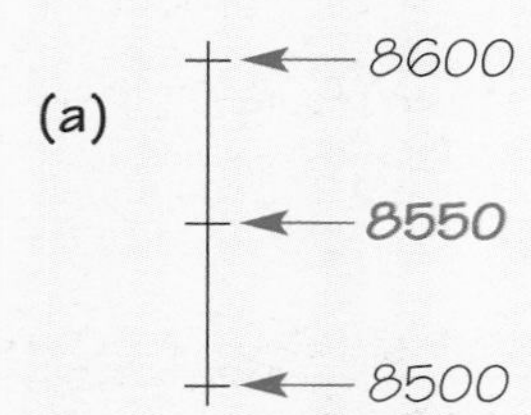

8550 is exactly halfway between 8500 and 8600

8550 is 8600, to the nearest 100

Halfway, so round upwards.

(b)

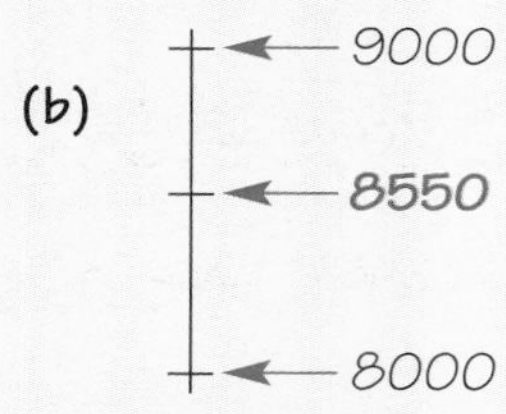

8550 is nearer to 9000 than 8000

(8500 is halfway)

8550 is 9000, to the nearest 1000

EXERCISE 2A

1 Round these numbers to the nearest 10.

(a) 38 (b) 592 (c) 245

(d) 3192 (e) 24 385

2 Round these numbers to the nearest 100.

(a) 634 (b) 4271 (c) 6850

(d) 25 351 (e) 165 387

3 Round these numbers to the nearest 1000.

(a) 8734 (b) 15 397 (c) 243 591

(d) 423 500 (e) 400 526

4 As you drive into Hartley and Eastpool you see these signs.

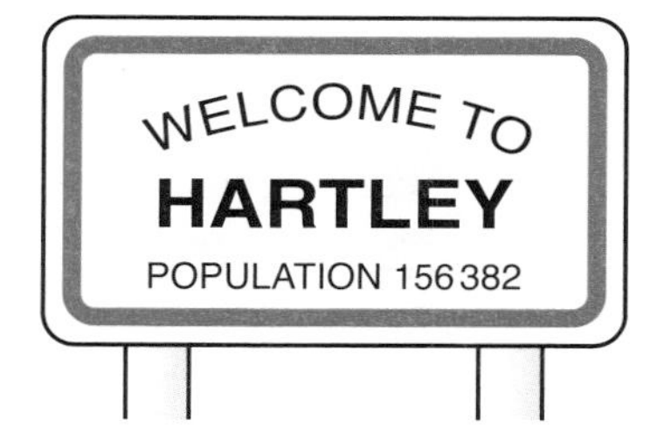

(a) Round each population to the nearest

(i) ten (ii) hundred (iii) thousand

(iv) ten thousand (v) hundred thousand.

(b) Do you think it is sensible to put the exact number of the population on the signs? Give a reason for your answer.

5 A warehouse has this stock.

Item	Number in stock
1cm washers	2 354 816
3 m copper pipe	14 535
90° elbow bends	8783
20 mm straight connectors	12 494

(a) Round each number in stock to the nearest thousand.

(b) Would it be sensible for the stock manager to round the numbers to the nearest thousand? Give a reason for your answer.

6 A newspaper quoted the number of spectators attending a football match as 34 500 to the nearest hundred.

Which of the statements could be true and which must be false?

(a) There were 34 526 at the match.

(b) There were 34 406 at the match.

(c) The attendance was 34 450.

(d) A total of 33 951 people were there.

(e) There were 34 549 spectators.

Rounding sensibly

In some practical problems where the answer is not a whole number, you need to decide whether to round up or down.

EXAMPLE 3

Renuka needs 63 tiles to finish tiling her bathroom.
The tiles are sold in boxes of 12.
How many boxes will she need?

Remember that $5 \times 12 = 60$ and $6 \times 12 = 72$

Number of boxes $= 63 \div 12 = 5.25$

She needs to buy 6 boxes. (She will have 9 tiles left.)

You must round up or she will not have enough tiles to finish the job.

EXAMPLE 4

Kim is packing eggs into boxes. Each box holds 6 eggs.
He has 125 eggs. How many boxes can he fill?

Round down as only whole boxes are filled.

Number of boxes $= 125 \div 6 = 20$ remainder 5

He can fill 20 boxes, with 5 eggs left over.

EXERCISE 2B

1 185 pupils are going on a school trip. Each bus can take 38 pupils. How many buses are needed?

2 Cakes are packed into boxes of 6. I have 74 cakes. How many boxes can I fill?

3 A school sells 108 tickets for a play. The audience sits in rows of 14 chairs. How many full rows of chairs are needed?

4 A room has a wall area of $270\,\text{m}^2$. Amina wants to paint the walls. She knows that 1 litre of paint covers $35\,\text{m}^2$ of wall.

(a) How many litres of paint does Amina need?

(b) The paint comes in 3 litre cans. How many cans will Amina need?

5 A factory makes 5 million small chocolate eggs.

(a) Each packet contains 24 eggs. How many packets will they have?

(b) 40 packets are packed into boxes. How many full boxes will they have?

(c) 18 boxes are packed into a carton. How many cartons will they have?

(d) A truck can carry 54 cartons. How many trucks are needed to deliver all the cartons at the same time?

Rounding to a given number of decimal places

Some calculations do not give an exact answer.

The question may ask you to give the answer to a number of **decimal places (d.p.)**.

Your final answer must have only as many decimal places as the question asks for, no more and no less.

7.56429651

An answer may fill the whole calculator display.

To round to a number of decimal places.

1 Count the number of decimal places you want, to the right of the decimal point.

2 Look at the *next* digit.

If it is *5 or more* you *round up* the digit in the previous decimal place,

If it is *4 or less* you *leave* the previous decimal digit as it is.

Round each of these numbers to 2 decimal places (2 d.p.).

(a) 7.2056 (b) 0.32491 (c) 12.698

(a) 7.20|56 = 7.21 (2 d.p.)

(b) 0.32|491 = 0.32 (2 d.p.)

(c) 12.69|8 = 12.70 (2 d.p.)

The *next* digit is 5, so round the 0 *up* to 1.

The *next* digit is 4, so leave the 2 *as it is.*

The *next* digit is 8, so round the 9 *up*. This makes 10, so 'carry' to the next column.

You need to write the zero as the question asks for 2 d.p. 6 changes to 7

EXAMPLE 6

Write 8.14973 to (a) 1 d.p. (b) 2 d.p. (c) 3 d.p.

(a) 8.1|4973 = 8.1 (1 d.p.)

(b) 8.14|973 = 8.15 (2 d.p.)

(c) 8.149|73 = 8.150 (3 d.p.)

The *next* digit is 4, so leave 1 *as it is.*

The *next* digit is 9, so round the 4 *up* to 5.

The *next* digit is 7, so round the 9 *up*. This makes 10, so 'carry' 1 to the next column. 4 changes to 5

Write the 0 because you must have 3 d.p. The answers of 8.15 in (b) and 8.150 in (c) are *not* the same, since 8.15 is accurate to 2 d.p. and 8.150 is accurate to 3 d.p.

EXERCISE 2C

1 Round these numbers to 1 decimal place.

(a) 5.83 (b) 7.39 (c) 2.15 (d) 5.681
(e) 4.332 (f) 15.829 (g) 11.264 (h) 17.155
(i) 145.077 (j) 521.999

2 Round to 2 decimal places.

(a) 3.259 (b) 6.542 (c) 0.877 (d) 0.031
(e) 11.055 (f) 4.007 (g) 3.899 (h) 2.3093
(i) 0.0009 (j) 5.1299

3 Round to 3 decimal places (d.p.).

(a) 1.2546 (b) 5.2934 (c) 4.1265 (d) 0.0007
(e) 0.000 08

4 Round 15.1529 to

(a) 1 d.p. (b) 2 d.p. (c) 3 d.p.
(d) the nearest whole number.

5 Tom bought a new CD for $14.75.
What is this

(a) to the nearest dollar (b) to the nearest 10 cents?

6 The mass of a dog is given as 5.625 kg.

Write this to

(a) the nearest kg (b) 1 d.p. (c) 2 d.p.

7 In a race, the time for the winner was given as 15.629 seconds. Write this
   (a) to the nearest second.
   (b) to 1 d.p.
   (c) to 2 d.p.

   The time for second place was 15.634 seconds.
   (d) Round this as you did for (a), (b) and (c) above. What do you notice?
   (e) Write a comment about rounding results of races.

Rounding to a given number of significant figures

Sometimes you are asked to round a value to a number of **significant figures (s.f.)**.

The most significant figure is the *first non-zero* figure reading from the left.

Significant figures are 'important' figures. The most significant figure in a number is the one with the greatest place value.

To round to a number of significant figures

1 Start from the most significant figure and count the number of figures that you want.

2 Look at the next digit.

   If it is 5 or more you round up the previous digit.

   If it is 4 or less you leave the previous digit as it is.

3 Use zero **place holders** to locate the decimal point and indicate place value.

This is *very* important.

EXAMPLE 7

Round (a) 240 to 1 significant place (1 s.f.)
   (b) 192 to 1 s.f.
   (c) 152 to 2 s.f.

(a) 2|40 = 200 (1 s.f.)

The most significant figure is 2. The *next* digit is 4, so leave the 2 *as it is*. Use zeros as place holders.

(b) 1|92 = 200 (1 s.f.)

The most significant figure is 1. The *next* digit is 9, so round 1 *up* to 2. Use zeros as place holders.

(c) 1|5|2 = 150 (2 s.f.)

The *next* digit is 2 so leave the 5 *as it is*. Use zeros as place holders.

EXAMPLE 8

Round 26 818 to (a) 1 s.f. (b) 2 s.f. (c) 3 s.f.

(a) 2|6818 = 30 000 (1 s.f.)

(b) 26|818 = 27 000 (2 s.f.)

(c) 26|8|18 = 26 800 (3 s.f.)

The most significant figure is 2.
The *next* digit is 6, so you round 2 *up* to 3.
Use zeros as place holders.

The *next* digit is 8 so round the 6 *up* to 7.
Use zeros as place holders.

The *next* digit is 1, so leave the 8 *as it is.*
Use zeros as place holders.

EXAMPLE 9

Write 0.0070386 to (a) 1 s.f. (b) 2 s.f. (c) 3 s.f.

(a) 0.007|0386 = 0.007

(b) 0.0070|386 = 0.0070

(c) 0.00703|86 = 0.00704

The most significant figure is the *first non-zero* digit reading from the left. Here it is 7.
The *next* digit is 0, so leave the 7 *as it is.*

The *next* digit is 3, so leave the 0 *as it is.* Write the 0 after the 7 because it is the second significant figure.

The *next* digit is 8, so round the 3 *up* to 4.

In (a), (b) and (c) the zeros after the decimal point but before the 7 locate the decimal point and indicate the place value of the 7.

EXERCISE 2D

1 Round these numbers to 1 significant figure.

(a) 384 (b) 2862 (c) 19 473
(d) 257 394 (e) 84 693 (f) 4.86

2 Round to 2 s.f.

(a) 612 (b) 6842 (c) 32 841
(d) 153 945 (e) 267 149 (f) 5.32

3 Round these numbers to 1 s.f.

(a) 0.037 (b) 0.042 (c) 0.0059
(d) 0.0035 (e) 0.509

4 Round to 2 s.f.

(a) 0.574 (b) 0.000 382 (c) 0.001 92
(d) 0.203 (e) 0.001 99

5 The population of Harpool was given as 293 465. What is this rounded to

(a) 1 s.f. (b) 2 s.f.?

6 The attendance at some football matches is given below.

Team A	Team B	Team C	Team D	Team E
48 354	21 924	11 589	5937	2051

(a) Write each number rounded to 1 s.f.

(b) Round each number to 3 s.f.

(c) Which attendance figures, (a) or (b), do you think a newspaper would give for each team? Give a reason for your answer.

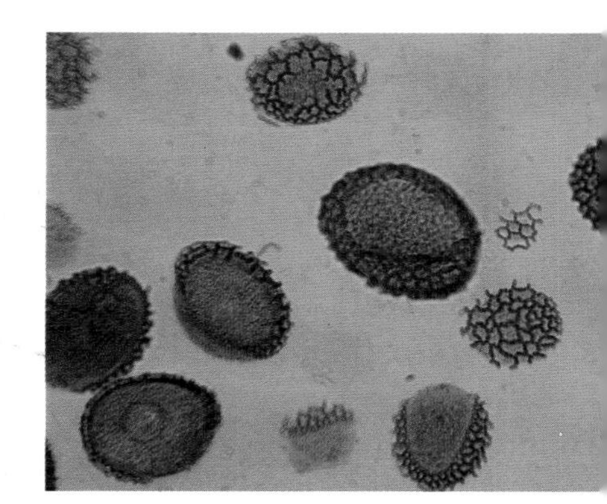

7 The size of a pollen grain is given as 0.0385 mm. What is this to 2 s.f.?

8 Round 15.045 correct to

(a) 1 s.f. (b) 2 s.f. (c) 3 s.f.

2.2 Estimating using approximations

When you do a calculation on a calculator, do you always believe the answer?

What if you press a wrong button? Would you realise what you had done?

You can check your answer using **approximations**.

To **estimate** an approximate answer

1 Round all of the numbers in the calculation to 1 significant figure.

2 Do the calculation using these approximations.

In the examination you might be asked to find estimates to calculations by using approximations.

EXAMPLE 10

Use approximations to estimate the answers to these calculations.

(a) $\frac{19 \times 59}{31}$ (b) 6.89×3.375 (c) $38.45 \div 7.56$

(d) $\frac{181.03 \times 0.48}{8.641}$ (e) $\frac{6.85^2}{2.19 \times 11.7}$

Continued ▼

(a) $\frac{19 \times 59}{31} \approx \frac{20 \times 60}{30} = \frac{1200}{30} = 40$

(b) $6.89 \times 3.375 \approx 7 \times 3 = 21$

(c) $38.45 \div 7.56 \approx 40 \div 8 = 5$

(d) $\frac{181.03 \times 0.48}{8.641} \approx \frac{200 \times 0.5}{9} = \frac{100}{9} \approx 11$

(e) $\frac{6.85^2}{2.19 \times 11.7} \approx \frac{7^2}{2 \times 10} = \frac{49}{20} = 2.5$

$\approx$ means 'approximately equal to'

All 1 s.f. approximations and the estimated answers are shown in blue.

$11 \times 9 = 99$, which is close enough.

49 is almost 50, and it makes the calculation easier.

If you work out the calculations from Example 10, you will see that they are close to the approximate answers.

(a) $\frac{19 \times 59}{31} = 36.161\,29...$ approximate answer $= 40$

(b) $6.89 \times 3.375 = 23.253\,75$ approximate answer $= 21$

(c) $38.45 \div 7.56 = 5.0859...$ approximate answer $= 5$

(d) $\frac{181.03 \times 0.48}{8.641} = 10.0560...$ approximate answer $= 11$

(e) $\frac{6.85^2}{2.19 \times 11.7} = 1.8312...$ approximate answer $= 2.5$

EXERCISE 2E

1 Use approximation to estimate the answers to these calculations.

(a) 67×53 (b) $11 \times 38 \times 12$

(c) $157 \div 47$ (d) $\frac{103 \times 32}{12 \times 11}$

(e) $\frac{426 \times 4982}{38 \times 51}$ (f) $\frac{276 \times 36}{114 \times 3}$

2 Calculate the actual answers to question 1. Check to see how close your estimates are.

3 Estimate the answers to these.

(a) 4.38×5.12 (b) $17.3 \div 3.925$

(c) 148.2×9.604 (d) $\frac{6.432 \times 3.618}{3.978 + 4.2}$

(e) $\frac{462.3 \times 0.48}{(4.852)^2}$ (f) $\frac{(5.3)^2 \times 7.759}{192 \times 4.2}$

4 Calculate the answers to question 3. Check to see how close your estimates are.

5 A milkman travels 32.5 kilometres each day delivering milk. He works 27 days each month. Estimate how many kilometres he travels each month.

6 On a holiday tour there were 23 coaches. Each coach held 48 people. How many people went on the tour?

7 For a project a workshop needs a total of 73.23 m of metal rod. They have 24 pieces of metal rod, each 3.75 m long. Use estimation to decide if they have enough rod for the project.

8 A theatre has 43 rows of 38 seats and 22 rows of 11 seats. Estimate the number of seats in the theatre.

9 There are 1025 paperclips in a packet. There are 144 packets in a box. 36 boxes are packed in a carton. Estimate the number of paperclips in a carton.

10 A car can travel 1.6 km in 58.6 seconds. Estimate how far it can travel in 1 hour.

EXAMINATION QUESTIONS

1 An athlete's time for a race was 43.78 seconds.
Write this time correct to
(a) one decimal place, [1]
(b) one significant figure. [1]

(CIE paper 1, Nov 2002)

---

2 In 1950 the population of Switzerland was 4 714 900.
Write this correct to 3 significant figures. [1]

(CIE Paper 2, Jun 2003)

---

3 Work out $2.7 \times 8.3 \div (12 - 2.7)$, writing down
(a) your full calculator display, [1]
(b) your answer to two decimal places. [1]

(CIE Paper 1, Nov 2003)

4

$$218 \div 39$$

(a) (i) Write both numbers in the calculation above correct to one significant figure. [1]
(ii) Use your answer to part (i) to estimate the value of the calculation. [1]
(b) Use your calculator to find the value of the calculation correct to two significant figures. [1]

(CIE Paper 1, Jun 2004)

---

5 The area of a small country is 78 133 square kilometres.
Write this area correct to 1 significant figure. [1]

(CIE Paper 2, Jun 2004)

---

6 Write $\frac{5}{9}$ as a decimal correct to two decimal places. [2]

(CIE Paper 1, Nov 2004)

7

$$\frac{8.95 - 3.05 \times 1.97}{2.92}$$

(a) (i) Write the above expression with each number rounded to one significant figure. [1]

(ii) Use your answer to find an **estimate** for the value of the expression. [1]

(b) Use your calculator to work out the value of the **original** expression.
Give your answer correct to 2 decimal places. [1]

(CIE Paper 1, Jun 2005)

---

8 **(a)** Write 0.48 correct to 1 significant figure. [1]

(b) (i) Find an approximate answer for the sum

$$9.87 - 5.79 \times 0.48$$

by rounding each number to 1 significant figure. Show your working. [1]

(ii) Use your calculator to find the exact answer for the sum in **part (b)(i)**.
Write down all the figures on your calculator. [1]

(CIE Paper1, Nov 2005)

---

9 Write the number 2381.597 correct to

(a) 3 significant figures, [1]

(b) 2 decimal places, [1]

(c) the nearest hundred. [1]

(CIE Paper 2, Nov 2005)

Chapter 3

Fractions

This chapter will show you how to

- ✔ find equivalent fractions and write a fraction in its simplest form
- ✔ put fractions in order of size
- ✔ find a fraction of a quantity
- ✔ use improper fractions and mixed numbers
- ✔ use the four rules for fractions
- ✔ convert between fractions, decimals and percentages
- ✔ recognise rational and irrational numbers
- ✔ handle terminating and recurring decimals
- ✔ write terminating and recurring decimals as fractions

3.1 Understanding fractions

This cake has been cut into six equal pieces.

One of the pieces has been eaten.

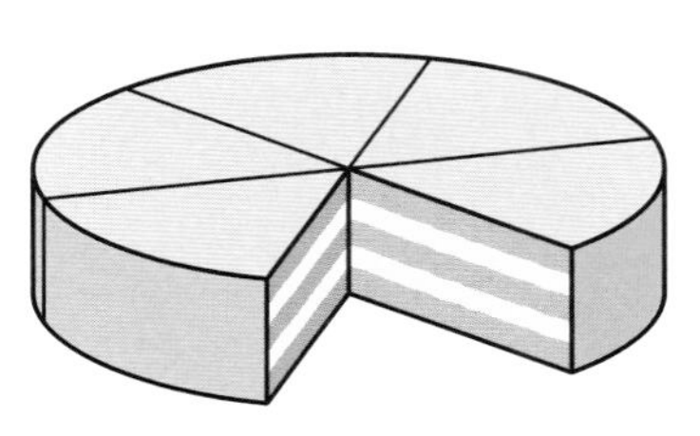

A fraction has a top number called the **numerator** and a bottom number called the **denominator**.

You say that $\frac{1}{6}$ (one sixth) of the cake has been eaten.

When the numerator is smaller than the denominator the fraction is a **proper fraction**.

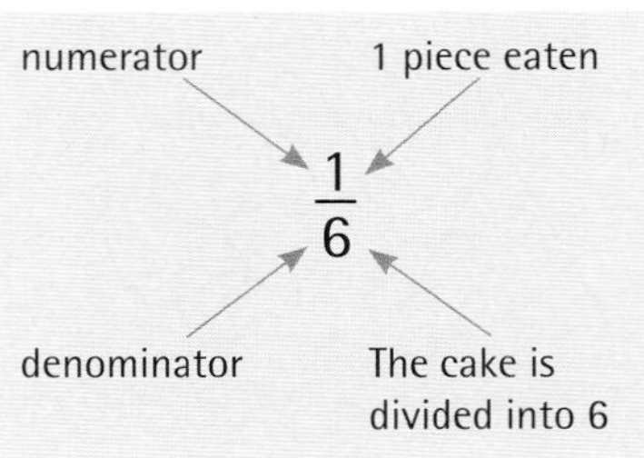

A proper fraction is always part of a whole.

You can use pictures to show proper fractions.

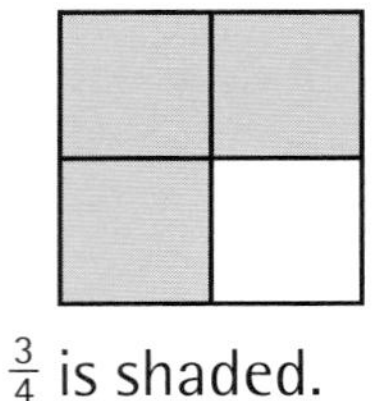

4 equal parts
3 parts shaded

$\frac{3}{4}$ is shaded.

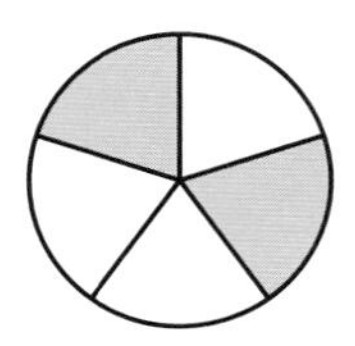

5 equal parts
2 parts shaded

$\frac{2}{5}$ is shaded.

EXAMPLE 1

What fraction of this diagram is shaded green?

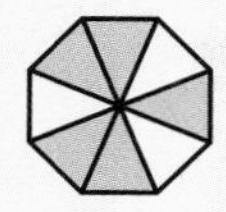

8 equal parts
5 parts are shaded green

$\frac{5}{8}$ is shaded green.

You can use fractions to represent information given in words.

EXAMPLE 2

Hassan has 45 DVDs.

10 are sport DVDs, 15 are music DVDs and the rest are film DVDs.

Write each of these as fractions of his collection.

Numerator = number of 'Sport' or 'Music' or 'Film'.
Denominator = total number of DVDs.

Sport = $\frac{10}{45}$ Music = $\frac{15}{45}$ Film = $\frac{20}{45}$

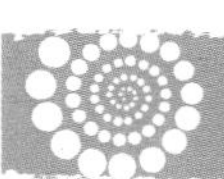

EXERCISE 3A

1 For each diagram write
   (i) the fraction shaded
   (ii) the fraction unshaded.

(a) 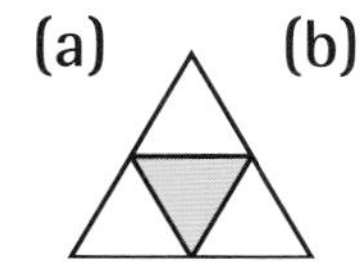 (b)  (c) 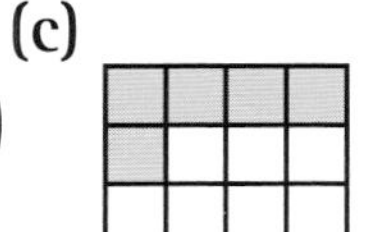 (d) 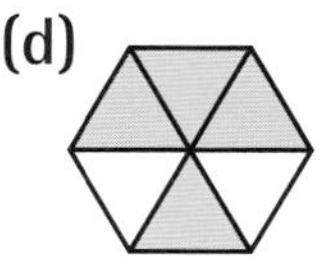 (e) 

2 For each of the following, draw a diagram to show the fraction shaded.
   (a) $\frac{1}{2}$ (b) $\frac{3}{4}$ (c) $\frac{1}{3}$
   (d) $\frac{3}{5}$ (e) $\frac{7}{10}$

3 There are 17 animals in a room. There are 9 dogs, 6 cats and 2 rabbits. What fraction of the animals are
   (a) dogs (b) cats (c) rabbits?

4 In a vehicle survey there were 12 cars and 9 trucks.

(a) How many vehicles were surveyed altogether?

(b) What fraction of the vehicles were cars?

(c) What fraction were trucks?

5 This is how Joti spent his day.

Activity	Sleeping	School	Eating	TV	Football	Homework
Hours	10	7	1	2	3	1
Fraction						

(a) Copy and complete the table to show the fraction of the day he spent on each activity.

(b) Make a similar table for the way you spend your day.

6 A box of sweets contains 6 with green wrappers, 4 with red wrappers and 3 with yellow wrappers. What fraction of the sweets have

(a) green wrappers (b) red wrappers

(c) blue wrappers?

Equivalent fractions

These three diagrams are exactly the same size and have exactly the same amount shaded.

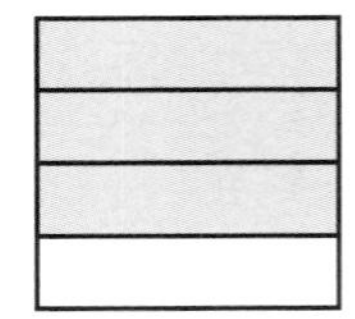

$\frac{3}{4}$ is shaded

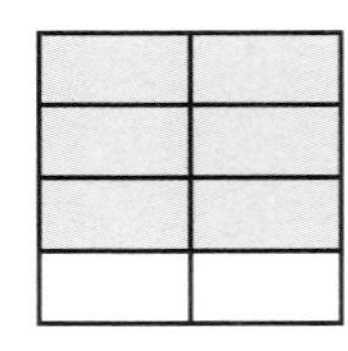

$\frac{6}{8}$ is shaded

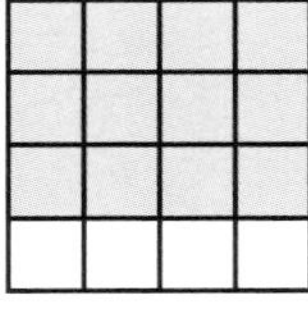

$\frac{12}{16}$ is shaded

The diagrams show that $\frac{3}{4} = \frac{6}{8} = \frac{12}{16}$.

They are called **equivalent fractions**.

There is a connection between these three fractions.

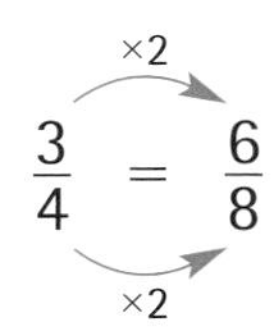

×2

$\frac{6}{8} = \frac{12}{16}$

×2

×4

$\frac{3}{4} = \frac{12}{16}$

×4

You could also write it like this.

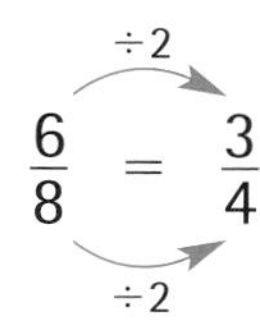

$$\frac{6}{8} \overset{\div 2}{=} \frac{3}{4} \qquad \frac{12}{16} \overset{\div 2}{=} \frac{6}{8} \qquad \frac{12}{16} \overset{\div 4}{=} \frac{3}{4}$$

To find equivalent fractions, multiply or divide the numerator and denominator by the same number.

When there is no number that divides exactly into the numerator and denominator, a fraction is in its **simplest form**.

No number divides exactly into both 3 and 4, so $\frac{3}{4}$ is the simplest form of this fraction.

EXAMPLE 3

Find three more fractions that are equivalent to $\frac{2}{9}$.

(i) Multiply numerator and denominator by 2 to get

$$\frac{2}{9} \overset{\times 2}{=} \frac{4}{18}$$

(ii) Multiply numerator and denominator by 3 to get

$$\frac{2}{9} \overset{\times 3}{=} \frac{6}{27}$$

(iii) Multiply numerator and denominator by 10 to get

$$\frac{2}{9} \overset{\times 10}{=} \frac{20}{90}$$

You can choose any multiplier to give an equivalent fraction, as long as you multiply both the numerator and the denominator.

EXAMPLE 4

Fill in the missing numbers to make equivalent fractions.

$$\frac{2}{\square} = \frac{6}{9} = \frac{\square}{30} = \frac{\square}{90} = \frac{14}{\square}$$

Start with the fraction you know, $\frac{6}{9}$ in this example.

$$\frac{6}{9} = \frac{2}{\boxed{3}} \quad (\div 3)$$

$$\frac{2}{3} = \frac{\boxed{20}}{30} \quad (\times 10)$$

$$\frac{6}{9} = \frac{\boxed{60}}{90} \quad (\times 10)$$

$$\frac{2}{3} = \frac{14}{\boxed{21}} \quad (\times 7)$$

You can use the fraction you were given or any of the answers you have worked out.

EXAMPLE 5

Write these fractions in their simplest form.

(a) $\frac{8}{10}$ (b) $\frac{36}{48}$ (c) $\frac{40}{72}$

(a) $\frac{8}{10} = \frac{4}{5}$ ($\div 2$) Answer $= \frac{4}{5}$

(b) $\frac{36}{48} = \frac{18}{24} = \frac{9}{12} = \frac{3}{4}$ ($\div 2$, $\div 2$, $\div 3$) Answer $= \frac{3}{4}$

You could work out (b) more quickly if you spotted that both 36 and 48 divide exactly by 4 or 6 or 12.

(c) $\frac{40}{72} = \frac{20}{36} = \frac{10}{18} = \frac{5}{9}$ ($\div 2$, $\div 2$, $\div 2$) Answer $= \frac{5}{9}$

In (c), you might spot that both 40 and 72 divide exactly by 4 or 8.

Even if you do not spot the quickest way, you will get the correct answer if you follow the rules.

EXERCISE 3B

1 Find two equivalent fractions for each of these.

(a) $\frac{1}{2}$ (b) $\frac{1}{4}$ (c) $\frac{1}{3}$ (d) $\frac{1}{5}$ (e) $\frac{1}{7}$

2 Find three equivalent fractions for each of these.

(a) $\frac{2}{3}$ (b) $\frac{2}{5}$ (c) $\frac{3}{4}$ (d) $\frac{3}{7}$ (e) $\frac{3}{10}$

3 Which of these fractions are equivalent to $\frac{2}{5}$?

$\frac{8}{20}$ $\frac{15}{25}$ $\frac{18}{45}$ $\frac{16}{30}$ $\frac{14}{35}$

4 Copy and complete these equivalent fractions.

(a) $\frac{1}{3} = \frac{\square}{6} = \frac{\square}{12} = \frac{5}{\square} = \frac{10}{\square}$

(b) $\frac{3}{5} = \frac{6}{\square} = \frac{\square}{15} = \frac{\square}{50} = \frac{36}{\square}$

(c) $\frac{\square}{7} = \frac{8}{14} = \frac{16}{\square} = \frac{12}{\square} = \frac{\square}{35}$

(d) $\frac{25}{\square} = \frac{\square}{12} = \frac{5}{6} = \frac{15}{\square} = \frac{\square}{36}$

(e) $\frac{20}{\square} = \frac{\square}{81} = \frac{12}{\square} = \frac{4}{9} = \frac{\square}{36}$

5 Write each of these fractions in its simplest form.

(a) $\frac{2}{4}$ (b) $\frac{6}{12}$ (c) $\frac{3}{9}$ (d) $\frac{6}{8}$ (e) $\frac{12}{20}$

(f) $\frac{10}{15}$ (g) $\frac{16}{20}$ (h) $\frac{15}{25}$ (i) $\frac{12}{16}$ (j) $\frac{14}{21}$

(k) $\frac{8}{28}$ (l) $\frac{27}{36}$ (m) $\frac{45}{60}$ (n) $\frac{45}{54}$ (o) $\frac{13}{52}$

6 Which of these fractions are not in their simplest form?

$\frac{2}{4}$ $\frac{7}{8}$ $\frac{4}{13}$ $\frac{5}{11}$ $\frac{12}{21}$ $\frac{9}{20}$ $\frac{13}{39}$ $\frac{5}{24}$

7 Sumala said that $\frac{16}{60}$ is equivalent to $\frac{1}{4}$. Is she right? Give a reason for your answer.

8 $\frac{3}{8}$ of the Gobi desert is without water. Write down three fractions equivalent to $\frac{3}{8}$.

9 Karl ate $\frac{15}{30}$ of an apple pie and Bavna ate $\frac{8}{32}$. Write these fractions in their simplest form. Who had more of the apple pie, Karl or Bavna? Explain your answer.

10 In a garden $\frac{2}{12}$ is grass, $\frac{3}{9}$ is flower beds and $\frac{9}{18}$ is a vegetable plot.

(a) Write each fraction in its simplest form.

(b) Draw a rectangle split into 6 equal parts to represent the garden. Shade in and label the fractions for the grass, flower beds and vegetable plot.

Ordering fractions

You can see that $\frac{5}{8}$ is bigger than $\frac{3}{8}$.

When fractions have the *same* denominator, the one with the bigger numerator is the bigger fraction.

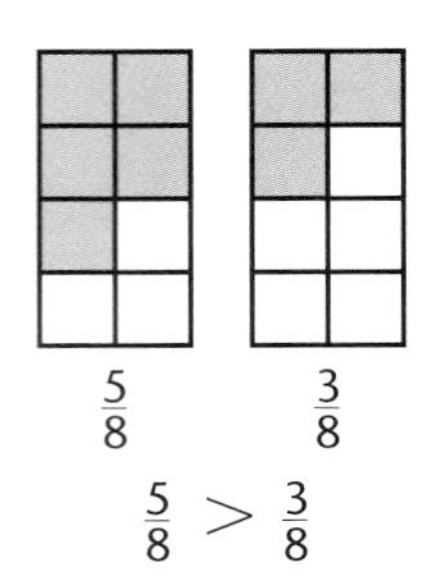

> means 'is greater than'

Which fraction is bigger, $\frac{4}{5}$ or $\frac{3}{4}$?

These fractions do *not* have the same denominator.

It is difficult to compare them.

You can use equivalent fractions to help you to decide which is bigger.

Look for a number that is in the 'times-tables' of both denominators. 20 is the smallest number in both the 4-times and the 5-times tables. It is called the **lowest common multiple (LCM)**. You will meet this again in Chapter 13.

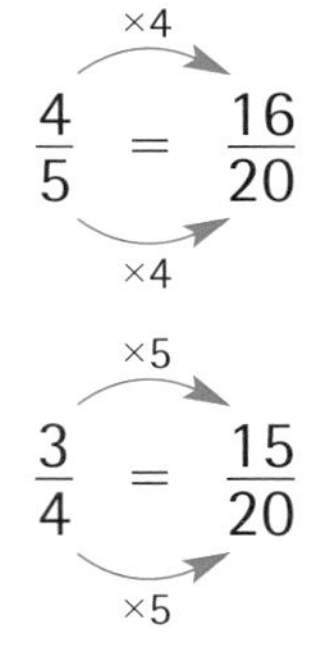

$\frac{16}{20}$ is bigger than $\frac{15}{20}$, so $\frac{4}{5}$ is bigger than $\frac{3}{4}$.

You can write $\frac{4}{5} > \frac{3}{4}$.

To order fractions, use equivalent fractions to write them with the same denominator, then compare them.

EXAMPLE 6

Put these fractions in order, smallest first $\frac{1}{3}, \frac{7}{20}, \frac{3}{10}$.

> 60 is the smallest number you can use for the denominator.

The smallest number that 3, 20 and 10 all divide into is 60.

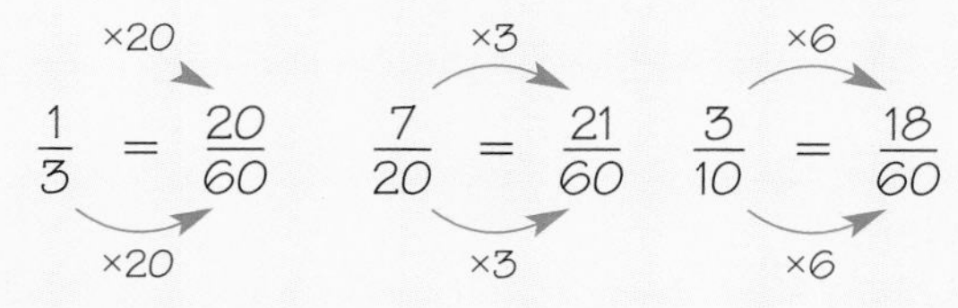

> Multiply both top and bottom by the same number.

The order is $\frac{18}{60}, \frac{20}{60}, \frac{21}{60}$ or $\frac{3}{10}, \frac{1}{3}, \frac{7}{20}$

EXERCISE 3C

1 In each question, write down the larger fraction.

(a) $\frac{3}{5}, \frac{1}{5}$ (b) $\frac{2}{6}, \frac{5}{6}$ (c) $\frac{4}{7}, \frac{3}{7}$ (d) $\frac{7}{10}, \frac{9}{10}$ (e) $\frac{7}{11}, \frac{5}{11}$

(f) $\frac{1}{2}, \frac{3}{4}$ (g) $\frac{4}{5}, \frac{7}{10}$ (h) $\frac{2}{3}, \frac{5}{6}$ (i) $\frac{3}{4}, \frac{3}{8}$ (j) $\frac{1}{2}, \frac{7}{12}$

2 Fill in each of these using > or < or =.

> < means less than
> \> means more than

(a) $\frac{3}{2}$ □ $\frac{2}{5}$ (b) $\frac{3}{4}$ □ $\frac{1}{4}$ (c) $\frac{2}{7}$ □ $\frac{5}{7}$ (d) $\frac{3}{8}$ □ $\frac{5}{8}$

(e) $\frac{5}{9}$ □ $\frac{4}{9}$ (f) $\frac{2}{4}$ □ $\frac{1}{2}$ (g) $\frac{3}{8}$ □ $\frac{1}{4}$ (h) $\frac{1}{5}$ □ $\frac{3}{10}$

(i) $\frac{3}{15}$ □ $\frac{1}{5}$ (j) $\frac{5}{24}$ □ $\frac{1}{8}$ (k) $\frac{5}{6}$ □ $\frac{2}{3}$ (l) $\frac{3}{5}$ □ $\frac{7}{10}$

(m) $\frac{3}{8}$ □ $\frac{9}{24}$ (n) $\frac{11}{20}$ □ $\frac{3}{5}$ (o) $\frac{5}{6}$ □ $\frac{27}{30}$ (p) $\frac{2}{3}$ □ $\frac{3}{4}$

(q) $\frac{3}{4}$ □ $\frac{3}{5}$ (r) $\frac{3}{5}$ □ $\frac{2}{3}$ (s) $\frac{4}{5}$ □ $\frac{5}{6}$ (t) $\frac{7}{9}$ □ $\frac{5}{7}$

3 Put these fractions in order, placing the smallest first.

> Put each fraction into its simplest form first.

(a) $\frac{3}{5}, \frac{1}{5}, \frac{2}{5}$ (b) $\frac{2}{3}, \frac{1}{6}, \frac{5}{6}$ (c) $\frac{1}{4}, \frac{3}{4}, \frac{5}{8}$ (d) $\frac{9}{10}, \frac{3}{5}, \frac{7}{10}$

(e) $\frac{1}{4}, \frac{1}{2}, \frac{3}{8}$ (f) $\frac{3}{5}, \frac{1}{2}, \frac{4}{10}$ (g) $\frac{5}{12}, \frac{2}{3}, \frac{3}{4}$ (h) $\frac{2}{3}, \frac{7}{27}, \frac{5}{9}$

(i) $\frac{2}{5}, \frac{3}{10}, \frac{4}{15}$ (j) $\frac{4}{5}, \frac{7}{12}, \frac{3}{4}$

4 Ali has $\frac{9}{40}$ of a bowl of cherries. Zainah has $\frac{18}{72}$ of the cherries and Lin has $\frac{7}{30}$. Place them in order from smallest to largest to find who has the largest fraction of cherries.

3.2 Finding a fraction of a quantity

In mathematics, 'of' means multiply so

$$\begin{aligned}\tfrac{3}{5} \text{ of } 80 &= \tfrac{3}{5} \times 80 \\ &= \frac{3 \times 80}{5} \\ &= \tfrac{240}{5} \\ &= 48\end{aligned}$$

Another way to do this is to work out $\frac{1}{5}$ of 80 first, then multiply by 3 because $\frac{3}{5} = 3 \times \frac{1}{5}$.

$$\begin{aligned}\tfrac{1}{5} \text{ of } 80 &= \tfrac{1}{5} \times 80 \\ &= \tfrac{80}{5} \\ &= 16\end{aligned}$$

So $\frac{3}{5}$ of $80 = 3 \times 16 = 48$

You can see that both methods give 48 for the answer. You can choose whichever one you prefer.

EXAMPLE 7

A factory employs 216 people.

$\frac{5}{12}$ of them are men.

(a) How many men work in the factory?

(b) How many women work in the factory?

(a) $$\begin{aligned}\tfrac{5}{12} \text{ of } 216 &= \tfrac{5}{12} \times 216 \\ &= \frac{5 \times 216}{12} \\ &= \frac{1080}{12} \\ &= 90\end{aligned}$$

90 men work in the factory.

(b) So the number of women $= 216 - 90$
$= 126$

EXERCISE 3D

1 Work out

(a) $\frac{1}{4}$ of 24 metres (b) $\frac{2}{3}$ of £36
(c) $\frac{3}{5}$ of 60 kg (d) $\frac{7}{9}$ of 36 litres
(e) $\frac{7}{10}$ of 40 minutes (f) $\frac{5}{8}$ of 48 pages.

2 Work out the following.

(a) $\frac{5}{6}$ of 162 km (b) $\frac{7}{11}$ of 121 days
(c) $\frac{13}{15}$ of \$630 (d) $\frac{8}{9}$ of \$153
(e) $\frac{3}{4}$ of 450 g (f) $\frac{2}{3}$ of 5 litres

3 Which is bigger?

(a) $\frac{4}{5}$ of 15 or $\frac{2}{3}$ of 24 (b) $\frac{3}{10}$ of 200 or $\frac{7}{8}$ of 96
(c) $\frac{1}{3}$ of 240 or $\frac{1}{4}$ of 440 (d) $\frac{3}{8}$ of 104 or $\frac{2}{3}$ of 69

4 Habib is awake for $\frac{5}{8}$ of a day. How many hours is this?

Change a day into hours first.

5 Hannah spends $\frac{3}{4}$ of her pocket money at the shops and saves the rest. One week she gets \$10 pocket money.

(a) How much does she spend?
(b) How much does she save?

6 Amit's grandfather says he can either have $\frac{2}{3}$ of \$75 or $\frac{3}{8}$ of \$120 for his birthday. Which one should he choose?

7 A man earns \$420.80 per week. He has to pay $\frac{3}{20}$ of this in tax. What will he have left?

3.3 Improper fractions and mixed numbers

This is a proper fraction. $\frac{3}{8}$ (3 — numerator; 8 — denominator)

The numerator is smaller than the denominator.

$\frac{9}{4}$ is an **improper fraction**. The numerator is greater than the denominator.

It can be represented in picture form as

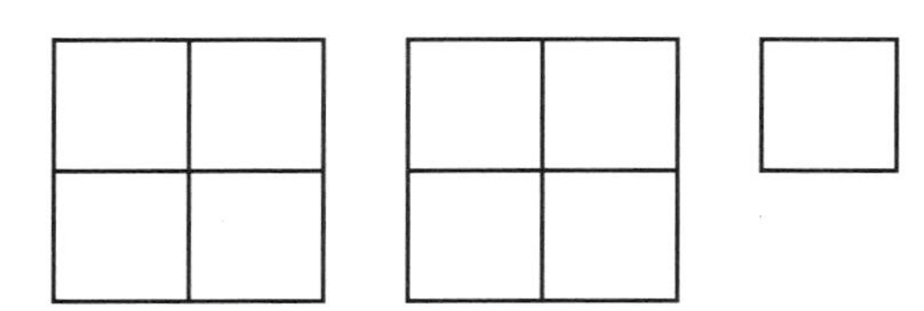

You can see that this is the same as $2\frac{1}{4}$ squares.

$2\frac{1}{4}$ is a **mixed number** because it contains a whole number part (2) and a fractional part ($\frac{1}{4}$).

You need to be able to write an improper fraction as a mixed number and also write a mixed number as an improper fraction.

Writing an improper fraction as a mixed number

For example, to change $\frac{9}{4}$ to $2\frac{1}{4}$

1 Divide the numerator by the denominator and write down the whole number part of your answer.

2 The remainder gives you the fractional part. The denominator is the same as the denominator of the improper fraction.

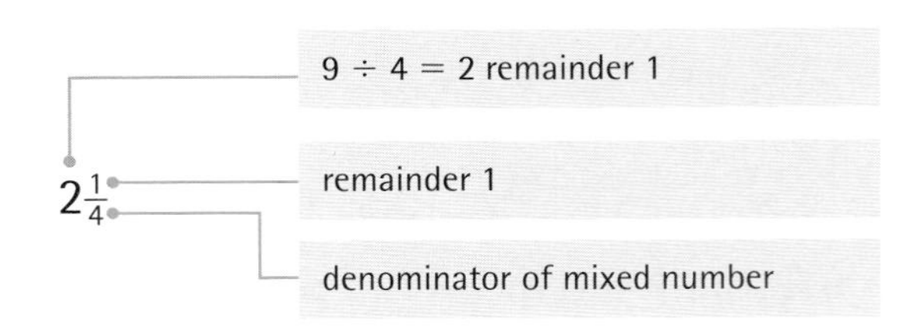

EXAMPLE 8

Write these improper fractions as mixed numbers.

(a) $\frac{7}{2}$ (b) $\frac{20}{3}$ (c) $\frac{29}{6}$

(a) $7 \div 2 = 3$ remainder 1, so $\frac{7}{2} = 3\frac{1}{2}$

(b) $20 \div 3 = 6$ remainder 2, so $\frac{20}{3} = 6\frac{2}{3}$

(c) $29 \div 6 = 4$ remainder 5, so $\frac{29}{6} = 4\frac{5}{6}$

EXERCISE 3E

1 Write these improper fractions as mixed numbers.

(a) $\frac{9}{2}$ (b) $\frac{13}{2}$ (c) $\frac{14}{3}$

(d) $\frac{18}{5}$ (e) $\frac{15}{4}$ (f) $\frac{21}{5}$

(g) $\frac{27}{4}$ (h) $\frac{19}{6}$ (i) $\frac{30}{7}$

2 Write these improper fractions as mixed numbers.

(a) $\frac{38}{5}$ (b) $\frac{47}{4}$ (c) $\frac{100}{3}$

(d) $\frac{48}{5}$ (e) $\frac{61}{6}$ (f) $\frac{37}{7}$

(g) $\frac{43}{8}$ (h) $\frac{51}{7}$ (i) $\frac{40}{9}$

It is important not to use a calculator when you see this symbol. You may be asked not to use one in a question in the examination.

3 Copy and fill in the missing numbers.

(a) $\frac{29}{2} = 14\frac{\square}{2}$ (b) $\frac{92}{3} = 30\frac{\square}{3}$ (c) $\frac{43}{4} = \square\frac{3}{4}$

(d) $\frac{47}{3} = 15\frac{\square}{3}$ (e) $\frac{104}{5} = \square\frac{4}{5}$ (f) $\frac{79}{6} = \square\frac{1}{6}$

(g) $\frac{55}{8} = \square\frac{7}{8}$ (h) $\frac{45}{7} = 6\frac{\square}{7}$ (i) $\frac{41}{9} = 4\frac{\square}{9}$

4 In each case

(i) write the improper fraction as a mixed number.

(ii) write your mixed number in its simplest form.

(a) $\frac{14}{4}$ (b) $\frac{25}{10}$ (c) $\frac{16}{6}$

(d) $\frac{30}{4}$ (e) $\frac{27}{6}$ (f) $\frac{30}{9}$

(g) $\frac{34}{6}$ (h) $\frac{42}{9}$ (i) $\frac{100}{12}$

5 Change the improper fractions into mixed numbers then write each pair of numbers using $>$ or $<$.

(a) $\frac{17}{5}$ $\frac{11}{2}$ (b) $\frac{10}{3}$ $\frac{23}{9}$

(c) $\frac{15}{4}$ $\frac{25}{6}$ (d) $\frac{13}{2}$ $\frac{43}{8}$

(e) $\frac{83}{10}$ $\frac{153}{20}$ (f) $\frac{24}{4}$ $\frac{125}{25}$

6 Write these improper fractions in order of size, smallest first.

(a) $\frac{15}{4}, \frac{17}{3}, \frac{25}{6}$ (b) $\frac{12}{5}, \frac{9}{3}, \frac{13}{3}$ (c) $\frac{25}{4}, \frac{18}{5}, \frac{30}{7}$

(d) $\frac{45}{6}, \frac{27}{5}, \frac{35}{8}$ (e) $\frac{31}{7}, \frac{32}{8}, \frac{33}{9}$ (f) $\frac{50}{7}, \frac{25}{4}, \frac{37}{9}$

Writing a mixed number as an improper fraction

For example, to change $2\frac{1}{4}$ to $\frac{9}{4}$

1 Multiply the whole number part by the denominator of the fractional part. For example, 2 is the same as $\frac{8}{4}$.

2 Add the fractional part of your answer to the result of step 1.

$\frac{8}{4} + \frac{1}{4} = \frac{9}{4}$

EXAMPLE 9

Write these mixed numbers as improper fractions.

(a) $1\frac{3}{8}$ (b) $4\frac{2}{7}$ (c) $10\frac{4}{9}$

(a) $1 \times 8 = 8$ $\frac{8}{8} + \frac{3}{8} = \frac{11}{8}$

(b) $4 \times 7 = 28$ $\frac{28}{7} + \frac{2}{7} = \frac{30}{7}$

(c) $10 \times 9 = 90$ $\frac{90}{9} + \frac{4}{9} = \frac{94}{9}$

EXERCISE 3F

1 Write these mixed numbers as improper fractions.

(a) $2\frac{1}{2}$ (b) $4\frac{1}{3}$ (c) $4\frac{1}{5}$ (d) $3\frac{3}{5}$

(e) $5\frac{1}{6}$ (f) $5\frac{4}{5}$ (g) $3\frac{3}{4}$ (h) $2\frac{5}{6}$

2 Write these mixed numbers as improper fractions.

(a) $7\frac{3}{5}$ (b) $9\frac{3}{4}$ (c) $21\frac{3}{5}$ (d) $5\frac{2}{7}$

(e) $6\frac{3}{8}$ (f) $7\frac{4}{9}$ (g) $17\frac{5}{6}$ (h) $10\frac{7}{9}$

3 Sabina ate $2\frac{1}{4}$ oranges. Alise ate $1\frac{3}{4}$ oranges.
How many orange quarters did each girl eat?

4 Joshua walks $1\frac{2}{5}$ km to school. Talil walks $\frac{8}{5}$ km to school.
Who has further to walk?

5 On Saturday the temperature at noon was $18\frac{1}{2}$ °C. On Sunday the temperature at noon was $\frac{35}{2}$ °C. Which day was hotter at noon? Explain why.

6 Erich says that $33\frac{1}{3}$ is bigger than $\frac{100}{3}$. He is wrong.
Explain why.

7 Copy and complete with the correct symbols, $<$, $>$ or $=$.

(a) $3\frac{1}{2}$ ☐ $\frac{7}{2}$ (b) $\frac{17}{5}$ ☐ $2\frac{4}{5}$ (c) $\frac{11}{4}$ ☐ $3\frac{1}{4}$

(d) $1\frac{8}{9}$ ☐ $\frac{16}{9}$ (e) $2\frac{4}{11}$ ☐ $\frac{28}{11}$ (f) $\frac{300}{50}$ ☐ 6

Change both numbers into improper fractions or mixed numbers to compare them.

3.4 The four rules for fractions

You can add, subtract, multiply and divide fractions.

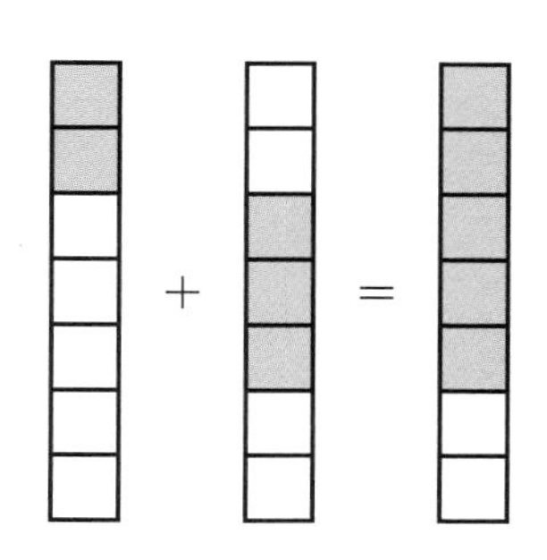

Adding fractions

When fractions have the *same* denominator it is easy to add them. You simply add the numerators.

For example, $\frac{2}{7} + \frac{3}{7} = \frac{5}{7}$

Notice that you do *not* add the denominators.

When fractions do *not* have the same denominator you need to use equivalent fractions to change one or both fractions so that the denominators are the same.

Equivalent fractions are in Section 3.1

For example, $\frac{2}{5} + \frac{3}{10}$

$\frac{2}{5}$ can be written as $\frac{4}{10}$

Multiplying both numerator and denominator by 2.

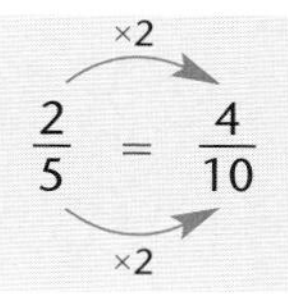

So the addition becomes $\frac{4}{10} + \frac{3}{10} = \frac{7}{10}$

To add fractions, you need to change them to equivalent fractions with the same denominator.

EXAMPLE 10

Work out $\frac{2}{3} + \frac{4}{5}$.

The denominators are 3 and 5. Look for a number that is in the 3-times and 5-times tables. 15 is in both.

$$\frac{2}{3} + \frac{4}{5} = \frac{10}{15} + \frac{12}{15}$$

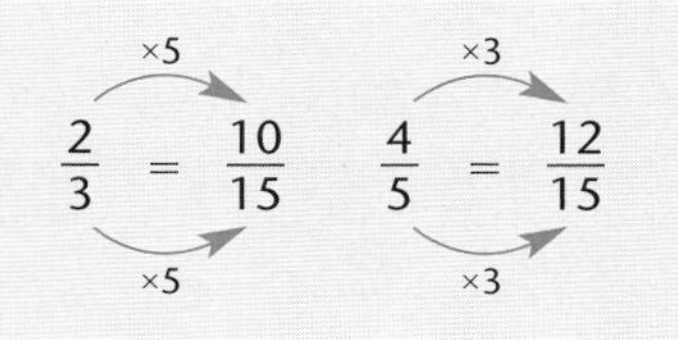

$$= \frac{22}{15}$$

$$= 1\frac{7}{15}$$

See Section 3.3, Improper fractions and mixed numbers.

EXAMPLE 11

Work out $3\frac{1}{4} + 1\frac{5}{6}$.

Method 1

$$3\frac{1}{4} + 1\frac{5}{6} = \frac{13}{4} + \frac{11}{6}$$

Write mixed numbers as improper fractions first.

$$= \frac{39}{12} + \frac{22}{12}$$

Multiplying by 3 and by 2 to get a denominator of 12 in each fraction.

$$= \frac{61}{12}$$

$$= 5\frac{1}{12}$$

Always give your answer as a mixed number.

Continued ▼

Method 2

$3\frac{1}{4} + 1\frac{5}{6} = 3 + 1 + \frac{1}{4} + \frac{5}{6}$

$= 4 + \frac{3}{12} + \frac{10}{12}$

$= 4 + \frac{13}{12}$

$= 4 + 1\frac{1}{12}$

$= 5\frac{1}{12}$

Add whole numbers together.

Multiply to get a common denominator of 12.

Convert improper fraction to mixed number and add.

EXERCISE 3G

1 Work these out.

(a) $\frac{3}{8} + \frac{3}{8}$ (b) $\frac{4}{9} + \frac{7}{9}$ (c) $\frac{3}{11} + \frac{10}{11}$

2 Work these out.

(a) $\frac{3}{4} + \frac{1}{3}$ (b) $\frac{2}{5} + \frac{1}{2}$ (c) $\frac{2}{3} + \frac{3}{5}$

(d) $\frac{4}{5} + \frac{3}{4}$ (e) $\frac{1}{3} + \frac{5}{6}$ (f) $\frac{1}{6} + \frac{3}{8}$

3 Work these out, giving your answers as mixed numbers.

(a) $1\frac{1}{4} + 2\frac{2}{3}$ (b) $1\frac{1}{6} + 2\frac{1}{3}$ (c) $1\frac{3}{5} + 1\frac{1}{10}$

(d) $2\frac{4}{5} + 1\frac{1}{4}$ (e) $2\frac{3}{10} + 1\frac{5}{6}$ (f) $1\frac{3}{8} + 2\frac{2}{3}$

4 After a party I have $\frac{2}{3}$ of one cake and $\frac{1}{6}$ of another identical cake left over. How much cake do I have altogether?

5 I have a piece of rope $3\frac{1}{3}$ metres long and another piece $2\frac{1}{4}$ metres long. What length of rope do I have altogether?

6 Juanita travelled by bus for $2\frac{3}{4}$ hours and then by train for $3\frac{1}{2}$ hours. How long was her journey?

7 A box has a mass of $\frac{1}{8}$ kg. Its contents have a mass of $2\frac{2}{3}$ kg. What is the total mass of the box and its contents?

Subtracting fractions

To subtract fractions, you need to change them to equivalent fractions with the same denominator.

You use the same method as for adding fractions.

EXAMPLE 12

Work these out

(a) $\frac{9}{11} - \frac{3}{11}$ (b) $\frac{7}{8} - \frac{11}{16}$

(c) $\frac{9}{10} - \frac{2}{3}$ (d) $3\frac{1}{4} - 1\frac{5}{6}$

(a) $\frac{9}{11} - \frac{3}{11} = \frac{6}{11}$

The denominators are the same, so you can subtract.

(b) $\frac{7}{8} - \frac{11}{16} = \frac{14}{16} - \frac{11}{16}$

Writing $\frac{7}{8}$ as $\frac{14}{16}$

$= \frac{3}{16}$

(c) $\frac{9}{10} - \frac{2}{3} = \frac{27}{30} - \frac{20}{30}$

Using a denominator of 30 and writing equivalent fractions.

$= \frac{7}{30}$

(d) $3\frac{1}{4} - 1\frac{5}{6} = \frac{13}{4} - \frac{11}{6}$

Convert to improper fractions first. These are the same fractions as in Example 11.

$= \frac{39}{12} - \frac{22}{12}$

$= \frac{17}{12}$

$= 1\frac{5}{12}$

Write the answer as a mixed number.

EXERCISE 3H

1 Work these out.

(a) $\frac{5}{8} - \frac{3}{8}$ (b) $\frac{5}{6} - \frac{1}{6}$ (c) $1\frac{3}{5} - \frac{4}{5}$

2 Work these out.

(a) $\frac{7}{10} - \frac{3}{5}$ (b) $\frac{8}{9} - \frac{5}{6}$ (c) $\frac{3}{4} - \frac{2}{5}$

3 Work these out.

(a) $4\frac{1}{3} - 2\frac{3}{4}$ (b) $2\frac{1}{3} - 1\frac{4}{5}$ (c) $3\frac{1}{8} - 1\frac{2}{3}$

4 I had $1\frac{2}{3}$ bars of chocolate but ate $\frac{3}{4}$ of a bar after lunch. How much do I have left?

5 I cut $1\frac{3}{4}$ m of copper pipe from a length $3\frac{1}{2}$ m long. How much copper pipe do I have now?

6 At my party we drank $3\frac{2}{3}$ litres of cola. There were $4\frac{1}{2}$ litres of cola to start with. How much cola remained?

7 In a garden, $\frac{2}{5}$ is grass and $\frac{1}{3}$ is paving.

(a) How much of the garden is grass and paving together?

(b) How much of the garden is left?

8 Kai spends $\frac{1}{3}$ of his allowance at the cinema and $\frac{1}{6}$ on a magazine. What fraction of his allowance does he have left?

Multiplying fractions

To multiply two fractions you multiply the numerators, and multiply the denominators.

Change mixed numbers to improper fractions before multiplying.

Writing a fraction in its simplest form is in Example 5.

You need to write the answer in its simplest form.

EXAMPLE 13

Work out (a) $\frac{4}{7} \times \frac{3}{5}$ (b) $\frac{2}{3} \times 5$ (c) $1\frac{1}{2} \times \frac{4}{5}$

(a) $\frac{4}{7} \times \frac{3}{5} = \frac{4 \times 3}{7 \times 5} = \frac{12}{35}$

Multiply the numerators and the denominators.

(b) $\frac{2}{3} \times 5 = \frac{2}{3} \times \frac{5}{1} = \frac{2 \times 5}{3 \times 1}$

$= \frac{10}{3}$

$= 3\frac{1}{3}$

Write 5 as $\frac{5}{1}$. Then multiply numerators and denominators. The answer is written in its simplest form.

(c) $1\frac{1}{2} \times \frac{4}{5} = \frac{3}{2} \times \frac{4}{5}$

$= \frac{3 \times 4}{2 \times 5}$

$= \frac{12}{10}$

$= 1\frac{2}{10}$

$= 1\frac{1}{5}$

First write $1\frac{1}{2}$ as an improper fraction.

Write the answer as a mixed number and then in its simplest form.

EXAMPLE 14

Molly buys $1\frac{1}{2}$ metres of material to make a skirt.

She only uses $\frac{3}{4}$ of the material.

How many metres of material does she use?

Remember that 'of' means multiply.

$$\frac{3}{4} \text{ of } 1\frac{1}{2} = \frac{3}{4} \times \frac{3}{2}$$

Write $1\frac{1}{2}$ as an improper fraction.

$$= \frac{3 \times 3}{4 \times 2}$$

$$= \frac{9}{8}$$

$$= 1\frac{1}{8}$$

Write the final answer as a mixed number.

Molly uses $1\frac{1}{8}$ metres of material.

EXERCISE 3I

1 Work these out.

(a) $\frac{3}{5} \times \frac{2}{7}$ (b) $\frac{1}{4} \times \frac{5}{9}$ (c) $\frac{1}{2} \times \frac{1}{2}$

(d) $\frac{1}{4} \times \frac{1}{4}$ (e) $\frac{4}{5} \times \frac{5}{7}$ (f) $\frac{3}{8} \times \frac{5}{9}$

2 Work these out.

(a) $\frac{2}{3}$ of $\frac{3}{8}$ (b) $\frac{5}{7}$ of $\frac{14}{15}$ (c) $\frac{3}{4}$ of $\frac{4}{5}$

'of' means multiply.

3 Work these out.

(a) $1\frac{1}{5} \times 3\frac{1}{3}$ (b) $1\frac{1}{3} \times 1\frac{1}{8}$ (c) $1\frac{5}{19} \times 1\frac{7}{12}$

(d) $2\frac{1}{7} \times 1\frac{2}{5}$ (e) $1\frac{1}{4} \times 1\frac{4}{6}$ (f) $2\frac{1}{2} \times 1\frac{3}{4} \times \frac{3}{10}$

Convert the mixed numbers to improper fractions.

4 Find the area of a rectangular photograph $9\frac{1}{4}$ cm long and $9\frac{1}{5}$ cm wide.

Area of rectangle = length × width.

5 One pizza has a mass of $\frac{2}{5}$ kg. What is the mass of $2\frac{1}{2}$ pizzas?

6 A rectangular path is $3\frac{1}{8}$ metres long and $\frac{2}{5}$ metres wide. What is its area?

7 It takes $3\frac{1}{3}$ minutes to fill a bag with sand. How long will it take to fill $10\frac{1}{2}$ bags?

Reciprocals

The reciprocal of $\frac{3}{4}$ is $\frac{4}{3}$ and the reciprocal of $\frac{4}{3}$ is $\frac{3}{4}$.

The reciprocal of 6 is $\frac{1}{6}$, and the reciprocal of $\frac{1}{6}$ is 6.

To find the reciprocal of a fraction, you **invert** it (turn it upside down).

Remember that $6 = \frac{6}{1}$

When you multiply a number by its reciprocal, you always get 1.

Note 0 has no reciprocal.

EXAMPLE 15

Find the reciprocal of (a) $\frac{5}{7}$ (b) $2\frac{2}{3}$.

(a) $\frac{7}{5}$

Check $\frac{5}{7} \times \frac{7}{5} = \frac{35}{35} = 1$

(b) $2 \times 3 = 6$ $\frac{6}{3} + \frac{2}{3} = \frac{8}{3}$

Reciprocal of $\frac{8}{3}$ is $\frac{3}{8}$

Check $\frac{8}{3} \times \frac{3}{8} = \frac{24}{24} = 1$

Change all mixed numbers into improper fractions before you find the reciprocal.

EXERCISE 3J

1 Invert these fractions.

(a) $\frac{2}{3}$ (b) $\frac{4}{5}$ (c) $1\frac{2}{7}$ (d) $2\frac{5}{9}$

2 Find the reciprocals of these fractions.

(a) $\frac{3}{7}$ (b) $\frac{9}{2}$ (d) 9 (d) $3\frac{4}{5}$

(f) $2\frac{5}{8}$ (g) $\frac{17}{3}$ (h) 18 (i) $10\frac{3}{7}$

Dividing fractions

To divide two fractions

1. **invert** (turn upside down) the fraction you are dividing by
2. change the division sign to a multiplication sign.

So $\frac{1}{3} \div \frac{2}{5} = \frac{1}{3} \times \frac{5}{2} = \frac{5}{6}$

$4 \div \frac{1}{2}$ means how many halves in 4?

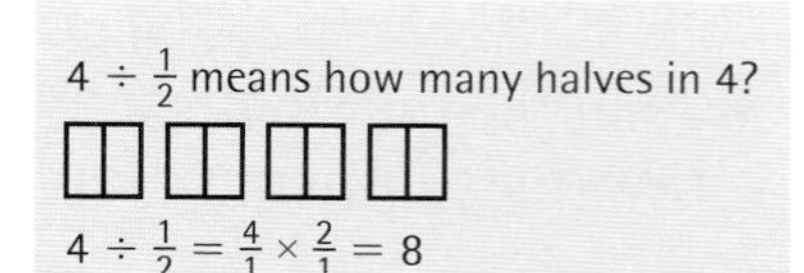

$4 \div \frac{1}{2} = \frac{4}{1} \times \frac{2}{1} = 8$

You turn the second fraction upside down. This is the fraction you are dividing by.

If the division involves mixed numbers, change these to improper fractions first.

EXAMPLE 16

Work out (a) $\frac{7}{8} \div 3$ (b) $\frac{2}{9} \div \frac{3}{4}$ (c) $1\frac{3}{4} \div \frac{5}{6}$

(a) $\frac{7}{8} \div 3 = \frac{7}{8} \div \frac{3}{1}$

$= \frac{7}{8} \times \frac{1}{3}$

$= \frac{7}{24}$

Write 3 as $\frac{3}{1}$, then use the rule for division.

(b) $\frac{2}{9} \div \frac{3}{4} = \frac{2}{9} \times \frac{4}{3}$

$= \frac{2 \times 4}{9 \times 3}$

$= \frac{8}{27}$

(c) $1\frac{3}{4} \div \frac{5}{6} = \frac{7}{4} \div \frac{5}{6}$

Write $1\frac{3}{4}$ as an improper fraction.

$= \frac{7}{4} \times \frac{6}{5}$

$= \frac{42}{20}$

$= \frac{21}{10}$

$= 2\frac{1}{10}$

Write the answer as a mixed number in its simplest form.

EXERCISE 3K

1 Work these out.

(a) $\frac{1}{4} \div 3$ (b) $\frac{2}{5} \div \frac{3}{8}$ (c) $7 \div \frac{2}{5}$

(d) $\frac{3}{4} \div 12$ (e) $\frac{3}{4} \div \frac{3}{8}$ (f) $\frac{5}{9} \div \frac{2}{3}$

2 Work these out.

(a) $1\frac{1}{2} \div \frac{3}{4}$ (b) $1\frac{2}{3} \div 1\frac{4}{6}$ (c) $5\frac{1}{4} \div 3\frac{1}{2}$

3 To wrap a parcel takes $\frac{3}{5}$ m of string. How many parcels can I wrap with $7\frac{1}{2}$ m of string?

4 $3\frac{1}{2}$ litres of orange juice is poured into $\frac{1}{3}$ litre glasses. How many full glasses are there?

5 How many pieces of ribbon $\frac{3}{8}$ m long can I cut from a piece $2\frac{1}{4}$ m long?

6 Share $2\frac{1}{4}$ cakes between 6 people. How much will they each receive? Give your answer in its simplest form.

3.5 Fractions, decimals and percentages

You will need to know

- about place value
- how to multiply and divide by 10, 100, 1000

Changing a percentage into a fraction or a decimal

There is more on percentages in Chapter 14.

Percentage (%) means 'out of a hundred'.

So if you score 60 marks out of 100 in a mathematics test your mark can be written as $\frac{60}{100}$ or as 60%.

To change a percentage to a fraction, you write it as a fraction with a denominator of 100.

So $78\% = \frac{78}{100}$, $43\% = \frac{43}{100}$, $7\% = \frac{7}{100}$

Always use a denominator of 100 to change a percentage into a fraction.

When you have written a percentage as a fraction with denominator 100, you can change it to its simplest form.

EXAMPLE 17

Write these percentages as fractions in their simplest form.

(a) 58% (b) 95% (c) 8% (d) $12\frac{1}{2}\%$

(a), (b) and (c) use the 'simplifying' skills from Section 3.1.

(a) $58\% = \frac{58}{100} = \frac{58}{100} = \frac{29}{50}$ (÷2 top and bottom)

(b) $95\% = \frac{95}{100} = \frac{95}{100} = \frac{19}{20}$ (÷5 top and bottom)

(c) $8\% = \frac{8}{100} = \frac{8}{100} = \frac{2}{25}$ (÷4 top and bottom)

(d) $12\frac{1}{2}\% = \frac{12\frac{1}{2}}{100} = \frac{12\frac{1}{2}}{100} = \frac{25}{200} = \frac{1}{8}$ (×2 top and bottom, then ÷25 top and bottom)

Multiply by 2 first to make the numerator a whole number. Don't stop at $\frac{25}{200}$, you must simplify as much as possible.

You already know how to divide by 100 using decimals. You move the digits 2 places to the right.

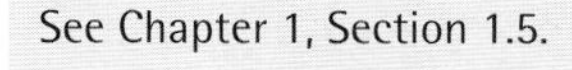

See Chapter 1, Section 1.5.

You can use this to change a percentage to a decimal.

$78\% = \frac{78}{100} = 0.78$

$\frac{78}{100}$ means $78 \div 100$

To change a percentage to a decimal

1 Write it as a fraction with a denominator of 100.

2 Divide the numerator by 100 (move the digits 2 places to the right).

EXAMPLE 18

Change these percentages to decimals.

(a) 43% (b) 7%

(a) $43\% = \frac{43}{100} = 0.43$

(b) $7\% = \frac{7}{100} = 0.07$

EXERCISE 3L

1 Write these percentages as fractions in their simplest form.

(a) 50% (b) 20% (c) 40% (d) 10%
(e) 18% (f) 26% (g) 38% (h) 16%
(i) 25% (j) 36% (k) 65% (l) 96%

2 Copy and complete each of these.

(a) $2\frac{1}{2}\% = \frac{\square}{40}$ (b) $3\frac{1}{3}\% = \frac{\square}{30}$ (c) $17\frac{1}{2}\% = \frac{7}{\square}$

(d) $3\frac{3}{4}\% = \frac{3}{\square}$ (e) $11\frac{2}{3}\% = \frac{7}{\square}$ (f) $37\frac{1}{2}\% = \frac{3}{\square}$

(g) $5\frac{5}{8}\% = \frac{\square}{160}$ (h) $14\frac{2}{7}\% = \frac{\square}{7}$ (i) $9\frac{3}{8}\% = \frac{\square}{32}$

3 Copy and complete with the correct sign, $<$, $>$ or $=$.

(a) $75\% \; \square \; \frac{3}{4}$ (b) $15\% \; \square \; \frac{1}{20}$ (c) $56\% \; \square \; \frac{13}{25}$

(d) $84\% \; \square \; \frac{19}{25}$ (e) $7\frac{1}{2}\% \; \square \; \frac{7}{40}$ (f) $66\frac{2}{3}\% \; \square \; \frac{2}{3}$

4 Write these percentages as decimals.

(a) 69% (b) 43% (c) 25% (d) 89%
(e) 50% (f) 60% (g) 10% (h) 5%
(i) 8% (j) 57.5% (k) 17.5% (l) 2.5%

5 Write these percentages as
(i) fractions in their simplest form
(ii) decimals.

(a) 30% (b) 90% (c) 15% (d) 85%
(e) 2% (f) 8% (g) 123% (h) 12.5%

6 Two friends were comparing test results. Yvan got 70% and Jacques got $\frac{12}{20}$. By changing them both into fractions in their simplest form, find which one scored the highest in the test.

7 Which of these percentages is equivalent to $\frac{9}{25}$?

35% 36% 40%

Changing a decimal or a fraction into a percentage

Changing a decimal into a percentage is the reverse of changing a percentage to a decimal.

$$85\% = \frac{85}{100} = 0.85$$

(÷100 from left to right; ×100 from right to left)

So $0.85 = (0.85 \times 100)\% = 85\%$

To multiply by 100, move the digits 2 places to the left. See Section 1.5.

EXAMPLE 19

Change into percentages.

(a) 0.075 (b) 1.64

(a) $0.075 = (0.075 \times 100)\% = 7.5\%$
(b) $1.64 = (1.64 \times 100)\% = 164\%$

$1.64 > 1$, so the answer is more than 100%.

To change a decimal to a percentage, multiply by 100.

To change a fraction into a percentage you multiply by 100.

You can go directly from a fraction to a % like this

$\frac{2}{5} \times 100 = \frac{200}{5} = 40\%$

EXAMPLE 20

Write these fractions as percentages.

(a) $\frac{9}{10}$ **(b)** $\frac{31}{40}$

(a) $\frac{9}{10} = (\frac{9}{10} \times 100)\% = 90\%$

(b) $\frac{31}{40} = (31 \div 40 \times 100)\% = 77.5\%$

When using a calculator, you can either use the fraction button as in part (a) or the division button as in (b).

To change a fraction to a percentage, multiply by 100.

Now that you can change between percentages, fractions or decimals, you can make comparisons between all three.

EXAMPLE 21

Write these in order of size, smallest first.

$\frac{2}{5}$ 0.39 45% $\frac{7}{20}$

$\frac{2}{5} = 0.4$

$45\% = \frac{45}{100} = 0.45$

$\frac{7}{20} = 0.35$

Since $0.35 < 0.39 < 0.4 < 0.45$

The order is $\frac{7}{20} < 0.39 < \frac{2}{5} < 45\%$

A calculator has been used to change the quantities into decimals.

EXERCISE 3M

1 Write these decimals as percentages.

(a) 0.32 (b) 0.79 (c) 2.39 (d) 0.125

2 Write these fractions as decimals and then as percentages.

(a) $\frac{7}{10}$ (b) $\frac{4}{5}$ (c) $\frac{3}{4}$ (d) $\frac{3}{8}$

3 Write these as percentages and then put them in order from smallest to largest.

$\frac{1}{10}$ $\frac{87}{100}$ $\frac{2}{5}$ $\frac{8}{25}$

4 Write the following fractions as percentages.

(a) $\frac{17}{34}$ (b) $\frac{15}{60}$ (c) $\frac{25}{125}$ (d) $\frac{50}{32}$

5 Copy and complete the following table of equivalent fractions, decimals and percentages.

Percentage	Fraction	Decimal
60%		
		0.48
	$\frac{3}{10}$	
		1.75
5%		

6 Yvette was comparing her test results. She had $\frac{16}{20}$ in History, 75% in Mathematics and $\frac{28}{40}$ in English. By changing all the results into percentages, put her results in order from lowest to highest.

Change both into percentages to compare them.

7 In a sale you can buy a TV with '15% off' or one with a '$\frac{1}{4}$ off' the original price. Which one is the best offer?

8 Write these in order from smallest to largest.

(a) 64% $\frac{7}{10}$ 0.625 $\frac{14}{25}$

(b) 0.438 $\frac{9}{20}$ 42% 0.4

9 Arrange these in order of size starting with the smallest.

(a) 37% 0.68 0.27 **(b)** 42% 38% 0.45

(c) 80% $\frac{9}{10}$ $\frac{7}{10}$ **(d)** 34% $\frac{38}{100}$ $\frac{13}{50}$

(e) $\frac{6}{25}$ 0.25 23%

Changing a decimal into a fraction

To write a decimal as a fraction you need to look at the place value of the **last** significant figure.

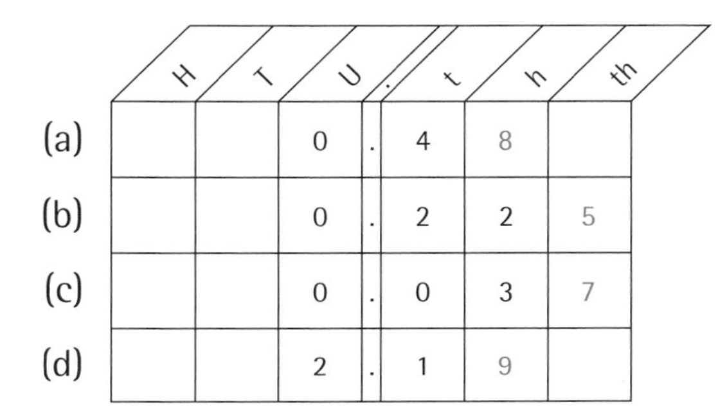

For significant figures see Section 2.1. For place value see Section 1.1.

In **(a)** the last significant figure is in the hundredths column.

So $0.48 = \frac{48}{100}$

$\frac{48}{100}$ simplifies to $\frac{12}{25}$

'hundredths' have denominator 100.

$\frac{48}{100} = \frac{12}{25}$ (÷4 top and bottom)

In **(b)** the last significant figure is in the thousandths column.

So $0.225 = \frac{225}{1000}$

'thousandths' have denominator 1000.

$$\frac{225}{1000} = \frac{9}{40}$$ (÷25 top and bottom)

In **(c)** the last significant figure is in the thousandths column.

So $0.037 = \frac{37}{1000}$

$\frac{37}{1000}$ cannot be simplified further.

(d) is a mixed number – it has a whole number part and a fractional (decimal) part. The last significant figure of the decimal part is in the hundredths column, so $2.19 = 2\frac{19}{100}$.

Decimals such as 0.48, 0.225 and 0.037 are **terminating decimals**. They end, or terminate, rather than going on for ever.

EXERCISE 3N

1 Change the following decimals into fractions.
Give each answer in its simplest form.

(a) 0.3 (b) 2.7 (c) 0.09
(d) 0.007 (e) 0.75 (f) 0.025
(g) 0.45 (h) 0.625 (i) 2.05
(j) 16.125 (k) 4.105 (l) 23.24

3.6 Definitions of rational and irrational numbers

All numbers in our number system are either rational or irrational.

A **rational** number is one that can be written in the form $\frac{a}{b}$ where a and b are integers and $b \neq 0$.
An **irrational** number is one that cannot be written in this way.

You will see more of these later in the chapter.

$\neq$ means 'not equal to'.

All integers are rational. For example $7 = \frac{7}{1}$ and $-8 = \frac{-8}{1}$

All fractions are rational. For example $\frac{1}{4}, \frac{5}{2}, \frac{8}{11}$

Some decimals are rational. For example $2.79 = \frac{279}{100}$,

$0.3 = \frac{3}{10}$

Some, as you will see, are not.

3.7 Terminating and recurring decimals

All rational numbers can be written as decimals, some of which are **terminating** decimals and some of which are not.

Terminating means that the decimal ends – there is a finite number of decimal places.

Here are two examples of terminating decimals

$\frac{7}{2} = 3.5 \qquad \frac{1}{8} = 0.125$

To change a fraction into a decimal you can do the division on your calculator.

For example, $\frac{5}{8} = 0.625$

Only those fractions that have a denominator with prime factors of *only* 2 and/or 5 can be converted to terminating decimals.

Fractions that do not satisfy this condition convert to decimals that repeat and do not terminate.

Look at these fractions written as decimals.

$\frac{1}{3} = 0.333\,333\,\ldots \qquad \frac{5}{12} = 0.416\,666\,\ldots$

$\frac{8}{11} = 0.727\,272\,\ldots \qquad \frac{3}{7} = 0.428\,571\,428\,571\,42\,\ldots$

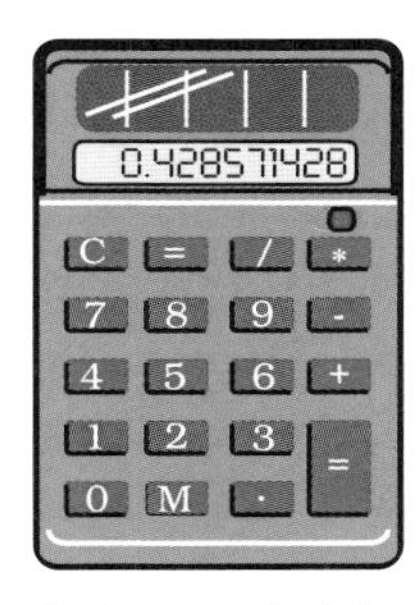

This is what your calculator display will show when you work out 3 ÷ 7.

When you work out the decimal answer on your calculator, the decimal places fill the whole of the calculator display.

If you do the division by short or long division methods you will see that the decimals never end.

The decimals do not terminate – they are called **recurring** decimals.

Recurring decimals have either a repeating digit or a repeating pattern of digits.

Because they never end, you need to have a way of writing them which makes the repeating pattern clear.

When a single digit repeats, you put a dot over this digit.

$\frac{1}{3} = 0.333\,333\ldots$ is written as $0.\dot{3}$

$\frac{5}{12} = 0.416\,666\ldots$ is written as $0.41\dot{6}$

When more than one digit repeats, you put a dot over the first digit of the pattern and a dot over the last digit of the pattern.

$\frac{8}{11} = 0.727272\ldots$ is written as $0.\dot{7}\dot{2}$

$\frac{3}{7} = 0.42857142857142\ldots$ is written as $0.\dot{4}2857\dot{1}$

Some fractions give decimals where the pattern does not start repeating immediately.

$\frac{47}{110} = 0.4272727\ldots$ which is written as $0.4\dot{2}\dot{7}$

EXERCISE 30

1 Use any method to change these fractions into decimals. Then say if they are terminating or recurring decimals.

(a) $\frac{2}{3}$ (b) $\frac{4}{5}$ (c) $\frac{4}{9}$ (d) $\frac{5}{11}$ (e) $\frac{7}{8}$

(f) $\frac{7}{12}$ (g) $\frac{6}{25}$ (h) $\frac{3}{16}$ (i) $\frac{5}{6}$ (j) $\frac{5}{7}$

2 Without converting these fractions into decimals, decide which of the denominators indicate terminating decimals and which indicate recurring decimals.

(a) $\frac{3}{8}$ (b) $\frac{5}{12}$ (c) $\frac{7}{16}$ (d) $\frac{9}{11}$ (e) $\frac{11}{64}$

(f) $\frac{13}{20}$ (g) $\frac{12}{22}$ (h) $\frac{17}{24}$ (i) $\frac{19}{125}$ (j) $\frac{21}{128}$

For a terminating decimal, the prime factors of the denominator of the fraction can only be 2 or 5.

3 Convert the following fractions into decimals, and write down how many numbers there are in the recurring patterns.

(a) $\frac{7}{9}$ (b) $\frac{8}{11}$ (c) $\frac{5}{7}$ (d) $\frac{14}{27}$ (e) $\frac{142}{111}$

4 Investigate when happens when you multiply

(a) a fraction that gives a terminating decimal by another fraction that gives a terminating decimal. Will your answer always be a terminating decimal?

(b) a fraction that gives a recurring decimal by another fraction that gives a recurring decimal. Will your answer always be a recurring decimal?

(c) a fraction that gives a terminating decimal by another fraction that gives a recurring decimal. Will the answer always be terminating, recurring or can't you tell?

5 Write out the full calculator display for the following recurring decimals, and the number of digits in the recurring pattern. Use them to answer the following questions.

$\frac{1}{3}$ $\frac{1}{6}$ $\frac{1}{7}$ $\frac{1}{9}$ $\frac{1}{12}$ $\frac{1}{13}$

(a) Is the number of digits in any of the recurring patterns larger than the denominator?

(b) How many recurring digits are there in the decimals of fractions with an even number denominator?

(c) Is there a connection between the denominators of the fractions with 1 recurring digit?

(d) Is there any connection between the denominators of the fractions with 6 recurring digits?

(e) Do your observations work with other fractions that give recurring decimals? Test them out on at least another four fractions each. What can you say about finding patterns from a small set of numbers?

3.8 Converting recurring decimals into fractions

To turn a decimal into a fraction is easy if the decimal terminates. Look at the place value of the last significant digit and write the fraction with this denominator. Simplify the fraction if possible.

2 decimal places means hundredths, 3 decimal places means thousandths, and so on.

For example, $0.48 = \frac{48}{100} = \frac{12}{25}$ $\quad 0.225 = \frac{225}{1000} = \frac{9}{40}$

When the decimal is a recurring decimal you need a special technique to find the equivalent fraction.

EXAMPLE 22

Write each of these recurring decimals as fractions in their simplest form.

(a) 0.888 88...

(b) 0.454 545...

(c) 0.610 810 810 8...

Continued ▼

(a) Let $x = 0.8888...$

$10x = 8.8888...$

Subtracting $9x = 8$

Dividing both sides by 9 $x = \frac{8}{9}$

(b) Let $x = 0.454545...$

$100x = 45.454545...$

Subtracting $99x = 45$

Dividing both sides by 99 $x = \frac{45}{99}$

Dividing top and bottom by 9 $x = \frac{5}{11}$

(c) Let $x = 0.610\,810\,810\,8...$

$1000x = 610.810\,810\,810$

Subtracting $999x = 610.2$

Dividing both sides by 999 $x = \frac{610.2}{999}$

Multiplying top and bottom by 10 $x = \frac{6102}{9990}$

Dividing top and bottom by 9 $x = \frac{678}{1110}$

Dividing top and bottom by 6 $x = \frac{113}{185}$

Example 1(c) can be done another way

Let $x = 0.610\,810\,810\,8...$

then $10x = 6.108108...$

and $10\,000x = 6108.108108...$

Subtracting $9990x = 6102$

Dividing both sides by 9990 $x = \frac{6102}{9990}$

$x = \frac{113}{185}$

When 1 digit recurs multiply both sides by 10.
When 2 digits recur multiply both sides by 100.
When 3 digits recur multiply both sides by 1000.

You must not leave your answer as $\frac{610.2}{999}$ because a fraction should consist of whole numbers.
Notice that there is still quite a lot of work to do to get the answer into its simplest form.
You will not score full marks if you do not simplify fully.

This method requires *two* multiplications at the start but then has the advantage that the *recurring decimal pattern is the same.*
When you subtract you automatically get whole numbers on the top and bottom of your fraction.
Both methods give you the same answer.

You can choose whichever method you prefer, but if you choose the second one you must remember to multiply x by multiples of 10 which will give you the same pattern of recurring decimals.

EXERCISE 3P

1 Write these recurring decimals using the dot notation.

(a) 0.666 666... (b) 0.111 111... (c) 0.733 333...
(d) 0.592 222... (e) 0.161 616... (f) 0.959 595...
(g) 0.123123123... (h) 0.175175... (i) 0.9642642642...
(j) 0.052 929 29... (k) 0.357 235 723 572...
(l) 0.303 130 313 031...

2 Write down the first 10 digits of these recurring decimals.

(a) $0.\dot{5}$ (b) $0.\dot{2}\dot{7}$ (c) $0.\dot{6}\dot{7}$ (d) $0.7\dot{1}\dot{5}$
(e) $0.\dot{5}1\dot{3}$ (f) $0.3\dot{2}7\dot{5}$ (g) $0.10\dot{3}9\dot{2}$ (h) $0.7\dot{8}05\dot{2}$

3 Change these fractions to decimals. Write

(i) the full calculator display.
(ii) the recurring decimal using the 'dot' notation.

(a) $\frac{2}{9}$ (b) $\frac{1}{6}$ (c) $\frac{7}{9}$ (d) $\frac{4}{15}$ (e) $\frac{5}{33}$
(f) $\frac{7}{18}$ (g) $\frac{7}{11}$ (h) $\frac{5}{12}$ (i) $\frac{5}{22}$ (j) $\frac{11}{24}$

4 Write these fractions as recurring decimals using the 'dot' notation.

(a) $\frac{1}{3}$ (b) $\frac{2}{3}$ (c) $\frac{7}{9}$
(d) $\frac{4}{11}$ (e) $\frac{5}{6}$ (f) $\frac{45}{55}$
(g) $\frac{5}{54}$ (h) $1\frac{1}{6}$ (i) $3\frac{7}{12}$
(j) $8\frac{3}{18}$ (k) $2\frac{16}{36}$ (l) $3\frac{1}{7}$

5 Find the fractions that are equivalent to the following terminating decimals. Express the fractions in their simplest terms.

(a) 0.45 (b) 0.375 (c) 0.005

6 Find the fractions that are equivalent to the following recurring decimals.

(a) 0.5555... (b) 0.8888...

(c) 0.212 121... (d) 0.454 545...

(e) 2.636 363... (f) 0.320 320 320...

(g) 0.624 624 624... (h) 0.157 815 78...

(i) 0.123 512 35... (j) 0.166 66...

(k) 0.045 454 5... (l) 6.042 042 042...

(m) 2.161 616... (n) 15.08 888 8...

(o) 4.032 532 5325...

3.9 Types of number

π $\sqrt{2}$ tan 10° 0.62 $2^{0.4}$ 1.59 3×10^{-2}

cos 20° −12 sin 60° $\frac{1}{2}$ 7 −6 60%

4^6 $1\frac{1}{4}$ 120 108% $\sqrt{-8}$ $0.\dot{1}4285\dot{7}$ 6×105

37 121 8 0.1268 $0.\dot{1}$ $\frac{7}{6}$ $\sqrt{4}$

You have studied many of the following types of number already and this section will show you how they are connected to each other. You will meet the other types of number listed below as you continue with your studies.

Topic	Example	Page
Whole numbers (integers)	1, 25, 1000	1
Negative integers	−2, −21, −29, −2090	19
Decimals	0.62, 1.59, 183.62	27
Fractions	$\frac{1}{2}$, $2\frac{1}{4}$, $23\frac{3}{4}$	60
Indices	2^5, 5^{-1}, 10^{-6}	182
Standard Form	3 × 10–2, 6.134 × 107	191
Percentages	60%, 109%, 4.06%	375
Rational numbers	All of the above	86, 359
Irrational numbers	π, tan 10°, cos 20°, sin 60°, $\sqrt{45}$	86, 544
Real numbers	All rational and irrational numbers	All

It is useful to know that, other than in the table, there is only one other type of number. The square roots of all negative numbers are called imaginary or complex numbers. This type of number is not in your syllabus.

The number tree that follows is included to help you understand how they all fit together.

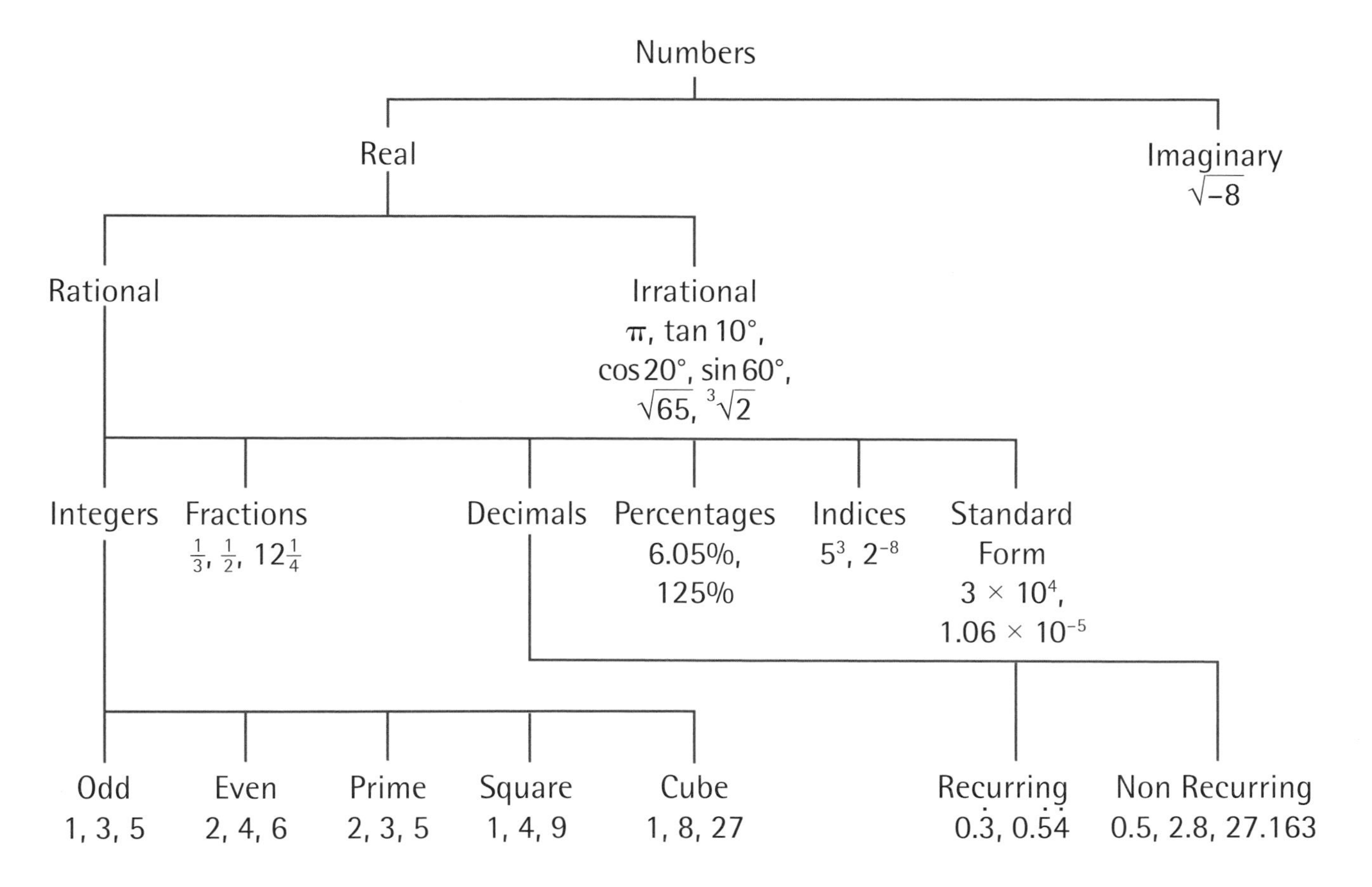

EXAMINATION QUESTIONS

1 Write down $\frac{1}{6}$ and 17%. Put one of the symbols $<$, $=$, or $>$ between $\frac{1}{6}$ and 17% to make a correct statement. [1]

(CIE Paper 1, Jun 2000)

---

2 **Showing all your working**, calculate $1\frac{1}{4} \div \frac{2}{3} - 1\frac{1}{3}$. [2]

(CIE Paper 2, Nov 2000)

---

3 Write

(a) $\frac{27}{50}$ as a decimal, [1]

(b) $\frac{83}{1000}$ as a percentage. [1]

(CIE Paper 1, Jun 2001)

---

4 Copy and fill in the missing numbers in the following statement.

$\frac{99}{132} = \frac{33}{\square} = \frac{\square}{12}$. [2]

(CIE Paper 1, Jun 2001)

---

5 **Showing your working**, calculate $1 - \frac{1}{3} \times \frac{1}{4}$. [2]

(CIE Paper 1, Nov 2001)

---

6 Last week, Mr and Mrs Hernandez spent $120.
This was $\frac{3}{8}$ of their earnings
What were their earnings? [2]

(CIE Paper 1, Nov 2001)

---

7 $\frac{5}{6}$ 8.3% 0.825 0.83 $\frac{33}{40}$

Five values are listed above. Find

(a) which two are equal, [1]

(b) which one is the smallest, [1]

(c) which one is the largest. [1]

(CIE Paper 1, Nov 2001)

---

8 Write as a decimal

(a) $\frac{7}{20}$, [1]

(b) 127%. [1]

(CIE Paper 1, Jun 2002)

---

9 Write 24% as a fraction in its lowest terms. [2]

(CIE Paper 1, Nov 2002)

---

10 Ian and Joe start to dig a garden. They both dig at the same rate.

(a) When they are half-way through the job, what fraction of the garden has Ian dug? [2]

(b) Keith arrives to help. All three dig at the same rate until the job is finished.

(i) What fraction of the garden did Ian dig after Keith arrived? [2]

(ii) What fraction of the garden did Ian dig altogether? [2]

(CIE Paper 3, Nov 2002)

---

11 **(a)** Write in order of size, smallest first 0.68 $\frac{33}{50}$ 67% [1]

(b) Convert 0.68 into a fraction in its lowest terms. [1]

(CIE Paper 1, Jun 2003)

---

12 **Show all your working** in the following calculations.

The answers are given so it is only your working that will be given marks.

(a) $\frac{1}{2} + \frac{2}{3} = 1\frac{1}{6}$, [2]

(b) $1\frac{1}{5} \times 1\frac{3}{4} = 2\frac{1}{10}$. [2]

(CIE Paper 1, Jun 2003)

---

13 Work out each of the following as a decimal.

(a) (i) 28% [1]

(ii) $\frac{275}{1000}$ [1]

(iii) $\frac{2}{7}$ [1]

(b) Write 28%, 275/1000 and $\frac{2}{7}$ in order of size, smallest first. [1]

(CIE Paper 1, Nov 2003)

---

14 Write as a fraction in its lowest terms

(a) 75%, [1]

(b) 0.07. [1]

(CIE Paper 1, Jun 2004)

---

15 Without using a calculator, work out $2\frac{1}{4} \div \frac{1}{2}$ as a single fraction.

Show all your working. [2]

(CIE Paper 1, Jun 2004)

---

16 $\frac{3}{5} \div \frac{7}{10} = \frac{6}{7}$

Show how this calculation is done without using a calculator.

Write down the working. [2]

(CIE Paper 1, Nov 2004)

---

17 Anne took a test in Chemistry.

She scored 20 marks out of 50.

Work out her percentage mark. [1]

(CIE Paper 1, Jun 2005)

---

18 Alphonse spends $28 on food.

This amount is $\frac{4}{9}$ of his allowance.

Calculate his allowance. [2]

(CIE Paper 1, Jun 2005)

---

19 Work out $\frac{5}{6} - \frac{3}{8}$.
Give your answer as a fraction in its lowest terms.
You must show all your working. [2]

(CIE Paper 1, Jun 2005)

Chapter 4

Basic rules of algebra

This chapter will show you how to

- ✔ use letters to represent numbers
- ✔ write simple expressions using letters to represent unknown numbers
- ✔ simplify algebraic expressions by collecting like terms
- ✔ multiply out brackets (by multiplying a single term over a bracket)
- ✔ simplify expressions involving brackets

4.1 Using letters to write simple expressions

To solve problems you often have to use a letter to represent an **unknown** number.

Using letters in mathematics is called **algebra**.

Suppose you have a bag of sweets but you do not know how many sweets are in the bag.

You could use s to represent this unknown quantity.

If you now add 4 sweets to the bag you will have s sweets plus 4 sweets. You can write this as $s + 4$ sweets.

$s + 4$ is called an **expression** in terms of s.

EXAMPLE 1

Use algebra to write expressions for these.

(a) 3 more than x **(b)** 5 less than w **(c)** a added to b

(a) $x + 3$

(b) $w - 5$

(c) $a + b$

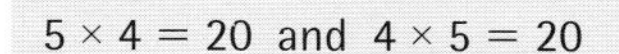

When you multiply two numbers the order in which you multiply them does not matter.

$5 \times 4 = 20$ and $4 \times 5 = 20$

It is the same when you are using algebra.

$3 \times x$ and $x \times 3$ are both the same and can be thought of as 3 lots of x, or $x + x + x$.

You write 3 lots of x as $3x$.

Always put the number first.

y times y, or $y \times y$, is written as y^2 ('y squared').

$d \div 4$ can be written as $\frac{d}{4}$.

You can also multiply two unknowns together.

$g \times h = gh$, $\quad x \times y = xy$

You leave out the '× sign'.

EXAMPLE 2

Use algebra to write expressions for these.

(a) A bag contains s sweets. 4 are taken out. How many sweets are left?

(b) $a + a + a + a$.

(c) I have 5 boxes of strawberries. Each has n strawberries. How many strawberries do I have altogether?

(d) x oranges are shared equally between three people. How many oranges does each one get?

(a) $s - 4$

(b) $a + a + a + a = 4a$

(c) $5 \times n = 5n$

(d) $x \div 3 = \frac{x}{3}$

Remember to put the number first.

EXAMPLE 3

Write expressions for the area of these shapes.

(a) a rectangle of length l and width w

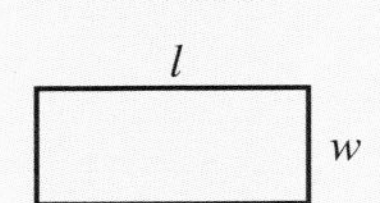

(b) a square of side b

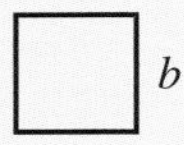

Area of a rectangle = length × width.

(a) Area $= l \times w$

$= lw$

(b) Area $= b \times b$

$= b^2$

EXERCISE 4A

1 Use algebra to write these expressions.

(a) 5 more than x (b) 3 less than w

(c) 8 more than m (d) 12 less than d

(e) 6 added to x (f) y subtract 2

(g) p added to 4 (h) 1 taken away from a

(i) x added to y (j) r take away t

2 Write these expressions using algebra.

(a) $g + g + g + g$ (b) $r + r + r + r + r$

(c) $h + h + h + h + h + h$ (d) $t + t + t$

3 Write an algebraic expression for each of these.

(a) 3 lots of y (b) z divided by 3

(c) k divided by 4 (d) f times 8

(e) n multiplied by 10 (f) 12 divided by x

(g) a multiplied by b (h) g multiplied by g

4 Use algebra to write expressions for these.
Use x to stand for the number.

(a) Choose a number and add 2

(b) Choose a number and multiply it by 7

(c) Choose a number and take 5 away

(d) Choose a number and double it

(e) Choose a number and halve it

(f) Choose a number, multiply it by 3, then subtract 2

(g) Choose a number, divide it by 4, then add 5

(h) Choose a number and multiply it by itself.

5 Write expressions for the area of these shapes.

(a) square, side a

(b) square, side x

(c) rectangle, width 4 and length l

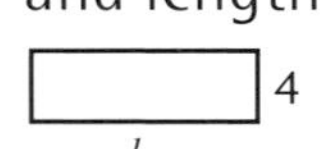

(d) rectangle, width x and length y

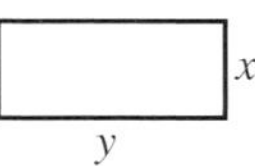

6 Write an expression for the total cost, in dollars, of

(a) 4 CDs at x dollars each

(b) 6 DVDs at y dollars each

(c) d CDs at \$4 each

(d) t DVDs at \$13 each

(e) 5 CDs at j dollars each and 8 DVDs at k dollars each

(f) a CDs at \$3 each and b DVDs at \$11 each

7 Hayley has 4 bags of beads. Each bag has n beads in it.

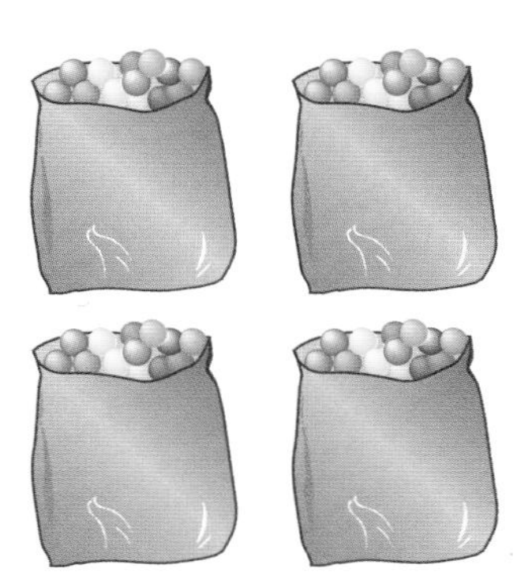

(a) How many beads does she have altogether?

(b) Hayley puts 2 more beads in each bag.
How many beads are in each bag now?

(c) How many beads does she have altogether now?

(d) Hayley gives 3 beads from one bag to her friend.
How many beads are left in this bag?

4.2 Simplifying algebraic expressions

You will need to know

- how to add and subtract positive and negative numbers

You use the same rules for algebra as you do when working with numbers.

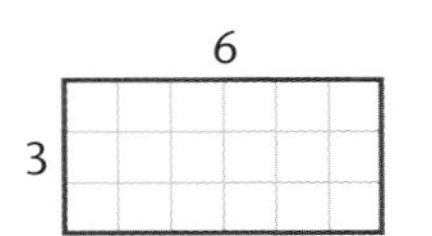

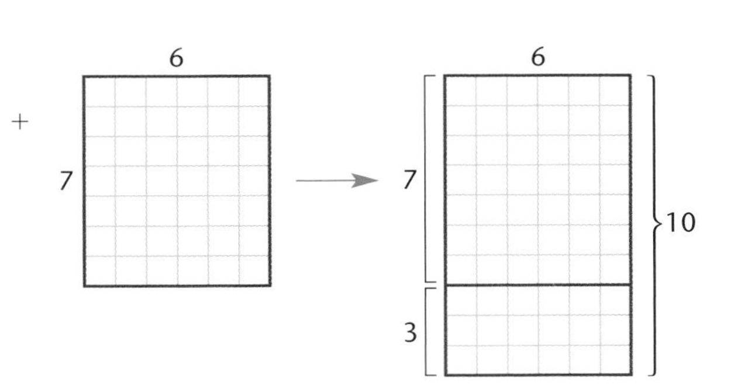

3 lots of 6, added to 7 lots of 6, is the same as 10 lots of 6

$(3 \times 6) + (7 \times 6) = 10 \times 6$

because $3 + 7 = 10$

You can do the same when you have a letter standing for an unknown quantity.

Remember that $3n$ means $3 \times n$.

$$3n + 7n = 10n$$

$$5x + x = 6x$$

x means $1x$ (you don't need to write the 1).

The same works when you subtract (take away).

10 sixes, take away 7 sixes, is the same as 3 sixes.

$(10 \times 6) - (7 \times 6) = 3 \times 6$
because $10 - 7 = 3$

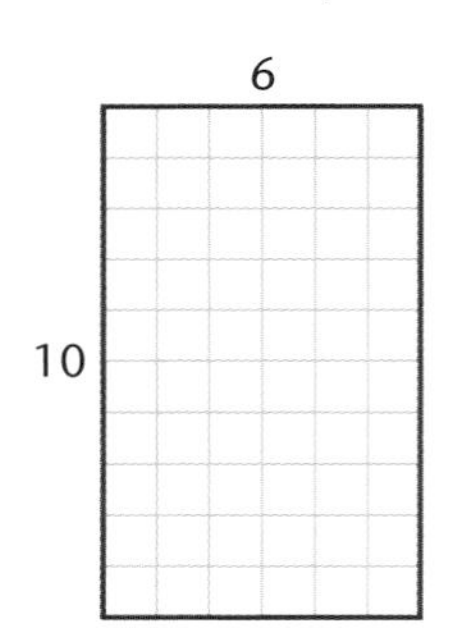

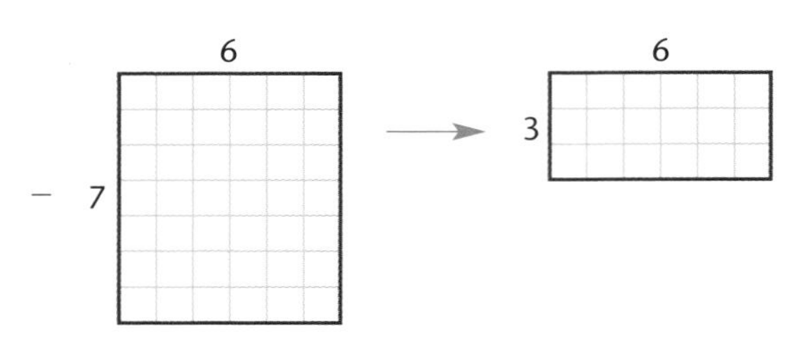

When you have a letter it stands for an unknown quantity

$$10n - 7n = 3n$$

$$6x - x = 5x$$

These answers are written in a shorter form than the original expression.

To **simplify** an expression, you write it in as short a form as possible.

EXAMPLE 4

Write each of these expressions in a shorter form.

(a)

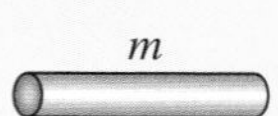

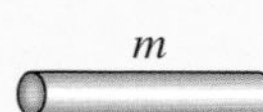

What is the total length of pipe?

(b) $3h + 6h + 2h$

(c) $6b + 5b - 7b$

(d) $6x - 2x + 8x - 10x$

(e) t [4] $+ t$ [5] $+ t$ [3]

What is the total area?

(a) $m + m + m = 3m$ — Remember to write the number first.

(b) $3h + 6h + 2h = 11h$ — $3 + 6 + 2 = 11$

(c) $6b + 5b - 7b = 11b - 7b$

$= 4b$ — $6 + 5 = 11$, $11 - 7 = 4$

Continued ▼

(d) $6x - 2x + 8x - 10x = 6x + 8x - 2x - 10x$
$= 14x - 12x$
$= 2x$

Just as with numbers you can add and subtract the terms in any order as long as each term keeps its own sign.
$6 - 2 + 8 - 10 = 2$

(e) $4t + 5t + 3t = 12t$

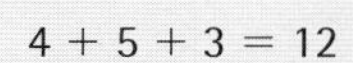

$4 + 5 + 3 = 12$

EXERCISE 4B

1 Write each of these expressions in a shorter form.

(a) 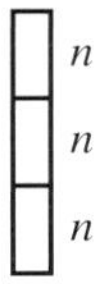

What is the total height?

(b) 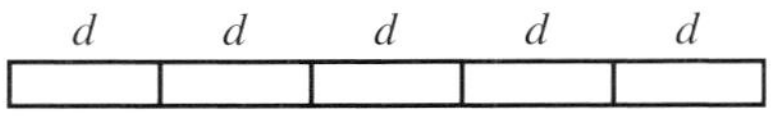

What is the total length?

(c) $a + a + a + a + a + a$

(d) $g + g$

(e) 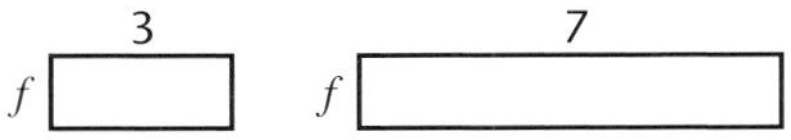

What is the total area?

(f) 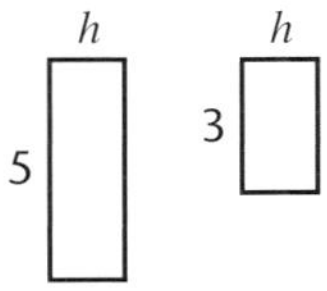

What is the total area?

(g) $5c + c$

c means $1c$.

(h) $4t + 5t + 3t$

(i) $x + 6x + 3x$

(j) $5l + l + 8l + 2l$

2 Write each of these in a longer form.

(a) Length

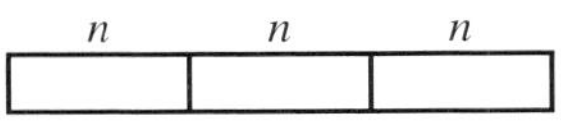

(b) Area

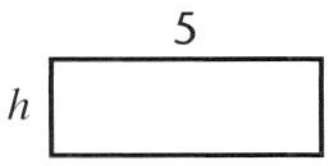

(c) $8y$ (d) $4r$ (e) $2w$ (f) $6v$

3 Simplify these algebraic expressions.

(a) $10b - 7b$ (b) $6y - y$ (c) $12z - 8z$
(d) $8t - 2t$ (e) $6j + 5j - 7j$ (f) $5u + 3u - 6u$
(g) $h + h + h$ (h) $7t - 5t + 3t$ (i) $6x - x - 3x$
(j) $4r - r + 5r - 2r$

4 In these rectangles the sum of the expressions in each row, each column, and in the two diagonals, is the same. Copy and complete each of these rectangles.

(a)

		$6x$
	$5x$	
$4x$	$3x$	

(b)

$5y$	$8y$	$11y$
		$9y$

In this square $4x + 5x + 6x = 15x$, so each row, column and diagonal must add up to $15x$.

4.3 Collecting like terms

$3a + 6b + 5a$ This expression has three **terms.**

$3a$ is one term. $6b$ is the second term. $5a$ is the third term.

Terms which use the same letter are called **like terms**.

$3a$ and $5a$ are like terms in the expression $3a + 6b + 5a$.

Terms which use the same combination of letters are also called like terms.

$6xy + 5x^2 - 2xy + 3x^2 + 2x$

$6xy$ and $2xy$ are like terms,

$5x^2$ and $3x^2$ are like terms,

$3x^2$ and $2x$ are not like terms.

You can simplify algebraic expressions by collecting like terms together.

$$3a + 6b + 5a = 3a + 5a + 6b$$
$$= 8a + 6b$$

Rearrange so that like terms are next to each other.

For example,

$$6xy + 5x^2 - 2xy + 3x^2 + 2x = 6xy - 2xy + 5x^2 + 3x^2 + 2x$$
$$= 4xy + 8x^2 + 2x$$

You keep the + or − sign with the term so the − stays with $2xy$, and the term is $-2xy$.

EXAMPLE 5

Simplify these expressions by collecting like terms.

(a) $2a + 7b + 3a$ (b) $4p + 5 + 3p - 2$

(c) $6f + 5g - 4f + g$ (d) $4ab - 6a + 3ab - 2a$

(a) $2a + 7b + 3a = 2a + 3a + 7b$
$= 5a + 7b$

(b) $4p + 5 + 3p - 2 = 4p + 3p + 5 - 2$
$= 7p + 3$

Collect the terms in p, and collect the terms which are just numbers.

(c) $6f + 5g - 4f + g = 6f - 4f + 5g + g$
$= 2f + 6g$

Remember to keep the − sign with the $4f$.

(d) $4ab - 6a + 3ab - 2a = 4ab + 3ab - 6a - 2a$
$= 7ab - 8a$

$4ab$ and $3ab$ are like terms, $-6a$ and $-2a$ are like terms.

EXAMPLE 6

Write an expression for the perimeter of this rectangle, in its simplest form.

$3x + 2$

$2x - 1$

To work out the perimeter of a shape you need to add together the lengths of all its sides.

$3x + 2 + 3x + 2 + 2x - 1 + 2x - 1$

$= 3x + 3x + 2x + 2x + 2 + 2 - 1 - 1$
$= 10x + 2$

EXERCISE 4C

1 Simplify these expressions by collecting like terms.

(a) $2c + 7d + 3c$ (b) $3m + 4r + 2m$

(c) $6x + 4y + x$ (d) $4a + 8b + 6a$

(e) $2q + 6 + 3q + 2$ (f) $7p + 2 + 5p + 3$

(g) $5j + 4 + 6j + 1$ (h) $6w + 7 + 2w + 3$

x means $1x$.

2 Simplify these expressions by collecting like terms.

(a) $5x + 4y - 2x$ (b) $7a + 3b - 5a$

(c) $8k + 3m - 4k$ (d) $12h + 7j - 4h$

(e) $4q + 5 + 3q - 2$ (f) $6p + 7 + 2p - 3$

(g) $5t + 2 - 3t + 1$ (h) $9z + 4 - 8z + 6$

3 Write expressions for the perimeter of these shapes.
Write each expression in its simplest form.

(a) a square, side $3a$

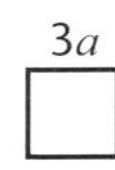

(b) a square, side $x + 1$

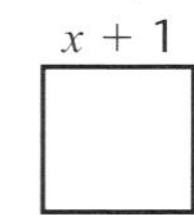

(c) a rectangle, width y and length $2x$

y

$2x$

(d) a rectangle, width $2x$ and length $3x + 1$

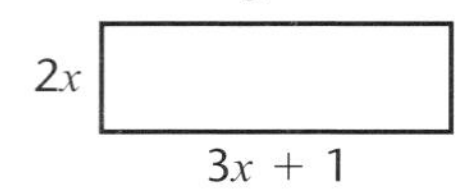

4 Simplify these expressions.

(a) $6a + 5b + 4a + 2b$
(b) $4m + 3r + 2m + 2r$
(c) $2x + 3y + 3x + 5y$
(d) $5q + 8r + 6q + r$
(e) $6k + 5l - 4k + y$
(f) $7v + 2w - 5v - 3w$
(g) $4y + 4z - 3y - z$
(h) $9c - 7d - 8c - 6d$

Keep the − sign with the $4k$.

5 Simplify these expressions.

(a) $5x + 3x + 2y + 2x + 4y$
(b) $6p + 7q + 2p + 3p - q$
(c) $7g + 3h - 4g + g - 2h$
(d) $7t + 7n - 4t - t + 2n - 3n$
(e) $4a + 6b - 2a + 3c + 2b - 4c$
(f) $5j + 6k + j + 4l - 2k + 3l$
(g) $8d + 5e + 6f - 3d + e - 4f$
(h) $6x - 2y - 5x + 8z - 3y - 4z$

6 Simplify these expressions.

(a) $4ab + 6a + 3ab - 2a$
(b) $6x^2 + 5x + 4x + 2x^2$
(c) $7t^2 + 4 - 3t^2 + 1$
(d) $6xy + 5x^2 - 2xy + 3x^2 + 2x$
(e) $9xy + 5x - 6xy + 6x^2 + 2x$
(f) $4ab + 6a + 2ab - 5b - 3ab + a - 2b$

7 In each rectangle the sum of the expressions in each row, column, and the two diagonals, is the same.
Copy and complete each of these rectangles.

(a)

$6a + 7b$		
$a + 6b$		
$8a + 2b$		$4a + 3b$

(b)

$3a + 2b$		$8a - 2b$
	$5a + b$	
$2a + 4b$		

(c)

$7a + b + 2c$		
$2a + 5b - 2c$	$10a - 2b + 5c$	$3a + 3b$

4.4 Multiplying algebraic terms

You will need to know

- how to multiply positive and negative numbers

Remember,

$3 \times x = x \times 3 = 3x \qquad y \times y = y^2 \qquad g \times h = gh$

You can simplify complicated multiplications.

For example,

$3f \times 4g = 12fg$

This is because $3f \times 4g = 3 \times f \times 4 \times g$
$= 3 \times 4 \times f \times g$
$= 12 \times fg$
$= 12fg$

$f \times 4 = 4 \times f$

To multiply algebraic terms

1 multiply the numbers,

2 multiply the letters.

EXAMPLE 7

Simplify these expressions.

(a) $2 \times 5b$ (b) $3c \times 2d$ (c) $3x \times 4x$

Multiply the numbers first, then multiply the letters.

(a) $2 \times 5b = 2 \times 5 \times b$
$= 10 \times b$
$= 10b$

(b) $3c \times 2d = 3 \times c \times 2 \times d$
$= 3 \times 2 \times c \times d$
$= 6 \times cd$
$= 6cd$

$x \times x = x^2$

(c) $3x \times 4x = 3 \times x \times 4 \times x$
$= 3 \times 4 \times x \times x$
$= 12 \times x^2$
$= 12x^2$

EXERCISE 4D

Simplify these expressions.

Use Example 7 to help.

$h = 1h$.

1 $2 \times 5k$	2 $3 \times 6b$	3 $6 \times 2x$
4 $4a \times 5$	5 $2h \times 7$	6 $3m \times 4$
7 $3a \times 2b$	8 $4c \times 3d$	9 $6p \times 7q$
10 $6g \times h$	11 $x \times 5y$	12 $7j \times 8k$
13 $3t \times 4t$	14 $6x \times 7x$	15 $5a \times 6a$
16 $4n \times n$	17 $6c \times 4d$	18 $x \times 7x$

Expanding brackets

You can work out 6×34 by thinking of it as '6 lots of 30' + '6 lots of 4'.

$6 \times 34 = 6 \times (30 + 4)$
$= 6 \times 30 + 6 \times 4$
$= 180 + 24$
$= 204$

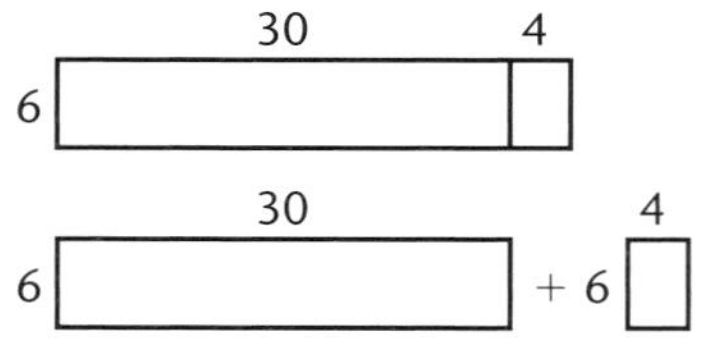

Multiply the 30 and the 4, by 6.

Then add the two answers together.

This is called **expanding** the **brackets**.

It is also sometimes called 'multiplying out the brackets' or 'removing the brackets'.

When you expand brackets you multiply each term inside the brackets by the term outside the bracket.

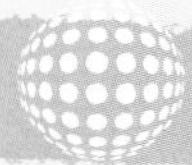

EXAMPLE 8

Expand the brackets to find the value of these expressions.

(a) $4(50 + 7)$ **(b)** $6(30 - 2)$

(a) $4(50 + 7) = 4 \times 50 + 4 \times 7$
$= 200 + 28$
$= 228$

(b) $6(30 - 2) = 6 \times 30 - 6 \times 2$
$= 180 - 12$
$= 168$

The minus sign means take '6 lots of 2' away from '6 lots of 30'.

We often use brackets in algebra.

$6(x + 4)$ means $6 \times (x + 4)$

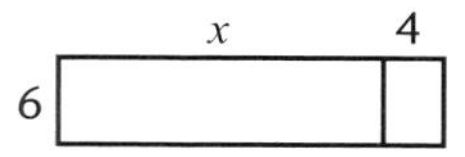

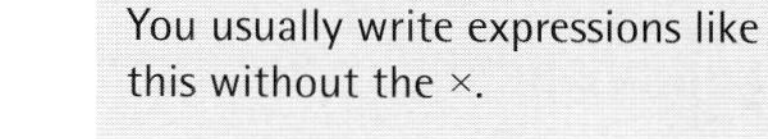
You usually write expressions like this without the ×.

You have to multiply each term in the brackets by 6.

The same as in the 6 × 34 example above.

$6(x + 4) = 6 \times x + 6 \times 4$
$= 6x + 24$

Remember you write $6 \times x$ as $6x$.

EXAMPLE 9

Simplify the following by multiplying out the brackets.

(a) $5(a + 6)$ **(b)** $2(x - 8)$ **(c)** $3(2c - d)$

Multiply each term inside the bracket by the term outside the bracket.

(a) $5(a + 6) = 5 \times a + 5 \times 6$
$= 5a + 30$

A common mistake is to forget to multiply the second term in the bracket.

(b) $2(x - 8) = 2 \times x - 2 \times 8$
$= 2x - 16$

(c) $3(2c - d) = 3 \times 2c - 3 \times d$
$= 6c - 3d$

$3 \times 2c = 2c + 2c + 2c = 6c$, or
$3 \times 2c = 3 \times 2 \times c$
$= 6 \times c$
$= 6c$

EXERCISE 4E

1 Expand the brackets to find the value of these expressions.

(a) $2(50 + 7)$ (b) $5(40 + 6)$
(c) $4(60 + 8)$ (d) $6(70 + 3)$
(e) $3(40 - 2)$ (f) $7(50 - 4)$
(g) $6(50 - 1)$ (h) $8(40 - 3)$

Use Example 8 to help.

2 Write an expression for the area of each shape.

(a) A rectangle of height 5 divided into two parts of widths m and 3.

Simplify by removing the brackets.

(b)

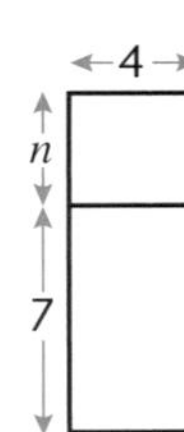

3 Simplify the following by multiplying out the brackets.

(a) $5(p + 6)$ (b) $3(a + 5)$ (c) $7(k + 2)$
(d) $4(m + 9)$ (e) $5(7 + f)$ (f) $2(8 + q)$
(g) $2(a + b)$ (h) $5(l + y)$ (i) $8(g + h + i)$
(j) $4(u + v + w)$

Multiply each term inside the bracket by the term outside the bracket.

4 Simplify the following by removing the brackets.

(a) $2(y - 8)$ (b) $3(x - 5)$ (c) $6(b - 4)$
(d) $7(d - 8)$ (e) $2(7 - l)$ (f) $4(8 - n)$
(g) $5(a - b)$ (h) $2(x - y)$ (i) $7(4 + p - q)$
(j) $8(a - b + 6)$

5 Expand the brackets in these expressions.

(a) $3(2c + 6)$ (b) $4(3m + 2)$ (c) $5(4t + 3)$
(d) $6(4y + 9)$ (e) $4(3e + f)$ (f) $2(5p + q)$
(g) $3(2a - b)$ (h) $6(3c - 2d)$ (i) $2(m - 4n)$
(j) $7(2x + y - 3)$ (k) $6(3a - 4b + c)$
(l) $4(2u - 5v - 3w)$

$3 \times 2c = 3 \times 2 \times c = 6c$

6 Simplify these expressions by removing the brackets.

(a) $2(x^2 + 3x + 2)$ (b) $3(x^2 + 5x - 6)$
(c) $2(a^2 - l + 2)$ (d) $4(y^2 - 3y - 10)$

Multiply each term inside the bracket by the term outside.

7 Write down the pairs of cards that show equivalent expressions.

A	B	C	D
$4(x + 2y)$	$4x + 2y$	$2(4x + y)$	$4(2x - y)$
E	**F**	**G**	**H**
$8x - 8y$	$4x + 8y$	$8(x - y)$	$2x - 8y$
I	**J**	**K**	**L**
$8x + 2y$	$2(x - 4y)$	$2(2x + y)$	$8x - 4y$

Equivalent expressions are the same when the brackets are expanded.

Letters outside brackets

Sometimes the term outside the bracket includes a letter.

You expand the brackets in the same way.

You multiply each term inside the brackets by the term outside.

EXAMPLE 10

Expand the brackets in these expressions.

(a) $a(a + 4)$ (b) $x(2x - y)$ (c) $3k(2k + 5)$

(a) $a(a + 4) = a \times a + a \times 4$
$= a^2 + 4a$

(b) $x(2x - y) = x \times 2x - x \times y$
$= 2x^2 - xy$

(c) $3k(2k + 5) = 3k \times 2k + 3k \times 5$
$= 6k^2 + 15k$

$a \times a = a^2$

$x \times 2x = x \times 2 \times x$
$= 2 \times x \times x$
$= 2x^2$
$x \times y = xy$

$3k \times 2k = 3 \times 2 \times k \times k = 6k^2$
$3k \times 5 = 3 \times 5 \times k = 15k$

EXERCISE 4F

Expand the brackets in these expressions.

1 $b(b + 4)$	2 $a(a + 5)$	3 $k(k - 6)$
4 $m(m - 9)$	5 $a(2a + 3)$	6 $g(4g + 1)$
7 $p(2p + q)$	8 $t(t + 5w)$	9 $m(m + 3n)$
10 $x(2x - y)$	11 $r(4r - t)$	12 $a(a - 4b)$
13 $2t(t + 5)$	14 $3x(x - 8)$	15 $5k(k + l)$
16 $3a(2a + 4)$	17 $2g(4g + h)$	18 $5p(3p - 2q)$
19 $3x(2y + 5z)$	20 $4p(3p + 2q)$	

Use Example 10 to help.

Negative numbers outside brackets

If you have an expression like $-3(2x - 5)$, multiply both terms in the brackets by -3.

$-3(2x - 5) = -3 \times 2x + (-3) \times (-5)$
$= -6x + 15$

The term outside the bracket is negative.

−	×	+	=	−
+	×	−	=	−
−	×	−	=	+
+	×	+	=	+

$-3 \times 2x = -6x$
$-3 \times -5 = 15$

EXAMPLE 11

Expand these expressions.

(a) $-2(3t + 4)$ (b) $-3(4x - 1)$

(a) $-2(3t + 4) = -2 \times 3t + -2 \times +4$
$= -6t + -8$
$= -6t - 8$

Multiply both terms in the bracket by −2.

$-2 \times 3 = -6$
$-2 \times 4 = -8$

(b) $-3(4x - 1) = -3 \times 4x + -3 \times -1$
$= -12x + 3$

$-3 \times 4 = -12$
$(-3) \times (-1) = +3$

EXERCISE 4G

Expand these expressions.

1 $-2(3k + 4)$ 2 $-3(2x + 6)$
3 $-5(3n + 1)$ 4 $-4(3t + 5)$
5 $-3(4p - 1)$ 6 $-2(3x - 7)$
7 $-6(x - 3)$ 8 $-5(2x - 3)$

4.5 Adding and subtracting expressions with brackets

Adding

To add expressions with brackets you expand the brackets first, then collect like terms to simplify your answer.

EXAMPLE 12

Expand then simplify these expressions.

(a) $3(a + 4) + 2a + 10$ (b) $3(2x + 5) + 2(x - 4)$

(a) $3(a + 4) + 2a + 10 = 3a + 12 + 2a + 10$
$= 3a + 2a + 12 + 10$
$= 5a + 22$

Expand the brackets first.
Then collect like terms.

(b) $3(2x + 5) + 2(x - 4) = 6x + 15 + 2x - 8$
$= 6x + 2x + 15 - 8$
$= 8x + 7$

Expand both sets of brackets first.

Subtracting

To subtract an expression with brackets you expand the brackets first, then collect like terms.

For example,

$$3(2x + 3) - 2(x - 1) = 6x + 9 - 2x + 2$$
$$= 6x - 2x + 9 + 2$$
$$= 4x + 11$$

$3(2x + 3) = 6x + 9$
$-2(x - 1) = -2x + 2$

EXAMPLE 13

Expand then simplify these expressions.

(a) $3(2t + 1) - 2(2t + 4)$ (b) $8(x + 1) - 3(2x - 5)$

(a) $3(2t + 1) - 2(2t + 4) = 6t + 3 - 4t - 8$
$= 6t - 4t + 3 - 8$
$= 2t - 5$

(b) $8(x + 1) - 3(2x - 5) = 8x + 8 - 6x + 15$
$= 8x - 6x + 8 + 15$
$= 2x + 23$

Multiply both terms in the second bracket by -2.

Expand the brackets first.
Then collect like terms.

EXERCISE 4H

Expand then simplify these expressions.

1 $3(y + 4) + 2y + 10$
2 $2(k + 6) + 3k + 9$
3 $4(a + 3) - 2a + 6$
4 $3(t - 2) + 4t - 10$
5 $3(2y + 3) + 2(y + 5)$
6 $4(x + 7) + 3(x + 4)$
7 $3(2x + 5) + 2(x - 4)$
8 $2(4n + 5) + 5(n - 3)$
9 $3(x - 5) + 2(x - 3)$
10 $4(2x - 1) + 2(3x - 2)$
11 $3(2b + 1) - 2(2b + 4)$
12 $4(2m + 3) - 2(2m + 5)$
13 $5(2k + 2) - 4(2k + 6)$
14 $2(4p + 1) - 4(p - 3)$
15 $5(2g - 4) - 2(4g - 6)$
16 $2(w - 4) - 3(2w - 1)$

Use Examples 12 and 13 to help.

4.6 Factorising algebraic expressions

Factors

A **factor** is a number or letter which divides exactly into another term.

EXAMPLE 14

(a) If 2 is a factor of these terms, show it by rewriting the term.

A 12 B 7 C $6t$ D $10x$ E $5y^2$

(b) If x is a factor of these terms, show it by rewriting the term.

A $6t$ B $4x$ C x^2 D 10 E xy

(a) A $12 = 2 \times 6$
B Not a factor
C $6t = 2 \times 3t$
D $10x = 2 \times 5x$
E Not a factor

$5y^2 = 5 \times y \times y$

(b) A Not a factor
B $4x = 4 \times x$
C $x^2 = x \times x$
D Not a factor
E $xy = x \times y$

If two terms have the same factor, or a factor 'in common', you call this a **common factor**.

For example, 2 is a factor of $4x$
2 is a factor of 6
2 is a common factor of $4x$ and 6.

EXAMPLE 15

(a) Is 2 a common factor of these pairs of terms?
If yes, then show how you can rewrite the terms.

A $8t$ and 12 B $3x$ and 6 C $20z$ and 4

(b) Is 3 a common factor of these pairs of terms?
If yes, then show how you can rewrite the terms.

A $5x$ and 9 B $6p$ and 12 C $3t$ and 15

(a) A Yes $8t = 2 \times 4t$, $12 = 2 \times 6$
B No
C Yes $20z = 2 \times 10z$, $4 = 2 \times 2$

Continued ▼

(b) A No
B Yes $6p = 3 \times 2p$, $12 = 3 \times 4$
C Yes $3t = 3 \times t$, $15 = 3 \times 5$

EXERCISE 4I

1 If 2 is a factor of these terms, show by rewriting the term.
A 10 B 9 C $4t$ D $8x$ E $3y^2$

2 If x is a factor of these terms, show by rewriting the term.
A 10 B 9 C $4t$ D $8x$ E $3y^2$

3 If x is a factor of these terms, show by rewriting the term.
A $6f$ B $5x$ C x^2 D 12 E wx

4 Is x a common factor of these pairs of terms?
If yes, then show how you can rewrite the terms.
A $12y$ and 6 B $13n$ and 9 C $6q$ and 21

5 Is x a common factor of these pairs of terms?
If yes, then show how you can rewrite the terms.
A $4x^2$ and $2x$ B xy and y C xy and tx

Factorising expressions

Factorising an algebraic expression is the opposite of expanding brackets.

Factorising an expression means you write it as one term multiplied by a simpler expression.

For example $6x + 10$ can be written as $2(3x + 5)$
because $6x = 2 \times 3x$
and $10 = 2 \times 5$

2 is a factor of $6x$.
2 is also a factor of 10.
2 is a common factor of $6x$ and 10.

EXAMPLE 16

Copy and complete these factorised expressions.

(a) $3t + 15 = 3(\square + 5)$ (b) $4n + 12 = \square(n + 3)$

(a) $3t + 15 = 3(t + 5)$

$3 \times t = 3t$ and $3 \times 5 = 15$

(b) $4n + 12 = 4(n + 3)$

$4 \times n = 4n$ and $4 \times 3 = 12$

EXAMPLE 17

Factorise these expressions.

(a) $5a + 20$ (b) $4x - 12$ (c) $x^2 + 7x$

(a) $5a + 20 = 5 \times a + 5 \times 4$
$= 5(a + 4)$
$= 5(a + 4)$

5 is a factor of $5a$
$5a = 5 \times a$
5 is a factor of 20
$20 = 5 \times 4$
So 5 is a common factor of $5a$ and 20.

Check $5(a + 4) = 5 \times a + 5 \times 4$
$= 5a + 20$ ✓

Check your answer by expanding the brackets.

(b) $4x - 12 = 4 \times x - 4 \times 3$
$= 4(x - 3)$
$= 4(x - 3)$

2 is a common factor of $4x$ and 12, 4 is also a common factor of $4x$ and 12.
Use 4 because it is higher than 2.

(c) $x^2 + 7x = x \times x + x \times 7$
$= x(x + 7)$
$= x(x + 7)$

x is a common factor of x^2 and $7x$

EXERCISE 4J

Don't forget to check your answers by removing the brackets.

1 Copy and complete these factorised expressions.

Use Example 16 to help you.

(a) $3x + 15 = 3(\square + 5)$ (b) $5a + 10 = 5(\square + 2)$
(c) $2x - 12 = 2(x - \square)$ (d) $4m - 16 = 4(m - \square)$
(e) $4t + 12 = \square(t + 3)$ (f) $3n + 18 = \square(n + 6)$
(g) $2b - 14 = \square(b - 7)$ (h) $4t - 20 = \square(t - 5)$

2 Factorise these expressions.

Use Example 17 to help you.

(a) $5p + 20$ (b) $2a + 12$ (c) $3y + 15$
(d) $7b + 21$ (e) $4q + 12$ (f) $6k + 24$
(g) $5a + 5$ (h) $4g + 8$

$5 = 5 \times 1$

3 Factorise these expressions.

Use Example 17 to help you.

(a) $4t - 12$ (b) $3x - 9$ (c) $5n - 20$
(d) $2b - 8$ (e) $6a - 18$ (f) $7k - 7$
(g) $4r - 16$ (h) $6g - 12$

4 Factorise these expressions.

(a) $y^2 + 7y$ (b) $x^2 + 5x$ (c) $t^2 + 2t$
(d) $n^2 + n$ (e) $x^2 - 7x$ (f) $z^2 - 2z$
(g) $p^2 - 8p$ (h) $a^2 - a$

$n = n \times 1$

5 Factorise these expressions.

$6p = 2 \times 3p$

Use Example 17 to help you.

(a) $6p + 4$ (b) $4a + 10$ (c) $4t - 6$
(d) $8m - 12$ (e) $10x + 15$ (f) $6y - 9$
(g) $4a + 8b$ (h) $10p + 5q$

6 Write down the pairs of cards that show equivalent expressions.

A	B	C	D	E	F
$4a - 12$	$2(2a - 3)$	$a(a - 4)$	$3a + 6$	$a^2 + 2a$	$6a + 9$

G	H	I	J	K	L
$4(a - 3)$	$a(a + 2)$	$3(a + 2)$	$a^2 - 4a$	$3(2a + 3)$	$4a - 6$

EXAMINATION QUESTIONS

1 Simplify $3(a - 4) - 2(5 - a)$. [2]

(CIE Paper 1, Nov 2000)

---

2 Factorise completely $2x^2 - 6x$. [2]

(CIE Paper 1, Jun 2001)

---

3 Factorise completely $6mp - 12p$. [2]

(CIE Paper 1, Nov 2001)

---

4 Factorise completely $8y - 12ty$. [2]

(CIE Paper 1, Jun 2002)

---

5 **(a)** Simplify $7k - 3m - k - 2m$. [2]
(b) Pencils cost p cents each and erasers cost e cents each.
Farah buys 7 pencils and 3 erasers.
Write down the total cost on cents, in terms of p and e. [2]

(CIE Paper 3, Jun 2002)

---

6 Factorise $40a - 8b + 32c$. [2]

(CIE Paper 1, Nov 2002)

---

7 Bottles of water cost 25 cents each.
(a) Find the cost of 7 bottles of water. [1]
(b) Write down an expression in terms of b for the cost of b bottles in cents. [1]
(c) Change your answer to **(b)** into dollars. [1]

(CIE Paper 3, Jun 2003)

---

8 **(a)** Multiply out the brackets $5x(2x - 3y)$. [2]
(b) Factorise completely $6x^2 + 12x$. [2]

(CIE Paper 1, Nov 2003)

---

9 Factorise completely $4xy - 6xz$. [2]

(CIE Paper 1, Jun 2004)

---

10 **(a)** Simon thought of a number x.
He multiplied this number by 3 and then added 8.
Write down an expression in x for his answer. [2]
(b) Simplify $-8a + 7b - a - 2b$. [2]
(c) Factorise fully $6a - 9a^2$. [2]

(CIE Paper 3, Nov 2004)

---

11 **(a)** Expand the bracket and simplify the expression $7x + 5 - 3(x - 4)$. [2]
(b) Factorise $5x^2 - 7x$. [1]

(CIE Paper 1, Jun 2005)

---

12 Factorise $3xy - 2x$. [1]

(CIE Paper 1, Nov 2005)

Chapter 5

Equations and inequalities

This chapter will show you how to

- ✔ solve simple equations
- ✔ deal with equations that have negative, decimal or fractional answers
- ✔ solve equations combining two or more operations, or involving brackets
- ✔ write inequalities and represent them on a number line

5.1 Equations

Equations involving addition and subtraction

A **simple equation** is an equation involving an unknown, usually represented by a letter. A **linear equation** has no letters with powers.

$3x + 5 = 17$ is a linear equation.
$x^2 = 16$ is not.

In the equation the letter may be added to, subtracted from, multiplied by or divided by a whole number.

When you **solve** an equation you find the value of the unknown. Your aim is to get the unknown letter on its own on one side of the equation.

$5x + 73 = 198$
$x = ?$

To solve a linear equation you must always do the same to both sides of the equation. This means do the same to the expression on either side of the equals sign.

For example, if you add 6 to one side of an equation you must add 6 to the other side.

EXAMPLE 1

Solve the equation $b + 2 = 5$ to find the value of b.

2 has been added to b. To get b by itself you must 'undo' this operation. You perform the 'opposite' or **inverse** operation, so subtract 2.

Continued ▼

$b + 2 = 5$

$b + 2 - 2 = 5 - 2$

$b = 3$

Subtract 2 from both sides of the equation.

EXAMPLE 2

Solve the equation $p - 4 = 9$.

Look for the operation.

4 has been subtracted from p. The inverse operation is add 4.

Add 4 to both sides of the equation.

$p - 4 = 9$

$p - 4 + 4 = 9 + 4$

$p = 13$

Write your equations with the equals signs underneath one another.

Sometimes the unknown letter appears on the right-hand side of the equation.

EXAMPLE 3

Solve the equation $7 = c + 2$.

Subtract 2 from both sides of the equation.

$7 = c + 2$

$7 - 2 = c + 2 - 2$

$5 = c$

$c = 5$

You usually write the answer with the unknown letter (the **subject**) on the left-hand side of the equals sign. $5 = c$ is the same as $c = 5$.

Alternatively you could write the equation as $c + 2 = 7$ and solve it as in Example 1.

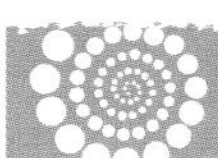

EXERCISE 5A

1 Solve each equation to find the value of the letter.

(a) $h + 4 = 6$ **(b)** $b + 1 = 12$ **(c)** $x + 9 = 10$

(d) $k - 5 = 2$ **(e)** $y - 9 = 11$ **(f)** $a - 20 = 30$

Make sure you treat both sides of the equation in the same way.

2 Find the value of the letter in each of these equations. Write the letter on the left-hand side of the equals sign in your answers.

(a) $4 = s + 3$ **(b)** $15 = d - 18$ **(c)** $6 = r + 4$

3 Find the value of the unknown letter in each of these equations.

(a) $5 + f = 32$ (b) $40 + w = 120$

(c) $14 = 8 + n$ (d) $0 = 3 + x$

4 Viewfone gave me an extra 25 free text messages. I now have 43 free text messages. How many did I have to start with? Write an equation and solve it.

Equations involving multiplication and division

Sometimes the unknown letter is multiplied or divided by a **whole number (integer).**

EXAMPLE 4

Solve the equation $4a = 12$.

$4a$ means $4 \times a$. The inverse operation is divide by 4.

In algebra you write $4a \div 4$ as $\frac{4a}{4}$.

$$4a = 12$$

$$\frac{4a}{4} = \frac{12}{4}$$

Divide both sides of the equation by 4.

$$a = 3$$

EXAMPLE 5

Solve the equation $\frac{d}{5} = 8$.

$\frac{d}{5}$ means $d \div 5$. The inverse operation of $\div 5$ is $\times 5$.

$$\frac{d}{5} = 8$$

$$\frac{d}{5} \times 5 = 8 \times 5$$

Multiply both sides of the equation by 5.

$$d = 40$$

EXERCISE 5B

1 Solve these equations.

Write the letter on the left-hand side of the equals sign in your answers.

(a) $3b = 9$ (b) $5t = 25$ (c) $7p = 21$
(d) $4g = 28$ (e) $9x = 36$ (f) $8n = 24$
(g) $10m = 130$ (h) $6y = 360$ (i) $7k = 49$
(j) $20 = 5h$ (k) $64 = 8r$ (l) $32 = 4a$

$64 = 8r$ is the same as $8r = 64$

2 Find the value of the letter in each of these equations.

(a) $\frac{t}{2} = 12$ (b) $\frac{y}{6} = 3$ (c) $\frac{f}{4} = 9$
(d) $\frac{r}{3} = 11$ (e) $\frac{q}{6} = 6$ (f) $\frac{g}{9} = 8$
(g) $\frac{s}{2} = 56$ (h) $\frac{b}{5} = 30$ (i) $\frac{n}{8} = 1$
(j) $4 = \frac{m}{5}$ (k) $16 = \frac{v}{3}$ (l) $100 = \frac{u}{9}$

Find the value of the unknown letter in each equation. Write the letter on the left-hand side of the equals sign in your answers.

(a) $8t = 48$ (b) $4 + p = 7$ (c) $\frac{r}{7} = 8$
(d) $15 = \frac{m}{4}$ (e) $45 = x - 35$ (f) $g - 100 = 550$
(g) $\frac{s}{12} = 5$ (h) $9d = 99$ (i) $42 = 7j$

4 It took James 20 minutes to plant one quarter of his seeds. How long will it take him to plant all of them? Write an equation and solve it.

Equations involving two operations

Some equations involve carrying out two **operations**.

In the equation $4x + 2 = 10$, the operations are

multiply by 4 then add 2

You solve the equation by 'undoing' the operations. The **inverse** operations are

subtract 2 then divide by 4

EXAMPLE 6

Solve the equation $3b + 2 = 8$.

The expression $3b + 2$ has been formed like this
multiply by 3 then add 2

So the inverse of this is
subtract 2 then divide by 3

$$3b + 2 = 8$$
$$3b + 2 - 2 = 8 - 2$$
$$3b = 6$$
$$\frac{3b}{3} = \frac{6}{3}$$
$$b = 2 \quad \text{Check } 3b + 2 = 3 \times 2 + 2 = 6 + 2 = 8$$

The order of this inverse process is very important.

First subtract 2 from both sides, then divide both sides by 3.

To check your answer try the value $b = 2$ in the expression.
You will cover substitution in more detail in Chapter 7.

EXAMPLE 7

Solve the equation $5g - 4 = 36$.

$$5g - 4 = 36$$
$$5g - 4 + 4 = 36 + 4$$
$$5g = 40$$
$$\frac{5g}{5} = \frac{40}{5}$$
$$g = 8$$

Check $5 \times 8 - 4 = 40 - 4 = 36$

g has been multiplied by 5 and then 4 has been subtracted. The inverse is add 4 and then divide by 5.

First add 4 to both sides and then divide both sides by 5.

EXAMPLE 8

Solve the equation $\frac{k}{4} + 3 = 7$.

The inverse of $\div 4$ is $\times 4$. The inverse of $+3$ is -3.

First do -3, then do $\times 4$

Look for the inverse operations and carry them out in the correct order.

Continued ▼

$$\frac{k}{4} + 3 = 7$$

$$\frac{k}{4} + 3 - 3 = 7 - 3$$

$$\frac{k}{4} = 4$$

$$\frac{k}{4} \times 4 = 4 \times 4$$

$$k = 16$$

Check $\frac{16}{4} + 3 = 16 \div 4 + 3 = 4 + 3 = 7$

First subtract 3 from both sides, then multiply both sides by 4.

Sometimes the whole of one side of an equation is divided by a number.

For example, $\frac{x + 2}{4} = 3$.

First multiply both sides of the equation by 4.

Operations $+2$ then $\div 4$

Inverse $\times 4$ then -2

$$\frac{x + 2}{4} \times 4 = 3 \times 4$$

$$x + 2 = 12$$

$$x = 10$$

EXAMPLE 9

Solve the equation $\frac{12 - 2n}{3} = 2$.

$$\frac{12 - 2n}{3} \times 3 = 2 \times 3$$

First multiply both sides of the equation by 3.

$$12 - 2n = 6$$

$$12 - 2n + 2n = 6 + 2n$$

This is now an equation like the ones in Examples 6 and 7. Next add $2n$ to both sides. This means that the value of $2n$ remains positive.

$$12 = 6 + 2n$$

$$12 - 6 = 6 - 6 + 2n$$

Subtract 6 from both sides.

$$6 = 2n$$

$$3 = n$$

Divide both sides by 2.

$$n = 3$$

Put n on the left-hand side of the equation.

EXERCISE 5C

1 Solve these equations.

(a) $3w + 2 = 14$ (b) $2u + 9 = 19$

(c) $5t - 4 = 36$ (d) $9m - 2 = 25$

(e) $8d + 3 = 19$ (f) $7g + 6 = 69$

(g) $4s + 5 = 17$ (h) $8b + 6 = 6$

(i) $20a - 30 = 70$

2 Find the value of the unknown letter in each equation.

(a) $\frac{r}{3} + 2 = 5$ (b) $\frac{d}{4} - 5 = 2$ (c) $\frac{f}{3} - 8 = 2$

(d) $\frac{x}{2} + 4 = 5$ (e) $\frac{d}{10} - 10 = 1$ (f) $\frac{a}{3} + 7 = 10$

3 Solve these equations.

(a) $\frac{3x - 2}{11} = 2$ (b) $\frac{6t + 5}{7} = 5$ (c) $\frac{4w + 9}{7} = 7$

(d) $\frac{14 - 3a}{2} = 4$ (e) $\frac{30 + 10n}{5} = 14$ (f) $\frac{22 + 8q}{9} = 6$

Solutions involving fractions and decimals

You will need to know

- how to change between fractions and decimals

So far you have solved equations with whole number (integer) answers.

Some equations have answers that are **fractions** or **decimals.**

The question may ask for the answer as a fraction or as a decimal. If not, you can give your answer in either form.

EXAMPLE 10

Solve the equation $9t + 7 = 19$. Give your answer as a **mixed number.**

A mixed number has a whole number part and a fractional part. $1\frac{3}{4}$ is a mixed number.

$$9t + 7 = 19$$
$$9t + 7 - 7 = 19 - 7$$
$$9t = 12$$
$$\frac{9t}{9} = \frac{12}{9}$$
$$t = 1\frac{3}{9}$$
$$t = 1\frac{1}{3}$$

First subtract 7 from both sides, then divide both sides by 9.

$\frac{3}{9} = \frac{1}{3}$ in its **lowest terms** or **simplest form.**

EXAMPLE 11

Solve the equation $5x - 6 = 3$. Give your answer as a decimal.

$$5x - 6 = 3$$
$$5x - 6 + 6 = 3 + 6$$
$$5x = 9$$
$$\frac{5x}{5} = \frac{9}{5}$$
$$x = 1.8$$

First add 6 to both sides, then divide both sides by 5.

$\frac{9}{5} = 9 \div 5 = 1.8$
See Section 3.5 Fractions, decimals and percentages.

EXERCISE 5D

1 Solve these equations. Give your answers as mixed numbers.

(a) $2b - 4 = 5$ (b) $7c + 4 = 13$ (c) $3y + 6 = 16$
(d) $7j - 13 = 6$ (e) $3g - 2 = 3$ (f) $8p - 2 = 10$

2 Solve these equations. Give your answers as decimals.

(a) $8k - 3 = 7$ (b) $4e - 5 = 1$ (c) $10i - 10 = 12$
(d) $5s + 9 = 40$ (e) $4m - 2 = 5$ (f) $2a + 5 = 20$

Solutions involving negative values

Some equations have **negative number** solutions.

EXAMPLE 12

Solve the equation $2u + 7 = 1$.

$$2u + 7 = 1$$
$$2u + 7 - 7 = 1 - 7$$
$$2u = -6$$
$$\frac{2u}{2} = \frac{-6}{2}$$
$$u = -3$$

First subtract 7 from both sides, then divide both sides by 2.

Dividing a negative number by a positive number gives a negative number answer. See Chapter 1 on rules for positive numbers.

EXAMPLE 13

Solve the equation $\frac{d}{2} - 4 = -5$.

$$\frac{d}{2} - 4 = -5$$
$$\frac{d}{2} - 4 + 4 = -5 + 4$$
$$\frac{d}{2} = -1$$
$$\frac{d}{2} \times 2 = -1 \times 2$$
$$d = -2$$

First add 4 to both sides, then multiply both sides by 2.

Multiplying a negative number by a positive number gives a negative number answer.

EXERCISE 5E

1 Solve these equations.

(a) $3k + 8 = 2$
(b) $5f + 12 = 7$
(c) $6m + 20 = 2$
(d) $10g - 5 = -55$
(e) $2w - 1 = -5$
(f) $4s - 4 = -12$
(g) $3u + 12 = 6$
(h) $5f + 13 = 3$
(i) $4v + 23 = 3$
(j) $19 + 2x = 7$
(k) $12 - 6a = 18$
(l) $3n + 7 = 28$

2 Find the value of the unknown letter in each equation.

(a) $\frac{m}{2} + 3 = 1$ (b) $\frac{q}{3} + 7 = 6$ (c) $\frac{x}{5} - 3 = -5$

(d) $\frac{b}{8} - 7 = -10$ (e) $20 + \frac{e}{4} = 17$ (f) $\frac{t}{6} + 10 = 4$

3 Sita gave $\frac{1}{20}$ of her wages, plus an extra \$2, to charity last week. She gave the charity \$18. Write an equation and solve it to find out her wages last week.

Equations with unknowns on both sides

Some equations have the unknown letter on both sides of the equation.

You can still solve the equation by treating both sides of the equation in the same way.

For example $5a - 2 = 2a + 4$.

It is useful to keep the letter on the side with the largest unknown, to keep it positive.

In this example, the left-hand side has $5a$ (the largest) and the right-hand side has $2a$. So you collect the a terms on the left-hand side.

To 'remove' the a terms on the right-hand side you subtract them. Remember to do the same to both sides.

$$5a - 2 = 2a + 4$$
$$5a - 2a - 2 = 2a - 2a + 4$$
$$3a - 2 = 4$$

Subtract $2a$ from both sides of the equation.

Now you have a simpler equation to solve.

See Example 6.

EXAMPLE 14

Solve the equation $3x + 1 = x + 7$

$$3x - x + 1 = x - x + 7$$
$$2x + 1 = 7$$
$$2x + 1 - 1 = 7 - 1$$
$$2x = 6$$
$$\frac{2x}{2} = \frac{6}{2}$$
$$x = 3$$

There are more xs on the left-hand side. Subtract x from both sides.

Subtract 1 from both sides, then divide by 2.

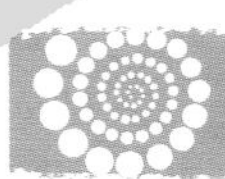

EXERCISE 5F

1 Solve these equations.

(a) $3x + 4 = 2x + 6$ (b) $5u + 9 = 3u + 17$

(c) $r + 2 = 5r - 10$ (d) $7p + 11 = 6p + 16$

(e) $4w - 6 = 3\mathit{1} + 1$ (f) $6b - 2 = 2b + 14$

2 Solve these equations.

(a) $4a + 6 = 2a + 2$ (b) $7m + 10 = 4m + 1$

(c) $8d - 2 = 3d - 12$ (d) $y - 5 = 6y + 15$

Equations involving brackets

Some equations involve **brackets**.

To solve an equation with brackets

1 **Expand** the brackets.

2 Solve by treating both sides of the equation in the same way.

Expanding brackets was covered in Section 4.4.

EXAMPLE 15

Solve $4(c + 3) = 20$

$$4(c + 3) = 20$$
$$4 \times c + 4 \times 3 = 20$$
$$4c + 12 = 20$$

First expand the brackets.

$$4c + 12 = 20$$
$$4c + 12 - 12 = 20 - 12$$
$$4c = 8$$
$$\frac{4c}{4} = \frac{8}{4}$$
$$c = 2$$

This equation is the now same type as the one in Example 6.

First subtract 12 from both sides, then divide both sides by 4.

EXAMPLE 16

Solve $5(3p - 4) = 10$.

$$5(3p - 4) = 10$$
$$5 \times 3p - 5 \times 4 = 10$$
$$15p - 20 = 10$$
$$15p - 20 + 20 = 10 + 20$$
$$15p = 30$$
$$\frac{15p}{15} = \frac{30}{15}$$
$$p = 2$$

Expanding the brackets.

First add 20 to both sides, then divide both sides by 15.

When there are two pairs of brackets, expand both brackets first before solving.

EXAMPLE 17

Solve $2(2m + 10) = 12(m - 1)$.

$$2(2m + 10) = 12(m - 1)$$
$$4m + 20 = 12m - 12$$
$$12 + 20 = 12m - 4m$$
$$32 = 8m$$
$$\frac{32}{8} = \frac{8m}{8}$$
$$4 = m$$
$$m = 4$$

Expand the brackets.

This equation has the unknown on both sides – see Example 14.

Subtract $4m$ from both sides.

Divide both sides by 8.

$4 = m$ is the same as $m = 4$.

EXERCISE 5G

1 Solve these equations.

(a) $4(g + 6) = 32$ (b) $7(k + 1) = 21$
(c) $5(s + 10) = 65$ (d) $2(n - 4) = 6$
(e) $3(f - 2) = 24$ (f) $7(v - 3) = 42$
(g) $4(m + 6) = 20$ (h) $2(w + 7) = 8$

2 Find the value of the letters in these equations.

(a) $7(2b + 2) = 56$ (b) $2(5r + 4) = 18$

(c) $4(5t + 2) = 48$ (d) $8(2v - 9) = 24$

(e) $9(3k - 3) = 0$ (f) $2(5y - 12) = 6$

(g) $3(4x + 15) = 9$ (h) $8(12 + 3f) = 48$

3 Solve these equations.

(a) $2(b + 1) = 8(2b - 5)$ (b) $5(4a + 7) = 3(8a + 9)$

(c) $6(x - 2) = 3(3x - 8)$ (d) $5(2p + 2) = 6(p + 5)$

(e) $9(3s - 4) = 5(4s - 3)$ (f) $4(10t - 7) = 3(6t - 2)$

(g) $4(2w + 2) = 2(5w + 7)$ (h) $3(3t - 2) = 7(t - 2)$

5.2 Using equations to solve problems

You can use equations to solve problems.

You will need to read the problem carefully then set up an equation and solve it using the methods introduced earlier.

EXAMPLE 18

The length of a rectangular garden is 10 m more than its width.

The perimeter of the garden is 48 m. Find the width of the garden.

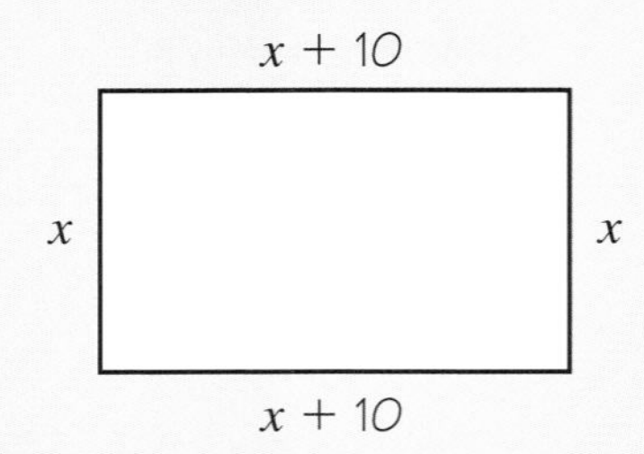

It is often useful to draw a diagram and label it.

The perimeter of the garden is

$x + (x + 10) + x + (x + 10) = 4x + 20$

This is equal to 48 m.

$4x + 20 = 48$

$4x = 28$

$x = 7$

The width of the garden is 7 m.

Use the letter x to represent the width of the garden.
The length is 10 m longer than the width so it must be $x + 10$.

Check by putting $x = 7$ into your original equation.
$7 + 17 + 7 + 17 = 48$ ✓

EXERCISE 5H

1 The length of a rectangular field is three times its width. If the perimeter of the field is 360 m find the width of the field.

Let the width of the field be x.

2 The sum of three consecutive whole numbers is 276. Find the three numbers.

Let the smallest of these numbers be x.
Then write the other two in terms of x.

3 Sarah is two years older than Susan, who is seven years older than Stephen. If their combined age is 61 years find the age of each person.

Let the age of youngest be x.

4 The perimeter of the triangle shown below is 22 cm.

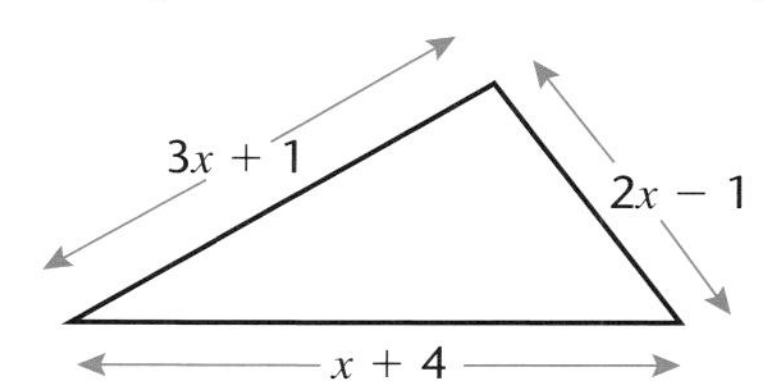

Formulate an equation in terms of x and solve it to find the lengths of the sides of the triangle.

5 Multiplying a number by 3 gives the same answer as adding 12 to it. What is the number?

Let the number be x.

6 AB is a straight line.

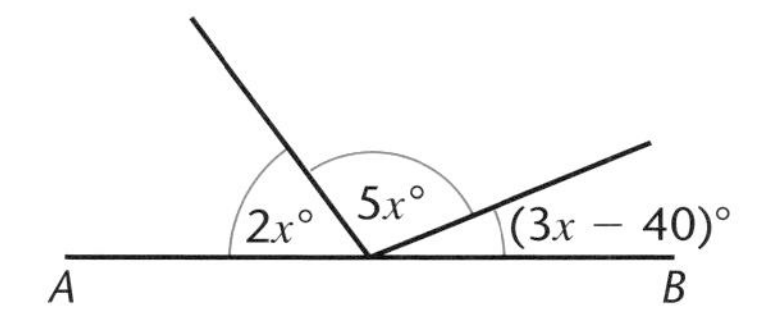

Angles on a straight line add up to 180°.

Find the value of x and the angles shown.

7 The opposite angles of a parallelogram are $(2a + 40)°$ and $(3a - 10)°$. Find the two angles.

Opposite angles in a parallelogram are equal.

8 A cricket bat costs \$15 more than a tennis racket. If the total price for a cricket bat and tennis racket is \$81 what is the cost of the cricket bat?

Let the price of the tennis racket be $\$x$.

5.3 Simultaneous equations

A linear equation such as $3x + 2 = 14$ has a unique solution, $x = 4$.

This is the only value of x that satisfies the equation.

These equations each contain two unknowns x and y.

$2x + y = 11$ (1)

$3x - y = 14$ (2)

An equation of a straight line is known as a linear equation. Each equation has an infinite number of co-ordinate pairs that satisfy it. Chapter 15 looks at linear equations and straight line graphs in a lot more detail. It includes finding points of intersection of straight lines.

Values of x and y that satisfy equation (1) may not satisfy equation (2).

For example, $x = 4$ and $y = 3$ satisfy equation (1) but they do not satisfy equation (2).

There are values of x and y that satisfy both equations at the same time, they are $x = 5$ and $y = 1$.

Solving two linear equations at the same time is called solving **simultaneous equations** .

You can solve simultaneous equations by the **method of elimination.** This eliminates one of the unknowns by adding or subtracting the equations, provided the x's or the y's cancel out when you add or subtract the equations.

The **coefficient** is the number in front of the x or y.

EXAMPLE 19

Solve the simultaneous equations.

$2x + y = 11$

$3x - y = 14$

Notice that the coefficients of y are 1 and -1 so adding the two equations will eliminate the y terms.

$2x + y = 11$ (1)

$3x - y = 14$ (2)

(1) + (2) $5x = 25$

so $x = 5$

Substitute $x = 5$ in equation (1)

$2 \times 5 + y = 11$

$10 + y = 11$

so $y = 1$

Check in (2) $3 \times 5 - 1 = 15 - 1 = 14$ ✓

The solutions are $x = 5$ and $y = 1$

When you have found the first solution, substitute it into one of the equations to work out the other solution.
It doesn't matter which equation you choose.

Check that the solutions you have found work for the other equation.

EXAMPLE 20

Solve the simultaneous equations.

$2x + 5y = 27$

$2x - 3y = 3$

$2x + 5y = 27$ (1)

$2x - 3y = 3$ (2)

(1) − (2) $8y = 24$

so $y = 3$

Substitute $y = 3$ in equation (1)

$2x + 5 \times 3 = 27$

$2x + 15 = 27$

$2x = 27 - 15$

$2x = 12$

so $x = 6$

Check in(2) $2 \times 6 - 3 \times 3 = 12 - 9 = 3$ ✓

The solutions are $x = 6$ and $y = 3$

When the coefficients (of x in this example) are exactly the same, subtracting the equations will eliminate one unknown.

Take great care when subtracting.
$5y - (-3y) = 5y + 3y$
$= 8y$

Substitute $y = 3$ into one of the equations.

Check that the solutions you have found work for the equation you did not choose before.

The rule is

If the x's or y's have the same numerical value but differ in sign ... ADD

If the x's or y's have the same numerical value and the same sign ... SUBTRACT

Sometimes you need to multiply one of the equations so that the coefficients of one of the terms become the same, as in the next example.

EXAMPLE 21

Solve the simultaneous equations.

$4g + 6h = 2$

$3g + 12h = 9$

	$4g + 6h = 2$	(1)
	$3g + 12h = 9$	(2)
(1) × 2 gives	$8g + 12h = 4$	(3)
	$3g + 12h = 9$	(2)
(3) − (2)	$5g = -5$	
so	$g = -1$	

The coefficients of g and h in (1) and (2) are not the same. To make the coefficient of h the same you multiply equation (1) by 2 giving a new equation (3).
Replace equation (1) with equation (3).
Do not alter equation (2).

Solve equations (2) and (3) by the method you used in the previous examples.

Substitute $g = -1$ in equation (1)

$4 \times (-1) + 6h = 2$

$-4 + 6h = 2$

$6h = 2 + 4$

$6h = 6$

$h = 1$

You can still substitute in equation (1) even though you replaced it with equation (3) to do the elimination.

Check in equation (2)

$3 \times -1 + 12 \times 1 = -3 + 12 \quad = 9$ ✓

Always check your solutions using the other equation.

The solutions are $g = -1$ and $h = 1$

You may have to multiply both equations before you can use the method of elimination. This is shown in Example 22.

EXAMPLE 22

Solve the simultaneous equations.

$2x - 5y = 13$

$3x + 2y = 10$

	$2x - 5y = 13$	(1)
	$3x + 2y = 10$	(2)
(1) × 3 gives	$6x - 15y = 39$	(3)
(2) × 2 gives	$6x + 4y = 20$	(4)

These equations can now be subtracted to eliminate the x terms.

Continued ▼

Alternatively

(1) × 2 gives $\quad 4x - 10y = 26 \quad$ (3)

(2) × 5 gives $\quad 15x + 10y = 50 \quad$ (4)

These equations can now be added to eliminate the y terms.

Whichever of these you choose you will obtain the solutions $x = 4$, $y = -1$.

EXAMPLE 23

The sum of two numbers is 60 and the difference between them is 26. What are the two numbers?

Let m and n be the two numbers.

First express the problem, using letters for the unknown numbers.

$$m + n = 60 \qquad (1)$$

$$m - n = 26 \qquad (2)$$

(1) + (2) $\quad 2m = 86$

$$m = \frac{86}{2}$$

$$m = 43$$

$$43 + n = 60 \qquad (1)$$

Substitute in equation (1).

$$n = 17$$

Check in equation (2)

$$43 - 17 = 26 \quad ✓$$

EXAMPLE 24

At a theme park, the cost for 4 adults and 5 children is \$156, and the cost for 2 adults and 3 children is \$86. How much is each ticket?

Let a = cost for 1 adult and c = cost for 1 child.

$$4a + 5c = 156 \qquad (1)$$

$$2a + 3c = 86 \qquad (2)$$

$$4a + 6c = 172 \qquad (3)$$

Multiply (2) × 2.

$$c = 172 - 156$$

Subtract (1) from (3).

$$c = 16$$

$$2a + 3 \times 16 = 86 \qquad (2)$$

$$2a = 86 - 48 = 38$$

$$a = 19$$

An adult ticket costs \$19 and a child's ticket \$16.

EXERCISE 5I

1 Solve the simultaneous equations.

(a) $3a + 2b = 14$
$a - 2b = 2$

(b) $5x - 3y = 21$
$3x + 3y = 3$

(c) $-3b + 4c = 2$
$3b - 7c = 1$

2 Solve the simultaneous equations.

(a) $4s - 2t = 14$
$3s - 2t = 9$

(b) $6u - 3v = 18$
$u - 3v = 8$

(c) $2p - 5q = 13$
$2p - 2q = 4$

3 Find the values of the two unknowns in the following.

(a) $x + 3y = 10$
$3x + 3y = 12$

(b) $4a + 2b = 7$
$2a + 2b = 4$

(c) $3p + 6q = 9$
$3p + q = -1$

4 Solve the following simultaneous equations.

(a) $2x + y = 5$
$x + 3y = 5$

(b) $3a + b = 5$
$a - 3b = 15$

(c) $x + 2y = 12$
$3x - 4y = 4$

(d) $5p - 3q = 1$
$-15p + 7q = 1$

(e) $3e - 2f = 5$
$5e - 6f = 3$

(f) $4x - 5y = 7$
$x - 3y = 7$

5 Solve these simultaneous equations.

(a) $2x - 3y = 3$
$5x + 2y = 17$

(b) $4x + 3y = 20$
$3x + 5y = 26$

(c) $5a - 2b = 9$
$3a + 7b = -11$

(d) $4g + 3h = 15$
$5g + 2h = 24$

(e) $2w + 5z = 43$
$9w - 4z = 8$

(f) $2x + 2y = 2$
$5x + 3y = -1$

6 Find two numbers whose sum is 116 and difference is 34.

7 The cost of 3 apples and 4 bananas is \$1.17 and the cost of 5 apples and 2 bananas is \$1.11. Find the cost of each type of fruit.

8 Solve these simultaneous equations.

(a) $3x - 2y = 0$
$5x - 7y = -22$

(b) $2a + 3b = 5$
$5a - 2b = -16$

(c) $5g + 3h = 23$
$2g + 4h = 12$

(d) $3w + 2z = 7$
$2w - 3z = 24$

(e) $5x - 7y = 27$
$3x - 4y = 16$

(f) $4x - 0.5y = 12.5$
$5x + 1.5y = 13.5$

5.4 Inequalities

An expression whose right-hand and left-hand sides are not equal is called an **inequality.**

You need to know these inequality symbols.

$<$ means less than

$>$ means greater than

$\leqslant$ means less than or equal to

$\geqslant$ means greater than or equal to

The symbol $\neq$ is used as shorthand for *not equal to.*

The pointed end points towards the smaller number. The open end points towards the larger number. You used the signs $<$ and $>$ in Chapter 1.

EXAMPLE 25

Write the correct inequality sign between these pairs of numbers.

(a) 3, 4 **(b)** 21, 56 **(c)** 9, 4 **(d)** 3, -5

(a) 3 is less than 4, $3 < 4$

(b) 21 is less than 56, $21 < 56$

(c) 9 is greater than 4, $9 > 4$

(d) 3 is greater than -5, $3 > -5$

For negative numbers you can use a **number line.**

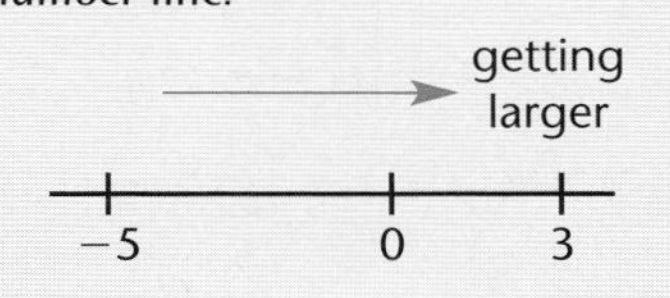

EXAMPLE 26

Write down the whole number values for each letter to make the inequality true.

(a) $n > 5$ and $n < 12$

(b) $b > 2$ and $b \leqslant 7$

(c) $x \geqslant -4$ and $x < 3$

(a) 6, 7, 8, 9, 10, 11 — The values of n are greater than 5 but less than 12.

(b) 3, 4, 5, 6, 7 — The values of b are greater than 2 but less than *or equal to* 7.

(c) $-4, -3, -2, -1, 0, 1, 2$ — The values of x must be greater than *or equal to* -4 but less than 3.

You can show inequalities on a number line.

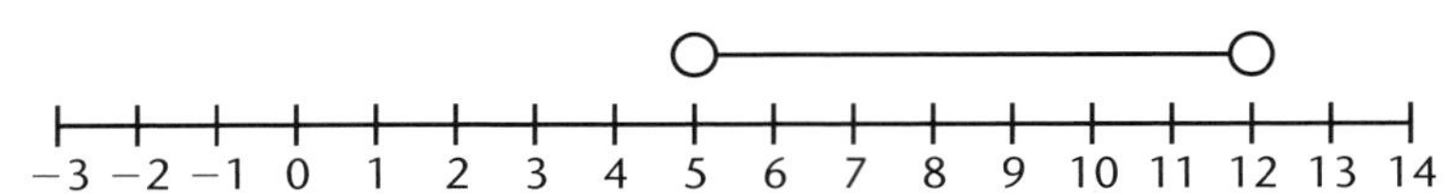

Use the symbol ○ when the end value is not included in the answer.

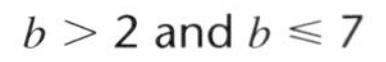

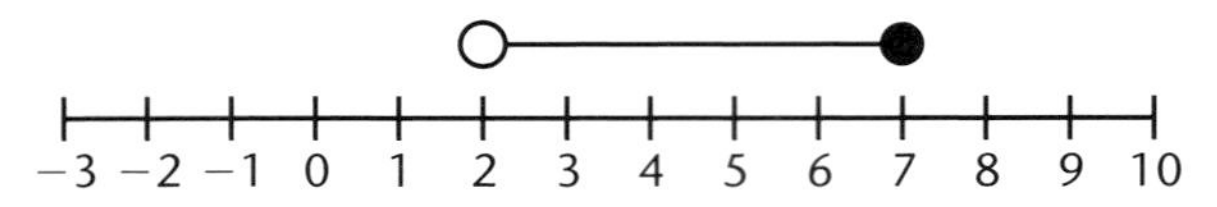

Use the symbol ● when the end value is included in the answer.

$x \geqslant -4$ and $x < 3$

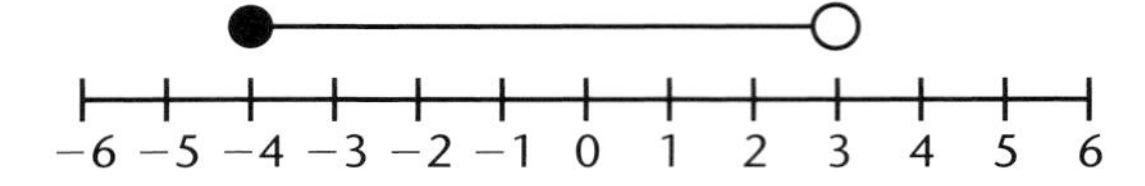

You can combine inequalities.
From the number lines you can see that

$n > 5$ and $n < 12 \rightarrow 5 < n < 12$

$b > 2$ and $b \leqslant 7 \rightarrow 2 < b \leqslant 7$

You say 'n is greater than 5 but less than 12' or that 'n lies between 5 and 12'.

$n > 5$ is the same as $5 < n$.

EXERCISE 5J

1 Write the correct inequality sign between these pairs of numbers in the order they are given.

(a) 5, 8 (b) 10, 5 (c) 27, 6
(d) 20, 80 (e) 2, −4 (f) 23, 21
(g) −5, −8 (h) 5.5, 6.4 (i) 12.7, 12.6
(j) 0, 0.01 (k) 0.01, 10.01 (l) 1112, 1121

2 Write down whether each statement is true (T) or false (F).

(a) $8 > 9$ (b) $12 < 17$ (c) $31 > 30$
(d) $5 \leqslant 4$ (e) $23 > 22.9$ (f) $0.01 > 0.001$
(g) $89 = 88$ (h) $45 \geqslant 44$ (i) $99.9 > 99.89$

3 Write down the whole number values for each letter to make the inequality true.

(a) $b > 4$ and $b < 9$
(b) $r \leqslant 12$ and $r > 7$
(c) $p > 0$ and $p \leqslant 5$
(d) $k \leqslant 100$ and $k > 94$
(e) $x > 22$ and $x \leqslant 3$
(f) $q \geqslant 29$ and $q < 24$

4 Write down the values for each letter, to 1 decimal place, to satisfy the inequalities.

(a) $h > 3.2$ and $h < 4.1$
(b) $t \leqslant 21.2$ and $t > 20.7$
(c) $y \leqslant 0.9$ and $y \geqslant 0.3$
(d) $m \geqslant 78.9$ and $m < 80.1$

5 Show each inequality on a separate copy of this number line.

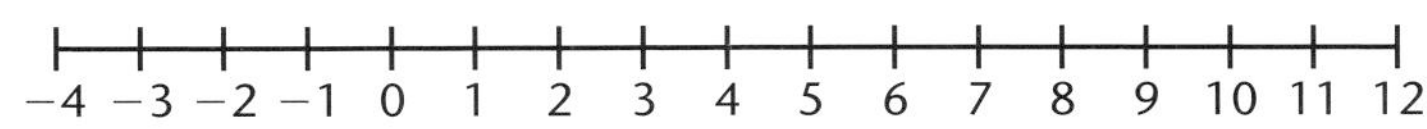

(a) $x > 4$
(b) $t > 2$ and $t < 11$
(c) $u \leqslant 12$ and $u > 6$
(d) $w \leqslant 9$ and $w > 0$
(e) $2 < j \leqslant 8$
(f) $8 \leqslant s < 12$
(g) $-3 \leqslant m < 6$
(h) $-4 \leqslant d \leqslant 0$

6 Write each pair of inequalities in questions 3 and 4 as a combined inequality. For example, write $b > 4$ and $b < 9$ as $4 < b < 9$.

EXAMINATION QUESTIONS

1 Solve the equation $\frac{3x + 2}{4} = 11$. [2]

(CIE Paper 1, Jun 2000)

---

2 Solve the equations
(a) $3x - 3 = 42$, [2]
(b) $5(x - 3) = 20$. [2]

(CIE Paper 1, Jun 2001)

3

$$y = 100 - 4x$$

Find the value of x when $y = 72$. [2]

(CIE Paper 3, Nov 2001)

---

4 The integer n is such that $-3 \leqslant n \leqslant 3$.
List all the possible values of n. [2]

(CIE Paper 1, Nov 2002)

---

5 Solve the equations
(a) $x - 7 = 9$,
(b) $2(y + 1) = 3y - 5$. [2]

(CIE Paper 1, Nov 2002)

---

6 Solve the equation $x + 4 = 3(2 - x)$. [3]

(CIE Paper 1, Jun 2003)

---

7 Solve the equations

(a) $4x - 5 = 31$, [2]

(b) $4(y - 5) = 36$. [2]

(CIE Paper 1, Jun 2004)

---

8 (a)

2x cm

The perimeter of the rectangle above is 36 cm. [2]

Find the vaue of x.

(b)

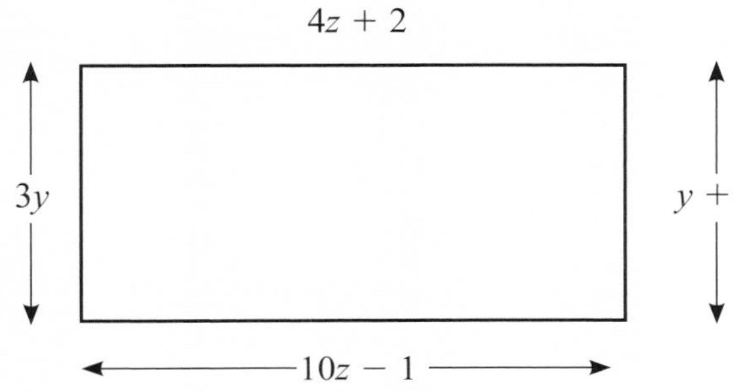

(i) In this rectangle $3y = y + 3$.
Solve this equation to find y. [2]

(ii) Write down an equation in z. [1]

(iii) Solve this equation to find z. [3]

(CIE Paper 3, Jun 2005)

---

9 Solve the equation $5x - 7 = 8$. [2]

(CIE Paper 1, Nov 2005)

Chapter 6

Angles, triangles and polygons

This chapter will show you how to

- ✔ describe a turn, recognise and measure angles of different types
- ✔ discover angle properties
- ✔ use bearings to describe directions
- ✔ investigate angle properties of triangles, circles, quadrilaterals and other polygons
- ✔ understand line and rotational symmetry

6.1 Angles

Turning

If you turn all the way round in one direction, back to your starting position, you make a **full turn** .

The minute hand of a clock moves a **quarter turn** from 3 to 6.

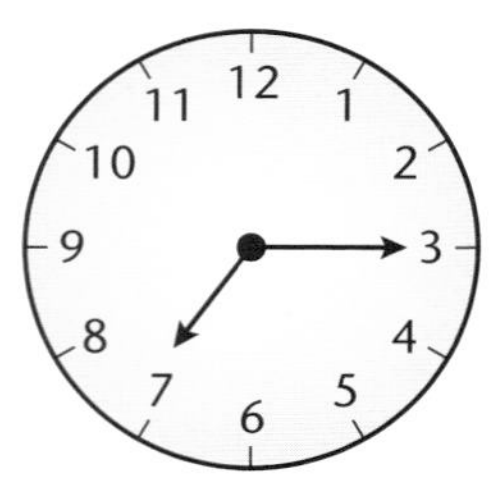

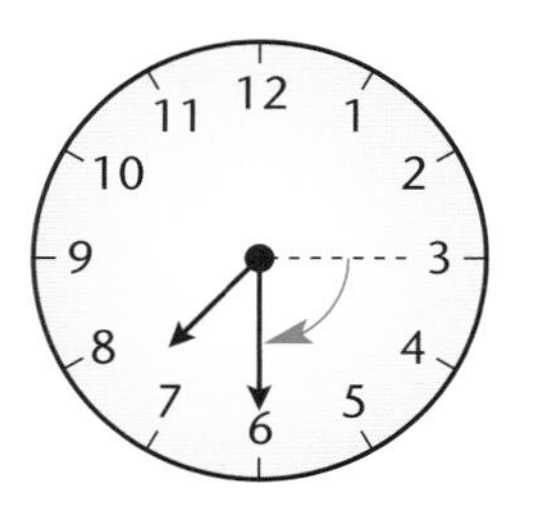

Minute hand makes $\frac{1}{4}$ turn.
Time taken $\frac{1}{4}$ hour.

The minute hand of a clock moves a **half turn** from 4 to 10.

Minute hand makes $\frac{1}{2}$ turn.
Time taken $\frac{1}{2}$ hour.

The hands of a clock turn **clockwise**.

This fairground ride turns **anti-clockwise** .

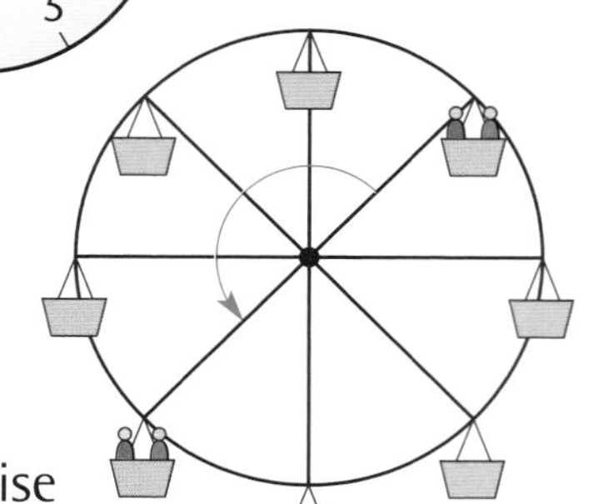

$\frac{1}{2}$ turn anti-clockwise

EXERCISE 6A

1 Describe the turn the minute hand of a clock makes between these times.

(a) 3 am and 3.30 am (b) 6.45 pm and 7 pm

(c) 2215 and 2300 (d) 0540 and 0710

Look at the clock examples on page 144.

2 Here is a diagram of a compass.

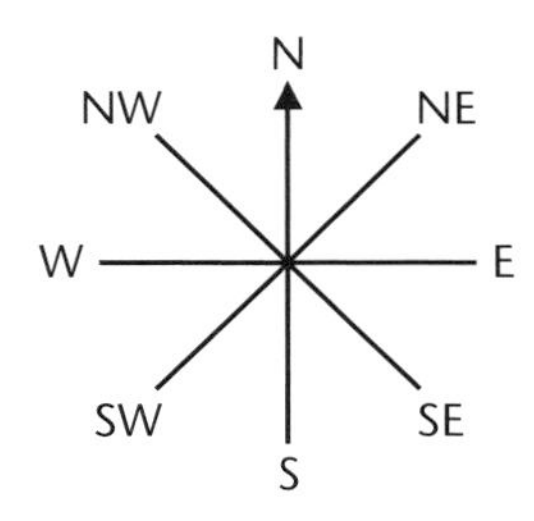

You are given a starting direction and a description of a turn.
What is the finishing direction in each case?

	Starting direction	Description of turn
(a)	N	$\frac{1}{4}$ turn clockwise
(b)	SE	$\frac{1}{4}$ turn anti-clockwise
(c)	SW	$\frac{1}{2}$ turn anti-clockwise
(d)	E	$\frac{3}{4}$ turn anti-clockwise
(e)	NE	$\frac{1}{4}$ turn clockwise
(f)	W	$\frac{1}{2}$ turn clockwise
(g)	NW	$\frac{3}{4}$ turn clockwise
(h)	S	$\frac{1}{4}$ turn anti-clockwise

Describing angles

An angle is a measure of turn. Angles are usually measured in degrees.
A complete circle (or full turn) is 360°.

The minute hand of a clock turns through 360° between 1400 (2 pm) and 1500 (3 pm).

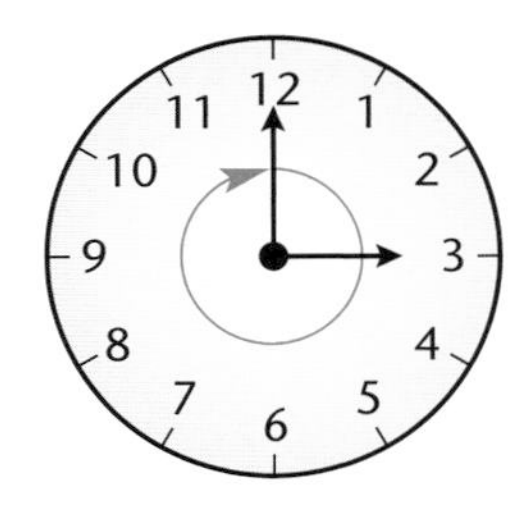

You will need to recognise the following types of angles.

acute		An angle between 0° and 90°, less than a $\frac{1}{4}$ turn.
right angle		An angle of 90°, a $\frac{1}{4}$ turn. A right angle is usually marked with this symbol.
obtuse		An angle between 90° and 180°, more than $\frac{1}{4}$ turn but less than $\frac{1}{2}$ turn.
straight line		An angle of 180°, a $\frac{1}{2}$ turn.
reflex		An angle between 180° and 360°, more than $\frac{1}{2}$ turn but less than a full turn.

You can describe angles in three different ways.

- 'Trace' the angle using capital letters. Write a 'hat' symbol over the middle letter $A\hat{B}C$

- Use an angle sign or write the word 'angle'
  ∠PQR or angle PQR

- Use a single letter.

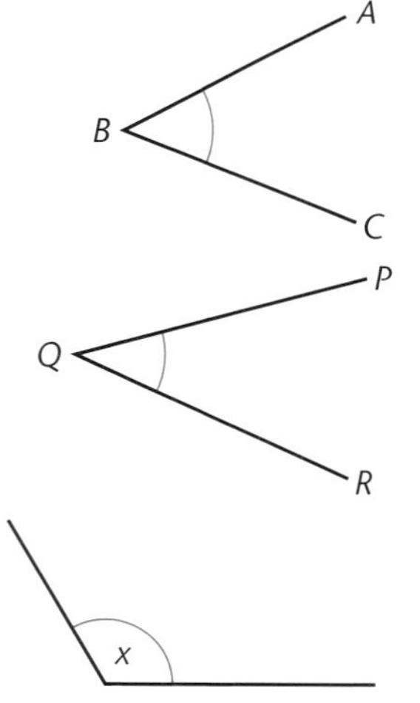

AB, BC, PQ and QR are called **line segments**.

The letter on the point of the angle always goes in the middle.

EXAMPLE 1

State whether these angles are acute, right angle, obtuse or reflex.

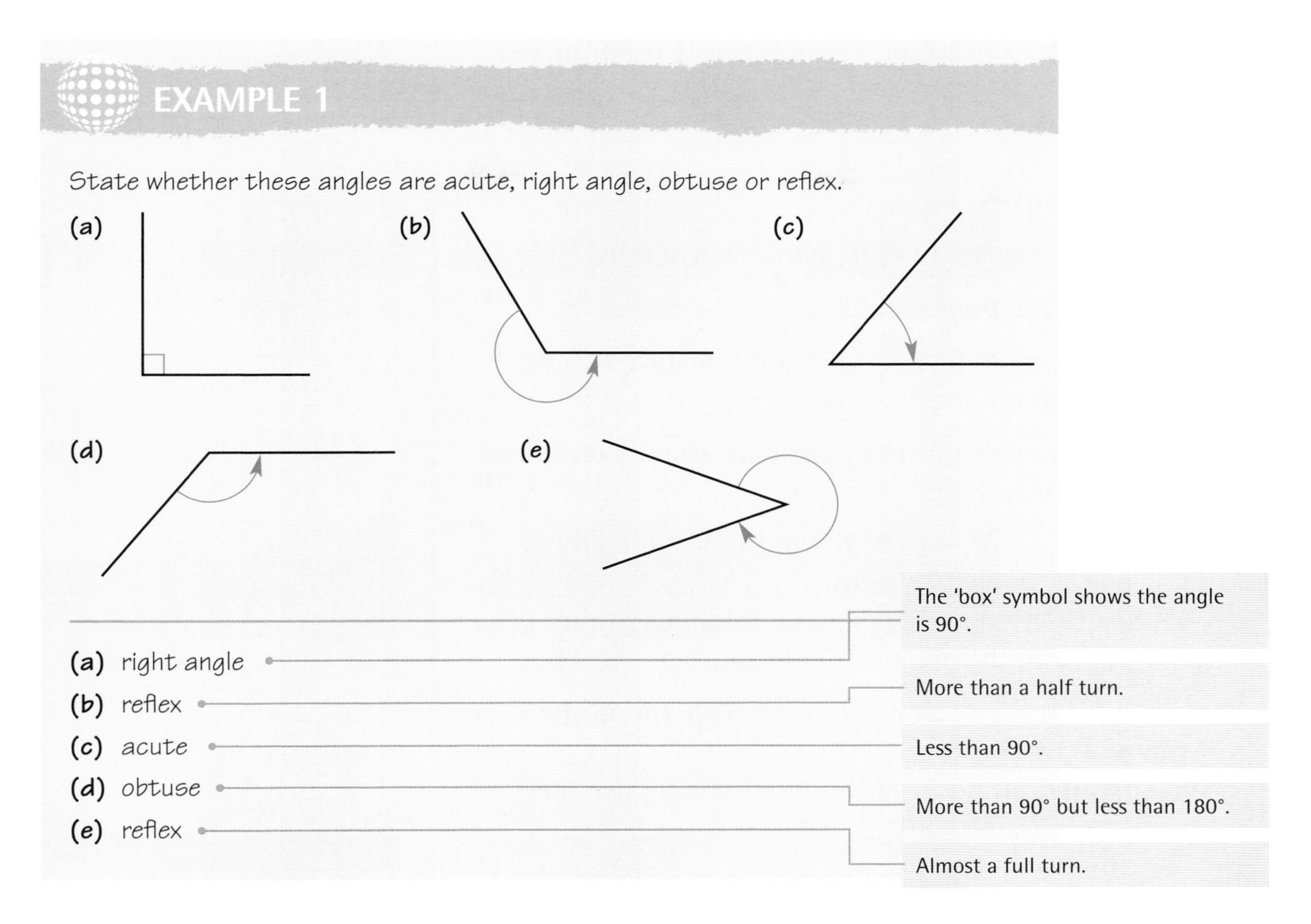

(a) right angle — The 'box' symbol shows the angle is 90°.
(b) reflex — More than a half turn.
(c) acute — Less than 90°.
(d) obtuse — More than 90° but less than 180°.
(e) reflex — Almost a full turn.

EXERCISE 6B

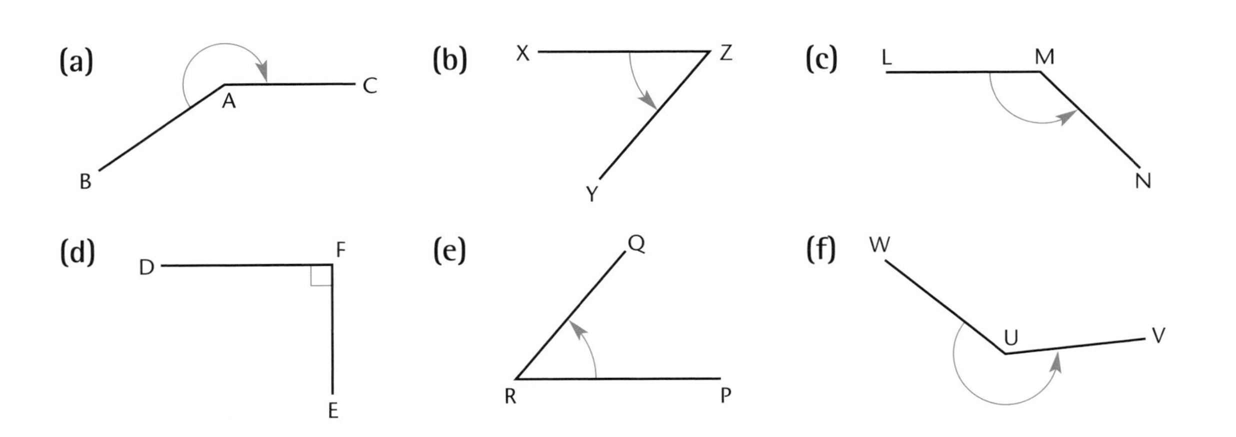

1 State whether these angles are acute, right angle, obtuse or reflex.

For example
$\angle GHK \quad \hat{GHK} \quad$ angle GHK

2 Describe each of the angles in question 1 using three letters.

Measuring angles

You use a **protractor** to measure angles accurately.

Follow these instructions carefully.

1 Estimate the angle first, so you don't mistake an angle of 30°, say, for an angle of 150°.

2 Put the centre point of the protractor exactly on top of the point of the angle.

3 Place one of the 0° lines of the protractor directly on top of one of the angle 'arms'.
If the line isn't long enough, draw it longer so that it reaches beyond the edge of the protractor.

4 Measure from the 0°, following the scale round the edge of the protractor.
If you are measuring from the *left-hand* 0°, use the *outside* scale.

Use your estimate to help you choose the correct scale.

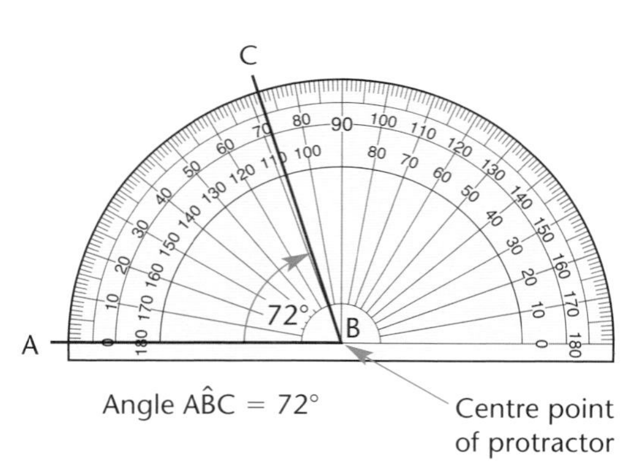

Measuring from the left-hand 0°.

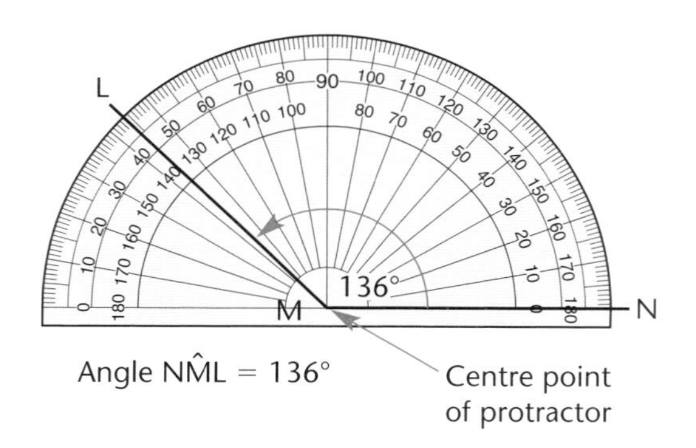

Measuring from the right-hand 0°.

If you are measuring from the *right-hand* 0°, use the *inside* scale.

5 On the correct scale, read the size of the angle in

EXERCISE 6C

degrees, where the other 'arm' cuts the edge of the protractor.

6 To measure a reflex angle (an angle that is bigger than 180°), measure the acute or obtuse angle, and subtract this value from 360°.

1 Measure these angles using a protractor.

(a)

(b)

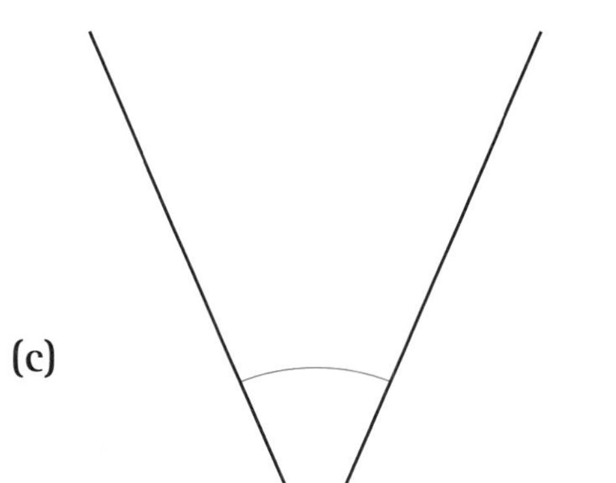

(c)

(d)

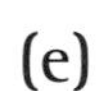

(e)

(f)

2 Draw these angles using a protractor. Label the angle in each case.

(a) angle PQR = 54°

(b) angle STU = 148°

(c) angle MLN = 66°

(d) angle ZXY = 157°

(e) angle DFE = 42°

(f) angle HIJ = 104°

For example

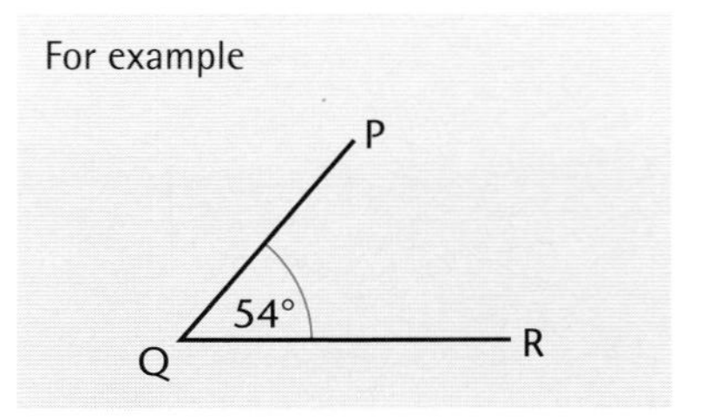

Angle properties

You need to know these angle facts.

Angles on a straight line add up to 180°.

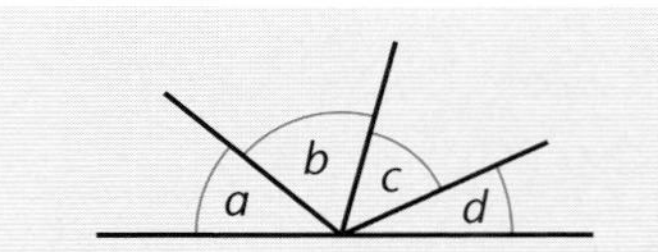

These angles lie on a straight line, so $a + b + c + d = 180°$.

Angles around a point add up to 360°.

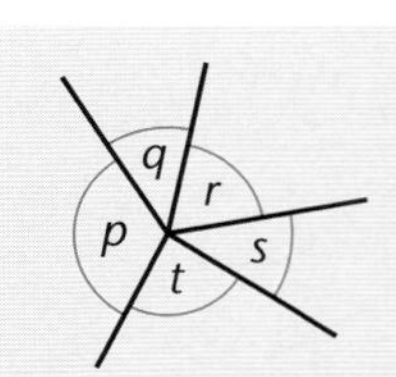

These angles make a full turn, so $p + q + r + s + t = 360°$.

Vertically opposite angles are equal.

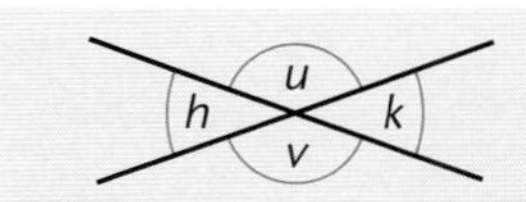

In this diagram $h = k$ and $u = v$.

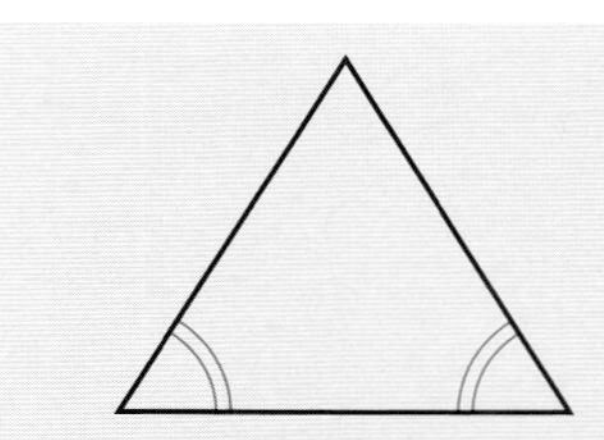

Equal angles are shown by matching arcs.

Perpendicular lines intersect at 90° and are marked with the right angle symbol.

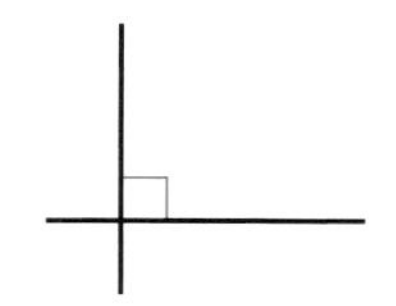

EXAMPLE 2

Calculate the size of the angles marked with letters.

(a)

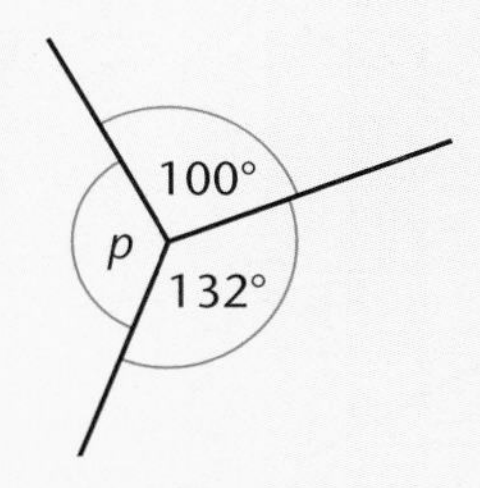

(b)

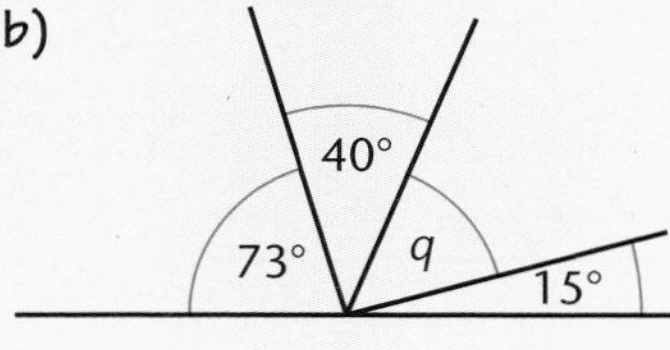

(c)

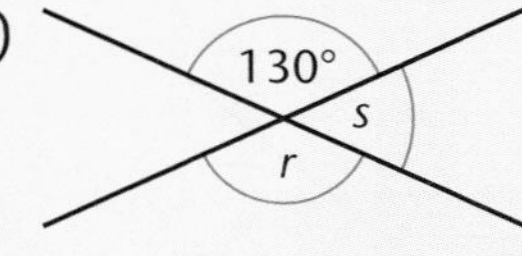

(d)

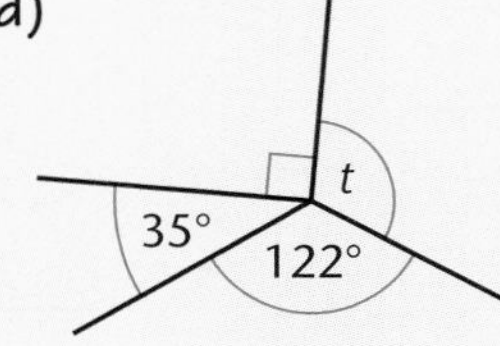

(a) $p + 100° + 132 = 360°$ (angles around a point)

$p = 360° - 100° - 132°$

$p = 128°$

Solve this equation to find *p*. Use Chapter 5 to help you.

(b) $q + 73° + 40° + 15° = 180°$ (angles on a straight line)

$q = 180° - 73° - 40° - 15°$

$q = 52°$

(c) $r = 130°$ (vertically opposite)

$s + 130° = 180°$ (angles on a straight line)

$s = 180° - 130° = 50°$

(d) $t + 122° + 35° + 90° = 360°$ (angles around a point)

$t = 360° - 122° - 35° - 90°$

$t = 113°$

means 90°.

EXERCISE 6D

Calculate the size of the angles marked with letters.

1

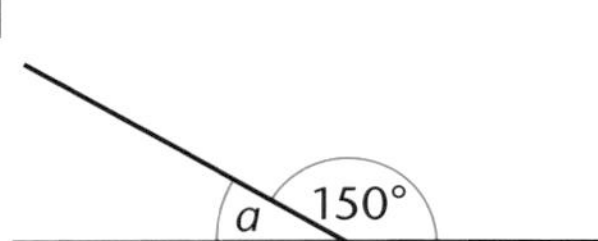

2

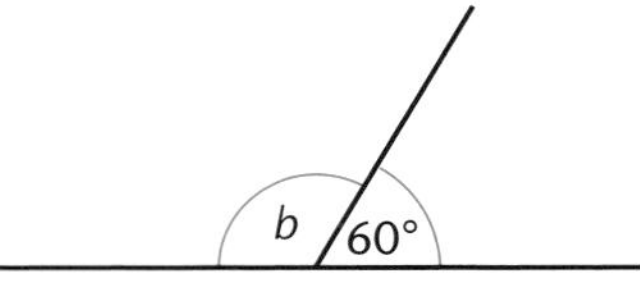

3

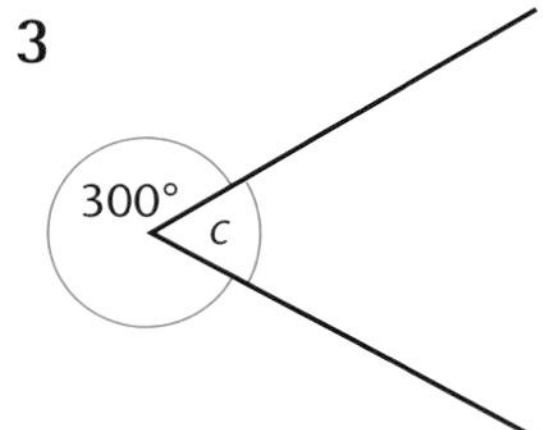

4

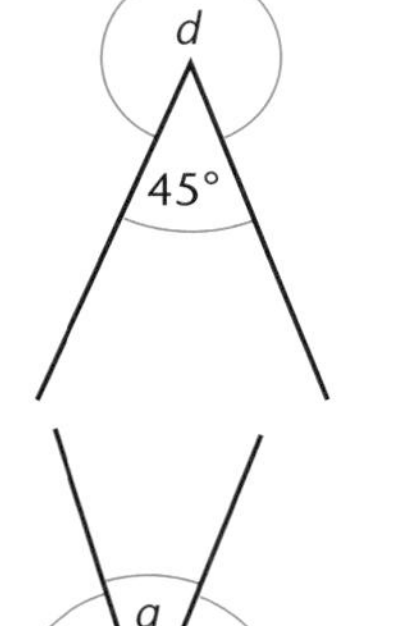

5

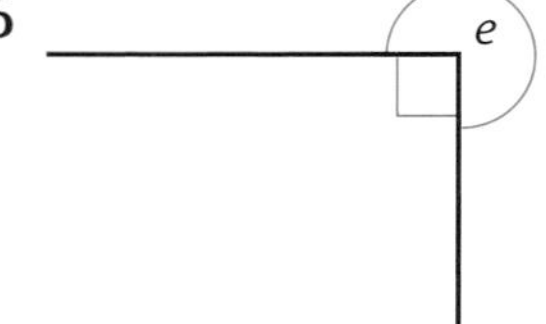

6

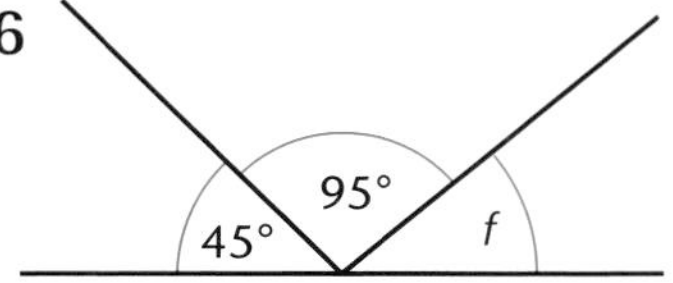

7

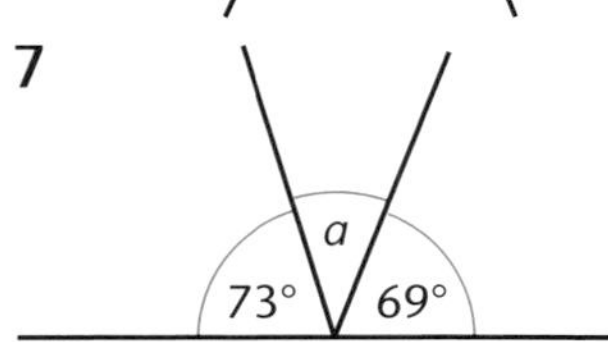

8

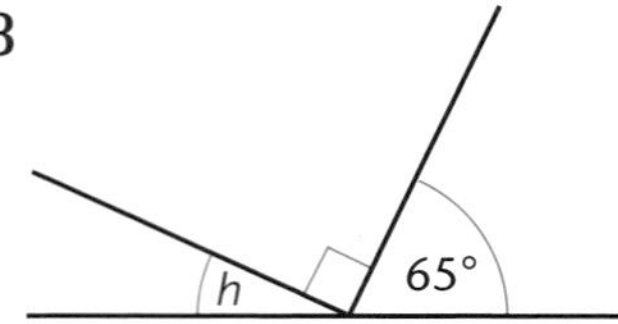

9

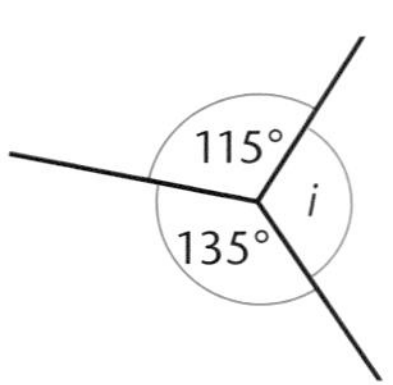

10

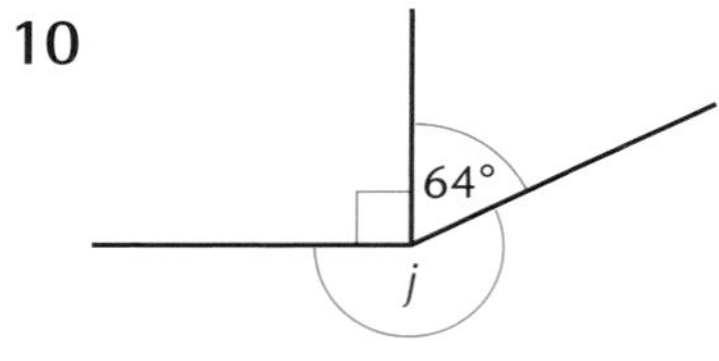

11

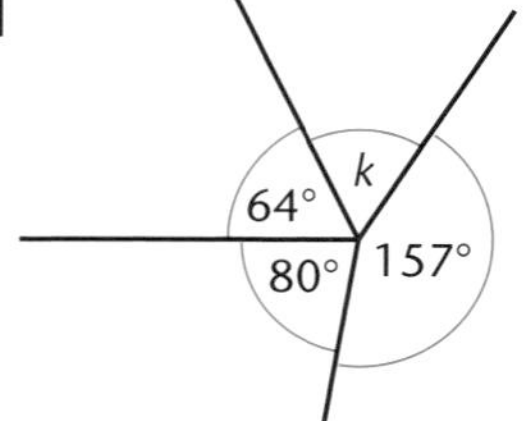

12

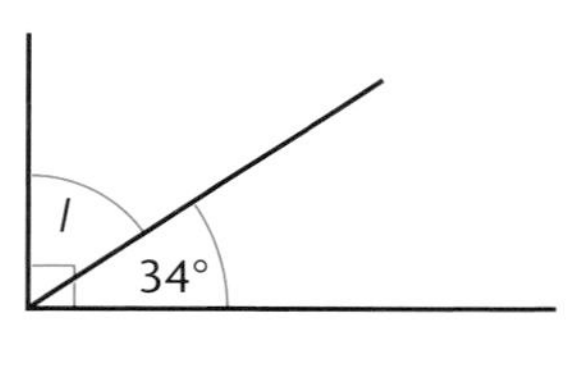

13

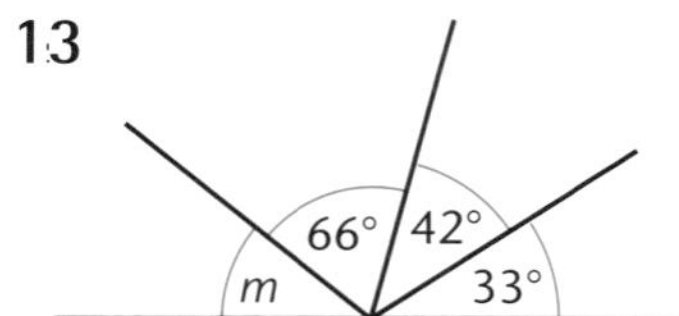

14

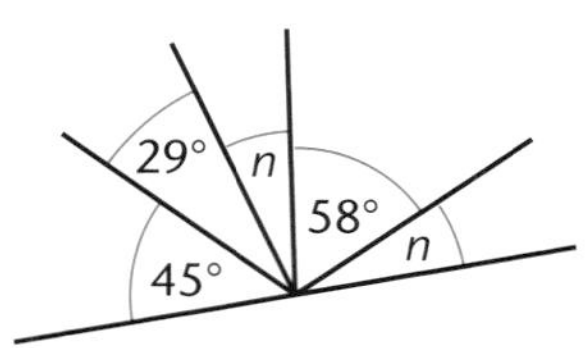

15

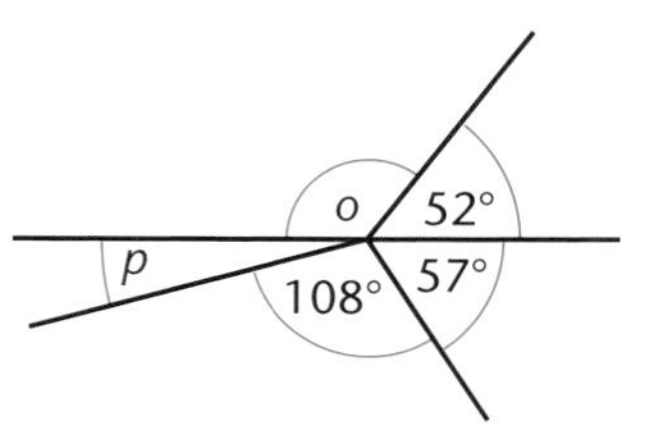

16

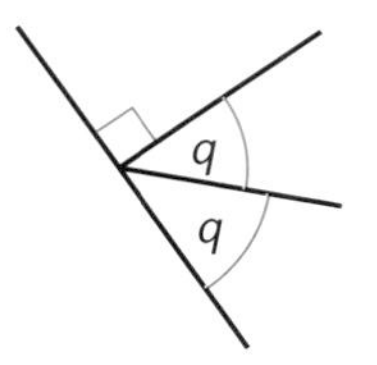

17

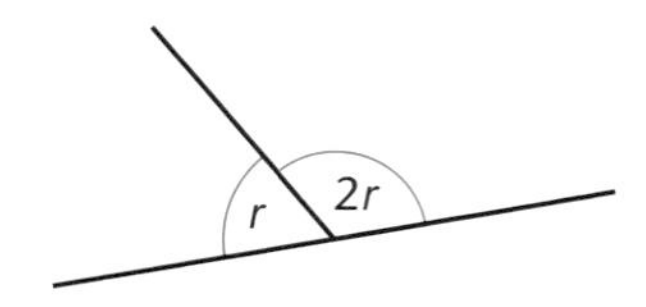

18

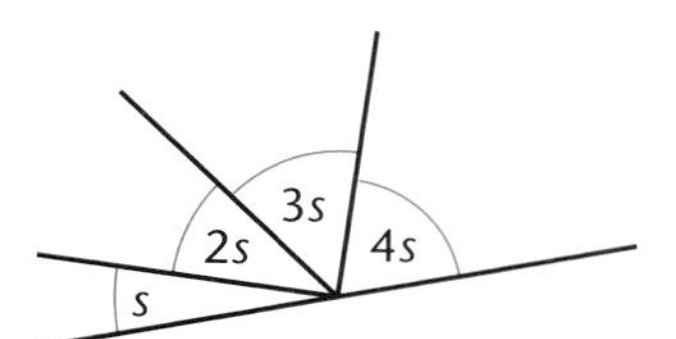

Angles in parallel lines

Parallel lines are the same distance apart all along their length. You can use arrows to show lines are parallel.

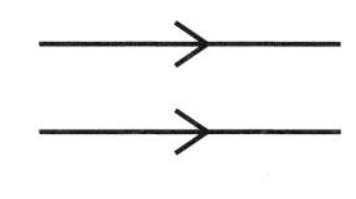

A straight line that crosses a pair of parallel lines is called a **transversal**.

A transversal creates pairs of equal angles.

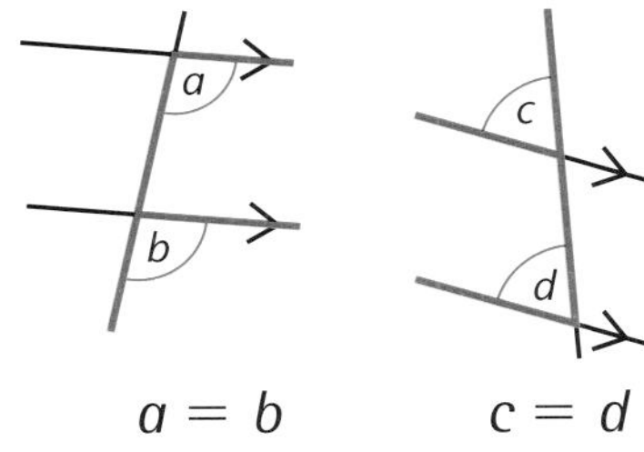

$a = b$ $\quad$ $c = d$

$\left.\begin{array}{l} a \text{ and } b \\ c \text{ and } d \end{array}\right\}$ are **corresponding** angles

The lines make an F shape.

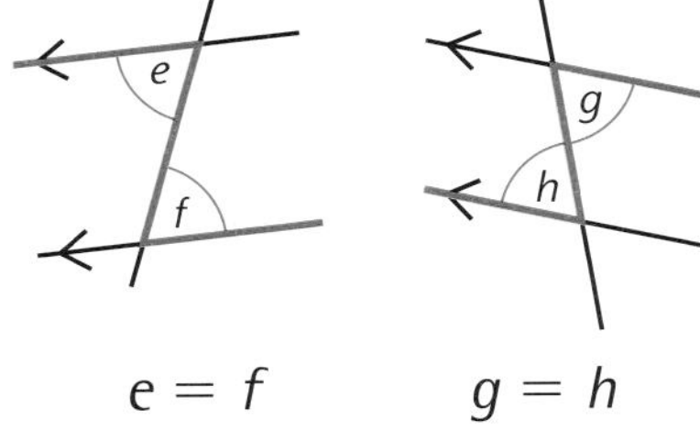

$e = f$ $\quad$ $g = h$

$\left.\begin{array}{l} e \text{ and } f \\ g \text{ and } h \end{array}\right\}$ are **alternate** angles.

The lines make a Z shape.

In this diagram the two angles are not equal, j is obtuse and k is acute.

$j + k = 180°$

The two angles lie on the *inside* of a pair of parallel lines. They are called **co-interior** angles or allied angles. Co-interior angles add up to 180°.

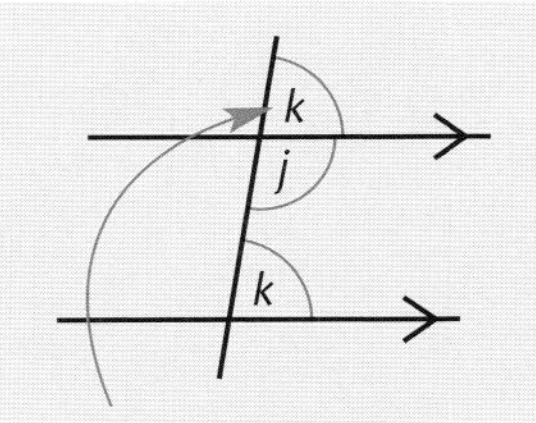

The angles marked k are corresponding angles.
$j + k = 180°$ (angles on a straight line).

Corresponding angles are equal.
Alternate angles are equal.
Co-interior angles add up to 180°.

EXAMPLE 3

Calculate the size of the angles marked with letters.

(a)

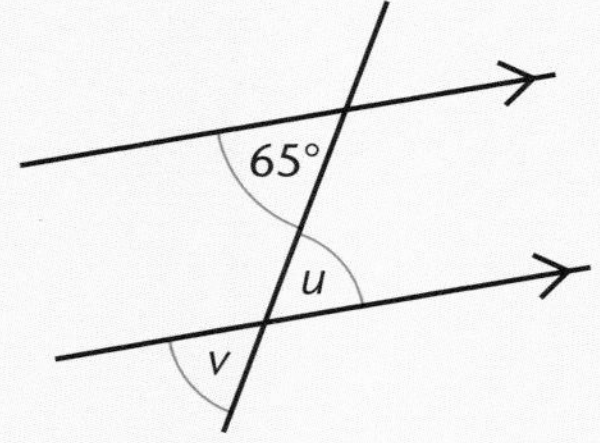

(b)

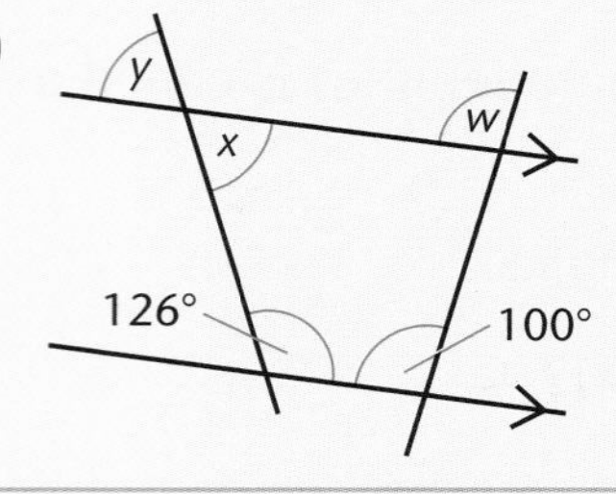

(a) Method A

$u = 65°$ (alternate)

$v = 65°$ (vertically opposite)

Method B

$v = 65°$ (corresponding)

$u = 65°$ (vertically opposite)

You could use method A or method B.

(b) $w = 100°$ (corresponding)

$x + 126° = 180°$ (co-interior angles)

$x = 180° - 126°$

$x = 54°$

$y = 54°$ (vertically opposite)

EXERCISE 6E

Calculate the size of the angles marked with letters.

1

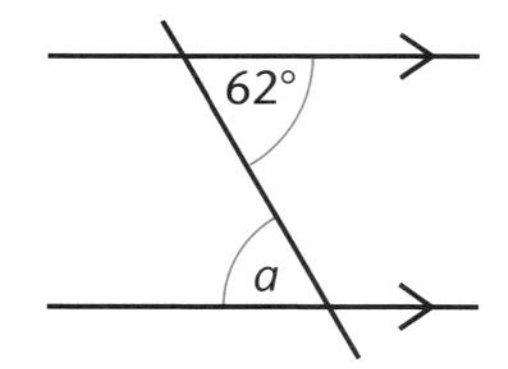

2

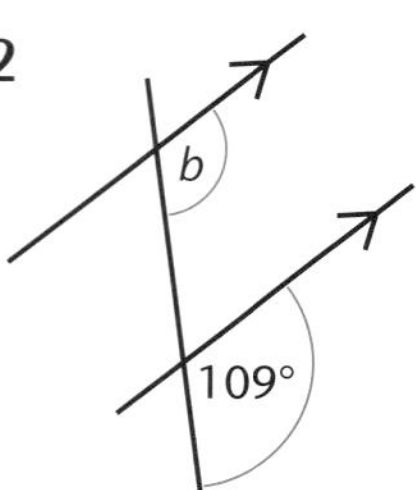

3

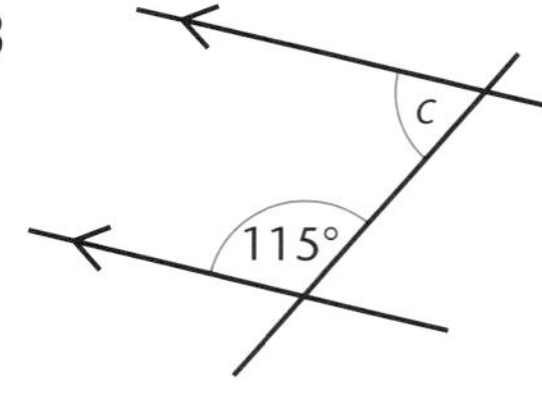

4

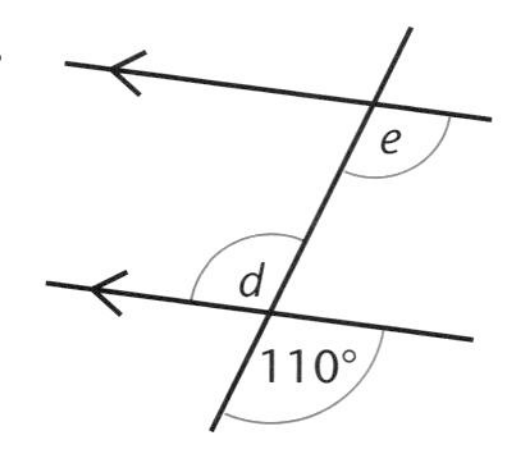

5

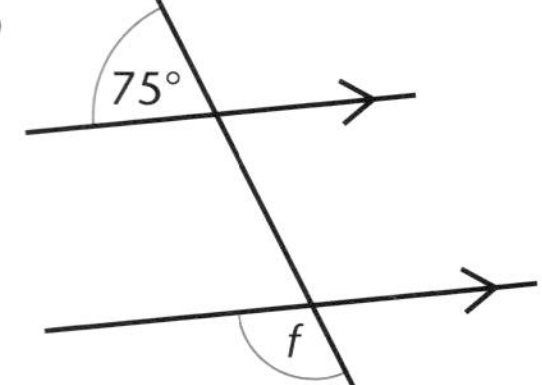

6

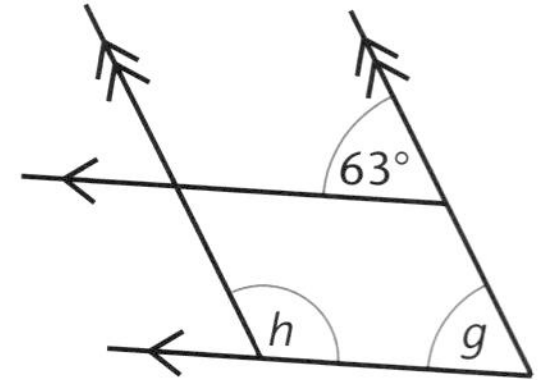

7

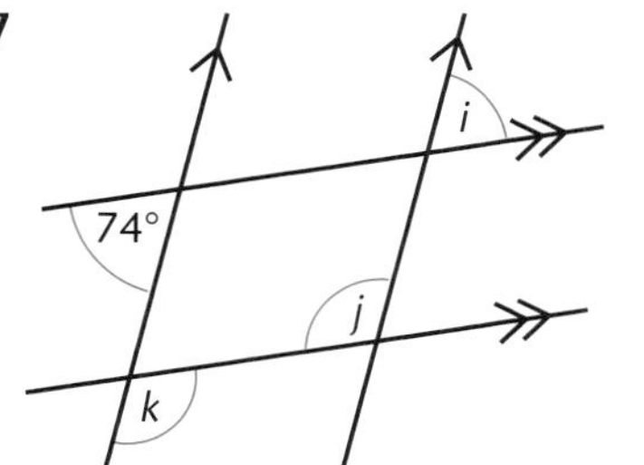

8

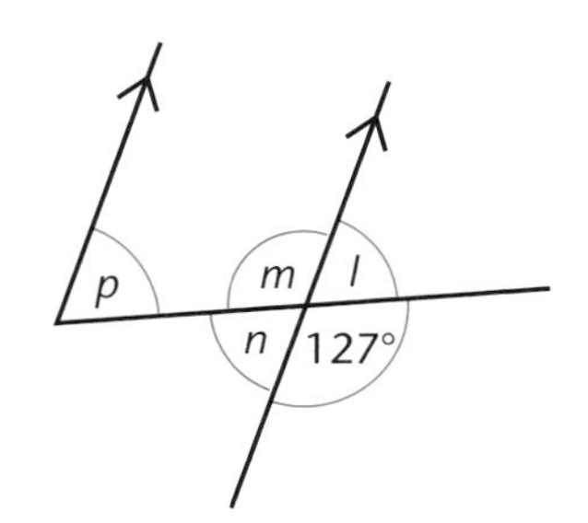

9

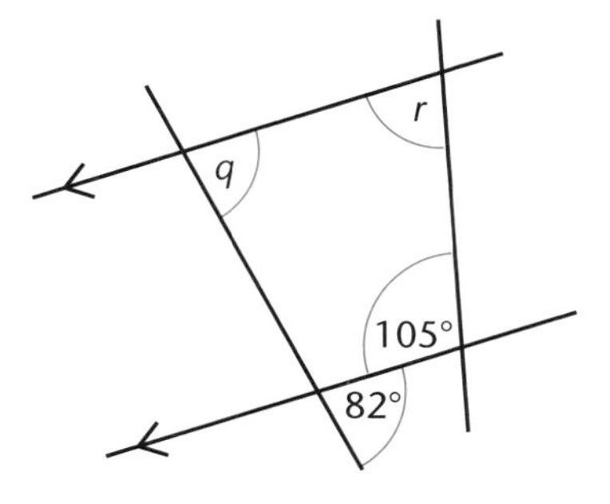

6.2 Three-figure bearings

You can use compass points to describe a direction, but in Mathematics we use three-figure **bearings**.

A three-figure bearing gives a direction in degrees.

It is an angle between 0° and 360°. It is always measured *from the north* in a *clockwise* direction.

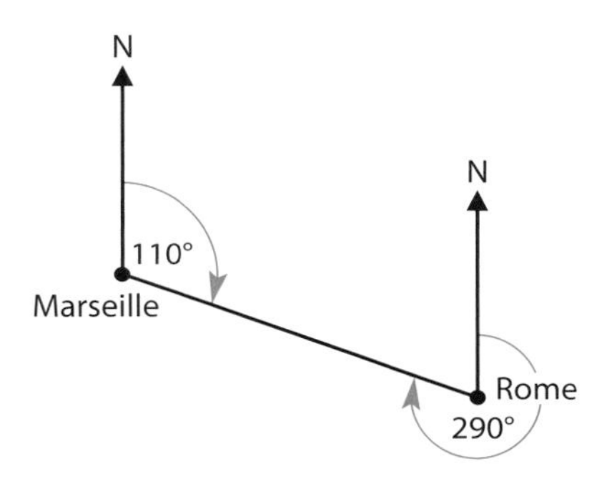

The diagram shows the cities of Marseille and Rome.
The bearing of Rome from Marseille is 110°.
The bearing of Marseille from Rome is 290°.

A bearing must always be written with *three figures*.

A bearing of 073°

You write 073° because the bearing must have three figures.

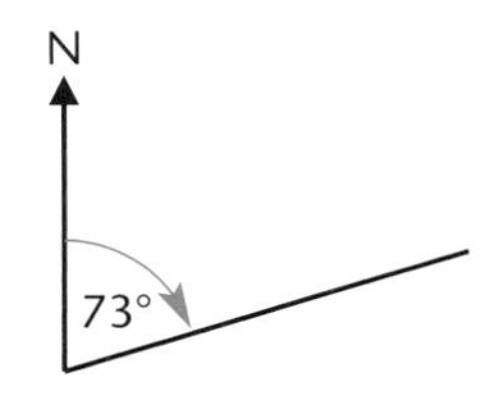

A bearing of 254°

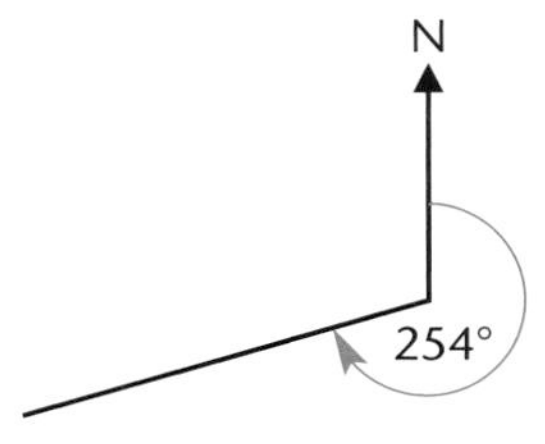

You need to be able to measure bearings accurately using a protractor.

Your answer needs to be within 1° of the correct value.

EXAMPLE 4

(a) Write down the bearing of Q from P.

(b) Work out the bearing of P from Q.

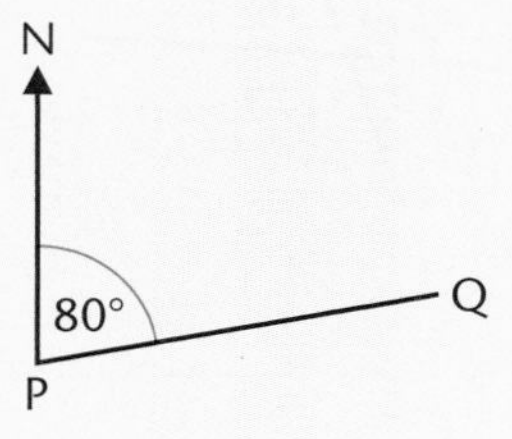

(a) 080°

If you stand at P and face north, you need to turn through 80° to face Q.

(b)

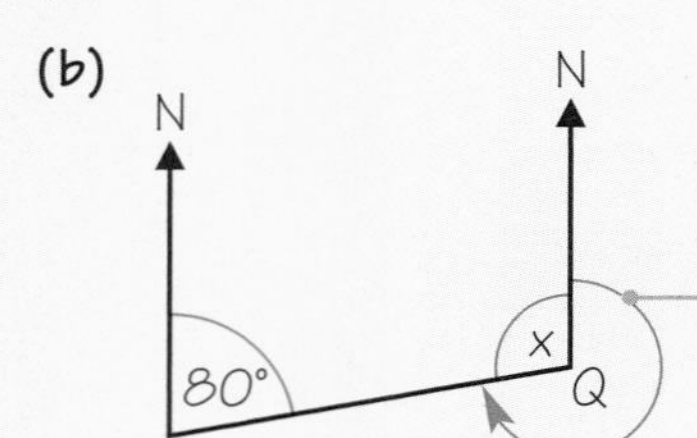

Draw in the north line at Q.

This angle is the bearing of P from Q.

The two north lines are parallel, so

$80° + x = 180°$ (co-interior angles)

$80° - 80° + x = 180° - 80°$

$x = 100°$

Bearing of P from Q $+ x = 360°$ (angles around a point)

Bearing of P from Q $+ 100° = 360°$

Bearing of P from Q $= 260°$.

EXERCISE 6F

1 For each diagram, write down the bearing of Q from P.

(a)

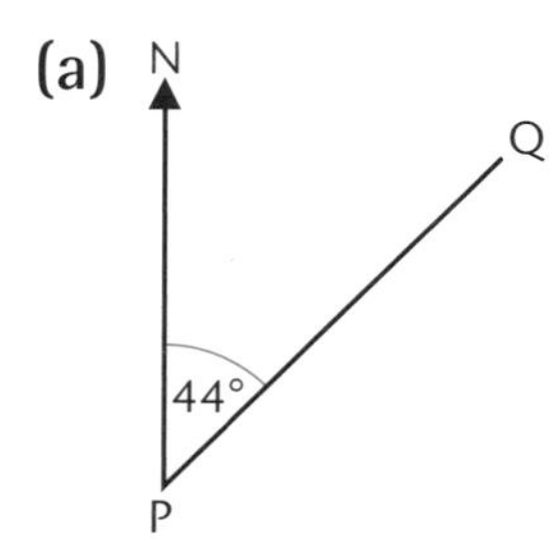

(b)

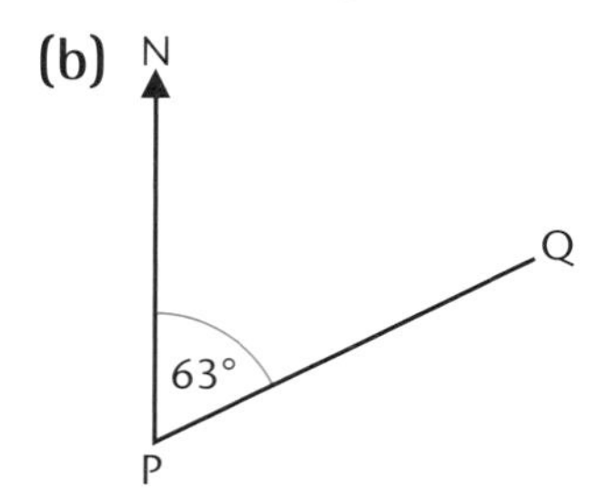

(c)

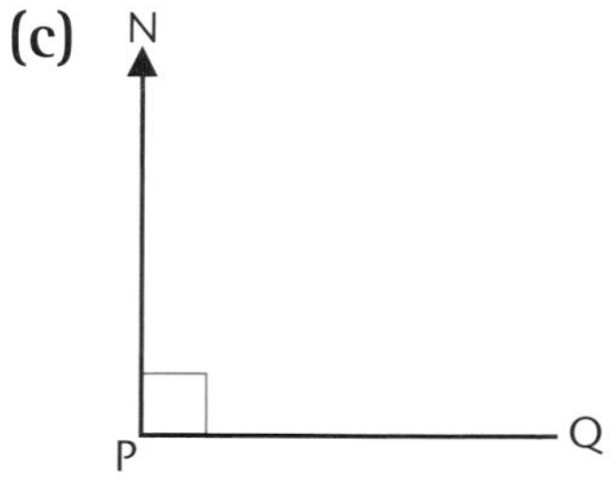

(d)

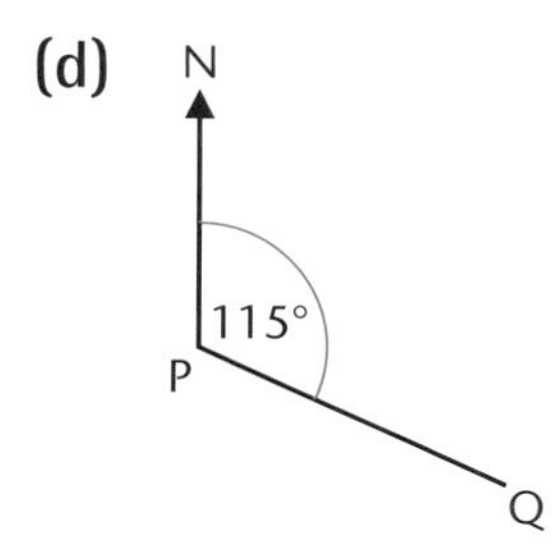

(e)

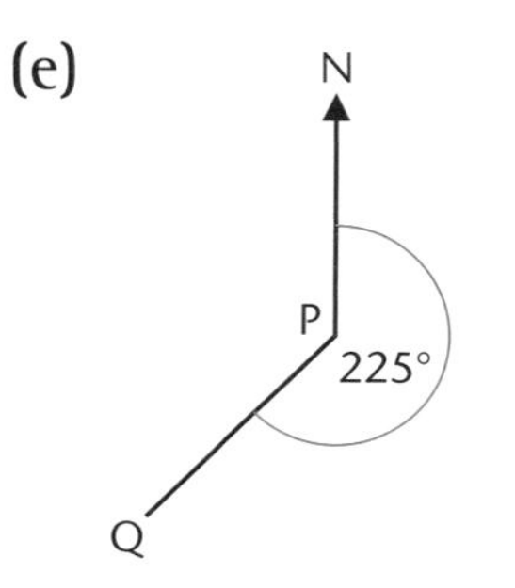

(f)

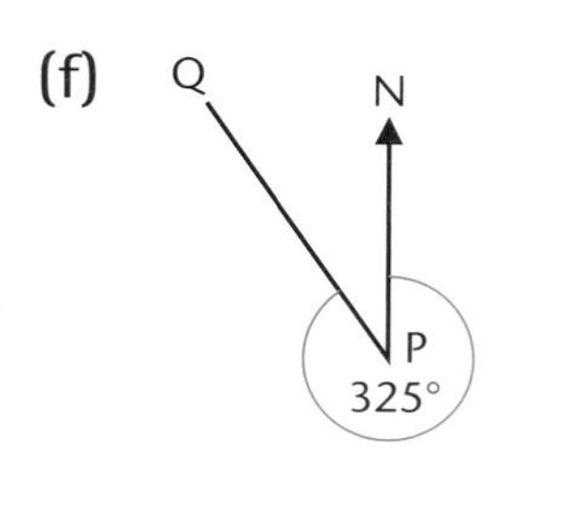

2 For each diagram work out the bearing of *B* from *A*.

(a)
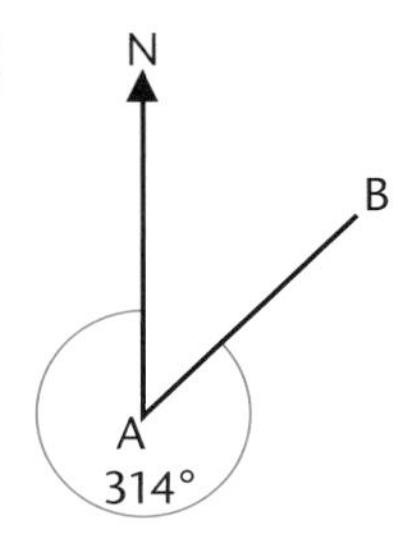

(b)
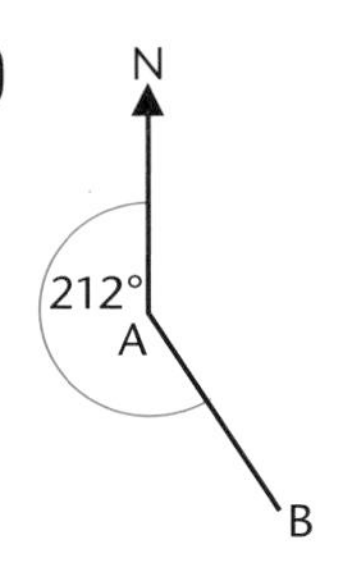

(c)
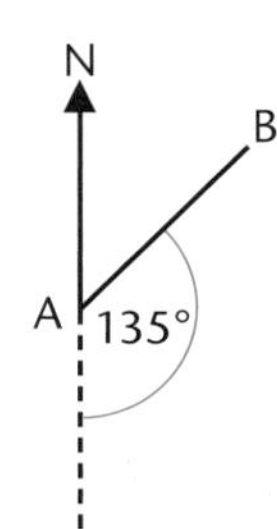

(d)
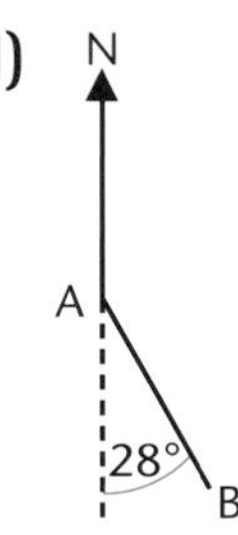

(e)
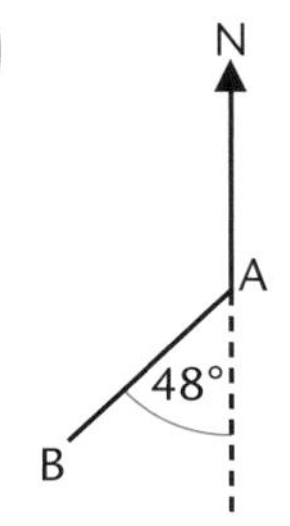

(f)
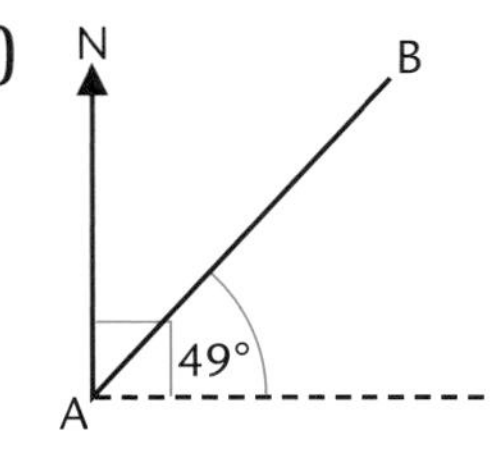

(g)
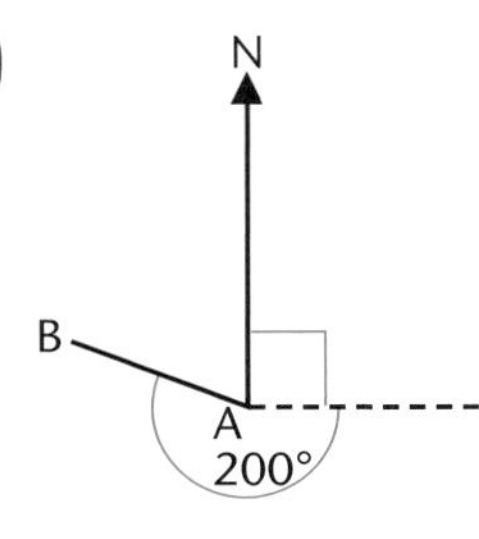

(h)
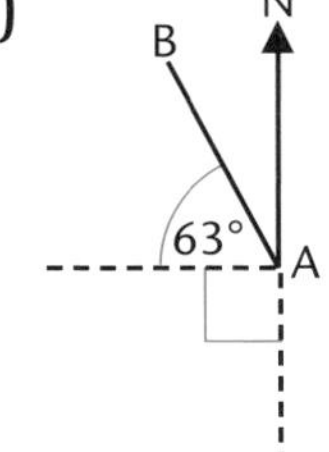

3 Draw accurate diagrams to show these three-figure bearings.

(a) 036° (b) 145° (c) 230° (d) 308°
(e) 074° (f) 256° (g) 348° (h) 115°

4 The diagram shows the peaks, P, L and M, of three mountains in the Himalayas.

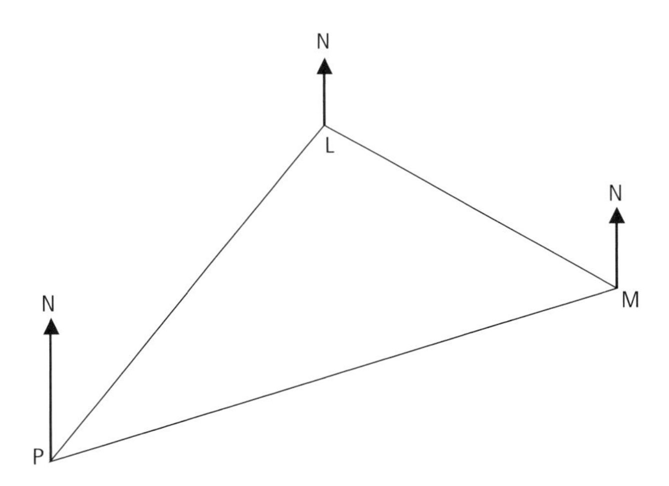

Use your protractor to find these bearings.

(a) *L* from *P* (b) *M* from *P* (c) *M* from *L*

5 In each diagram work out the bearing of *X* from *Y*.

(a)

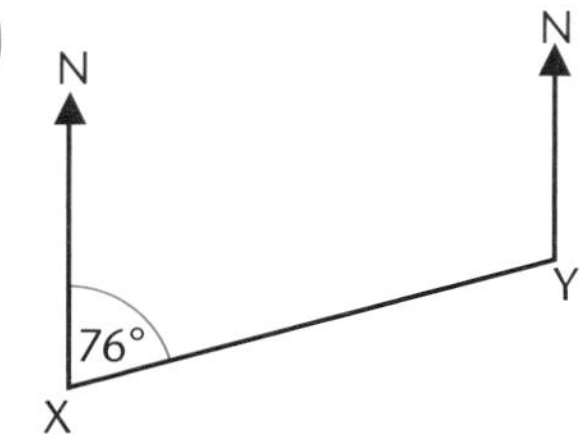

(b)

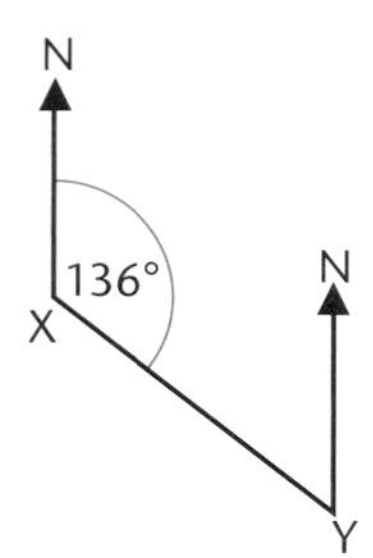

(c)

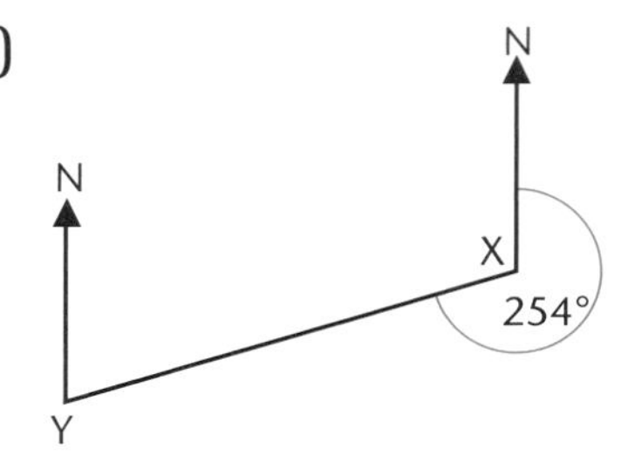

(d)

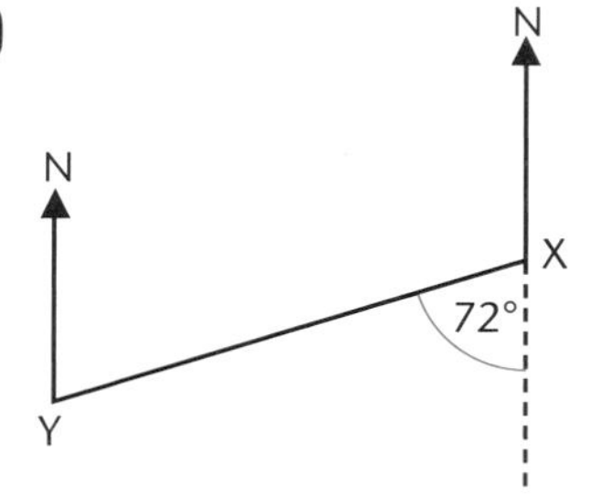

(e)

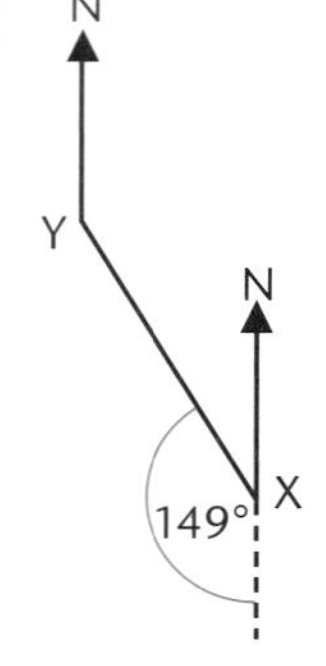

(f)

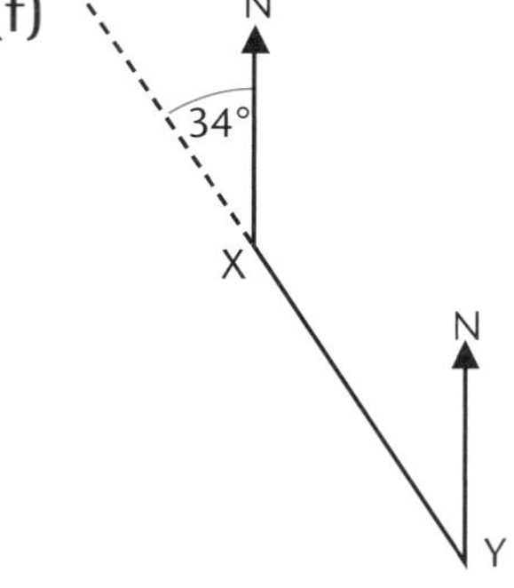

The bearing of *D* from *C* is

(a) 037° (b) 205° (c) 167° (d) 296°

6 In each case

- Sketch a diagram to show the bearing of *D* from *C*.
- Work out the bearing of *C* from *D*.

6.3 Triangles

Types of triangle

There are four types of triangle. They can be described by their properties.

Triangle	Picture	Properties
scalene		The three sides are different lengths. The three angles are different sizes.
isosceles	A B C	Two equal sides. AB = AC Two equal angles ('base angles') angle ABC = angle ACB.
equilateral	60° 60° 60°	All three sides are equal in length. All angles 60°.
right-angled	X Y Z	One of the angles is a right angle (90°). Angle XZY = 90°.

The marks across the sides of the triangles show which sides are equal.

Angles in a triangle

1 Draw a triangle on a piece of paper.

2 Mark each angle with a different letter or shade them different colours.

3 Tear off each corner and place them next to each other on a straight line.

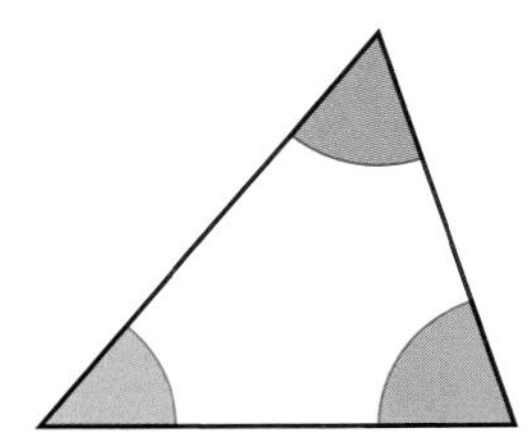

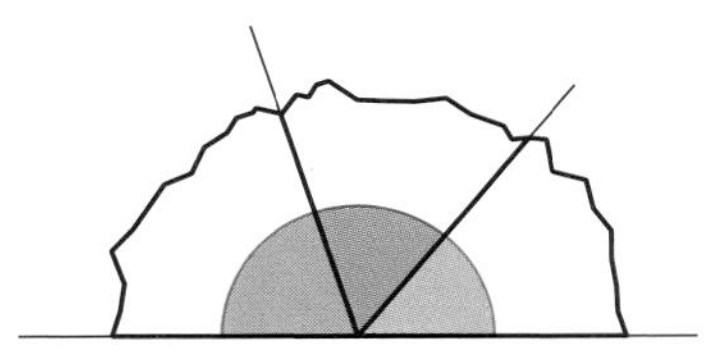

You will see that the three angles fit exactly onto the straight line.
So the three angles add up to 180°.

You are not *proving* this result, you are simply showing that it is true for the triangle you drew.

You can prove this for all triangles using facts about alternate and corresponding angles.

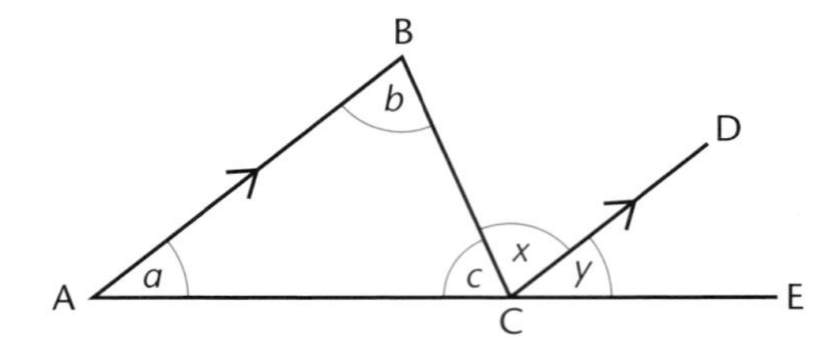

For the triangle ABC

1 Extend side AC to point E.

2 From C draw a line CD parallel to AB.

Let $\angle BCD = x$ and $\angle DCE = y$.

$x = b$ (alternate angles)

$y = a$ (corresponding angles)

$c + x + y = 180°$ (angles on a straight line at point C)

which means that $c + b + a = 180°$.

The sum of the angles of a triangle is 180°.

EXAMPLE 5

Calculate the size of the angles marked with letters.

(a)

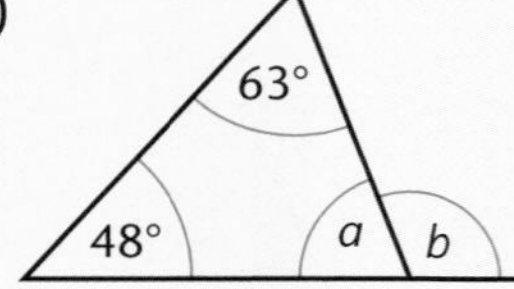

(b)

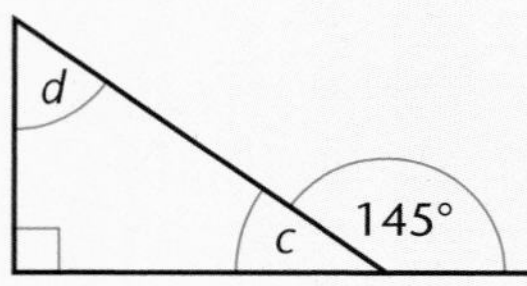

(c)

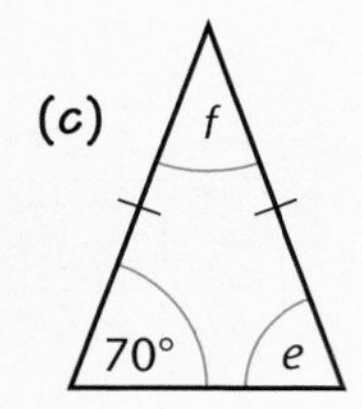

(a)

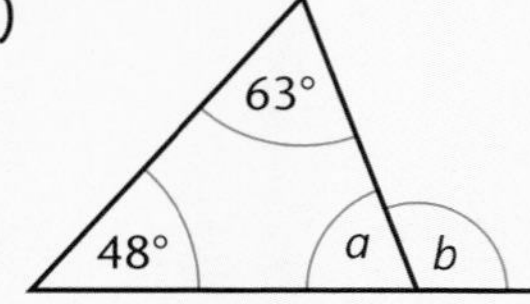

$a = 180° - 63° - 48°$

(angle sum of $\triangle$)

$a = 69°$

$b = 180° - 69°$

(angles on a straight line)

$b = 111°$

$\triangle$ means 'triangle'

Continued ▼

(b)

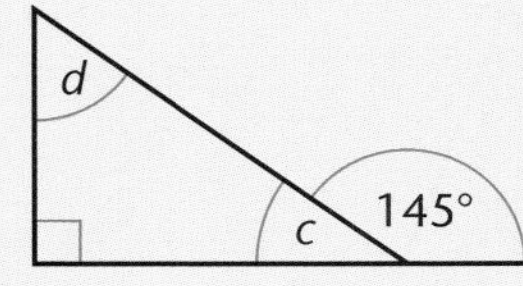

$c = 180° - 145°$
(angles on a straight line)
$c = 35°$
$d = 180° - 90° - 35°$
(angle sum of △)
$d = 55°$

(c)

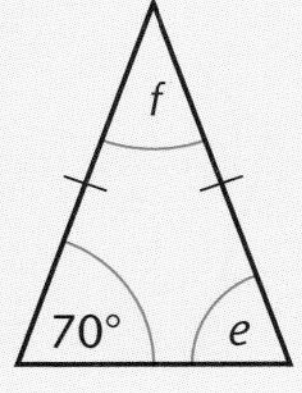

$e = 70°$ (base angles isosceles △)
$f = 180° - 70° - 70°$
(angle sum of △)
$f = 40°$

EXERCISE 6G

Calculate the size of the angles marked with letters.

1

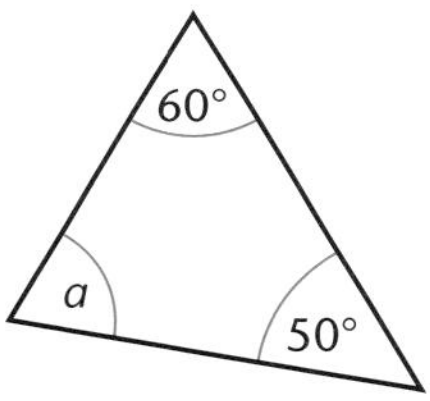

2

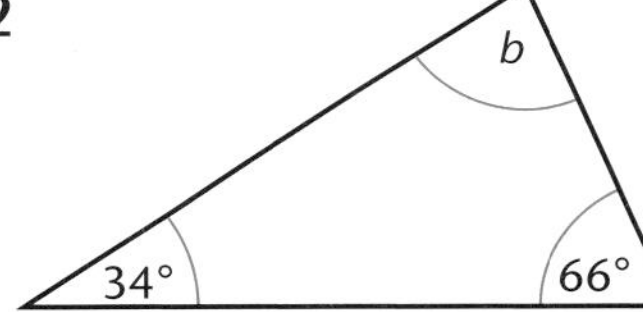

3

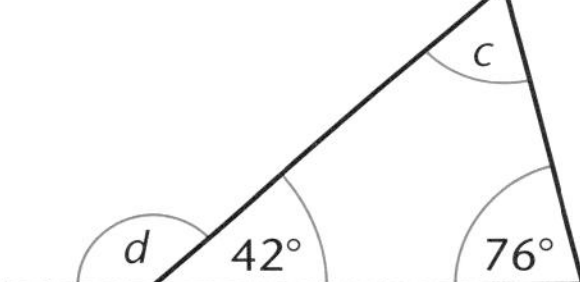

4

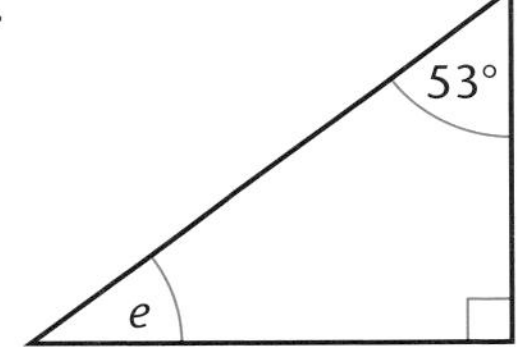

5

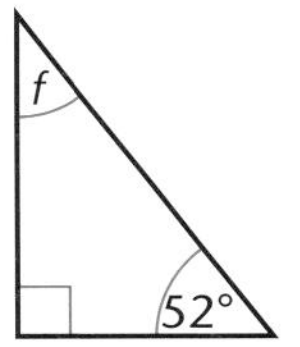

6

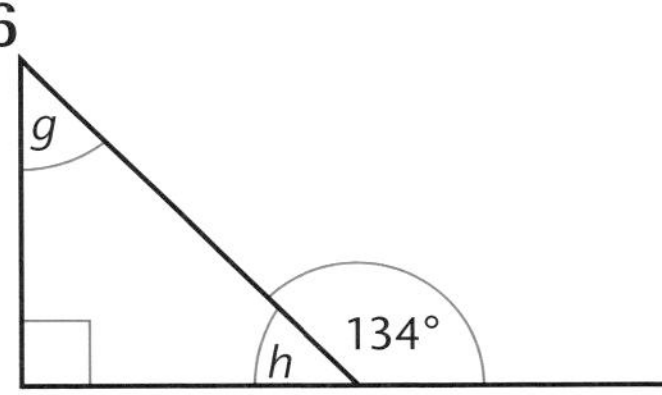

7

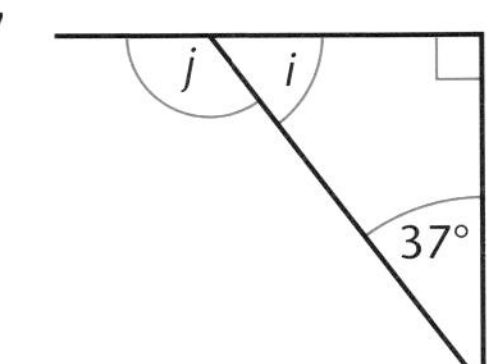

8

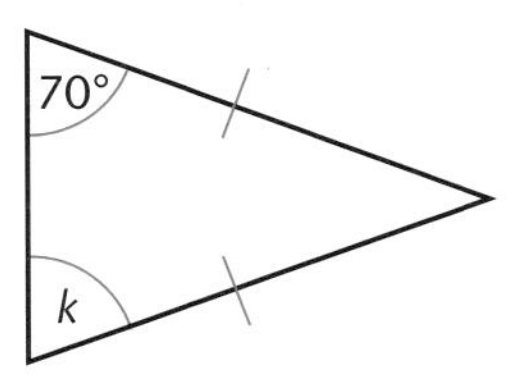

9

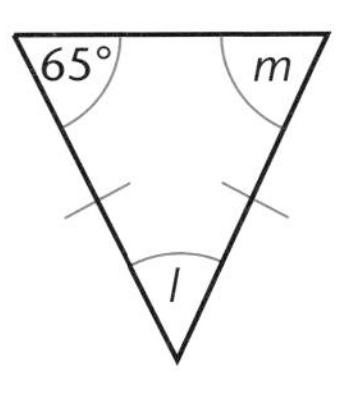

10

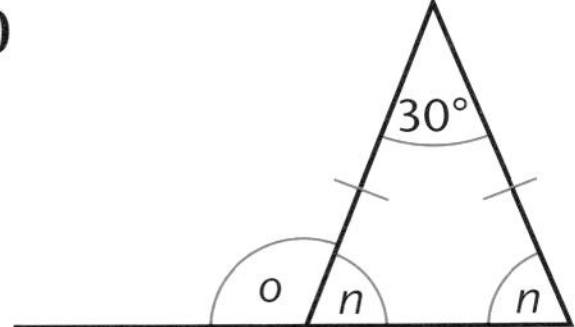

11

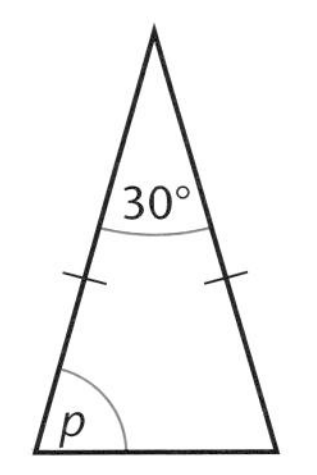

12

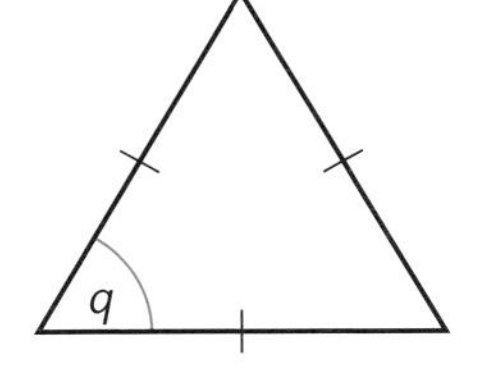

13

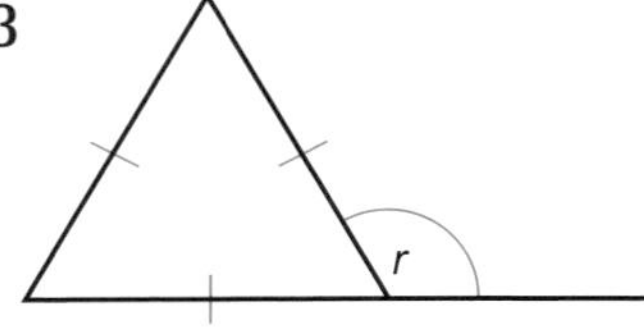

14

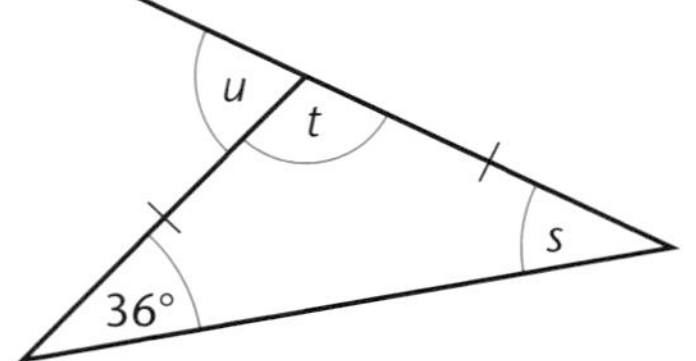

15

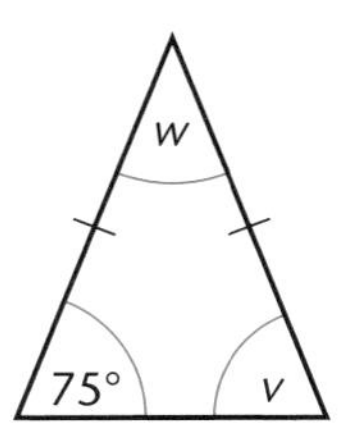

16

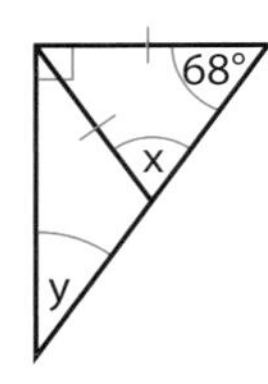

17

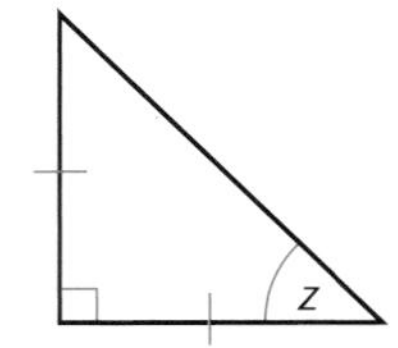

18

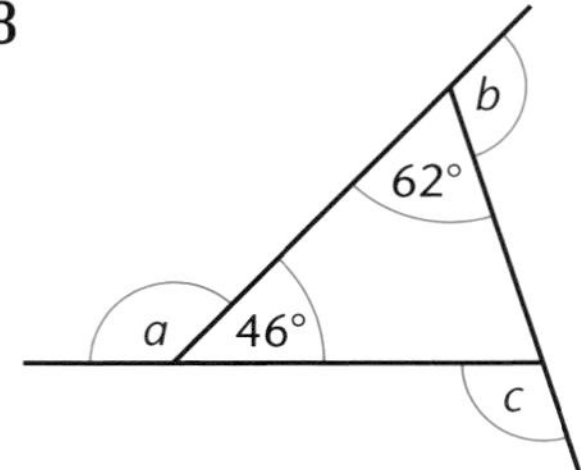

Interior and exterior angles in a triangle

The angles inside a triangle are called **interior** angles.

An **exterior** angle is formed by extending one of the sides of the triangle.

Angle BCE in this diagram is an exterior angle.

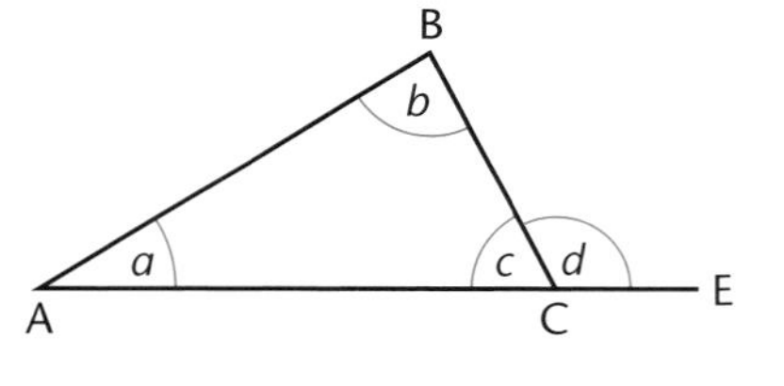

Let $BCE = d$

then $c + d = 180°$ (angles on a straight line)

but $c + b + a = 180°$ (angle sum of $\triangle ABC$)

which means that $d = a + b$

> In a triangle, the exterior angle is equal to the sum of the two opposite interior angles.

Sometimes you need to calculate the size of unmarked angles before you can calculate the ones you want. You can label the extra angles with letters.

> This helps make your explanations clear.

EXAMPLE 6

Calculate angles g and h.

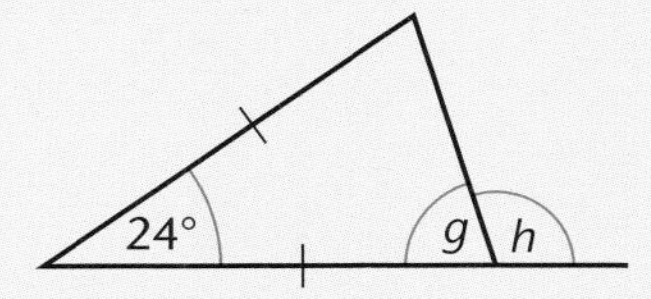

Let the third angle of the triangle $= x$.

$g + x + 24° = 180°$ (angle sum of △)
$g + x = 180° - 24°$
$g + x = 156°$
$g = x$ (base angles isosceles △)
so both g and $x = \frac{1}{2}$ of $156°$ (or $156° \div 2$)
$g = 78°$
$h = 180° - 78°$ (angles on a straight line)
$h = 102°$

$156 \div 2 = 78$.

EXAMPLE 7

Calculate angle i.

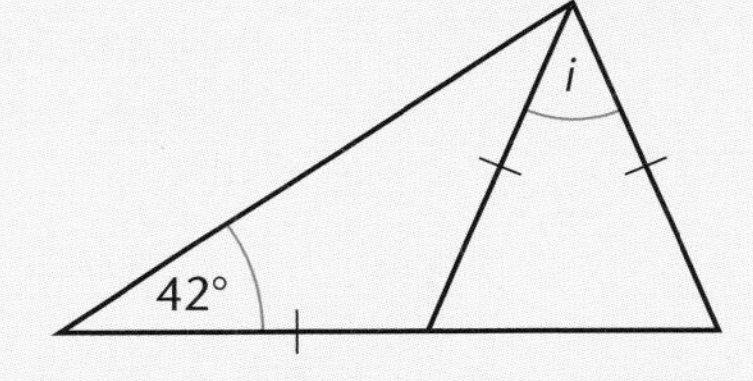

Label the unknown angles x, y and z.

Label the two triangles 1 and 2.

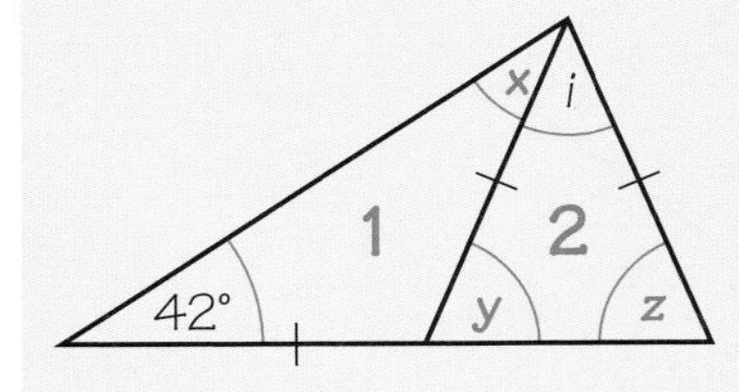

$x = 42°$
(base angles isosceles △ 1)
$y = 42° + 42°$
(exterior angle of △ 1)
$y = 84°$
$z = 84°$
(base angles isosceles △ 2)
$i = 180° - 84° - 84°$
(angle sum △ 2)
$i = 12°$

EXAMPLE 8

Calculate angle j.

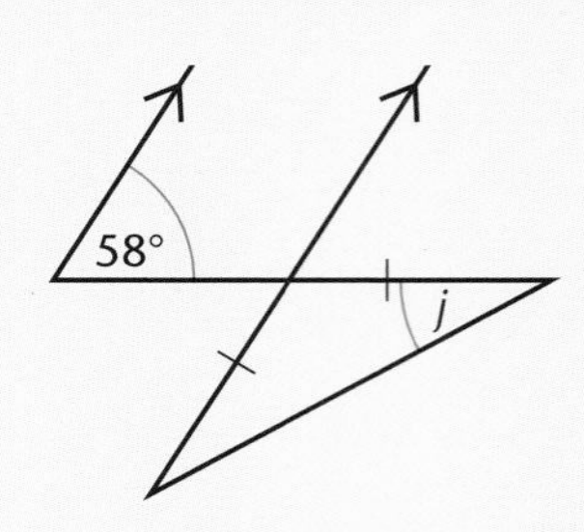

Label the 'missing' angles x, y and z.

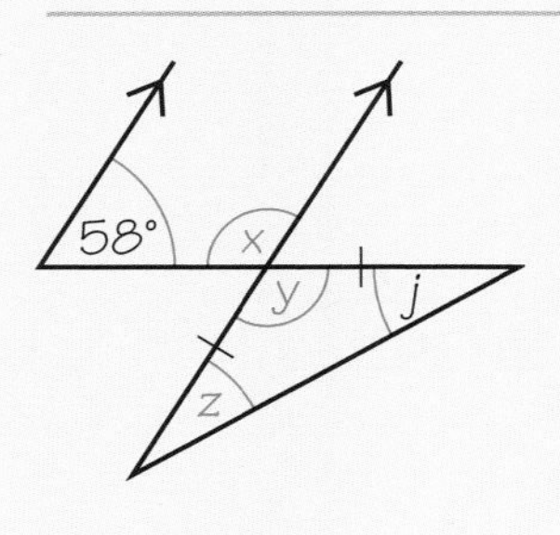

$x = 180° - 58°$
(co-interior angles, parallel lines)
$x = 122°$
$y = 122°$ (vertically opposite)
$z + j = 180° - 122°$ (angle sum of $\triangle$)
$z + j = 58°$
$z = j$ (base angles isosceles $\triangle$)
so both z and $j = \frac{1}{2}$ of $58°$ (or $58° \div 2$)
$j = 29°$

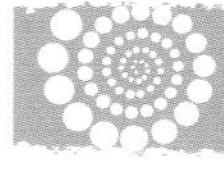

EXERCISE 6H

Copy the diagram and label any 'missing' angles.

Calculate the size of the angles marked with letters.

1

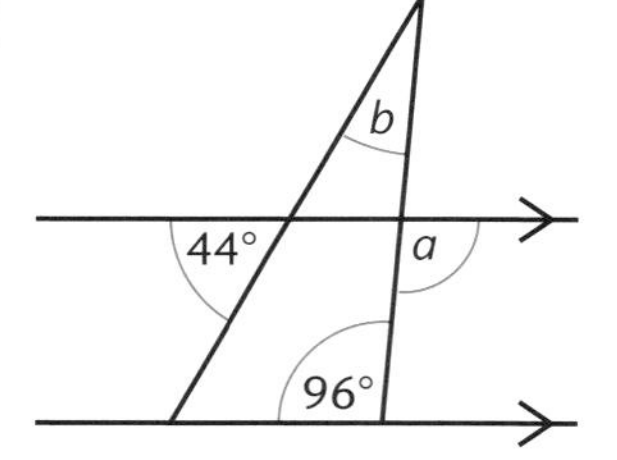

2

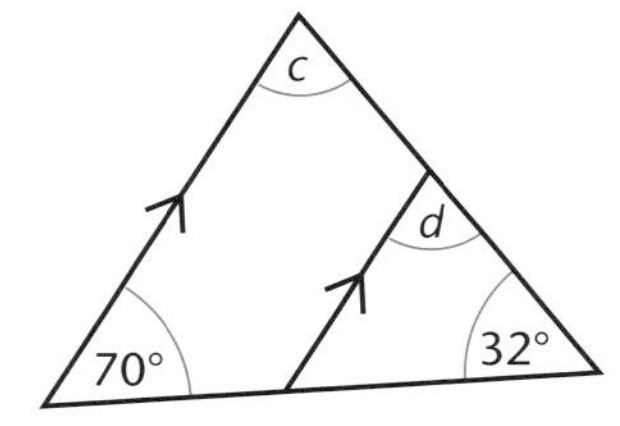

3

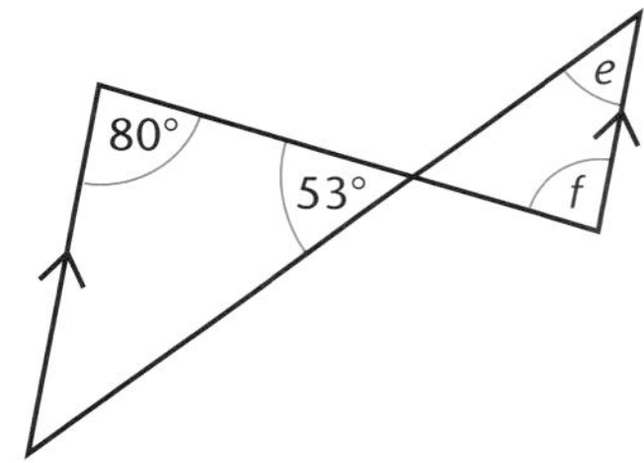

4

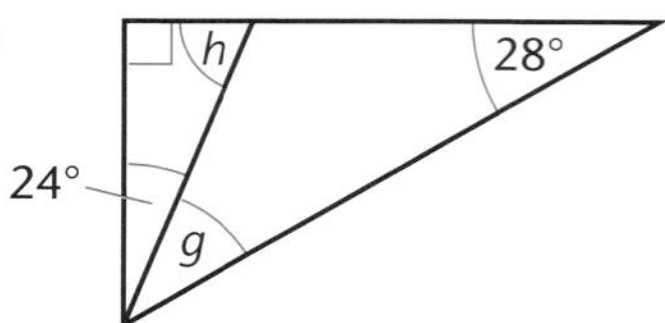

5

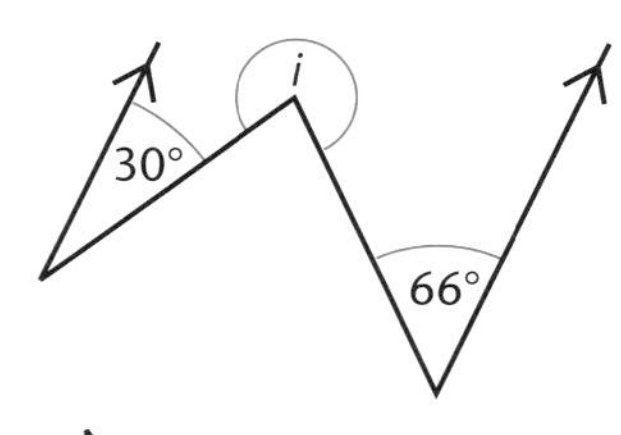

6

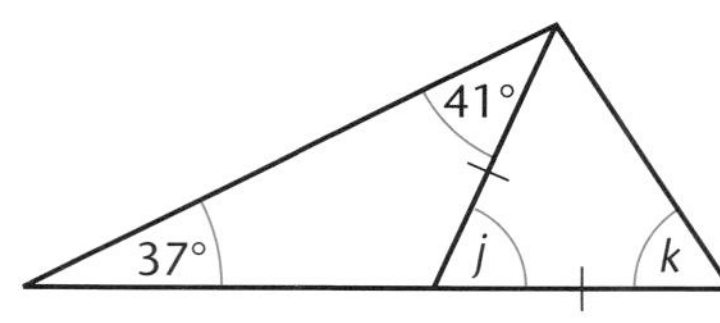

7

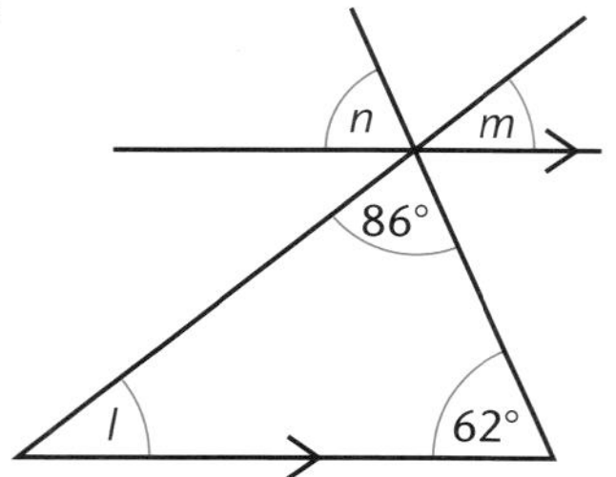

8

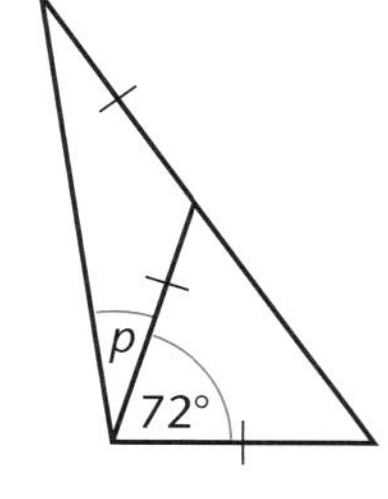

6.4 Circle properties

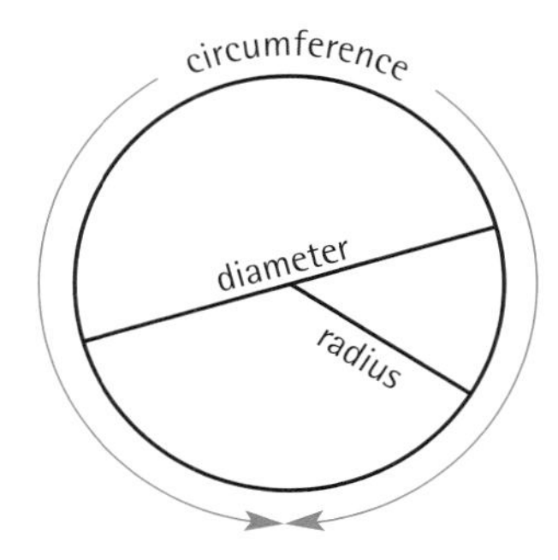

The diagram above shows some of the terms you will use when solving problems involving circles.

A **tangent** to a circle is a line that just touches the circumference at only one point of contact.

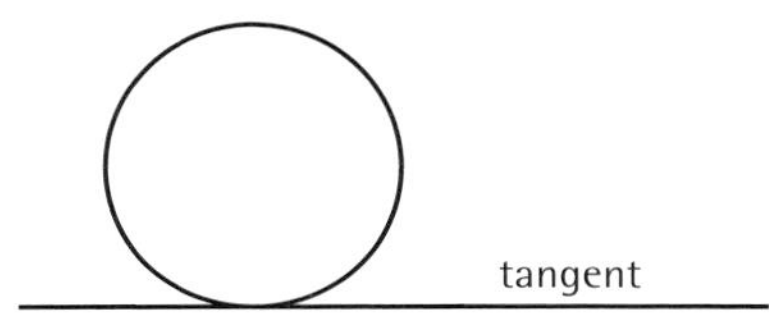

In this section we start with two important circle properties.

1 The angle between a tangent and a radius is 90°.

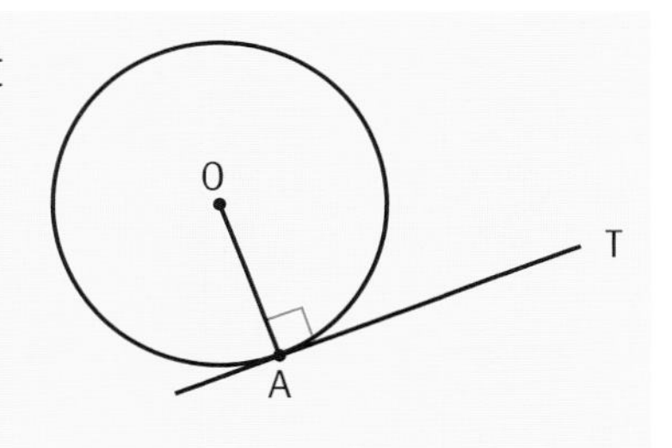

2 The angle in a semi-circle is a right angle.

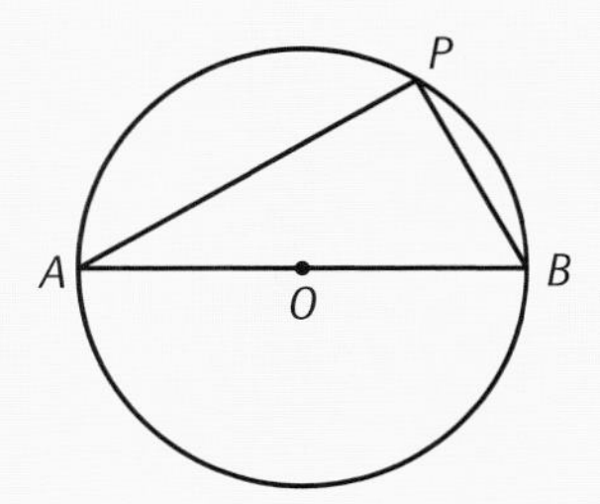

Angle $APB = 90°$

Note that AB is a diameter.

You need to remember these results because they are often needed for examination questions.

EXAMPLE 9

In the diagram, O is the centre of the circle and AB is a diameter.
Calculate the value of x, giving reasons for your answer.

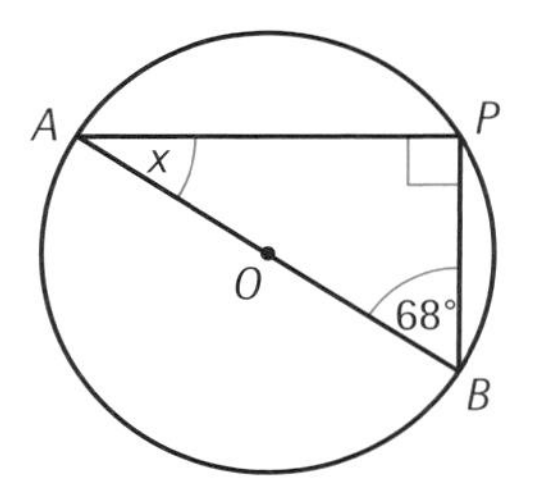

angle $APB = 90°$ (angle in a semi-circle)

$x + 68° + 90° = 180°$ (angle sum of triangle = 180°)

$x = 22°$

EXERCISE 6I

1

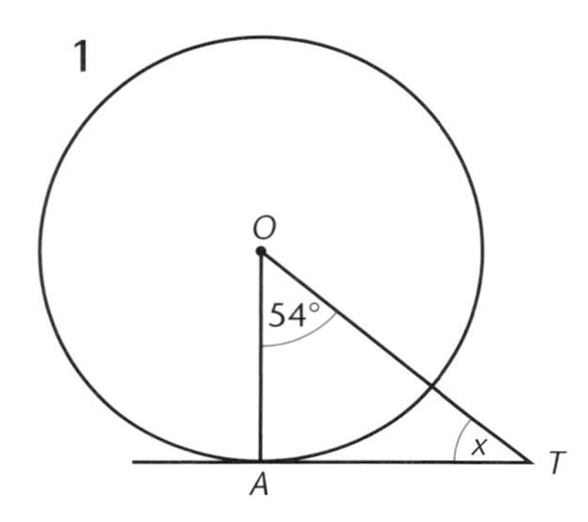

Find angle x.

2

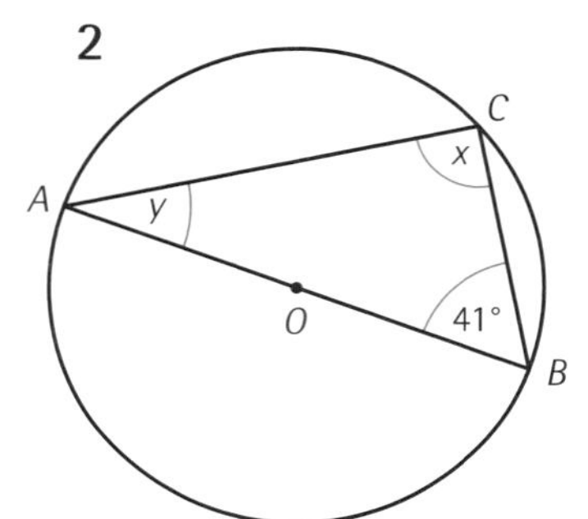

Find angles x and y.

3

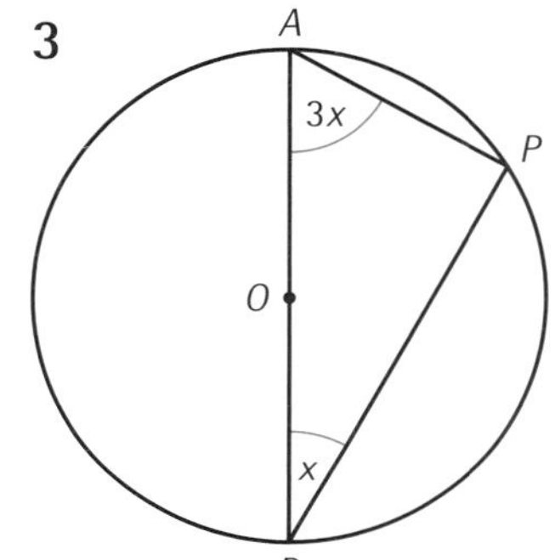

Find angle x.

4

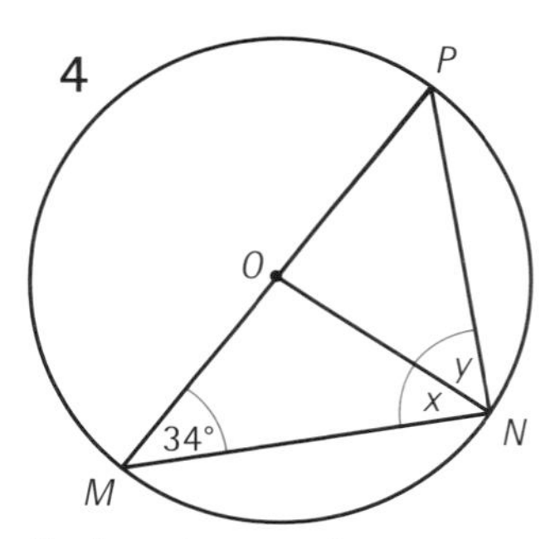

Find angles x and y.

5

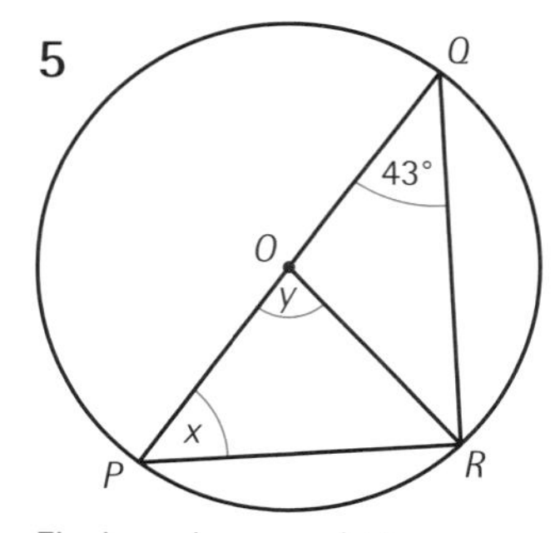

Find angles x and y.

6.5 Quadrilaterals and other polygons

Quadrilaterals

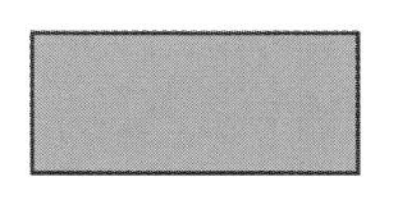

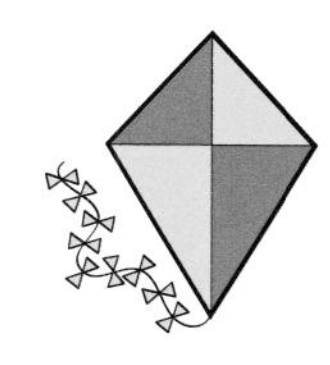

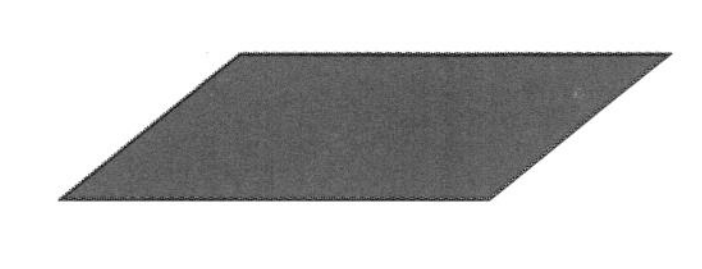

A **quadrilateral** is a 2-dimensional 4-sided shape.

This quadrilateral can be divided into 2 triangles

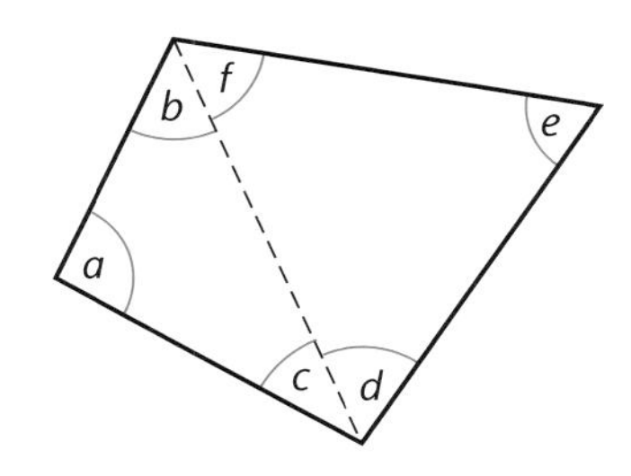

You can divide *any* quadrilateral into 2 triangles.

In each triangle, the angles add up to 180°,
so $a + b + c = 180°$
and $d + e + f = 180°$

The 6 angles from the 2 triangles add up to 360°.
$a + b + c + d + e + f = 360°$

a, $(b + f)$, e, $c + d$ are the angles of the quadrilateral.

In any quadrilateral the sum of the interior angles is 360°.

Some quadrilaterals have special names and properties.

Quadrilateral	Picture	Properties
square		Four equal sides. All angles 90°. Diagonals **bisect** each other at 90°.
rectangle		Two pairs of equal sides. All angles 90°. **Diagonals** bisect each other.

Bisect means 'cut in half'.

Quadrilateral	Picture	Properties
parallelogram		Two pairs of equal and parallel sides. Opposite angles equal. Diagonals **bisect** each other.
rhombus		Four equal sides. Two pairs of parallel sides. Opposite angles equal. **Diagonals** bisect each other at 90°.
trapezium		One pair of parallel sides.
kite		Two pairs of **adjacent** sides equal. One pair of opposite angles equal. One diagonal bisects the other at 90°.

Lines with equal numbers of arrows are parallel to each other.

Adjacent means 'next to'.

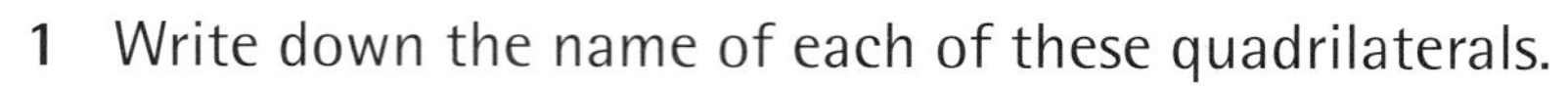

EXERCISE 6J

1 Write down the name of each of these quadrilaterals.

(a)  (b) 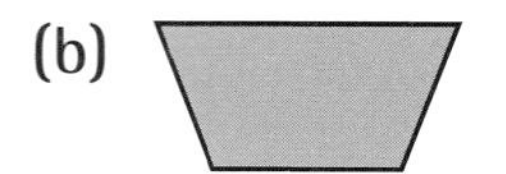 (c) 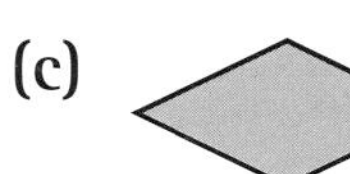

(d) 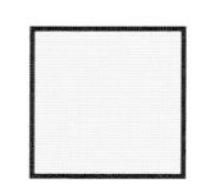 (e) 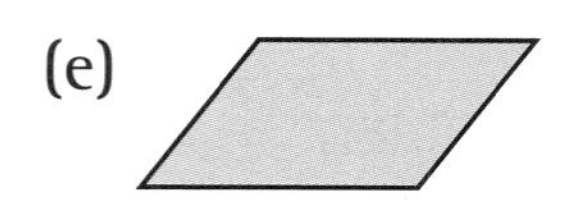  (f) 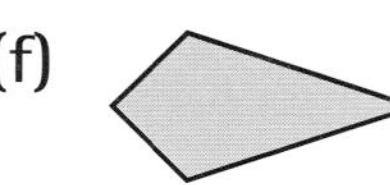

2 Write down the names of all the quadrilaterals with these properties

(a) diagonals which cross at 90°

(b) all sides are equal in length

(c) only one pair of parallel sides

(d) two pairs of equal angles but not all angles equal

(e) all angles are equal

(f) two pairs of opposite sides are parallel

(g) only one diagonal bisected by the other diagonal

(h) two pairs of equal sides but not all sides equal

(i) the diagonals bisect each other

(j) at least one pair of opposite sides are parallel

(k) diagonals equal in length

(l) at least two pairs of adjacent sides equal.

Polygons

A **polygon** is a 2-dimensional shape with many sides and angles.

Here are some of the most common ones.

Pentagon (5 sides)

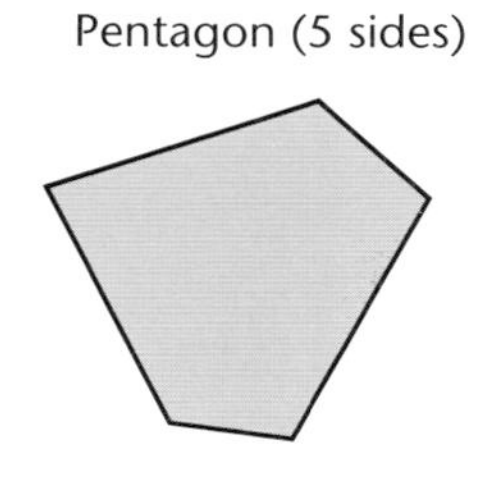

Hexagon (6 sides)

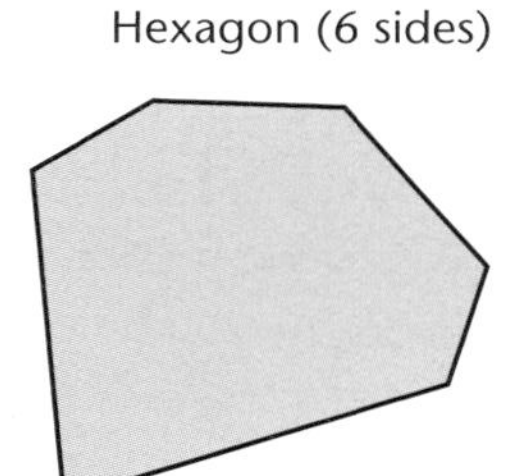

Octagon (8 sides)

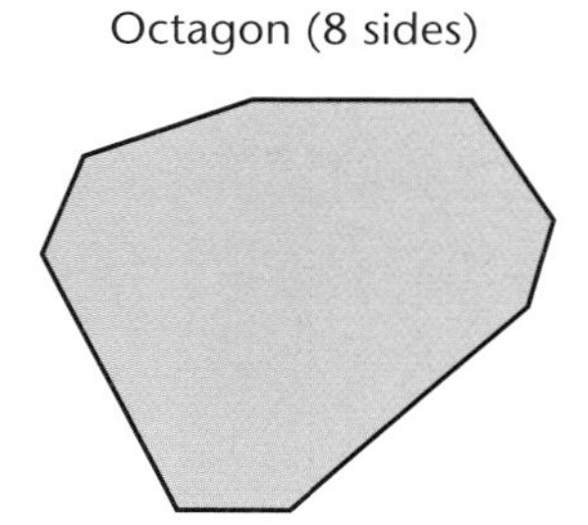

Polygon means 'many angled'.

A polygon with all of its sides the same length and all of its angles equal is called **regular**.

Regular pentagon

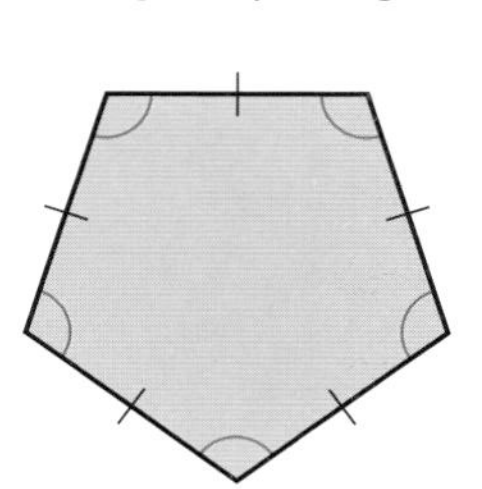

Regular hexagon

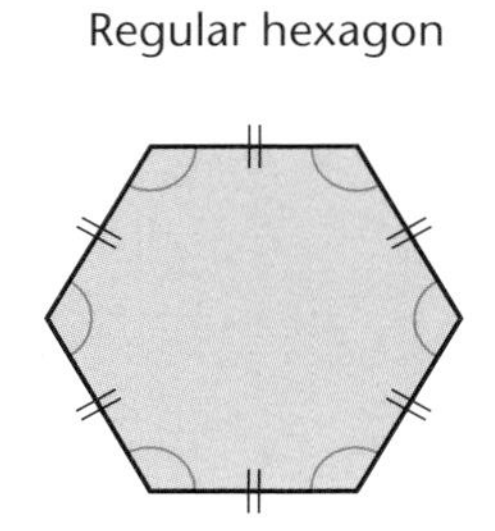

Regular octagon

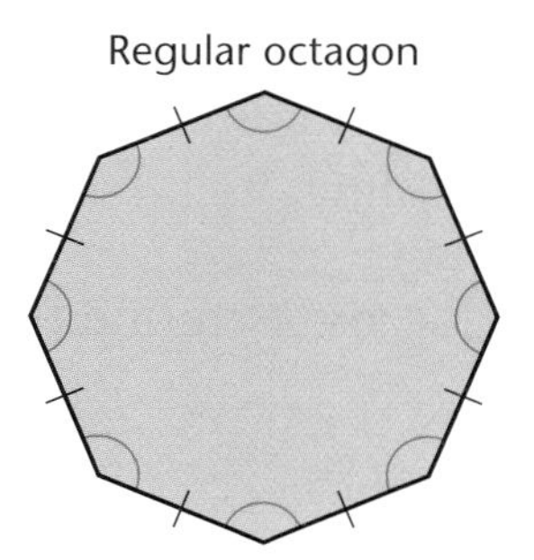

The sum of the exterior angles of any polygon is 360°.

You can explain this result like this.

Exterior angle

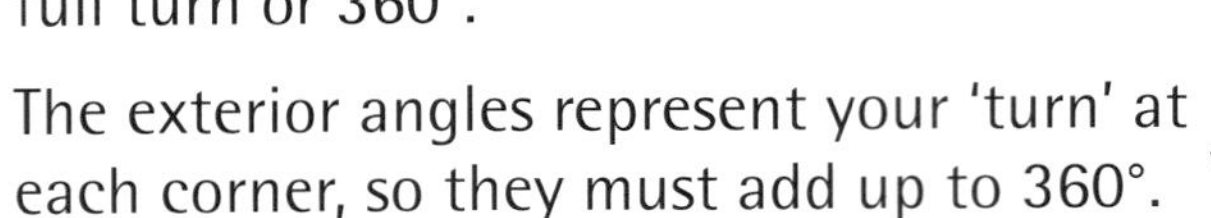

If you 'walk round' the sides of this hexagon until you get back to where you started, you complete a full turn or 360°.

The exterior angles represent your 'turn' at each corner, so they must add up to 360°.

On a regular hexagon, all 6 exterior angles are the same size.

$\frac{360°}{6} = \frac{360°}{\text{number of sides}}$

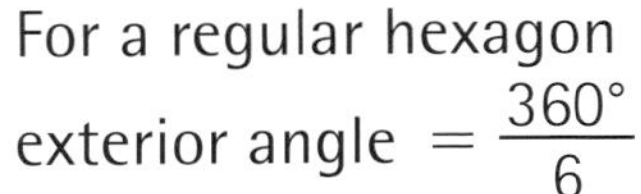

For a regular hexagon

$$\text{exterior angle} = \frac{360°}{6} = 60°$$

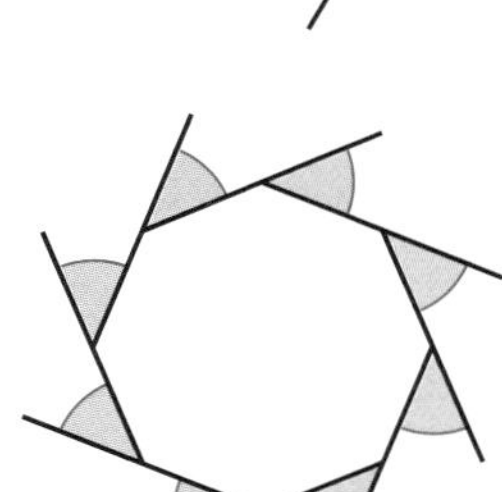

For a regular octagon

$$\text{exterior angle} = \frac{360°}{8} = 45°$$

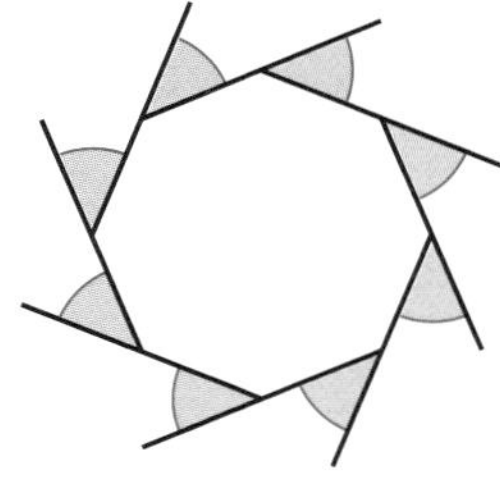

A regular octagon has 8 equal exterior angles.

For a regular polygon,

$$\text{exterior angle} = \frac{360°}{\text{number of sides}}$$

You can use this equation to work out the number of sides in a regular polygon.

For a regular polygon,

$$\text{number of sides} = \frac{360°}{\text{exterior angle}}$$

In a regular polygon with exterior angles of 18°

$$\text{number of sides} = \frac{360°}{18°} = 20$$

Once you know the exterior angle, you can calculate the interior angle.

At each **vertex**, the exterior and interior angles lie next to each other (adjacent) on a straight line.

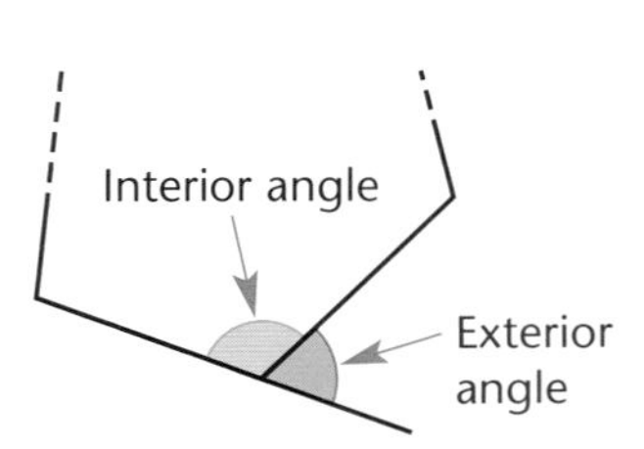

In a polygon, each pair of interior and exterior angles adds up to 180°.

So, for any polygon

interior angle = 180° − exterior angle

For example,

- regular hexagon interior angle = 180° − 60° = 120°
- regular octagon interior angle = 180° − 45° = 135°

In a regular polygon all the exterior angles are equal. So all the interior angles are equal.

When a polygon is *not* regular the interior angles could all be different sizes. You can find the *sum* of the interior angles.

These diagrams show how you can divide any polygon into triangles.

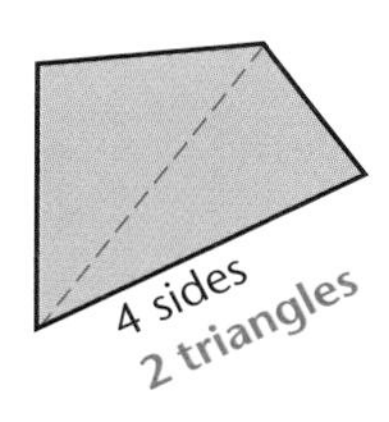

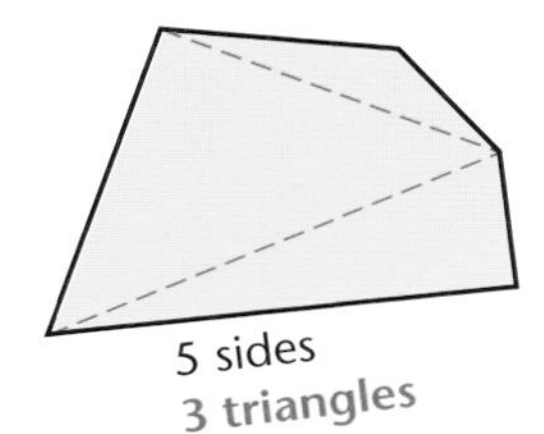

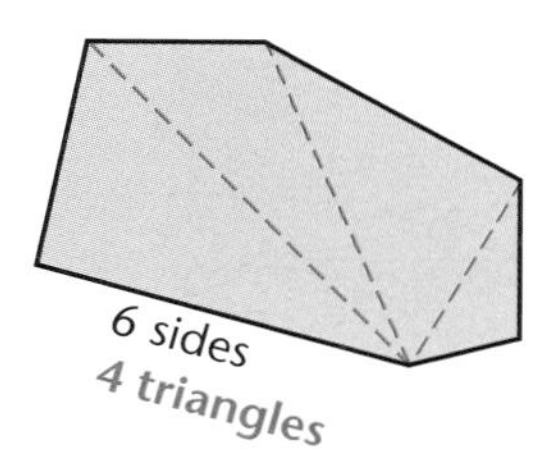

Draw dotted lines from the **vertices**.

You can see that the number of triangles is always 2 less than the number of sides of the polygon.

The sum of the interior angles for each polygon is given by the formula

$(n - 2) \times 180°$ where n is the number of sides of the polygon.

You should learn this formula.

Name of polygon	Number of sides	Number of triangles	Sum of interior angles
Triangle	3	1	1 × 180° = 180°
Quadrilateral	4	2	2 × 180° = 360°
Pentagon	5	3	3 × 180° = 540°
Hexagon	6	4	4 × 180° = 720°
Heptagon	7	5	5 × 180° = 900°
Octagon	8	6	6 × 180° = 1080°
Nonagon	9	7	7 × 180° = 1260°
Decagon	10	8	8 × 180° = 1440°

EXAMPLE 10

Calculate the size of the angles marked with letters in each of these diagrams.

You could use the fact that the sum of the angles of a quadrilateral is 360° to calculate q.

(a)

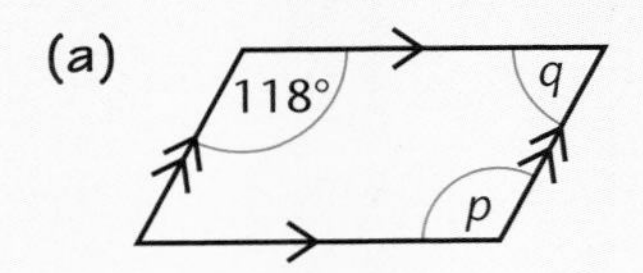

$p = 118°$ (opposite angles of parallelogram are equal

$q = 62°$ (p and q are co-interior angles, so $p + q = 180°$)

(b)

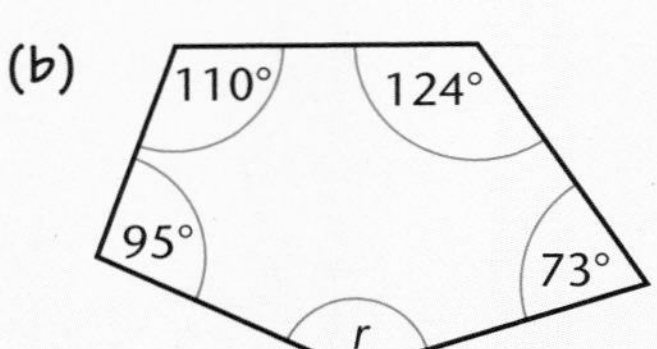

$r = 540° -$ (the sum of the other 4 angles)

$r = 540 - (95° + 110° + 124° + 73°)$

$r = 138°$

Sum of the interior angles of a pentagon (5 sides) = $3 \times 180° = 540°$.

(c)

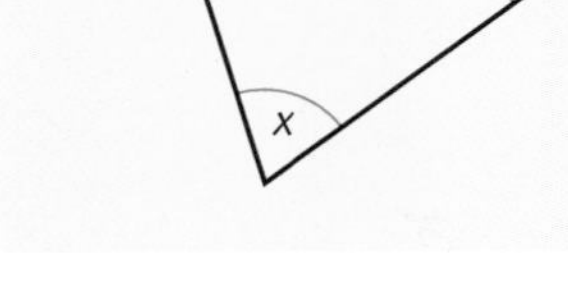

$x + x + 133° + 79° = 360°$

$2x = 360° - 133° - 79°$

$2x = 148°$

$x = 74°$

Sum of interior angles of quadrilateral = 360°.

EXAMPLE 11

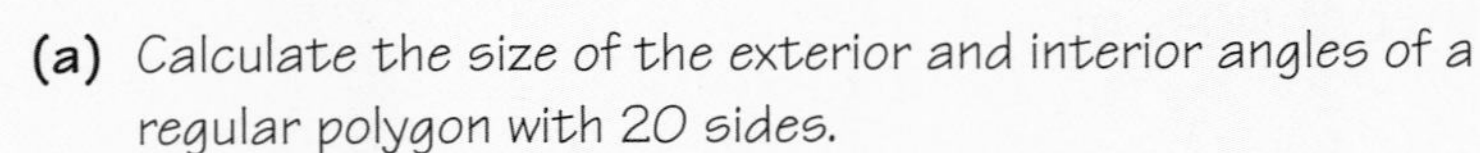

(a) Calculate the size of the exterior and interior angles of a regular polygon with 20 sides.

(b) How many sides has a regular polygon with interior angle 168°?

Exterior angle of regular polygon $= \dfrac{360°}{\text{number of sides}}$

(a)

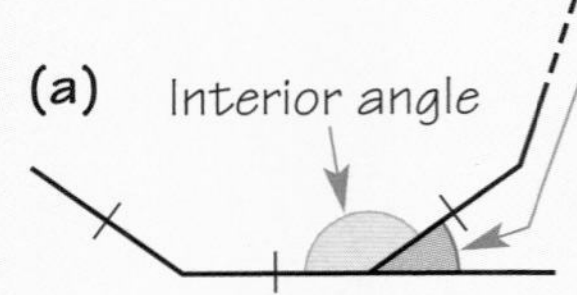

Exterior angle $= \dfrac{360°}{20} = 18°$

Interior angle $= 180° -$ exterior angle

$= 180° - 18°$

$= 162°$

(b) Exterior angle $= 180° -$ interior angle

$= 180° - 168°$

$= 12°$

Number of sides $= \dfrac{360}{12°} = 30$

EXERCISE 6K

Calculate the size of the angles marked with letters.

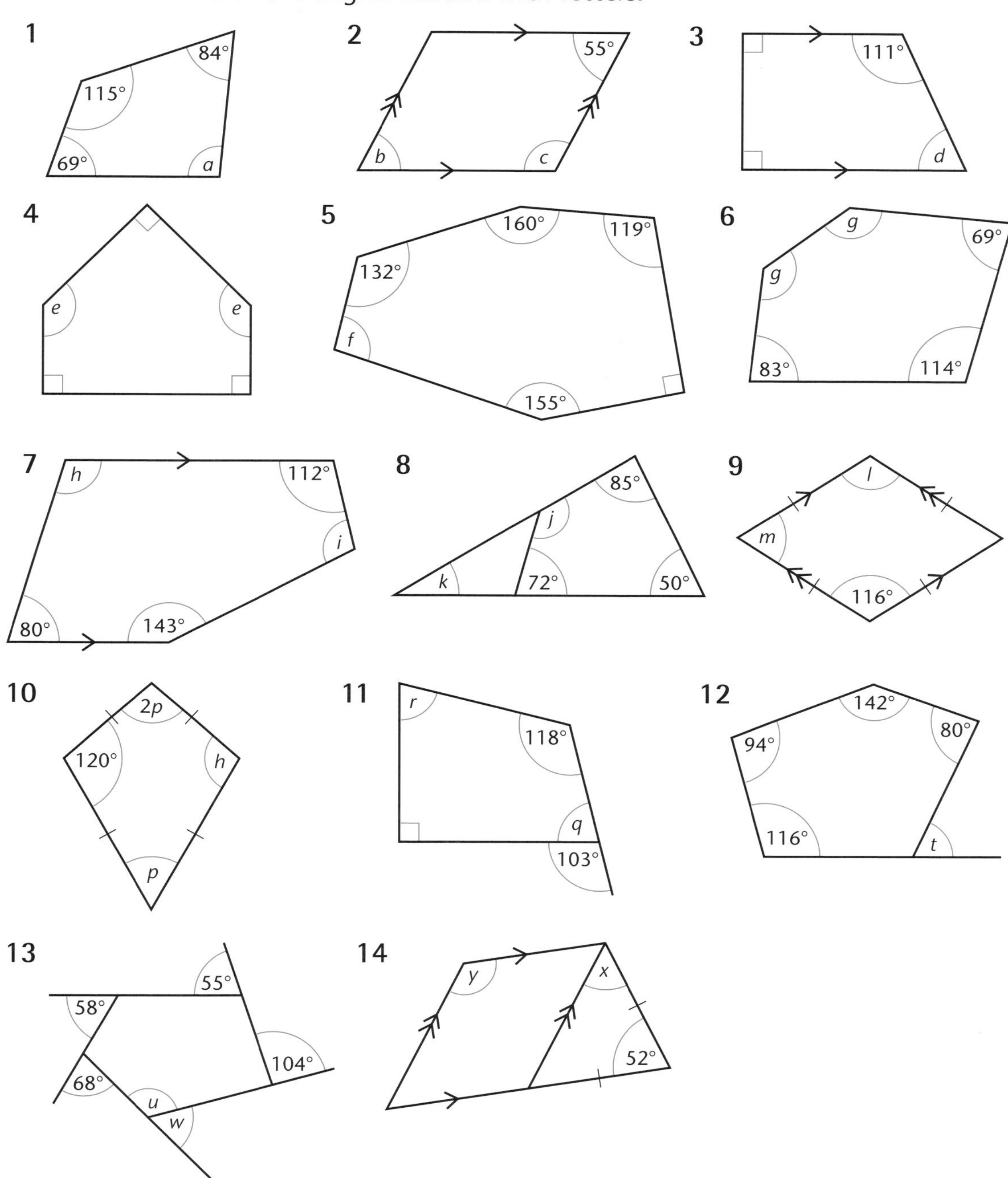

15 Calculate the interior angle of a regular polygon with 18 sides.

16 Can a regular polygon have an interior angle of 130°? Explain your answer.

6.6 Symmetry

Line symmetry

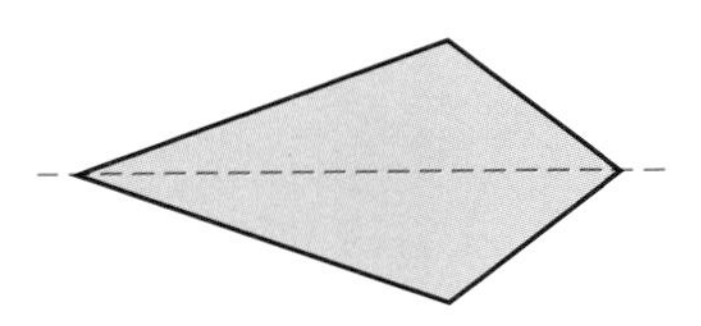

A kite is **symmetrical**.

If you fold it along the dashed line, one half fits exactly onto the other.

The dashed line is called a **line of symmetry**.

A line of symmetry divides a shape into two halves.
One half is the mirror image of the other.

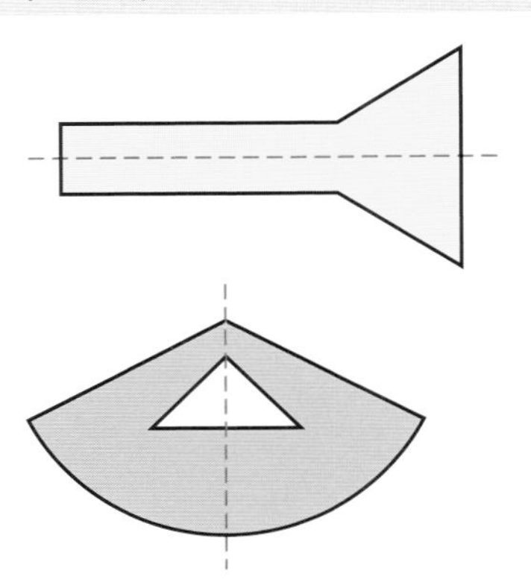

Some shapes have more than one line of symmetry.

Rectangle ... 2 lines of symmetry

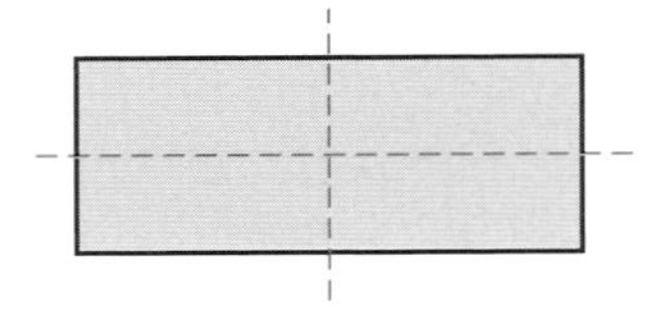

You can draw lines of symmetry with solid lines.

Equilateral triangle ... 3 lines of symmetry

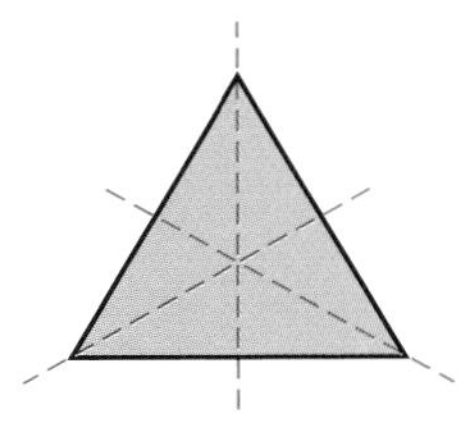

Regular hexagon ... 6 lines of symmetry

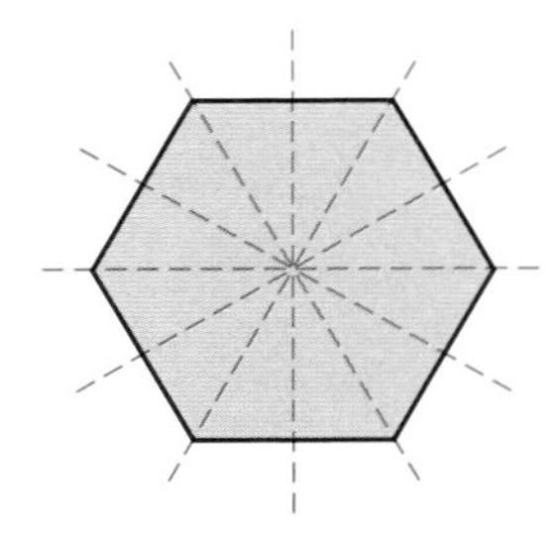

Some shapes have no lines of symmetry.

A parallelogram has no lines of symmetry.

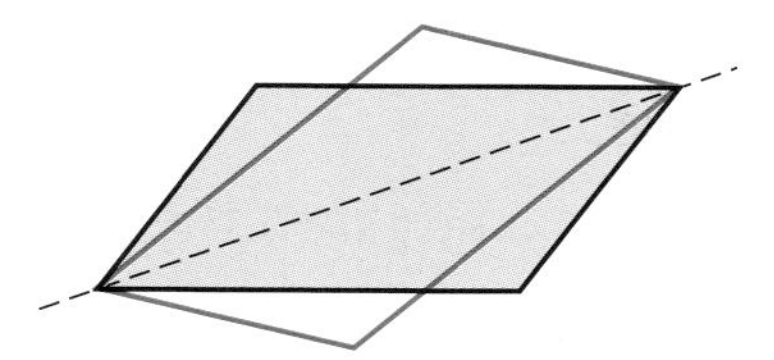

The dashed line is *not* a line of symmetry. If you reflect the parallelogram in it you get the red parallelogram.

These shapes have no lines of symmetry.

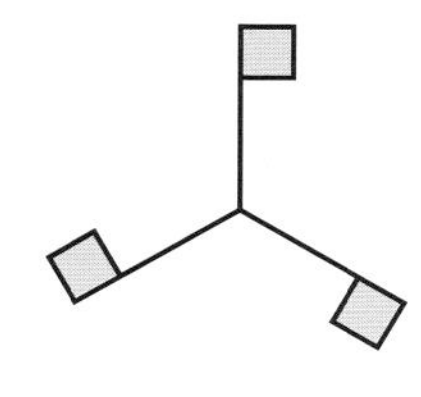

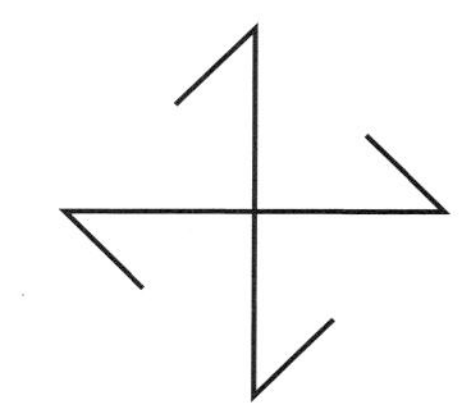

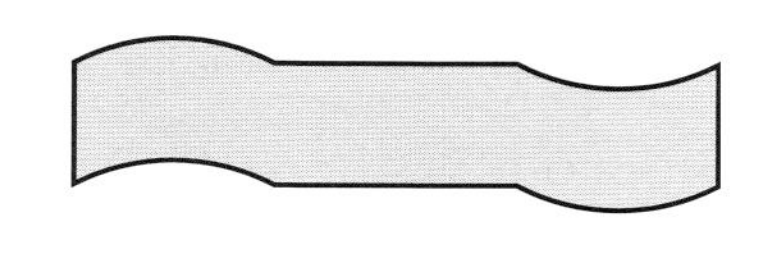

EXERCISE 6L

Copy these shapes and draw in all the lines of symmetry (if any).

1

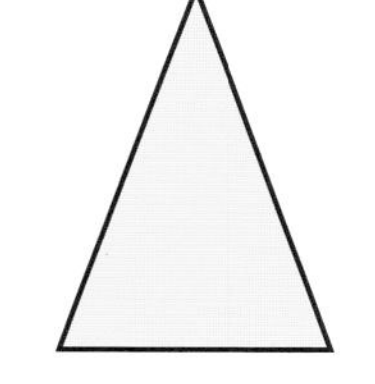

2

3

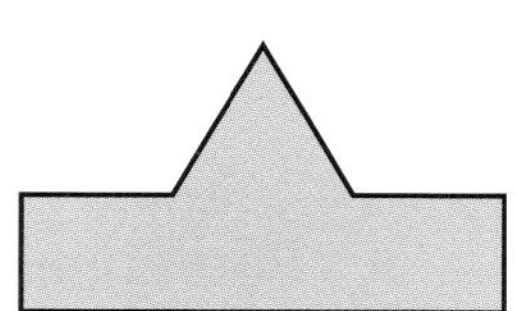

4

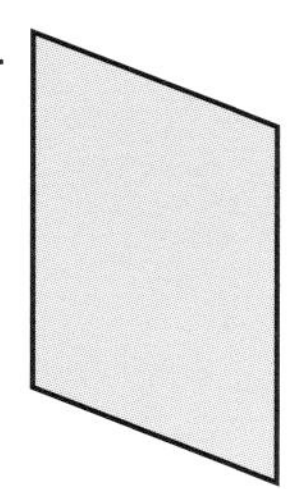

5

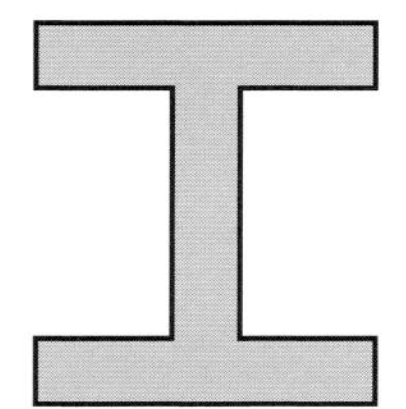

6

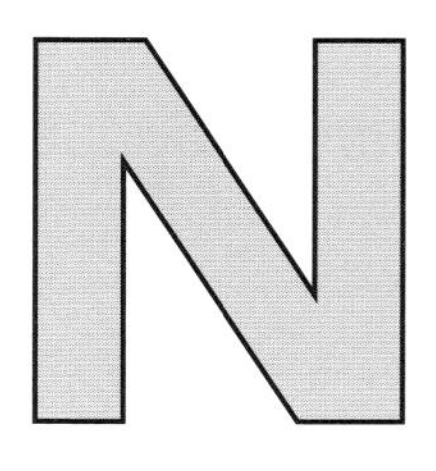

7 Copy and complete these grids so that they are symmetrical about the dashed line.

(a)

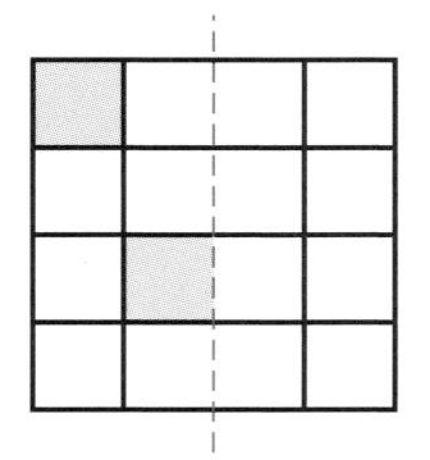

(b)

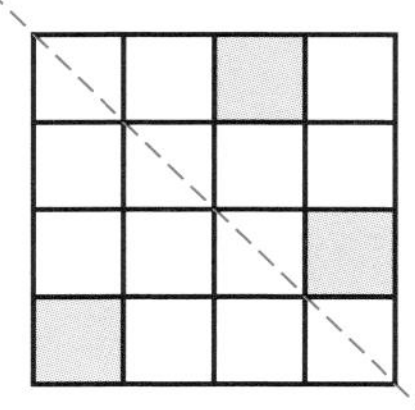

(c)

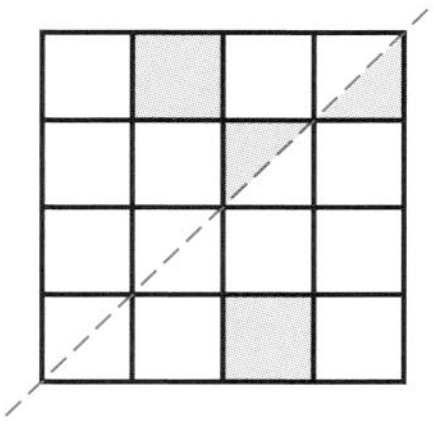

Rotational symmetry

Look again at the shapes before Exercise 6K.
They have no lines of symmetry but they have **rotational symmetry**.

You can turn them and they will fit exactly into their original shape again.

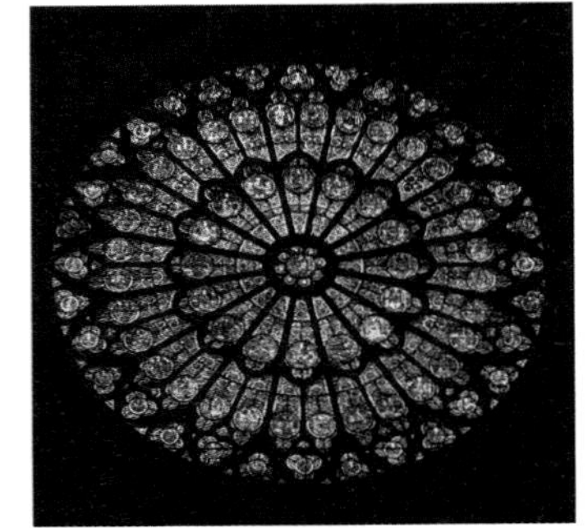

The **order** of rotational symmetry is the number of times a shape looks the same during one full turn.

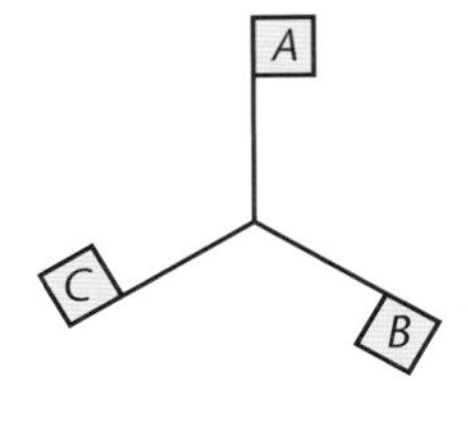

If *A* turns to any of the positions *A*, *B* or *C* the shape will look exactly the same.

Rotational symmetry of order 3.

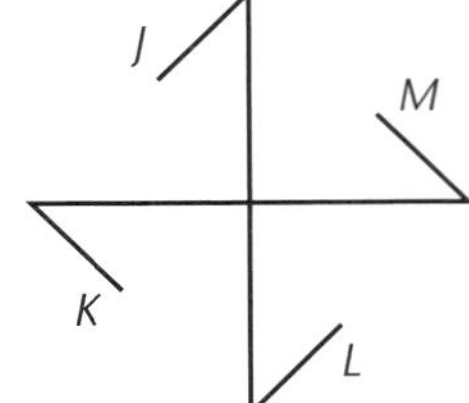

If *J* turns to *J*, *K*, *L* or *M* the shape will look exactly the same.

Rotational symmetry of order 4.

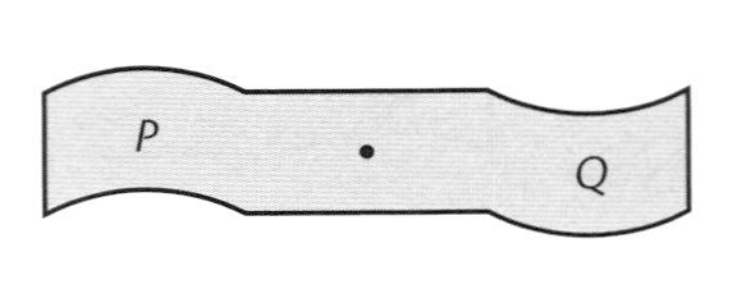

If *P* turns to positions *P* or *Q* the shape will look exactly the same.

Rotational symmetry of order 2.

Some shapes have line symmetry *and* rotational symmetry.

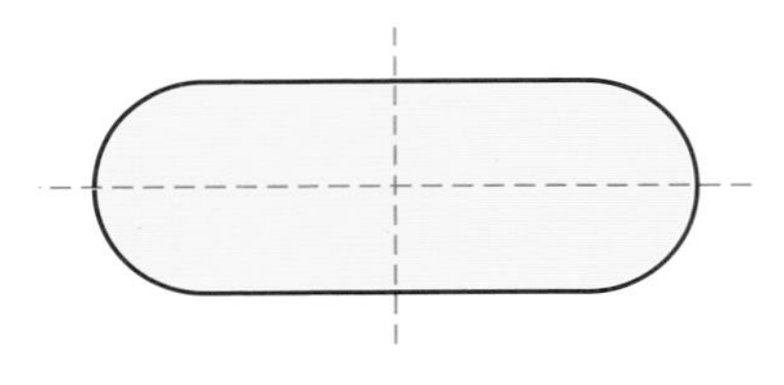

2 lines of symmetry.

Rotational symmetry of order 2.

3 lines of symmetry.

Rotational symmetry of order 3.

EXERCISE 6M

1 Which of these shapes has rotational symmetry.

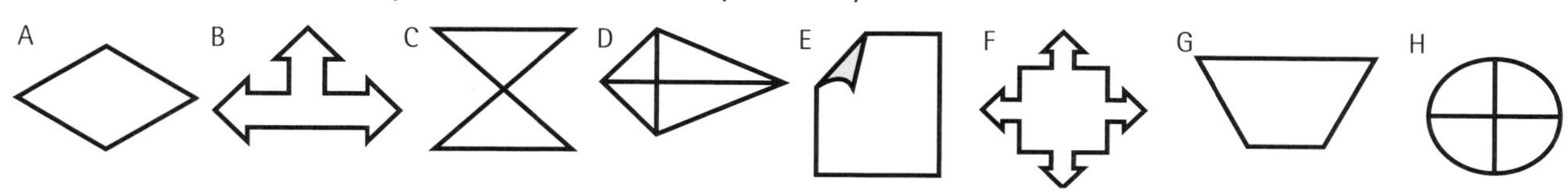

2 Write down the letters of the shapes that have rotational symmetry of order.

(a) 2 (b) 3 (c) 4 (d) 5

A 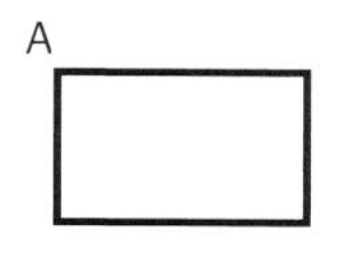 B C  D  E  F 

G 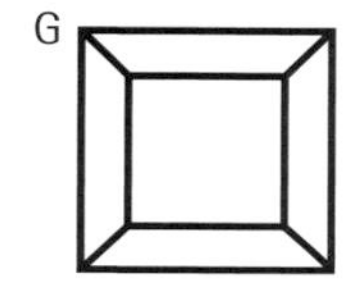 H 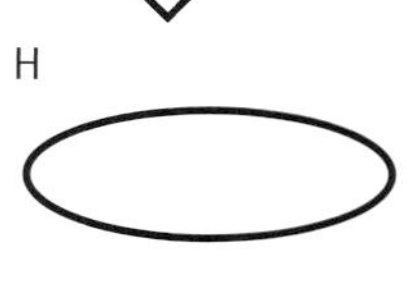 I  J 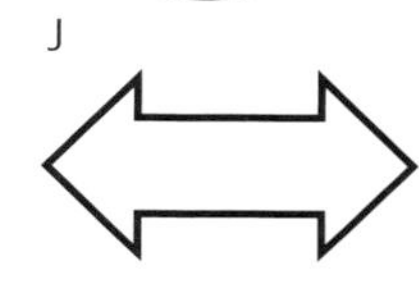 K 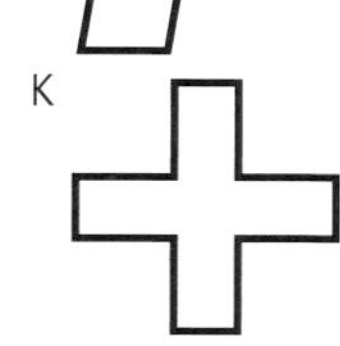 L 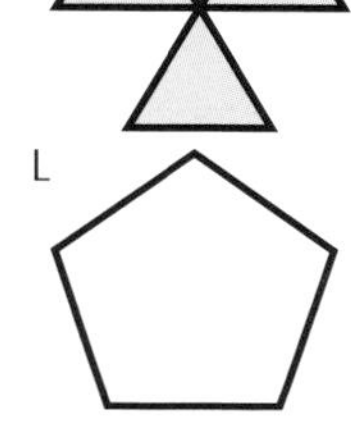

3 Copy and shade one more square in each grid to make shapes with rotational symmetry of order 2.

(a) (b) 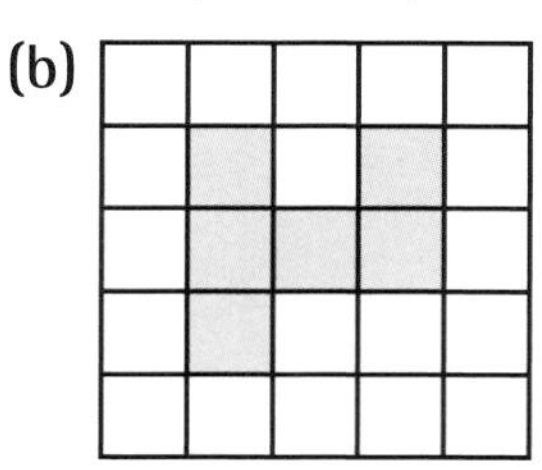 (c) 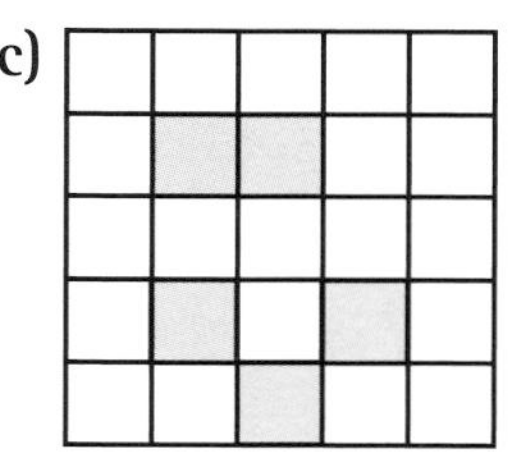

4 Copy and complete this table.

Name of shape		Order of rotational symmetry
		3
Square		
Pentagon		5

EXAMINATION QUESTIONS

1

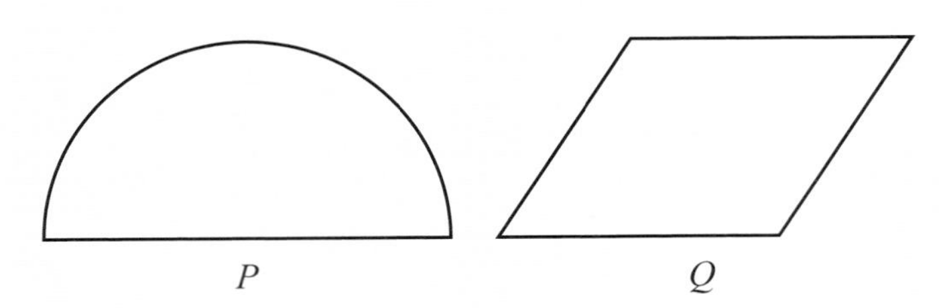

(a) Write down the number of lines of symmetry of shape
  (i) P, [1]
  (ii) Q. [1]
(b) Write down the order of rotational symmetry of shape Q. [1]

(CIE, Paper 1, Jun 2000)

2

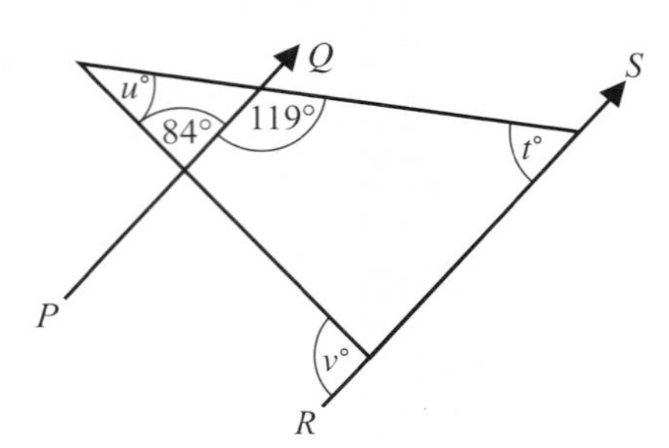

In the diagram the lines PQ and RS are parallel.
Calculate the values of t, u and v. [3]

(CIE Paper 1, Jun 2000)

3

MATHEMATICS

(a) In the word above
  (i) which letters have two lines of symmetry? [2]
  (ii) which letter has rotational symmetry but no line symmetry? [1]

(CIE Paper 1, Nov 2000)

4

North S North Not to scale

A B

A and *B* are two points on a coastline. *B* is directly East of *A*.
A ship *S* can be seen from both *A* and *B*.
The bearing from *S* to *A* is 054°. The bearing of *S* from *B* is 324°.

(a) Find the number of degrees in
(i) angle *SAB*, [1]
(ii) angle *SBA*. [1]

(b) Use your answer in **part (a)** to show that angle $ASB = 90°$. [1]

(c) The straight line distance *AB* is 6 kilometres.
Work out the distance of the ship
(i) from *A*, [2]
(ii) from *B*. [2]

(d) Later the ship is seen 2.5 kilometres from *A* with angle *ASB* still 90°. What is the bearing of the ship from A now? [3]
Give your answer correct to the nearest degree.

(CIE Paper 1, Nov 2000)

5

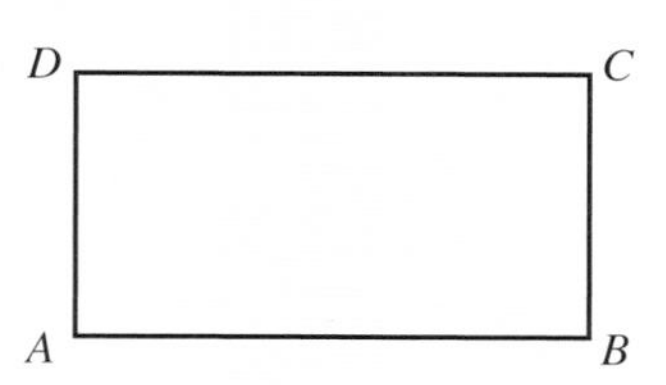

Copy and draw accurately the lines of symmetry of the rectangle *ABCD*. [1]

(CIE Paper 3, Nov 2001)

6

North

North

S

$140°$

T

Samira (S) and Tamara (T) walk towards each other.
Samira walks on a bearing of 140°.
Find the bearing on which Tamara walks. [2]

(CIE Paper 1, Nov 2001)

---

7 **(a)** Write down the name of the special quadrilateral which has rotational symmetry of order 2 but no line symmetry. [1]

(b) Draw, on a grid, a quadrilateral which has exactly one line of symmetry but no rotational symmetry.
Draw the line of symmetry on your diagram. [2]

(CIE Paper 1, Jun 2002)

---

8 **(a)**

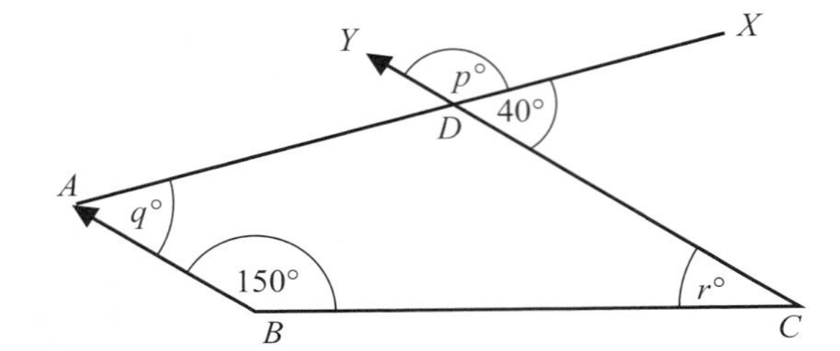

In the diagram, AX and CY are straight lines which intersect at d.
BA and CY are parallel.
Angle $CDX = 40°$ and angle $ABC = 150°$.

(i) Find p, q and r. [3]

(ii) What is the name of the special quadrilateral $ABCD$? [1]

(b) (i) A nonagon is a polygon with nine sides.
Calculate the size of an interior angle of a **regular** nonagon. [3]

(ii) Each angle of another regular polygon is 150°.
Calculate the number of sides of this polygon. [3]

(CIE Paper 3, Jun 2002)

9

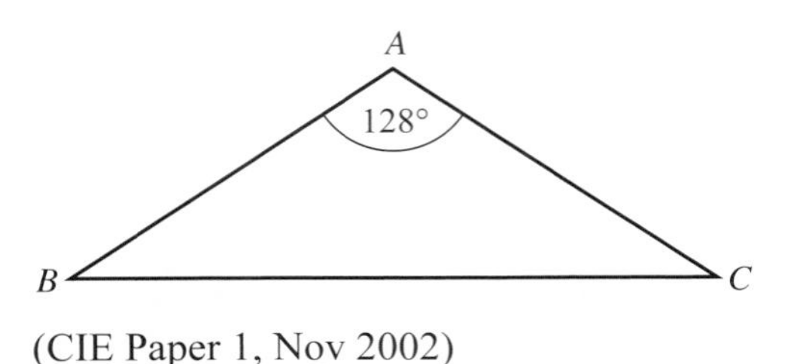

In the triangle ABC, $AB = BC$.

(a) What is the special name of this triangle? [1]

(b) Angle $BAC = 128°$. Work out angle ABC [2]

(CIE Paper 1, Nov 2002)

10 Find the size of one of the ten interior angles of a regular decagon. [3]

(CIE Paper 1, Nov 2003)

11

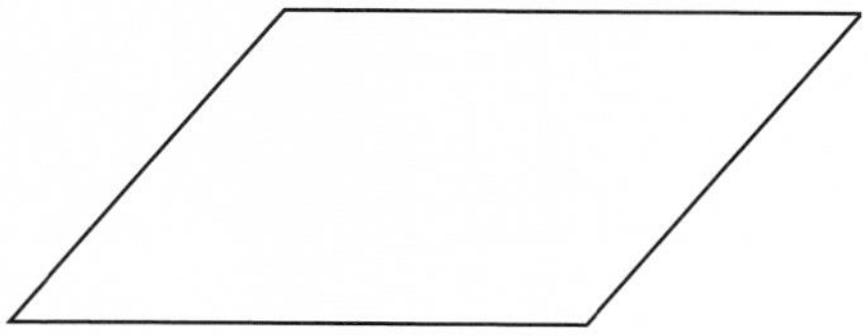

For the shape shown, write down
(a) the number of lines of symmetry, [1]
(b) the order of rotational symmetry. [1]

(CIE Paper 1, Jun 2004)

---

12 **(a)** (i) What is the special name given to a five-sided polygon? [1]
(ii) Calculate the total sum of the interior angles of a regular five-sided polygon. [2]
(iii) Calculate the size of one interior angle of a regular five-sided polygon. [1]

(CIE Paper 3, Jun 2004)

---

13 Reflex Right Acute Obtuse
Use one of the above terms to describe each of the angles given.
(a) 100°,
(b) 200°.

(CIE Paper 1, Nov 2004)

---

14 Write down the order of rotational symmetry of each of the following shapes.
(a) Equilateral triangle
(b) Rhombus

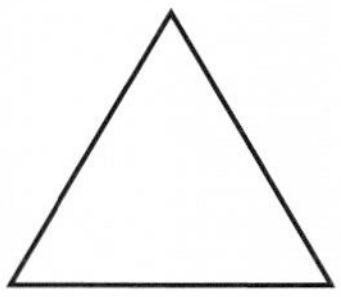

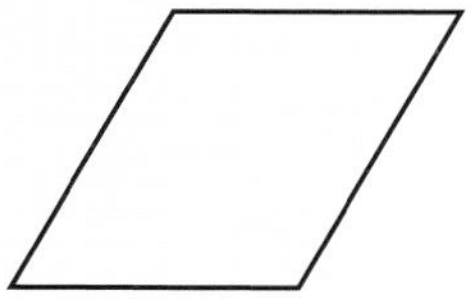

(CIE Paper 1, Nov 2004)

Chapter 7

Indices and formulae

This chapter will show you how to

- ✔ use index (power) notation and the index laws
- ✔ use formulae given in words or letters and symbols
- ✔ substitute numbers into formulae
- ✔ change the subject of a formula
- ✔ calculate using standard form

7.1 Indices

You will need to know

- **how to multiply positive and negative numbers**
- **how to use the power key on your calculator**
- **the correct order of operations**

Using index notation to simplify expressions

6^2 (6 to the power 2) is called '6 **squared**'.

$6^2 = 6 \times 6$

8^3 (8 to the power 3) is called '8 **cubed**'.

$8^3 = 8 \times 8 \times 8$

You can write $5 \times 5 \times 5 \times 5$ as 5^4.

You say '5 raised to the **power** of 4', or '5 to the power 4'. The 4 is called the **index**.

You can also use powers (or **indices**) in algebraic expressions where there are unknown or **variable** values.

A *variable* is a letter which stands for an unknown quantity.

You can write

- $d \times d \times d$ as d^3 ('d cubed', or 'd to the power 3').
- $y \times y \times y \times y \times y$ as y^5.

d^3 and y^5 are written in index notation.

EXAMPLE 1

Write these expressions using index notation.

(a) $d \times d \times d \times d \times d \times d$ (b) $2 \times x \times x \times x \times x$

(a) $d \times d \times d \times d \times d \times d = d^6$

(b) $2 \times x \times x \times x \times x = 2x^4$

It is all multiplied by 2 as well. Write the 2 at the front of the expression.

EXAMPLE 2

Simplify these expressions.

(a) $3 \times a \times a \times 5 \times a \times a \times a$ (b) $4t \times t \times 2t$

(c) $3x \times y \times x \times 2y \times y$

You can multiply the values in any order.

Multiply numbers first, then letters.

(a) $3 \times a \times a \times 5 \times a \times a \times a = 3 \times 5 \times a \times a \times a \times a \times a$
$= 15 \times a^5$
$= 15a^5$

$3 \times 5 = 15$ and $a \times a \times a \times a \times a = a^5$

$4t$ means $4 \times t$.

(b) $4t \times t \times 2t = 4 \times t \times t \times 2 \times t$
$= 4 \times 2 \times t \times t \times t$
$= 8t^3$

Multiply numbers first, then letters.

(c) $3x \times y \times x \times 2y \times y = 3 \times x \times y \times x \times 2 \times y \times y$
$= 3 \times 2 \times x \times x \times y \times y \times y$
$= 6 \times x^2 \times y^3$
$= 6x^2y^3$

Multiply numbers, then the letters. You usually write letters in alphabetical order.

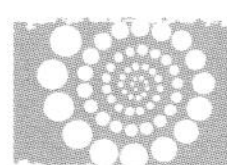

EXERCISE 7A

1 Write these expressions using index notation.

(a) $d \times d \times d \times d$

(b) $a \times a \times a \times a \times a \times a$

(c) $x \times x \times x \times x \times x$

(d) $y \times y \times y \times y \times y \times y \times y \times y$

(e) $b \times b$

(f) $m \times m \times m \times m$

(g) $p \times p \times p$

(h) $r \times r \times r \times r \times r \times r \times r$

Use index notation.

2 Simplify these expressions.

(a) $5 \times p \times p \times p \times p$

(b) $a \times 4 \times a \times a$

(c) $2 \times x \times 4 \times x \times x \times x$

(d) $f \times f \times 3 \times f \times 5 \times f \times f$

(e) $b \times 4 \times 2 \times b \times 1 \times b$

(f) $h \times h \times 6 \times h \times 5 \times 2 \times h \times h$

(g) $3 \times z \times 3 \times z \times 3 \times z$

(h) $5 \times r \times r \times 5 \times r \times r$

3 Simplify these using index notation.

(a) $b \times b \times c \times c \times c$

(b) $f \times g \times f \times f \times g \times f$

(c) $3 \times x \times x \times x \times 4 \times y$

(d) $5 \times d \times d \times 2 \times e \times e \times d \times d$

(e) $4 \times t \times 2 \times s \times s \times t \times s$

(f) $k \times k \times 3 \times j \times 5 \times k \times 2 \times j$

(g) $2 \times p \times p \times 2 \times q \times q \times q$

(h) $4 \times a \times a \times 4 \times b \times b$

4 Simplify these algebraic expressions using index notation.

(a) $2x \times 3x$

(b) $4y \times y \times 3y$

(c) $a \times 3a \times 2a \times a$

(d) $3b \times 2b \times 4b$

(e) $5n \times n \times n \times 4n \times 2n$

(f) $7k \times 3k \times 3k \times 2k$

(g) $3x \times 3x \times 3x$

(h) $6z \times z \times z \times 6z \times z \times z$

5 Simplify these algebraic expressions.

(a) $2a \times a \times 3b$

(b) $p \times 2q \times q \times 4p \times p$

(c) $3x \times 2y \times 3y$

(d) $5r \times 2s \times 3s \times 2r$

(e) $2d \times c \times 3d \times 7c$

(f) $5y \times 4 \times x \times 3x \times 2x$

(g) $4p \times r \times 4p \times r$

(h) $7r \times r \times s \times 7r \times r \times s$

6 Match each question (1 to 10) with an answer (a to j).

Questions	Answers
(1) $y \times y^2$	(a) y^4
(2) $y \times 5y$	(b) y^6
(3) $2y \times 3y$	(c) $5y^2$
(4) $y^2 \times y^2$	(d) $2y^3$
(5) $2y \times 4y$	(e) $4y^4$

Questions	Answers
(6) $4y \times y^3$	(f) $8y^3$
(7) $y^3 \times y^2$	(g) y^3
(8) $2y^2 \times y$	(h) $8y^2$
(9) $y \times y^2 \times y^3$	(i) y^5
(10) $2y \times 2y \times 2y$	(j) $6y^2$

Multiplying and dividing expressions involving indices

You will need to know

- how to simplify (cancel down) fractions

You can multiply different powers of the *same* variable.

EXAMPLE 3

Simplify

(a) $r^3 \times r^2$ (b) $4h^2 \times 5h^4$

(a) $r^3 \times r^2 = r \times r \times r \times r \times r$

$= r^5$

$r^3 = r \times r \times r$ and $r^2 = r \times r$

(b) $4h^2 \times 5h^4 = 4 \times h \times h \times 5 \times h \times h \times h \times h$

$= 4 \times 5 \times h \times h \times h \times h \times h \times h$

$= 20 \times h^6$

$= 20h^6$

Numbers first, then letters.

EXERCISE 7B PART 1

1 Simplify each of these.

(a) $a^3 \times a^2$ (b) $b^4 \times b^3$ (c) $c^2 \times c^4$ (d) $d^5 \times d^5$

2 Simplify each of these.

(a) $x^3 \times x$ (b) $y^2 \times y$ (c) $z \times z^4$ (d) $w \times w^3$

3 Simplify each of these.

(a) $2a^3 \times 3a^4$ (b) $5b^4 \times 3b^5$

(c) $4c^2 \times c^6$ (d) $3d^4 \times 3d^4$

Use Example 3 to help.

4 Can you see a quicker way to answer questions 1 to 3? Describe it, using an example to help.

EXAMPLE 4

Simplify (a) $d^2 \times d^3$ (b) $3a^4 \times 5a^3$ (c) $g^3 \times g \times g^2$

Multiply numbers first, then letters.

(a) $d^2 \times d^3 = d^{2+3}$
$= d^5$

(b) $3a^4 \times 5a^3 = 3 \times a^4 \times 5 \times a^3$
$= 3 \times 5 \times a^4 \times a^3$
$= 15 \times a^{4+3}$
$= 15a^7$

You can write the answer down without the lines of working.

(c) $g^3 \times g \times g^2 = g^3 \times g^1 \times g^2$
$= g^{3+1+2}$
$= g^6$

$g = g^1$
You do not usually write the index 1.

EXERCISE 7B PART 2

5 (a) $m^3 \times m^2$ (b) $a^4 \times a^3$ (c) $n^2 \times n^4$
(d) $u^5 \times u^5$ (e) $t^3 \times t^6$

6 (a) $d^3 \times d$ (b) $a^2 \times a$ (c) $r \times r^4$
(d) $e \times e^3$ (e) $t \times t^5$

7 (a) $2h^3 \times 3h^4$ (b) $5e^4 \times 3e^5$ (c) $4g^2 \times g^6$
(d) $3r \times 3r^4$ (e) $6e^3 \times 4e$

8 (a) $h^3 \times h^4 \times h^2$ (b) $e^4 \times e \times e^3$ (c) $4c^2 \times c^6 \times 6c$

You can divide different powers of the *same* variable.

EXAMPLE 5

Simplify (a) $a^5 \div a^2$ (b) $c^3 \div c^4$ (c) $10b^2 \div 5b^4$

In algebra you write $x \div y$ as $\frac{x}{y}$

(a) $a^5 \div a^2 = \frac{a \times a \times a \times a \times a}{a \times a}$

$= \frac{a \times a \times a \times {}^1\not{a} \times {}^1\not{a}}{{}^1\not{a} \times {}^1\not{a}}$

$= a \times a \times a$

$= a^3$

The top and bottom of the fraction are multiplications. You can cancel any terms which are on both the top and bottom.

$a \div a = 1$

Continued ▼

(b) $c^3 \div c^4 = \dfrac{c \times c \times c}{c \times c \times c \times c}$

$= \dfrac{{}^1\not{c} \times {}^1\not{c} \times {}^1\not{c}}{c \times {}^1\not{c} \times {}^1\not{c} \times {}^1\not{c}}$

$= \dfrac{1}{c}$

The numerator is now $1 \times 1 \times 1 = 1$.

(c) $10b^2 \div 5b^4 = \dfrac{10 \times b \times b}{5 \times b \times b \times b \times b}$

$= \dfrac{10 \times {}^1\not{b} \times {}^1\not{b}}{5 \times b \times b \times {}^1\not{b} \times {}^1\not{b}}$

Cancel the b terms.

$= \dfrac{10}{5 \times b \times b}$

$= \dfrac{10}{5b^2}$

Divide the 10 by 5.

$= \dfrac{2}{b^2}$

EXERCISE 7C PART 1

1 Simplify

(a) $p^5 \div p^2$ (b) $q^6 \div q^2$ (c) $r^4 \div r^3$ (d) $s^6 \div s^3$

2 Simplify each of these.

(a) $e^2 \div e^5$ (b) $f^2 \div f^6$ (c) $g^3 \div g^4$ (d) $h \div h^6$

Use Examples to help.

3 Simplify

(a) $5x^5 \div x^2$

(b) $8y^6 \div 4y^2$

(c) $12r^4 \div 3r^3$

(d) $2s^2 \div 10s^5$

4 Can you see a quicker way to answer questions 1 to 3? Describe it, using an example to help.

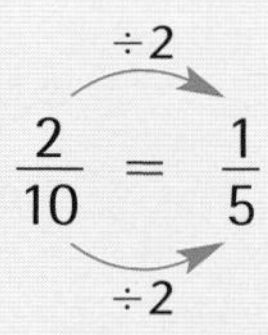

EXAMPLE 6

Simplify each of these.

(a) $y^6 \div y^2$ (b) $x^3 \div x^5$ (c) $12b^5 \div 4b^3$ (d) $\dfrac{t^4 \times t}{t^3}$

(a) $y^6 \div y^2 = y^{6-2}$

$= y^4$

You can have negative powers as well as positive ones.

(b) $x^3 \div x^5 = x^{3-5}$
$= x^{-2}$

(c) $12b^5 \div 4b^3 = \dfrac{12 \times b^5}{4 \times b^3}$

Numbers first, then letters.

$= \dfrac{12}{4} \times \dfrac{b^5}{b^3}$

$12 \div 4 = 3$
$b^5 \div b^3 = b^2$

$= 3 \times b^{5-3}$
$= 3 \times b^2$
$= 3b^2$

(d) $\dfrac{t^4 \times t}{t^3} = \dfrac{t^4 \times t^1}{t^3}$

Simplify the top first, $t^4 \times t^1 = t^5$.

$= \dfrac{t^5}{t^3}$

Then the division, $t^5 \div t^3 = t^2$.

$= t^{5-3}$
$= t^2$

EXERCISE 7C PART 2

5 (a) $a^5 \div a^2$ (b) $t^6 \div t^2$ (c) $e^4 \div e^3$
(d) $s^6 \div s^3$ (e) $t^2 \div t^2$

6 (a) $e^2 \div e^5$ (b) $f^2 \div f^6$ (c) $g^3 \div g^4$
(d) $h \div h^4$ (e) $w \div w^6$

7 (a) $5x^5 \div x^2$ (b) $6y^6 \div 2y^2$ (c) $12r^4 \div 4r^3$
(d) $8t^4 \div 4t$ (e) $12s^2 \div 6s^5$

EXAMPLE 7

Simplify each of these.

(a) $x^3 \div x^3$ (b) $18b^4 \div 3b^4$ (c) $5y^6 \div 20y^7$ (d) $\dfrac{3d^2 \times 4d^3}{6d^7}$

(a) $x^3 \div x^3 = x^{3-3}$
$= x^0$
$= 1$

$x^3 \div x^3 = \dfrac{x \times x \times x}{x \times x \times x}$
$= \dfrac{{}^1\cancel{x} \times {}^1\cancel{x} \times {}^1\cancel{x}}{{}_1\cancel{x} \times {}_1\cancel{x} \times {}_1\cancel{x}}$
$= 1$

(b) $18b^4 \div 3b^4 = \dfrac{18 \times b^4}{3 \times b^4}$

Numbers first, then letters.

$= \dfrac{18}{3} \times \dfrac{b^4}{b^4}$
$= 6 \times b^{4-4}$
$= 6 \times b^0$

$b^0 = 1$

$= 6 \times 1$
$= 6$

(c) $5y^6 \div 20y^7 = \frac{5y^6}{20y^7}$

$= \frac{5}{20} \times \frac{y^6}{y^7}$

$= \frac{5}{20} \times y^{6-7}$

$= \frac{5}{20} \times y^{-1}$

$= \frac{1}{4}y^{-1}$

Use $x^m \div x^n = x^{m-n}$.
So, $y^6 \div y^7 = y^{6-7}$
$= y^{-1}$

Simplify the fraction.
Divide top and bottom by 5.

(d) $\frac{3d^2 \times 4d^3}{6d^7} = \frac{3 \times 4 \times d^2 \times d^3}{6d^7}$

$= \frac{12 \times d^5}{6d^7}$

$= \frac{12}{6} \times \frac{d^5}{d^7}$

$= \frac{12}{6} \times d^{5-7}$

$= 2d^{-2}$

Use $x^m \div x^n = x^{m-n}$.
So, $d^5 \div d^7 = y^{5-7}$
$= d^{-2}$

Simplify the fraction.
$12 \div 6 = 2$

EXERCISE 7D

Simplify each of these.

1 (a) $h^4 \div h^4$ (b) $y^7 \div y^7$ (c) $a^5 \div a^5$
(d) $x^2 \div x^2$ (e) $t^3 \div t^3$ (f) $n^6 \div n^6$

2 (a) $6a^2 \div 3a^2$ (b) $8c^3 \div 4c^3$ (c) $12d^6 \div 4d^6$
(d) $15x^4 \div 5x^5$ (e) $20y \div 4y^2$ (f) $36h^7 \div 6h^8$
(g) $2b^3 \div 4b^3$ (h) $3w^5 \div 9w^5$ (i) $3k^4 \div 18k^6$
(j) $25m^4 \div 25m^4$ (k) $6d^2 \div 9d^2$ (l) $14g^3 \div 21g^3$

3 (a) $8b^2 \div 4b^3$ (b) $9d^3 \div 34^4$ (c) $10e^6 \div 5e^7$
(d) $18y^4 \div 6y^5$ (e) $21x \div 7x^2$ (f) $39c^7 \div 13c^8$
(g) $3t^3 \div 6t^6$ (h) $4h^5 \div 12h^7$ (i) $2m^4 \div 18m^7$
(j) $25n^4 \div 25n^6$ (k) $12w^2 \div 18w^6$ (l) $18f^3 \div 27f^3$

4 (a) $\frac{t^3 \times t^4}{t^7}$ (b) $\frac{2y^2 \times 3y^4}{y^6}$ (c) $\frac{4a^4 \times 5a}{a^5}$
(d) $\frac{7h^3 \times 5h^3}{h^6}$ (e) $\frac{3p^2 \times 6p^3}{2p^5}$ (f) $\frac{2b^2 \times 8b}{4b^3}$

5 (a) $\frac{5q^4 \times q^2}{10q^6}$ (b) $\frac{h^2 \times 4h^2}{20h^4}$ (c) $\frac{3k^5 \times 5k^3}{15k^8}$

(d) $\frac{3m^4 \times 2m}{12m^6}$ (e) $\frac{5c^4 \times 2c}{30c^7}$ (f) $\frac{2r^5 \times 3r^3}{7r^{12}}$

6 (a) $\frac{4q^5}{q^2 \times q^3}$ (b) $\frac{5h^7}{h^3 \times 5h^4}$ (c) $\frac{12y^6}{y^3 \times 3y^3}$

(d) $\frac{6p^3}{4p^3 \times 3p^4}$ (e) $\frac{6p^3 \times 4p^3}{2p^5 \times 3p^2}$ (f) $\frac{9a^4 \times 2a^3}{5a \times 6a}$

Laws of indices

Any value raised to the power 1 is equal to the value itself.

For example, $3^1 = 3$, $x^1 = x$.

You may have noticed some of the rules that make it easier to simplify expressions involving indices.

To multiply powers of the **same** number or variable, add the indices.

$$\begin{aligned} 4^3 \times 4^2 &= 4^{3+2} \\ &= 4^5 \\ x^3 \times x^2 &= x^{3+2} \\ &= x^5 \end{aligned}$$

$$\begin{aligned} 4^3 \times 4^2 &= 4 \times 4 \times 4 \times 4 \times 4 \\ &= 4^{3+2} \\ &= 4^5 \end{aligned}$$

In general $x^m \times x^n = x^{m+n}$

To divide powers of the **same** number or variable, subtract the indices.

$$\begin{aligned} 6^5 \div 6^3 &= 6^{5-3} \\ &= 6^2 \\ t^5 \div t^3 &= t^{5-3} \\ &= t^2 \end{aligned}$$

$$\begin{aligned} &6^5 \div 6^3 \\ &= \frac{6 \times 6 \times \cancel{6}^1 \times \cancel{6}^1 \times \cancel{6}^1}{\cancel{6}^1 \times \cancel{6}^1 \times \cancel{6}^1} \\ &= 6^{5-3} \\ &= 6^2 \end{aligned}$$

In general $x^m \div x^n = x^{m-n}$

Any value raised to the power $0 = 1$.

For example, $3^0 = 1$, $x^0 = 1$.

7.2 Standard Form

This is a special form for numbers which are either very large or very small. Your calculator will automatically change into this form when necessary.

Numbers greater than 1

Number	Calculation	Standard form
600	6 × 10 × 10	6 × 102
631	6.31 × 10 × 10	6.31 × 102
2000	2 × 10 × 10 × 10	2 × 103
2318.5	2.3185 × 10 × 10 × 10	2.3185 × 103
40 000	4 x 10 × 10 × 10 × 10	4 × 104

You can see from the pattern in the table that in standard form

- if there is a decimal point it is always after the first number
- the power of 10 tells you the size of the number

In the same way 40827.59 = 4.082759 x 10^4

Did you notice that the power of 10 is always one less than the number of digits before the decimal point in the original number?

Never write answers as 4.08275904 which is how your calculator may display numbers in standard form.

This part is always a number between 1 and 10.

This part is always a power of 10.

You MUST write the x sign between the two parts.

Now complete question 1 of Exercise 7E.

Numbers less than 1

To understand these numbers you need to remember from indices that

6 ÷ 10 = 6 x 10^{-1} 5 ÷ 102 = 5 x 10^{-2}

4 ÷ 102 = 4 x 10^{-3} 3 ÷ 104 = 3 x 10^{-4}

and so on. You will then see that very small numbers follow the same pattern as very large numbers.

Number	Calculation	Standard form
0.6	6 ÷ 10	6 × 10^{-1}
0.05	5 ÷ 10 ÷ 10	5 × 10^{-2}
0.0003	3 ÷ 10 ÷ 10 ÷ 10 ÷ 10	3 × 10^{-4}
0.071	7.1 ÷ 10 ÷ 10	7.1 × 10^{-2}
0.000 008	8 ÷10 ÷ 10 ÷ 10 ÷ 10 ÷ 10 ÷ 10	8 × 10^{-6}

Did you notice that the power of 10 is always negative and one more than the number of zeros between the decimal point and the first non-zero digit in the original number?

$$0.000917 = 9.17 \times 10^{-4}$$

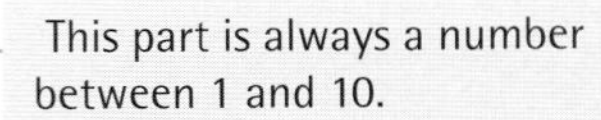
This part is always a number between 1 and 10.

This part is always a negative power of 10.

You MUST write the x sign between the two parts.

Now complete question 2 of Exercise 7E.

EXERCISE 7E

1 Write the following numbers in standard form.

(a) 3 000 000 (b) 7400 (c) 32 000
(d) 603 500 (e) 108 (f) 68
(g) 650.5 (h) 99.9

2 Write in standard form.

(a) 0.0005 (b) 0.006 (c) 0.4
(d) 0.00012 (e) 0.0717 (f) 0.0001975
(g) 0.9009 (h) 0.0010003

3 Match a number in the first column with a correct standard form number from the second column.

A	0.9
B	900 000
C	900
D	0.09
E	90
F	9000

G	9×10^3
H	9×10^2
I	9×10^5
J	9×10^1
K	9×10^{-1}
L	9×10^{-2}

4 Write these numbers in standard form.

(a) The diameter of the Earth is approximately 12 735 kilometres.

(b) The total mass of krill, a small sea shrimp, is estimated to be 650 million tonnes.

5 A single-celled organism is 2 tenths of a millimetre long. What is this in metres? Give your answer in standard form.

Changing between standard form and a decimal number

Standard form	Calculation	Number
9×10^5	$9 \times 10 \times 10 \times 10 \times 10 \times 10$	900 000
4.08×10^3	$4.08 \times 10 \times 10 \times 10$	4080
8.123×10^2	$8.123 \times 10 \times 10$	812.3
9×10^{-3}	$9 \div 10 \div 10 \div 10$	0.009
4.08×0^{-2}	$4.08 \div 10 \div 10$	0.0408
8.123×10^{-5}	$8.123 \div 10 \div 10 \div 10 \div 10 \div 10$	0.00008123

The table shows changing various numbers in standard form into decimal numbers.

Positive ⟶ Right

Negative ⟵ Left

Always put a zero in front of the decimal point when the number is small

0.000419

is much clearer than

.000419

EXERCISE 7F

1 Write these as decimal numbers.

(a) 5×10^4 (b) 3.8×10^3

(c) 6×10^{-3} (d) 7.26×10^9

(e) 8.492×10^{-2} (f) 4.37×10^6

(g) 1.006×10^{-4} (h) 6.2387×10^3

2 Write these numbers in standard form as decimal numbers.

(a) The distance from the Earth to the Sun is approximately 1.488×10^8 kilometres.

(b) The average width of an iris of an eye is 1×10^{-2} metres.

(c) A billion molecules of water has a mass of 3×10^{-11} grams.

3 These numbers are *not* in standard form. Rewrite them in standard form.

(a) 123×10^2 (b) 0.8×10^7

(c) 17×10^{-2} (d) 0.25×10^{-4}

(e) 18 million (f) $\frac{1}{8}$

(g) $36 \times 10^4 \times 0.006$ (h) $\sqrt{40 \times 10}$

4 Complete these calculations, giving your answer in standard form.

(a) $(6.4 \times 10^5) - (8.34 \times 10^4)$
(b) $(4.2 \times 10^2) + (5.6 \times 10^3)$
(c) $(3.9 \times 10^{-2}) + (4.2 \times 10^{-3})$
(d) $(8.2 \times 10^{-3}) - (6.1 \times 10^{-4})$
(e) $(4 \times 10^2) - (7.2 \times 10^{-1})$
(f) $(6.1 \times 10^3) + (5.7 \times 10^2) - (4.8 \times 10^{-1})$

Write these as decimal numbers before you do the calculation.

Write your answers in standard form.

5 Calculate the following, giving your answer in standard form.

(a) $(8.4 \times 10^3) \times (2 \times 10^4)$ (b) $(7.1 \times 10^5) \times (3 \times 10^2)$
(c) $(4 \times 10^{-5}) \times (5.2 \times 10^{11})$ (d) $(8.9 \times 10^{-2}) \times (3 \times 10^{-5})$
(e) $(1.4 \times 10^3) \div (7 \times 10^9)$ (f) $(3.6 \times 10^{-2}) \div (4 \times 10^{-5})$
(g) $(4 \times 10^6) \div (5 \times 10^7)$ (h) $(5.2 \times 10^2) \div (1.7 \times 10^{-2})$
(i) $(1.5 \times 10^3)^2$ (j) $(3.5 \times 10^2)^2$
(k) $\sqrt{(1.6 \times 10^3)}$ (l) $\sqrt{(2.25 \times 10^8)}$

6 On Jane's computer a picture of a dragonfly uses 7 832 450 bytes. Her picture of a bee uses 7.68×10^6 bytes. Which is larger and by how much?

7 Find the circumference of a circle radius 4×10^8 giving your answer in standard form.

8 Light travels 3×10^8 metres in 1 second. Find how far it travels in

(a) 1 minute (b) 1 hour.

9 Given that $p = 3.1 \times 10^5$ and $q = 1.05 \times 10^{-2}$ calculate, giving your answer in standard form.

(a) pq (b) $p + q$ (c) $\frac{p - q}{2}$ (d) p^2q
(e) $p \div q$ (f) $10q$ (g) $\frac{1}{2}pq$ (h) q^3

10 In 2008 an average of 2.56×10^5 passengers a week used an airport.
Calculate, in standard form, the average number of passengers using the airport in a year.

Calculators and standard form

To enter a number given in standard form on your calculator you can use the button marked EXP

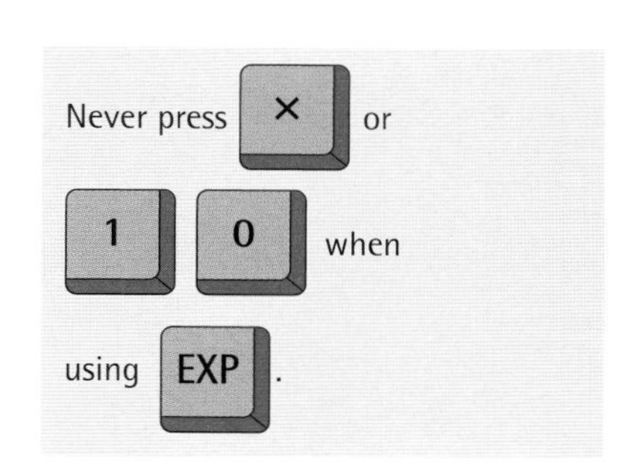

EXP stands for *exponent* which means 'a power of 10'.
So to enter 1.86×10^5 you press,

This is much quicker than pressing ...

... especially when you have quite a few numbers written in standard form in a calculation!

A word of warning!

When you use your calculator to obtain an answer that is either very big or very small, the number in the calculator display may or may not appear in standard form or be displayed to enough decimal places.

For example, to calculate the reciprocal of 2.3×10^8:

The reciprocal is $\frac{1}{2.3 \times 10^8}$

The reciprocal button is quickest way of doing this calculation.

The reciprocal button is

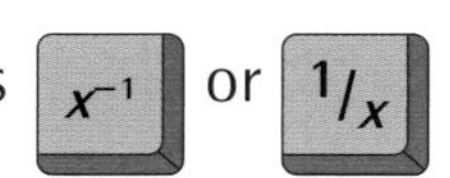

Press

(making sure your calculator is in the correct mode).

The answer given is $4.347\,826\,087^{-09}$
Written in standard form this is 4.35×10^{-9} (3 s.f.)

The choice of 'NORM 1~2 ?' does not just write an answer in standard form when you select 1, it also gives you all the decimal places. When you select 2 you may not see enough decimal places, or the answer may disappear off the display screen.

The reciprocal button is the quickest way of doing this calculation.

This may not give the answer to the required degree of accuracy ... it depends whether your calculator is in the correct mode.

To check, press the mode button until the word 'NORM' appears (this is usually option 3).
When you choose this you will see 'NORM 1 ~2 ?'
If you select 2 your answer will **not** be in standard form, so select 1.

Remember that calculators vary. Check how yours works.

EXERCISE 7G

1 Work out the following, giving your answers in standard form.

(a) $(2.2 \times 10^3) \times (4 \times 10^5)$ (b) $(3.3 \times 10^6) \times (3 \times 10^{-4})$

(c) $(8.4 \times 10^5) \div (2 \times 10^4)$ (d) $(5.6 \times 10^3) \div (4 \times 10^{-2})$

(e) $(4.2 \times 10^4) \times (5.3 \times 10^3)$ (f) $(1.6 \times 10^{-2}) \div (4 \times 10^{-4})$

2 This year it is estimated that, on average, each person in the United Kingdom will spend £480 making mobile phone calls. If the population is 5.4×10^7, what is the total amount spent on mobile phone calls? Give your answer in standard form.

3 The radius of the Earth is approximately 6.4×10^3 km. Estimate the volume of the Earth if we assume it is a sphere.

Volume of a sphere $= \frac{4}{3}\pi r^3$

4 If $a = 1.32 \times 10^7$ and $b = 4.28 \times 10^6$, find the following, giving your answers in standard form.

(a) $3b$ (b) ab (c) a^2 (d) $a + b$ (e) $b \div a$

5 If there are 6×10^9 people in the world and they eat 7×10^7 tonnes of fish per year, how much fish does each person eat?

6 The diameter of the Sun is approximately 1.392×10^6 km. Find its surface area using the formula $4\pi r^2$.

7 A rectangular picture measures 1.4×10^2 cm by 2.7×10^3 cm. What is

(a) the area (b) the perimeter of the picture?

8 The number of krill (a type of plankton) is estimated to be 6×10^{15} and their mass is about 6.5×10^8 tonnes.

(a) What is the mass of 1 krill in grams?

(b) If the volume of 1 million krill is 100 cm^3, what is the mass of one cubic cm of krill?

9 After the Sun, the next nearest star to the Earth is Proxima Centauri. It takes about 4.24 years for light from this star to reach the Earth. If light travels at 1.86×10^5 miles per *second*, estimate the distance of Proxima Centauri from Earth.

7.3 Using and writing formulae

You will need to know

- how to use the correct order of operations

Formulae is the plural of formula.

Using formulae given in words

Lisa has a job. The pay is $6 an hour.
How much is Lisa paid if she works for 20 hours?

You can write a **formula** to find her pay for any number of hours she works.

pay = number of hours worked × rate of pay

You can use the formula to answer the question.

For 20 hours work pay = 20 × $6
= $120

The number of hours Lisa works is a variable – it can change from day to day. Her total pay and her rate of pay are both variables as well.

A formula is a general rule that shows the relationship between quantities.

The quantities in a formula can vary in size.

These quantities are called **variables**.

You can use the formula to find Lisa's pay for any number of hours worked and any rate of pay.

EXAMPLE 8

To work out his pay Raol uses the formula

pay = hours worked × rate of pay + bonus

What is his pay for 10 hours' work at a rate of $5.50 an hour, with a bonus of $7?

Pay = hours worked × rate of pay + bonus
= 10 × $5.50 + $7
= $55 + $7
= $62

Remember the order of operations you do the multiplication before the addition.

EXERCISE 7H

1 To work out his pay Amit uses the formula

$$\text{pay} = \text{hours worked} \times \text{rate of pay}$$

(a) Work out his pay for 10 hours' work at $6 an hour.

(b) Work out his pay for 20 hours' work at $5.50 an hour.

2 Lauris uses this formula to work out the cost of stamps

$$\text{cost} = \text{number of stamps} \times \text{cost of one stamp}$$

(a) Work out the cost of 20 stamps at 30 cents each.

(b) Work out the cost of 15 stamps at 35 cents each.

(c) Lauris spends $6 on 25-cent stamps.
How many stamps does Lauris buy?

Use your equation solving skills from Chapter 5 to help.

3 To work out her pay Angharad uses the formula

$$\text{pay} = \text{hours worked} \times \text{rate of pay} + \text{bonus}$$

(a) What is her pay when she works 10 hours at a rate of $6.50 an hour, with a bonus of $7.50?

(b) What is her pay when she works 15 hours at a rate of $6 an hour, with a bonus of $8?

Use Example 8 to help you.

4 Use this formula

$$\text{Average speed} = \frac{\text{distance travelled}}{\text{time taken}}$$

to work out the average speed of these journeys.

(a) 100 kilometres from Munich to Salzburg in 2 hours.

(b) 180 kilometres by train in 3 hours.

(c) A sponsored walk of 12 kilometres that takes 4 hours.

(d) A marathon runner who takes 4 hours to run 26 kilometres.

Your answers will be in km/h (kilometres per hour).

Writing formulae using letters and symbols

You can use letters for the variables in a formula.

For example, for

pay = hours worked × rate of pay

you could write

$p = hr$

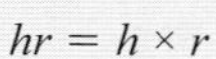

where

p is the pay,
h is the number of hours worked,
r is the rate of pay

You must say what each letter stands for.

You can put in numbers for the hours worked and the rate of pay, to calculate the pay.

EXAMPLE 9

The formula for the area of a rectangle is $A = lw$, where l is the length and w is the width.

Work out the value of A when $l = 8$ and $w = 4$.

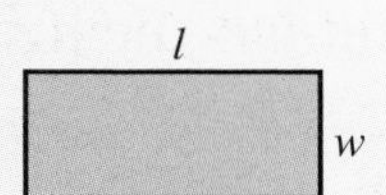

$A = lw$
$= l \times w$
$= 8 \times 4$
$= 32$

EXAMPLE 10

The perimeter of a rectangle is given by $P = 2l + 2w$, where l is the length and w is the width.

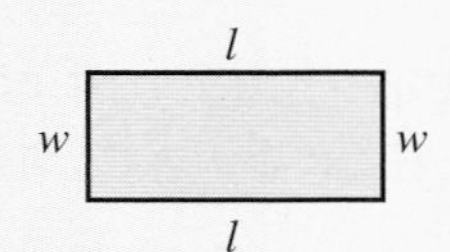

The perimeter is the distance around the outside.

Work out the perimeter when $l = 9$ and $w = 6$.

$P = 2l + 2w$
$= 2 \times 9 + 2 \times 6$
$= 18 + 12$
$= 30$

Order of operations
Multiplication before Addition.

EXERCISE 7I

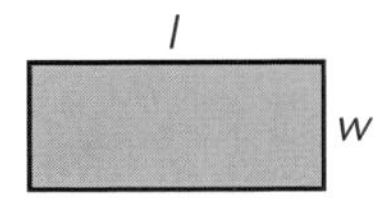

1 The formula for the area of a rectangle is $A = lw$, where l is the length and w is the width.
Work out the value of A when

(a) $l = 8$ and $w = 5$ (b) $l = 7$ and $w = 6$

(c) $l = 10$ and $w = 6$ (d) $l = 9$ and $w = 7$

2 The formula for the voltage, V, in an electrical circuit is $V = IR$, where I is the current and R is the resistance.
Work out the voltage when

(a) $I = 2$ and $R = 6$ (b) $I = 3$ and $R = 8$

(c) $I = 1.5$ and $R = 9$ (d) $I = 4$ and $R = 60$

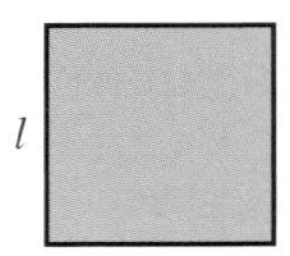

3 The perimeter of a square is given by $P = 4l$, where l is the length of the square.
Work out the value of P when

(a) $l = 12$ (b) $l = 25$ (c) $l = 3.5$ (d) $l = 5.2$

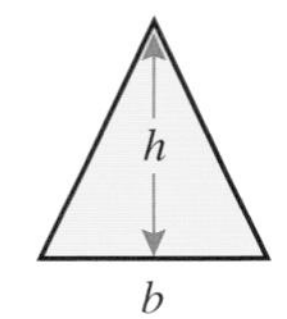

4 The formula for the area of a triangle is $A = \frac{1}{2}bh$, where b is the length of the base and h is the height. Work out the value of A when

(a) $b = 8$ and $h = 6$ (b) $b = 10$ and $h = 5$

(c) $b = 9$ and $h = 8$ (d) $b = 7$ and $h = 5$

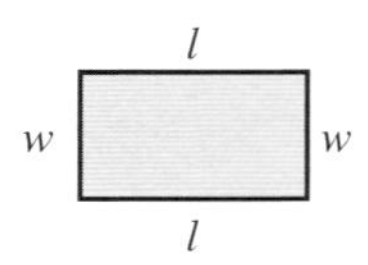

5 Ben uses the formula $P = 2l + 2w$ to work out the perimeter of a rectangle, where l is the length and w is the width. Work out the value of P when

(a) $l = 8$ and $w = 5$ (b) $l = 12$ and $w = 5$

(c) $l = 7$ and $w = 4.5$ (d) $l = 7.5$ and $w = 3.5$

Use Example 9 to help.

6 Jenna uses the formula $P = 2(l + w)$ to work out the perimeter of rectangles, where l is the length and w is the width.
Use her formula to work out the perimeter of each rectangle in question 5 to show that she gets the same answers as Ben.
You must show your working.

Order of operations
Work out the bracket before doing the multiplication.

Writing your own formulae

EXAMPLE 11

Alex buys x packets of sweets.
Each packet of sweets costs 45 cents.
Alex pays with a \$5 note.
Write a formula for the change, C, in cents, Alex should receive.

$C = 500 - 45x$

\$5 = 500 cents

The sweets cost 45 cents per packet so the cost, in cents, for x packets is $45x$.

EXERCISE 7J

1 Nilesh buys y packets of sweets.
Each packet of sweets costs 48 cents.
Nilesh pays with a \$5 note.
Write a formula for the change, C, in cents, Nilesh should receive.

2 Apples cost r cents each and bananas cost s cents each.
Sam buys 7 apples and 5 bananas.
Write a formula for the total cost, t, in cents, of these fruit.

3 To roast a chicken you allow 45 minutes per kg and then a further 20 minutes.
Write a formula for the time, t, in minutes, to roast a chicken that weighs w kg.

4 To roast lamb you allow 30 minutes plus a further 65 minutes per kg.
Write a formula for the time, t, in minutes, to roast a piece of lamb that weighs w kg.

5 A rectangle has a length of $3x + 1$ and a width of $x + 2$.
Write down a formula for the perimeter, p, of this rectangle.

7.4 Substitution

Substitution into simple expressions

You can **substitute** numbers for the variables in an expression.

Substitute means 'replace'. You replace each letter with a number.

Use the correct order of operations to do the calculations.

EXAMPLE 12

If $a = 4$, $b = 2$ and $c = 3$,
work out the value of these expressions.

(a) $3a$
(b) $5b + 4c$
(c) $ab - c$

(a) $3a = 3 \times 4$
$= 12$

Substitute the value 4 for a in the expression $3a = 3 \times a$

(b) $5b + 4c = 5 \times 2 + 4 \times 3$
$= 10 + 12$
$= 22$

(c) $ab - c = 4 \times 2 - 3$
$= 8 - 3$
$= 5$

$ab = a \times b$

EXERCISE 7K

If $a = 4$, $b = 2$ and $c = 3$,
work out the value of these expressions.

1	$2a$	2	$5b$	3	$6c$	4	$a + c$
5	$a + b + c$	6	$a - 3$	7	$3a - 5$	8	$5b + 6$
9	$4c - 7$	10	$5a + 4b$	11	$3b + 7c$	12	$8a + 5c$
13	$6a - c$	14	$5b - 2c$	15	$9c - 4a$	16	$ab + c$
17	$ac - b$	18	$ab + ac$	19	$bc - 2b$	20	abc

You can substitute positive and negative integers, fractions and decimals into expressions.

Use the rules for adding, subtracting, multiplying and dividing negative numbers.

EXAMPLE 13

If $x = 5$, $y = \frac{1}{2}$ and $z = -2$,
work out the value of these expressions.

(a) $3x + 4y$ (b) $6x + 2z$ (c) $4x - 6y + 3z$

(a) $3x + 4y = 3 \times 5 + 4 \times \frac{1}{2}$
$= 15 + 2$
$= 17$

(b) $6x + 2z = 6 \times 5 + 2 \times (-2)$
$= 30 + (-4)$
$= 30 - 4$
$= 26$

Adding -4 is the same as subtracting 4.

(c) $4x - 6y + 3z = 4 \times 5 - 6 \times \frac{1}{2} + 3 \times (-2)$
$= 20 - 3 + (-6)$
$= 20 - 3 - 6$
$= 11$

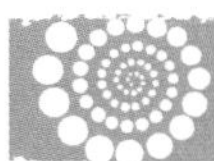

EXERCISE 7L

If $x = 5$, $y = \frac{1}{2}$ and $z = -2$, work out the value of these expressions.

1 $6x$
2 $2y$
3 $5z$
4 $3x + 4$
5 $4y + 7$
6 $3z + 14$
7 $3x + 10y$
8 $3x + 5y$
9 $4x + 3z$
10 $12y + 2z$
11 $7x - 8y + 4z$
12 $4x - 5y + 2z$

If $f = 6$, $g = 1.5$ and $h = -3$, work out the value of these expressions.

13 $7f$
14 $4g$
15 $6h$
16 $f + g$
17 $4g + h$
18 $3f + 5h$
19 $2f - 5g$
20 $fg + 8g + 3h$

Substitution into more complicated expressions

You can substitute values into expressions involving brackets and powers (indices).

EXAMPLE 14

If $a = 5$, $b = 4$ and $c = 3$,
work out the value of these expressions.

(a) $\frac{a+3}{2}$ (b) $3b^2 - 1$ (c) $\frac{5c+1}{b}$

(a) $\frac{a+3}{2} = \frac{5+3}{2}$
$= 8 \div 2$
$= 4$

(b) $3b^2 - 1 = 3 \times 4^2 - 1$
$= 3 \times 16 - 1$
$= 48 - 1$
$= 47$

Order of operations
indices ($4^2 = 16$)
then the multiplication (3×16)
then the subtraction ($48 - 1$)

(c) $\frac{5c+1}{b} = (5 \times 3 + 1) \div 4$
$= (15 + 1) \div 4$
$= 16 \div 4$
$= 4$

EXAMPLE 15

Evaluate (a) $f^3 + f$ when $f = 2.3$ (b) $2x^3$ when $x = -4$

Evaluate means 'work out the value of'.

(a) $f^3 + f = 2.3^3 + 2.3$
$= 14.467$

Use the power key on your calculator.

(b) $2x^3 = 2 \times x^3$
$= 2 \times (-4)^3$
$= 2 \times -64$
$= -128$

You need to know how to input a negative number in your calculator. You may have to enter $(-4)^3$ like this

EXERCISE 7M

If $r = 5$, $s = 4$ and $t = 3$,
work out the value of these expressions.

1 $\frac{r+3}{2}$ 2 $\frac{s+5}{3}$ 3 $\frac{t+7}{2}$

4 $3r^2 + 1$ 5 $4t^2 - 6$ 6 $2s^2 + r$

$3r^2 = 3 \times r^2 = 3 \times r \times r$

7 $4(5s + 1)$ 8 $t(r + s)$ 9 $5(2s - 3t)$

Order of operations brackets first.

10 $\frac{5t+1}{s}$ 11 $\frac{4r-2}{t}$ 12 $\frac{3s+t}{r}$

If $a = 5$, $b = 1.5$, $c = -2$, work out the value of these expressions.

13 $3a^2 + b$ **14** $c^2 + a$ **15** $2c^3 + b$

16 $10b + 2a^2$ **17** $4c^2 - a + b$ **18** $b^3 - c$

19 Copy and complete this table.

x	1	2	3	4	5
$x^2 + 2x$			15		

$3^2 + 2 \times 3 = 9 + 6 = 15$

20 Copy and complete this table.

x	3	4	3.5	3.7	3.8
$x^3 - x$		60			

$x^3 = x \times x \times x$
$4^3 - 4 = 64 - 4 = 60$

If $A = 6$, $B = -4$, $C = 3$ and $D = 30$, work out the value of these expressions.

21 $D(B + 7)$ **22** $A(B + 1)$ **23** $A^2 + 2B + C$

24 $\frac{2A + 3}{C}$ **25** $\frac{4B + D}{2}$ **26** $\frac{A^2 + 3B}{C}$

EXAMPLE 16

A formula for working out acceleration is

$$a = \frac{v - u}{t}$$

where v is the final velocity, u is the initial velocity and t is the time taken.
Work out the value of a when $v = 50$, $u = 10$, $t = 8$.

$$a = \frac{v - u}{t}$$
$$= \frac{50 - 10}{8}$$
$$= 40 \div 8$$
$$= 5$$

Work out the numerator first, then divide.

EXAMPLE 17

A formula for working out distance travelled is

$$s = ut + \frac{1}{2}at^2$$

where u is the initial velocity, a is the acceleration and t is the time taken.
Work out the value of s when $u = 3, a = 8, t = 5$.

Initial velocity = starting velocity.

$s = ut + \frac{1}{2}at^2$
$= 3 \times 5 + \frac{1}{2} \times 8 \times 5^2$
$= 15 + 4 \times 25$
$= 15 + 100$
$= 115$

$ut = u \times t$

Order of operations indices ($5^2 = 25$) then multiplication ($3 \times 5 = 15, \frac{1}{2} \times 8 = 4$) then addition ($15 + 100 = 115$).

EXERCISE 7N

Substitution into formulae

You can substitute values into a formula to work out the value of a variable.

1 Use the formula $a = \frac{v - u}{t}$ to work out the value of a when

(a) $v = 15, u = 3, t = 2$ (b) $v = 29, u = 5, t = 6$
(c) $v = 25, u = 7, t = 3$ (d) $v = 60, u = 10, t = 4$

2 The formula for the area of a trapezium is

$A = \frac{1}{2}(a + b)h$

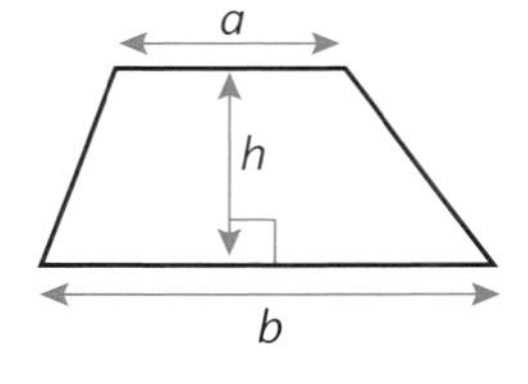

Work out the value of A when

(a) $a = 10, b = 6, h = 4$ (b) $a = 13, b = 9, h = 8$
(c) $a = 9, b = 6, h = 4$ (d) $a = 15, b = 10, h = 6$

3 Use the formula $s = ut + \frac{1}{2}at^2$ to work out the value of s when

Use Example 17 to help you.

(a) $u = 3, a = 10, t = 2$ (b) $u = 7, a = 6, t = 5$
(c) $u = 2.5, a = 5, t = 4$ (d) $u = -4, a = 8, t = 3$

4 Body Mass Index, b, is calculated using the formula

$$b = \frac{m}{h^2}$$

where m is mass in kilograms and h is height in metres.

Work out the value of b when

(a) $m = 70, h = 1.8$ (b) $m = 38, h = 1.4$

(c) $m = 85, h = 1.9$ (d) $m = 59, h = 1.7$

Round your answers to three significant numbers.

5 A formula for working out the velocity of a car is

$$v = \sqrt{(u^2 + 2as)}$$

where u is the initial velocity, a is the acceleration and s is the distance travelled.

Work out the value of v when

(a) $u = 3, a = 4, s = 5$ (b) $u = 6, a = 8, s = 4$

(c) $u = 9, a = 10, s = 2$ (d) $u = 7, a = 4, s = 15$

$\sqrt{}$ is the symbol for square root.

6 *Talkalot* calculates telephone bills using this formula

$$C = \frac{7.5n + 995}{100}$$

where C is the total cost (in \$) and n is the number of calls.

Work out the telephone bill for

(a) 84 calls (b) 156 calls (c) 328 calls.

Give your answers correct to the nearest cent.

7 My water company calculates water bills each quarter using this formula

$$A = 2.87V + 4.94$$

where A is the amount to pay (in \$) and V is the volume of water used (in m^3).

In January I used 10 m^3, in February I used 11 m^3 and in March I used 8 m^3.

(a) Work out my bill for the first quarter of the year.

(b) How many m^3 of water would I need to use to make my bill for the quarter more than \$100?

The first quarter is January, February and March.

7.5 Changing the subject of a formula

In the formula $v = u + at$ the variable v is called the **subject** of the formula.

A variable is a letter that can take different values.

The subject of a formula is always the letter on its own on one side of the equation. This letter only appears once in the formula.

P is the *subject* of the formula $P = 2l + 2w$. You can use the formula to find the value of P.

You can **rearrange** a formula to make a different variable the subject. This is called 'changing the subject' of a formula.

This uses equation solving skills (Chapter 5).

If you are given some values, you can substitute these before you rearrange.

EXAMPLE 18

A formula for working out the perimeter of a regular hexagon is $P = 6x$, where x is the length of each side.

Work out the value of x when $P = 48$.

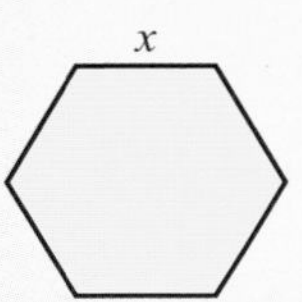

Substitute the value you know into the formula.

$$P = 6x$$
$$48 = 6x$$

Solve the equation to find x.

$$48 \div 6 = x$$
$$8 = x$$

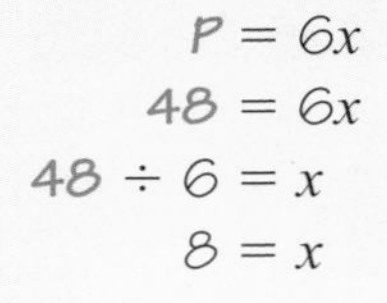

EXAMPLE 19

The perimeter of a rectangle is given by $P = 2y + 2w$, where y is the length and w is the width.

Work out the value of y when $P = 24$ and $w = 5$.

Substitute the values you know into the formula.

$$P = 2y + 2w$$
$$24 = 2y + 2 \times 5$$
$$24 = 2y + 10$$

Solve the equation to find y.

$$24 - 10 = 2y$$
$$14 = 2y$$
$$14 \div 2 = y$$
$$7 = y$$

You can use the same techniques to rearrange a formula when you don't know any of the values.

EXAMPLE 20

Rearrange $d = a + 8$ to make a the subject.

$$d = a + 8$$
$$d - 8 = a + 8 - 8$$
$$d - 8 = a$$
$$a = d - 8$$

You need to finish with a on its own on one side.

Subtract 8 from both sides as when solving an equation.

EXAMPLE 21

Rearrange $A = lw$ to make l the subject.

$$A = lw$$
$$\frac{A}{w} = \frac{lw}{w}$$
$$\frac{A}{w} = l \quad \text{or} \quad l = \frac{A}{w}$$

lw means $l \times w$.
l is multiplied by w so divide both sides by w to leave l on its own.

EXAMPLE 22

Make x the subject of the formula $y = 5x - 2$.

$$y = 5x - 2$$
$$y + 2 = 5x - 2 + 2$$
$$y + 2 = 5x$$
$$\frac{y + 2}{5} = \frac{5x}{5}$$
$$\frac{y + 2}{5} = x \quad \text{or} \quad x = \frac{y + 2}{5}$$

Add 2 to both sides to get $5x$ on its own.

Now divide both sides by 5 to leave x on its own.

EXERCISE 70

1 A formula for working out the perimeter of a regular hexagon is $P = 6l$, where l is the length of each side.

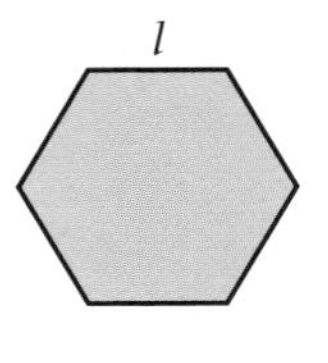

Use Example 18 to help you.

Work out the value of l when

(a) $P = 60$ (b) $P = 30$ (c) $P = 120$

2 The formula for the area of a rectangle is $A = lw$, where l is the length and w is the width.

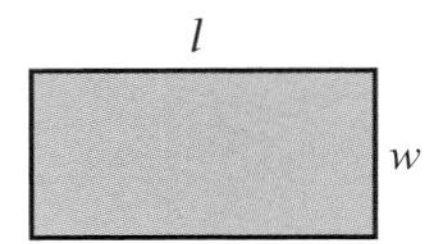

Work out the value of w when

(a) $A = 12$ and $l = 4$ (b) $A = 36$ and $l = 9$

(c) $A = 42$ and $l = 7$ (d) $A = 60$ and $l = 15$

3 The perimeter of a rectangle is given by

$$P = 2l + 2w$$

where l is the length and w is the width.

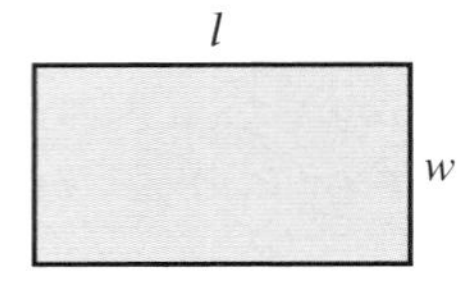

Use Example 19 to help.

Use the formula to

(a) find l when $P = 18$ and $w = 4$

(b) find l when $P = 32$ and $w = 7$

(c) find w when $P = 60$ and $l = 17$

(d) find w when $P = 50$ and $l = 13.5$

4 Use the formula $v = u + at$

(a) to find u when $v = 30$, $a = 8$ and $t = 3$

(b) to find u when $v = 47$, $a = 4$ and $t = 9$

(c) to find a when $v = 54$, $u = 19$ and $t = 7$

(d) to find t when $v = 60$, $u = 15$ and $a = 5$

(e) to find u when $v = 20$, $a = 7$ and $t = 4$

5 Rearrange each of these formulae to make a the subject.

(a) $d = a + 8$ (b) $t = a + 12$

(c) $k = a - 6$ (d) $w = a - 7$

6 Rearrange each of these formulae to make w the subject.

(a) $P = 4w$ (b) $a = 3w$

(c) $A = lw$ (d) $h = kw$

Use Example 22 to help you.

7 Make x the subject of each of these formulae.

(a) $y = 5x - 6$ (b) $y = 4x - 7$

(c) $y = 2x + 1$ (d) $y = 6x + 5$

8 The formula $F = 1.8C + 32$ can be used to convert degrees Celsius, $°C$, to degrees Fahrenheit, $°F$.

(a) Convert $15°C$ to degrees Fahrenheit.

(b) Rearrange the formula to make C the subject.

(c) Use your new formula to convert $68°F$ to $°C$.

(d) What is $82°F$ in degrees Celsius to the nearest degree? You must show your working.

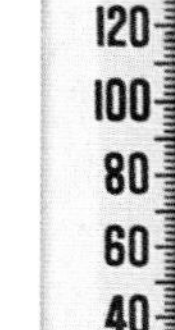

9 Make r the subject of each of these formulae.

(a) $p = 4r + 2t$ (b) $v = 7r + 4h$

(c) $w = 3r - 2s$ (d) $y = 6r - 5p$

10 A formula used to calculate velocity is

$$v = u + at$$

where u is the initial velocity, a is the acceleration and t is the time taken.

(a) Rearrange the formula to make a the subject.

(b) Rearrange the formula to make t the subject.

11 Rearrange these formulae to make a the subject.

(a) $b = \frac{1}{2}a + 6$ (b) $b = \frac{1}{2}a + 7$ (c) $b = \frac{1}{3}a - 1$

(d) $b = \frac{1}{4}a - 3$ (e) $b = 2(a + 1)$ (f) $b = 3(a - 5)$

EXAMINATION QUESTIONS

1 Find the value of $a - b - 2c$ when $a = -2$, $b = 7$ and $c = -11$. [1]

(CIE Paper 1, Jun 2000)

---

2 Write down the value of

(a) $\left(\frac{5}{2}\right)^{-2}$, [1]

(b) $\left(\frac{1}{3}\right)^{0}$. [1]

(CIE Paper 1, Jun 2000)

---

3 Work out $\left(\frac{0.07728}{27600}\right)^2$, giving your answer in standard form. [2]

(CIE Paper 1, Jun 2000)

---

4 The population of Argentina is 35 400 000.
Write this number in standard form. [1]

(CIE Paper 1, Nov 2000)

---

5 Work out $4^5 - 5^4$. [1]

(CIE Paper 1, Nov 2000)

---

6 A person's body mass index, is calculated by using the formula $I = \frac{M}{h^2}$.
where M is the mass in kilograms and h is the height in metres.

(a) Anne has a height of 1m 62 cm.

(i) Write 1m 62 cm in metres. [1]

(ii) Anne has a mass of 60 kilograms.
Calculate her body mass index. [3]

(b) Make M the subject of the formula $I = \frac{M}{h^2}$. [2]

(CIE Paper 3, Nov 2000)

---

7 The Earth is 1.5×10^8 kilometres from the Sun.
Light from the Sun takes 8 minutes to reach the Earth.
Jupiter is 7.78×10^8 kilometres from the Sun.
Work out how long it takes light from the Sun to reach Jupiter. [2]

(CIE Paper 1, Jun 2001)

---

8 Two coins are weighed.
The difference between their masses is 5.7×10^{-3} grams.
(a) Write 5.7×10^{-3} as a decimal. [1]
(b) Change 5.7×10^{-3} grams into milligrams. [1]

(CIE Paper 1, Nov 2001)

---

9 Find the exact value of
(a) 2^3, [1]
(b) 7^{-1}, [1]
(c) $\left(\frac{3}{2}\right)^{-2}$. [1]

(CIE Paper 1, Nov 2001)

10
$$y = 100 - 4x$$
(a) Find the value of y when $x = 20$. [1]
(b) Find the value of x when $y = 72$. [2]
(c) Make $\boldsymbol{x}$ the subject of the equation. [3]

(CIE Paper 3, Nov 2001)

---

11 Work out $48k^{10} \div 24k^8$ giving your answer in its simplest form. [2]

(CIE Paper 1, Jun 2002)

---

12 A spoon can hold 5 ml of medicine.
(a) Write 5 ml in litres. [1]
(b) Write your answer in standard form. [1]

(CIE Paper 1, Jun 2002)

---

13 Javed says that his eyes will blink 415 000 000 times in 79 years.
Write 415 000 000 in standard form. [1]

(CIE Paper 2, Jun 2002)

---

14 The population of Argentina is 3.164×10^7. Its area is 2.8×10^6 square kilometres.
Work out the average number of people per square kilometre in Argentina. [2]

(CIE Paper 1, Nov 2002)

---

15 The radius of the Earth at the equator is appoximately 6.4×10^6 metres.
Calculate the circumference of the Earth at the equator.
Give your answer in standard form, correct to 2 significant figures. [3]

(CIE Paper 2, Nov 2002)

16

$$T = 2\sqrt{n}.$$

(a) Find T when $n = 25$. [1]
(b) Make n the subject of the formula. [2]

(CIE Paper 1, Nov 2002)

17

$$F = \frac{300\,000}{l}$$

(a) Calculate the value of F when $l = 1500$. [1]
(b) Calculate the value of l when $F = 433$,
giving your answer to the nearest whole number. [3]

(CIE Paper 1, Nov 2002)

18

$$y = 4uv - 3v$$

Find the value of y when $u = -3$ and $v = -2$. [2]

(CIE Paper 1, Jun 2003)

---

19 Write down the value of $(1\frac{1}{2})^{-2}$ as a fraction. [2]

(CIE Paper 1, Jun 2003)

---

20 The perimeter P, of a triangle is given by the formula $P = 6x + 3$.
(a) Find the value of P when $x = 4$. [1]
(b) Find the value of x when $P = 39$. [2]
(c) Rearrange the formula to find x in terms of P. [2]

(CIE Paper 3, Nov 2003)

---

21 Write down the value of n in the following statements.
(a) $1500 = 1.5 \times 10^n$. [1]
(b) $0.00015 = 1.5 \times 10^n$. [1]

(CIE Paper 1, Nov 2003)

---

22 **(a)** $(\frac{1}{2})^x = \frac{1}{8}$ Write down the value of x. [1]
(b) $7^y = 1$ Write down the value of y. [1]

(CIE Paper 1, Jun 2004)

23

$$y = a + bc$$

(a) Find the value of y when $a = -3$, $b = 2$ and $c = 8$. [2]
(b) Make c the subject of the formula. [2]

(CIE Paper 1, Jun 2004)

---

24 In 1997 the population of China was 1.24×10^9.
In 2002 the population of China was 1.28×10^9.
Calculate the percentage increase from 1997 to 2003. [2]

(CIE Paper 2, Nov 2004)

---

25 When $x = 5$ find the value of
(a) $4x^2$, [1]
(b) $(4x)^2$. [1]

(CIE Paper 1, Nov 2004)

---

26 Simplify the following expressions
(a) $a^2 \times a^5$, [1]
(b) $b^4 \div b^3$. [1]

(CIE Paper 1, Nov 2004)

---

27 When $\boldsymbol{x} = -3$ find the value of $x^3 + 2x^2$. [2]

(CIE Paper 1, Jun 2005)

---

28 Work out 4^{-3} as a fraction. [2]

(CIE Paper 1, Jun 2005)

---

29 Make s the subject of the formula $p = st - q$. [2]

(CIE Paper 1, Jun 2005)

---

30 The formula for changing degrees Celsius (C) to degrees Fahrenheit (F) is $F = \frac{9}{5}C + 32$.

Use the formula to change 6 degrees Celsius to degrees Fahrenheit. [2]

(CIE Paper 3, Jun 2005)

Chapter 8

Measurements

This chapter will show you how to

- ✔ estimate length, volume (capacity) and mass
- ✔ calculate with time and read timetables
- ✔ use metric units of length, volume and mass and convert between them
- ✔ understand the accuracy of measurement
- ✔ handle the compound measures of speed and density

8.1 Estimating length, volume and mass

You need to be able to estimate length, volume (or capacity) and mass in everyday life.

To estimate a measurement, compare with a measurement you already know.

For example, compare with
- the length of a ruler (30 cm)
- the mass of a bag of sugar (1 kg)
- the capacity of a carton of juice (1 ℓ)

Estimating length

You can measure lengths and distances in these units.

kilometres (km), metres (m), centimetres (cm), millimetres (mm)

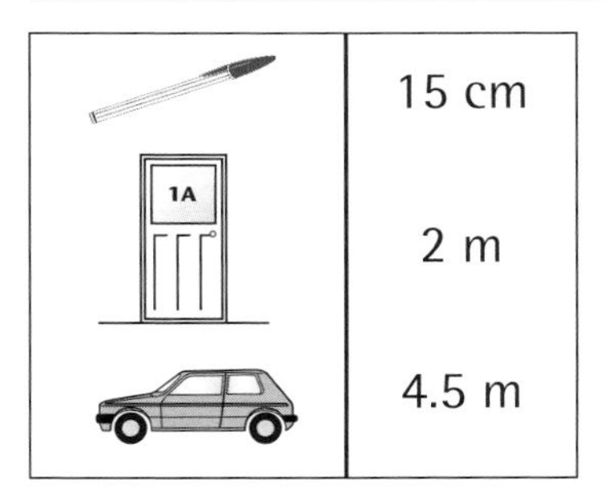

Estimating volume (capacity)

Volume or capacity is a measure of the amount of space inside a container, or the amount it can hold. You can measure it in these units.

The capacity of a bath is the volume of water it can hold.

litres (ℓ), centilitres (cℓ), millilitres (mℓ)

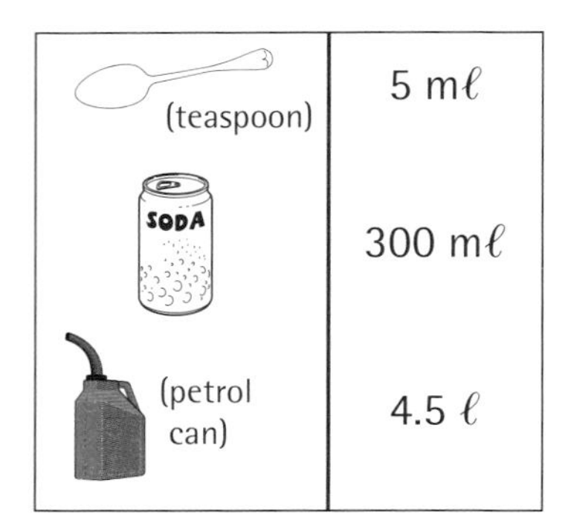

Estimating mass

You can measure mass in these units.

tonnes (t), kilograms (kg), grams (g), milligrams (mg)

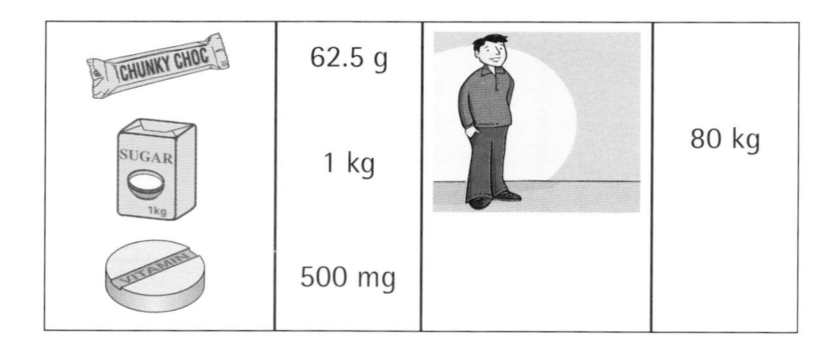

EXERCISE 8A

For each statement, give the most appropriate units of measurement.

1 The mass of your chair.

2 The distance from the door to the window in your classroom.

3 The capacity of a teaspoon.
4 The mass of this textbook.
5 The length of this textbook.
6 The mass of a Jumbo Jet.
7 The distance from Paris to London.
8 The amount of water in a swimming pool.

8.2 Reading scales

You need to be able to read different types of scales. Some of the most common are

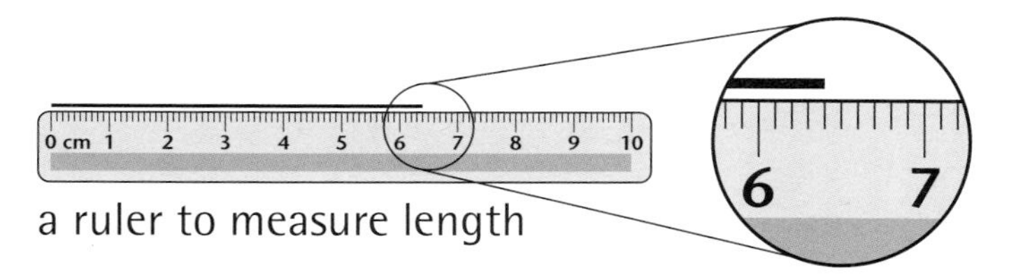

a ruler to measure length

There are 10 spaces between the 6 and the 7.
10 spaces = 1 cm.
1 space = 1 ÷ 10 = 0.1 cm.
The line shown is 6.4 cm long.

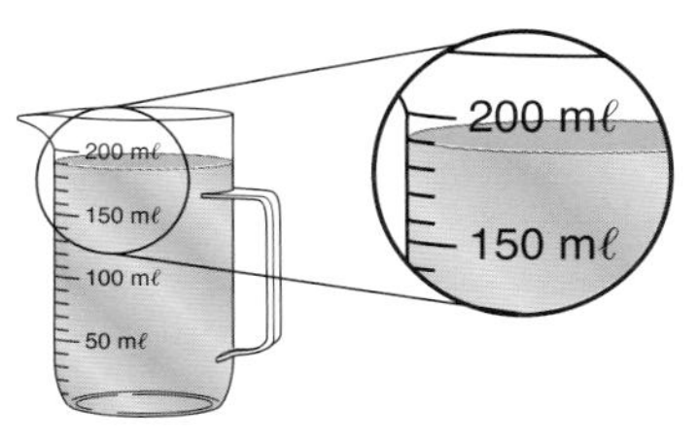

a measuring jug to measure capacity

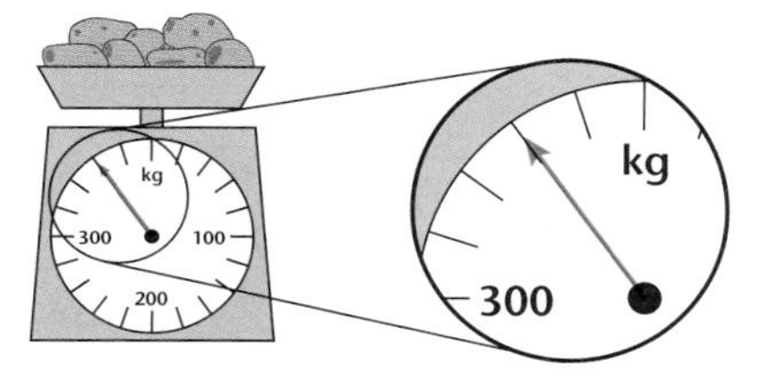

kitchen scales to weigh ingredients.

There are 5 spaces between the 150 and the 200.
5 spaces = 50 mℓ.
1 space = 50 ÷ 5 = 10 mℓ.
The scale shows a reading of 190 mℓ.

There are 5 spaces between the 300 and the 400.
5 spaces = 100 g.
1 space = 100 ÷ 5 = 20 g.
The scale shows a reading of 360 g.

To read a scale, you need to look at how it is marked.

Does it go up in 1s, 10s, 50s, 100s or something else?

How many spaces are there between the numbers?
Work out what one space represents.

EXERCISE 8B

1 What is the reading on the following scales?

(a)

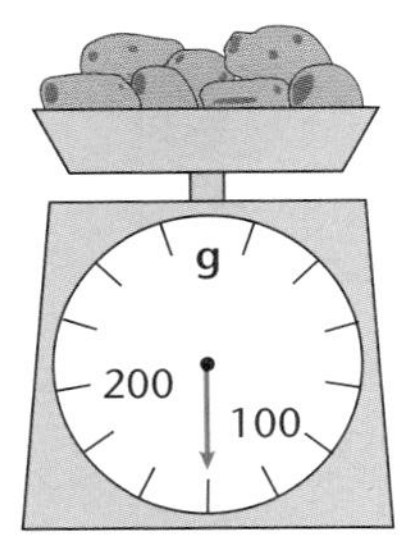

(b)

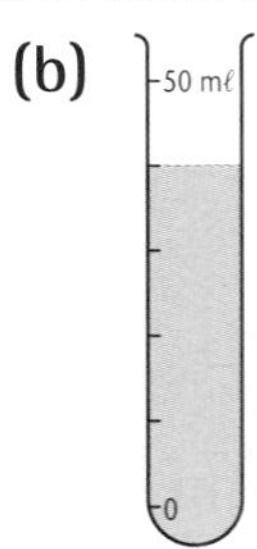

(c)

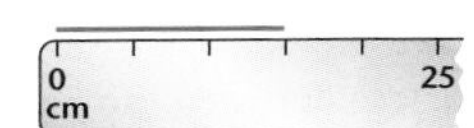

(d)

(e)

(f)

(g)

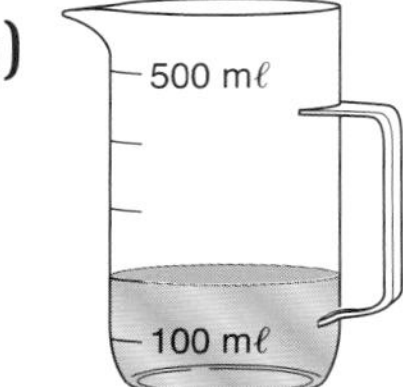

(h)

2 Copy the scales used in question 1 and mark on them

(a) 180 grams (b) 20 m, (c) 20 cm

(d) 5.75 m (e) 54 km/h (f) 62 kg

(g) 375 mℓ (h) $\frac{3}{4}$ full

Some scales have very few markings on them and you need to estimate a reading.

This speedometer scale is marked in tens. There are 2 spaces between each marked value, so each space represents 10 ÷ 2 = 5 km/h.

The pointer lies between 35 and 40. It is slightly nearer to 35 than to 40. A sensible estimate is 37 km/h.

EXERCISE 8C

1 Make a sensible estimate for each measurement shown.

(a) 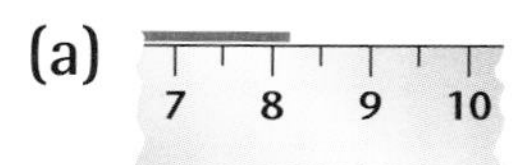

(b) 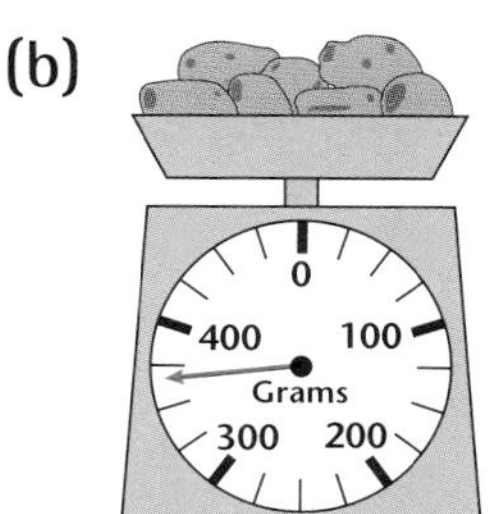

(c) 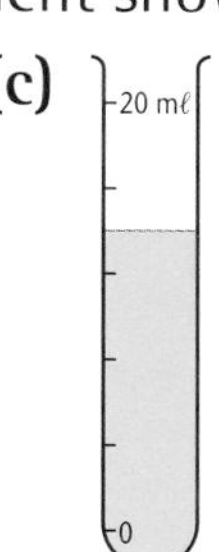

(d) 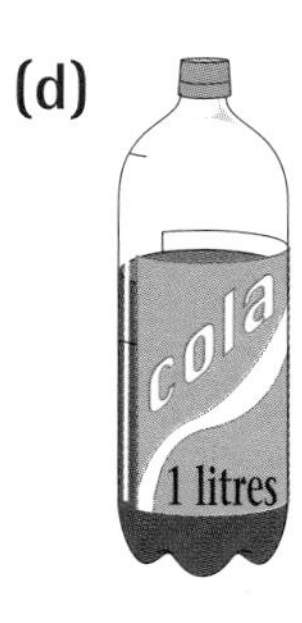

(e) 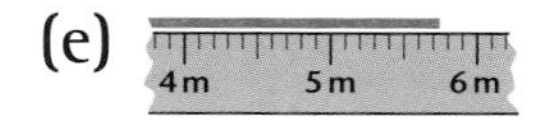

(f) 

(g) 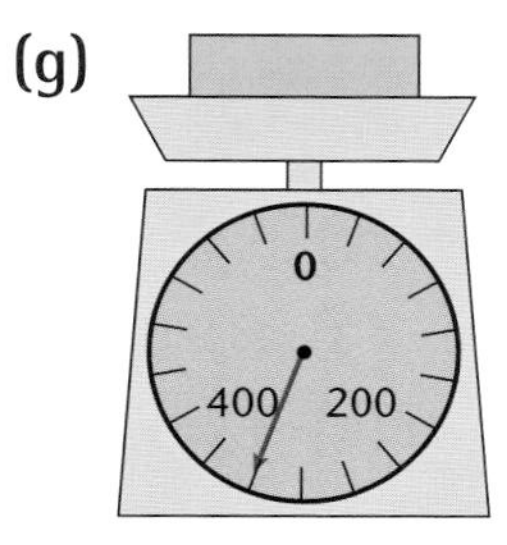

(h) 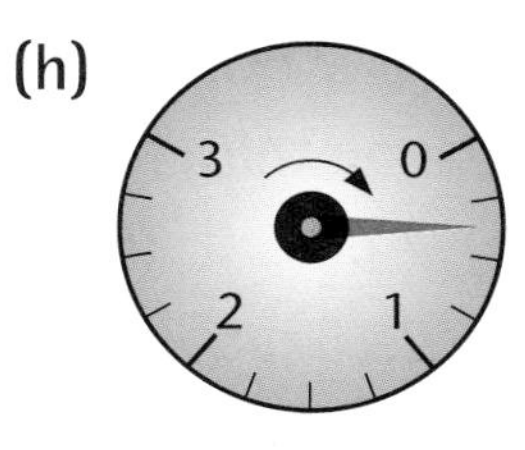

8.3 Time, dates and timetables

You will need to know

- how to write times in the 24-hour clock system
- how to write times in am or pm

12-hour and 24-hour clock

There are two ways of measuring time

- using **am** for morning and **pm** for afternoon
- using the **24-hour clock**

The '12-hour clock' means time in am and pm.

Most digital watches use the 24-hour clock.

am and pm	24-hour clock
12 midnight	0000 or 2400
7.15 am	0715
10 am	1000
12 noon (midday)	1200
3.40 pm	1540
9.55 pm	2155

0715: You say, 'quarter past seven in the morning'

1540: 'twenty to four in the afternoon'

2155: 'five to ten in the evening'

EXERCISE 8D

1 Write these times using **(i)** am and pm (the 12-hour clock) **(ii)** the 24-hour clock.

(a) twenty past six in the morning
(b) quarter to eleven in the morning
(c) five past three in the afternoon
(d) ten minutes to seven in the evening
(e) half past ten in the evening

2 Write these times as you would say them.

(a) 9.15 pm **(b)** 7.40 am **(c)** 1200 **(d)** 1430 **(e)** 0000

3 Change these times into 24-hour clock times.

(a) 7 pm	**(b)** 6 am	**(c)** 8.30 am	**(d)** 12.45 pm
(e) 2.15 pm	**(f)** 2.15 am	**(g)** 11.23 pm	**(h)** 4.55 pm
(i) 11.47 am	**(j)** 12.15 am	**(k)** 9.45 pm	**(l)** 2.20 am

4 Write these 24-hour clock times as am or pm times.

(a) 1100	**(b)** 1300	**(c)** 0600	**(d)** 1630
(e) 0920	**(f)** 0030	**(g)** 2315	**(h)** 1005
(i) 1545	**(j)** 0420	**(k)** 1822	**(l)** 2400

Calculating time

Take care when using a calculator to calculate times. You may fall into the trap of thinking that there are 100 minutes in *1 hour* instead of *60 minutes*.

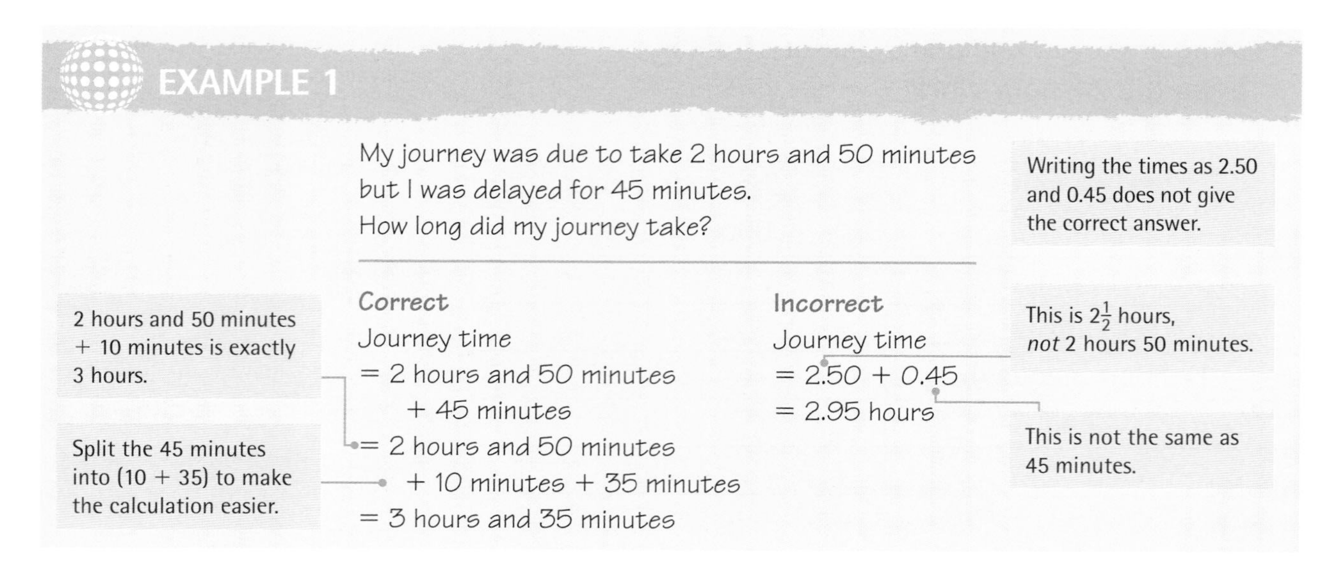

EXAMPLE 2

My plane leaves Changi airport, Singapore at 0723.
I have to be there $1\frac{1}{2}$ hours beforehand.
What time is this?

$\frac{1}{2}$ hour = 60 ÷ 2 = 30 mins.

30 − 23 = 7

Using a calculator to work out 7.23 − 1.5 gives a ridiculous answer!

Correct

1 hour before 0723 is 0623.

I need to be 30 minutes earlier than this.

23 minutes earlier is 0600.

So I must be there 7 minutes before 0600 which is 0553.

Incorrect

Arrival time = 7.23 − 1.5
= 5.73

I need to be there by 0573.

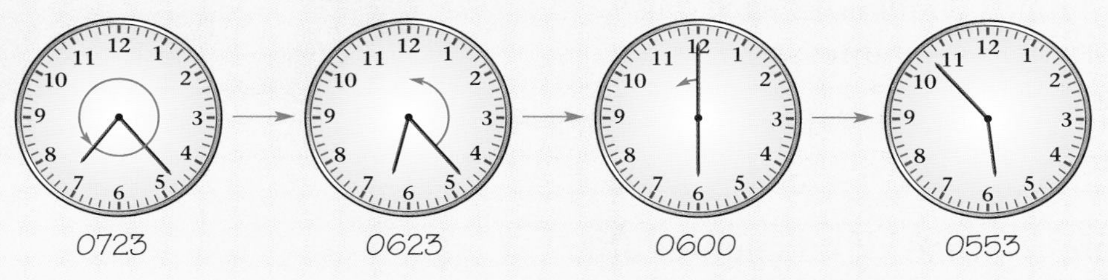

To calculate times you need to remember

60 seconds	=	1 minute
60 minutes	=	1 hour
24 hours	=	1 day
7 days	=	1 week
365 days	=	1 year
366 days	=	1 leap year

In some calculations, you might use 52 weeks = 1 year.
This is not quite true because
52 weeks = 52 × 7 days
= 364 days!

EXAMPLE 3

(a) Write 3 hours 49 minutes in minutes.
(b) Write 438 minutes in hours and minutes.

(a) 3 hours = 3 × 60 minutes = 180 minutes
So 3 hours 49 minutes = 180 + 49 minutes
= 229 minutes

(b) Try some numbers of hours
5 × 60 = 300   6 × 60 = 360
7 × 60 = 420   8 × 60 = 480

438 comes between 420 and 480 so it is more than 7 hours.
438 − 420 = 18, so 438 minutes = 7 hours 18 minutes

If you use your calculator to work out 438 ÷ 60 you will get an answer of 7.3.
What will you write as your answer?
7 hours 3 minutes?
7 hours 30 minutes?
Neither of these is correct.

If you do use your calculator to get 7.3 you can get the correct answer if you then do this

- The '7' is the whole number of hours.
- Work out 0.3×60 to change 0.3 hours into minutes.
- This gives 18 minutes.
- So 438 minutes = 7 hours 18 minutes.

Multiply by 60 to change hours into minutes.

EXERCISE 8E

1 The bus normally takes 1 hour and 52 minutes to travel from Barcelona to Tarragona. It was held up for 13 minutes in a traffic jam. How long did the journey take?

2 I usually work for 4 hours 30 minutes each day. On Fridays I work 1 hour 40 minutes less. How long do I work on a Friday?

3 Copy and complete the table.

55 minutes before	Time	45 minutes after
	0920	
	1135	
	1317	
	1648	
	2121	
	2352	

4 My journey to work takes 35 minutes by bus, plus 8 minutes' walk. I want to be at work by 8.25 am. What time should I catch the bus?

5 Write the following in the units given.

(a) 3 weeks in days
(b) 2 years in days (not leap years)
(c) 4 hours in seconds
(d) 1 week in hours
(e) 3 hours and 20 minutes in minutes
(f) $2\frac{1}{2}$ minutes in seconds
(g) 250 minutes in hours and minutes
(h) 343 minutes in hours and minutes
(i) 91 days in weeks
(j) 1 leap year in minutes

Handling dates

To work out dates, you need to know how many days there are in each month. This rhyme will help you to remember.

30 days have September,
April, June and November.
All the rest have 31
except in February alone
which has but 28 days clear
and 29 in each leap year.

EXAMPLE 4

Xavier is going on holiday on 5th August.
He is counting the days until he goes.
Today is 12th April. He counts this as 'day 1'.
How many days are there until he goes away?

April	M	T	W	T	F	S	S	
						1	2	
	3	4	5	6	7	8	9	
	10	11	12	13	14	15	16	There are 19 days left in April
	17	18	19	20	21	22	23	
	24	25	26	27	28	29	30	

May 31 days   June 30 days   July 31 days
August 5 days to the 5th
Total = 19 + 31 + 30 + 31 + 5 = 116 days

EXERCISE 8F

1 Use the calendar for April in Example 4 to answer the following.

(a) Today is Thursday the 6th April. What will be the date two weeks from today?

(b) If it is Wednesday the 5th today, what will be the date a week on Friday?

(c) My last day at work is the 20th April. I have 10 days' holiday. What is the date of the last day of my holidays?

(d) Today is Sunday 2nd April. My driving lesson is three weeks on Saturday. What date is this?

2 How many days are there between

(a) 25th November and the 3rd December?

(b) 19th March and the 8th April?

(c) 8th June and the 17th August?

(d) 20th July and the 16th October?

(e) 25th September and Christmas Day?

Include both start and end dates.

3 I finish my exams on 26th June. I do not start college until 11th September. How many days of holiday do I have?

4 I was born on 7th December 1995. My brother is 1 year 3 months and 4 days younger than me. When is my brother's birthday?

A calendar will help.

Timetables

Bus and train timetables help you plan a journey.
They list the places where the bus or train stops and the departure time for each of these places.
Times are usually shown as 24-hour clock times.

This train timetable shows a train timetable from Avignon to Antibes.

This train leaves Avignon at 1130.

Avignon	1103	1130	1157	1233	1259
Aix en Provence		1143		1246	
Marseille	1131	1202		1305	
Toulon	1202	1231	1250	1335	1354
St Raphael		1255		1401	
Cannes	1310	1323	1359	1450	1501
Antibes	1412	1443	1455	1546	1559

This is the 1401 from St Raphael.

The 1401 from St Raphael arrives in Antibes at 1546.

The times for any one train are written in a *column*. You read the time the train leaves each place in the *row* opposite the place name.

The 1130 train from Avignon stops at all the stations listed. It leaves Marseille at 1202 and Cannes at 1323. It arrives in Antibes at 1443.

The journey starts at 1130 and ends at 1443, this is a total time of 3 hours and 13 minutes.

1130 to 1430 is 3 hours.

If you live in St Raphael and want to go to Antibes you can either catch the 1255 train or the 1401 train. These are the only two trains that stop at St Raphael.

If you catch the 1401 train and arrive in Antibes at 1546 the journey time is 1 hour and 45 minutes.

EXAMPLE 5

Use the train timetable for Avignon to Antibes to answer these questions.

(a) Michel arrives at the train station at Toulon at 1 pm.
- **(i)** What time is the next train he could catch to Antibes?
- **(ii)** What is the earliest time he could arrive in Antibes?

(b) Which train from Avignon has the shortest journey time to Antibes?

(a) (i) 1 pm = 1300
The next train to Antibes after 1300 is the 1335 train.

Convert to 24-hour clock time.

(ii) The earliest time he could be in Antibes is 1546.

This is if he catches the first possible train (the 1335).

(b) The trains that stop the least number of times will take the shortest time.
These are the 1157 and 1259 trains.
The 1157 train gets in at 1455. The journey time is 2 minutes less than 3 hours, which is 2 hours and 58 minutes. The 1259 train gets in at 1559. The journey time is exactly 3 hours.
The 1157 train has the shortest journey time.

EXERCISE 8G

Use the train timetable for Avignon to Antibes for questions 1 and 2.

1 (a) How many trains on this timetable stop at Aix en Provence?
(b) What time does the 1103 from Avignon leave Cannes?
(c) What time does the 1305 from Marseille leave Avignon?
(d) What time will the 1250 from Toulon arrive in Antibes?
(e) How long does it take the 1131 from Marseille to get to Antibes?

2 Michelle has a job interview in Cannes at 3 pm.
(a) What is the last train from Avignon she can catch to get to Cannes before 3 pm?
(b) How long does this train take from Avignon to Cannes?
(c) It takes 15 minutes to get from Cannes station to the interview. Which train should Michelle catch? Give a reason for your answer.

3 This is a bus timetable from Chester to Ramsholt.

Chester	0845	0914	0934	0954		1614	1634	1654	1714	1739
Barton	0900	0920	0940	1000		1620	1640	1700	1720	1745
Holm	0911	0931	0951	1011	at	1631	1651	1711	1731	1756
Whitby	0920	0940	1000	1020	every	1640	1700	1720	1740	1805
Ellesmere (arr)	0926	0946	1006	1026	20	1646	1706	1726	1746	1811
Ellesmere (dep)	0929	0949	1009	1029	mins.	1649	1709	1729	1749	1814
Pooltown	0934	0954	1014	1034	until	1654	1714	1734	1754	1819
Overpool	0938	0958	1018	1038		1658	1718	1738	1758	1823
Ramsholt	0940	1000	1020	1040		1700	1720	1740	1800	1825

(a) What time does the 0914 from Chester arrive at Ramsholt?

(b) What time does the 1700 from Barton arrive in Overpool?

(c) How long do these journeys take
   (i) from Chester to Ramsholt (if you leave after 0900)
   (ii) from Whitby to Pooltown?

(d) I arrive in Ellesmere at 1646. What time did I catch the bus in Holm?

(e) In Chester, what time is the next bus after the 0954?

(f) How long does the bus stop in Ellesmere?

4 Use the bus timetable in question 3 for this question.
I live in Chester and I am meeting my friend in Barton at 4.45 pm.

(a) What is the latest bus I could catch?

We are going to the cinema in Ramsholt. The film starts at 6.15 pm.

(b) How long will the journey take?

(c) What is the time of the last bus we can catch?

(d) Another friend is getting on the bus at Ellesmere to travel with us. What time would he catch the bus?

(e) How long do we have to get from the bus to the cinema if the bus is on time?

8.4 Converting units

You will need to know

- how to multiply and divide whole numbers and decimals by 10, 100 and 1000.

Converting length, mass and capacity

You need to learn these conversions.

Units of length
10 mm = 1 cm   100 cm = 1 m
1000 mm = 1 m   1000 m = 1 km

1 cm^3 is shown by this cube whose sides are all 1 cm.

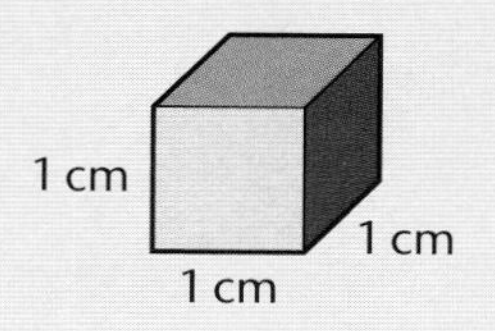

Units of capacity
100 cℓ = 1 ℓ   1000 mℓ = 1 ℓ

1 mℓ = 1 cm^3 = 1 cc.

Units of mass
1000 mg = 1 g   1000 g = 1 kg   1000 kg = 1 tonne

cm^3 and cc both stand for cubic centimetres.

For the units of length

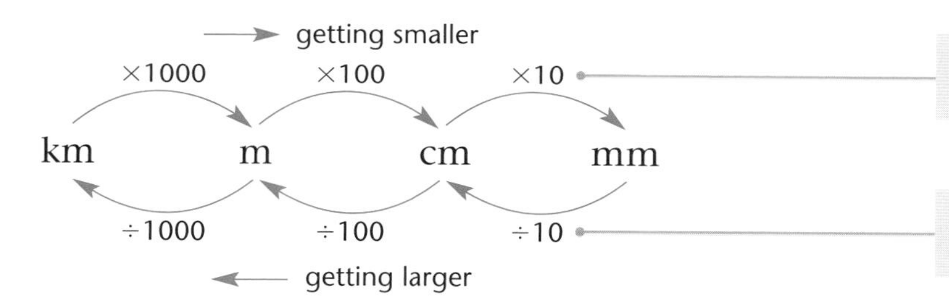

To change from larger units to smaller units you *multiply*.

To change from smaller units to larger units you *divide*.

EXAMPLE 6

(a) How many metres are there in 4 kilometres?
(b) How many millimetres are there in 3 metres?
(c) Change 160 millimetres into centimetres.
(d) Change 8 kilometres into centimetres.

(a) 4 km = 4 × 1000 m = 4000 m
(b) 3 m = 3 × 100 cm = 300 cm
300 cm = 300 × 10 mm = 3000 mm
(c) 160 mm = 160 ÷ 10 cm = 16 cm
(d) 8 km = 8 × 1000 m = 8000 m
8000 m = 8000 × 100 cm = 800 000 cm

Move digits to the left to multiply.

Move digits to the right to divide.

See Section 1.5 for a reminder of the rules.

You can use the same ideas to convert metric units of capacity and mass.

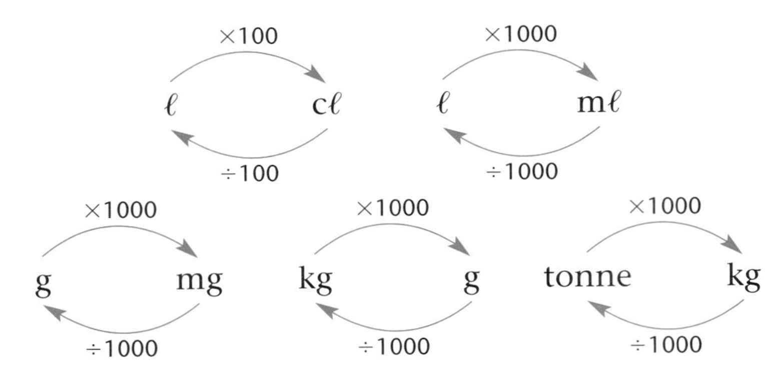

EXAMPLE 7

Change these volumes to litres.

(a) 4000 mℓ (b) 72 000 cℓ

(a) 4000 mℓ = 4000 ÷ 1000 ℓ = 4 ℓ

(b) 72000 cℓ = 72000 ÷ 100 ℓ = 720 ℓ

EXAMPLE 8

Write these masses in kilograms.

(a) 3 tonnes (b) 83 000 g

(a) 3 tonnes = 3 × 1000 kg = 3000 kg

(b) 83 000 g = 83 000 ÷ 1000 kg = 83 kg

A tonne is more than a kilogram. To change tonnes to kg you multiply.

EXERCISE 8H

1 (a) How many centimetres are there in 6 metres?

(b) How many metres are there in 8 kilometres?

2 Change these lengths into the units given.

(a) 3 metres → millimetres

(b) 5 metres → centimetres

(c) 12 cm → mm

(d) 4000 m → km

(e) 6000 mm → m

(f) 300 mm → cm

3 Change these masses into the units given.

(a) 5 tonnes → kilograms
(b) 8000 grams → kg
(c) 60 kg → grams
(d) 16 000 kg → tonnes

4 Change these capacities into the units given.

(a) 7000 mℓ → ℓ   (b) 800 cℓ → ℓ
(c) 30 cℓ → mℓ   (d) 5ℓ → cℓ

5 Find the missing values.

(a) 7000 mm = ___ m   (b) 50 mℓ = ___ cℓ
(c) 4 kg = ___ g   (d) 15 km = ___ m
(e) 5000 kg = ___ t   (f) 2000 mℓ = ___ ℓ

6 A small box of chocolates weighs 0.4 kg. I have 3 boxes. How many grams of chocolates do I have altogether?

7 How many 150 mℓ glasses can be filled from a 3-litre bottle?

Converting and comparing measurements

To compare measurements, they must be in the same units.

EXAMPLE 9

Change these lengths to centimetres.

(a) 43 mm   (b) 7.8 m   (c) 0.75 km
(d) 0.645 m   (e) 136 mm   (f) 9 mm

Which is the longest length? Which is the shortest?

(a) 43 mm = 43 ÷ 10 cm = 4.3 cm
(b) 7.8 m = 7.8 × 100 cm = 780 cm
(c) 0.75 km = 0.75 × 1000 m = 750 m
and 750 m = 750 × 100 cm = 75 000 cm
(d) 0.645 m = 0.645 × 100 cm = 64.5 cm
(e) 136 mm = 136 ÷ 10 cm = 13.6 cm
(f) 9 mm = 9 ÷ 10 cm = 0.9 cm

Continued ▼.

Look at Section 1.5 if you need help with × or ÷ by 10, 100, 1000.

The lengths are 4.3 cm 780 cm 75 000 cm 64.5 cm 13.6 cm 0.9 cm

The longest is 75 000 cm or 0.75 km.

The shortest is 0.9 cm or 9 mm.

For your answer, write the lengths with the units from the question.

EXAMPLE 10

Put these masses in order, smallest first.

(a) 950 g (b) 0.003 tonne (c) 2250 g (d) 2.16 kg

(a) 950 g

(b) 0.003 tonne = 0.003×1000 kg = 3 kg
and 3 kg = 3×1000 g = 3000 g

(c) 2250 g

(d) 2.16 kg = 2.16×1000 g = 2160 g

So $950\text{ g} < 2.16\text{ kg} < 2250\text{ g} < 0.003\text{ tonne}$

You can only compare if they are all written in the same units.
Always change them to the *smallest* unit.

$950\text{ g} < 2160\text{ g} < 2250\text{ g} < 3000\text{ g}$
Write the measurements in the units from the question.

EXERCISE 8I

1 Change these lengths to metres.

(a) 320 cm (b) 4500 mm (c) 4.2 km
(d) 0.485 km (e) 87 cm (f) 750 mm

Which is the longest length? Which is the shortest?

2 Change these masses to kilograms.

(a) 5640 g (b) 2.6 t (c) 800 g

Which is the smallest?

3 Change these capacities to centilitres.

(a) 163 mℓ (b) 9.3 ℓ (c) 0.7 ℓ

Which is the largest?

4 Put these lengths in order from smallest to largest.

450 cm 0.05 km 4620 mm 5.1 m

5 Two lengths of timber are the same price. One has a label saying 2450 mm. The other says 2.4 m. Which one is the best buy?

6 Place these bottles in order from the smallest volume to the largest volume of foam bath.

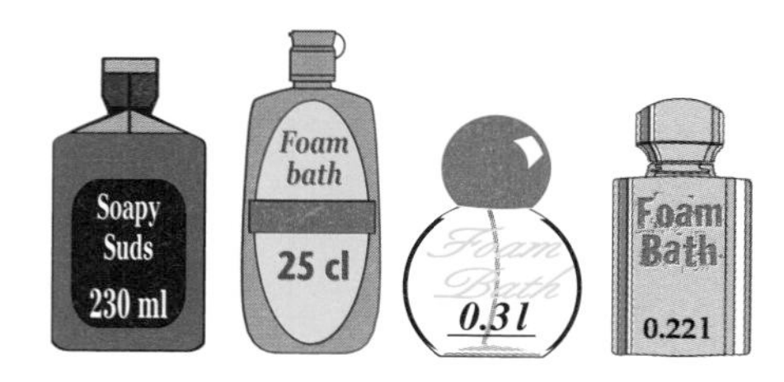

8.5 Converting areas and volumes

These two squares are the same size.

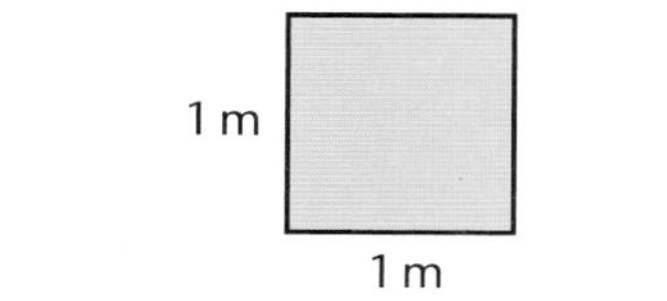

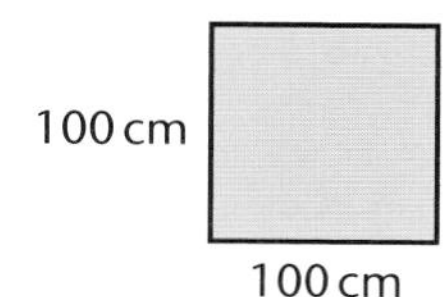

Area = 1 m × 1 m
Area = 1 m²

Area = 100 cm × 100 cm
Area = 10 000 cm²

Area is measured in **square** units cm², m² or km².

$1\ m^2 = 10\,000\ cm^2$

The most common mistake is to think that because there are 100 cm in 1 m there must be 100 cm² in 1 m². You can see that the correct answer is 10 000 cm² = 1 m².

In the same way $1\ km = 1000\ m$

so $1\ km^2 = 1000\ m \times 1000\ m$

$1\ km^2 = 1\,000\,000\ m^2$

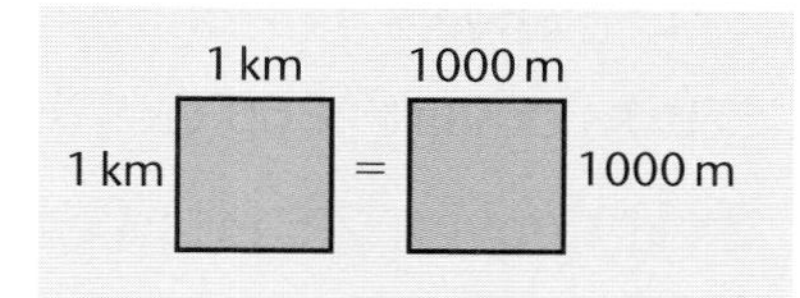

EXAMPLE 11

(a) Convert 20 m² to cm². (b) Convert 15 000 cm² to m².

(a) $20\ m^2 = 20 \times 10\,000\ cm^2 = 200\,000\ cm^2$

(b) $15\,000\ cm^2 = 15\,000 \div 10\,000\ m^2 = 1.5\ m^2$

You can use the same ideas for volumes.

> To change from larger to smaller units you *multiply*.

> Smaller unit → larger unit → *divide*.

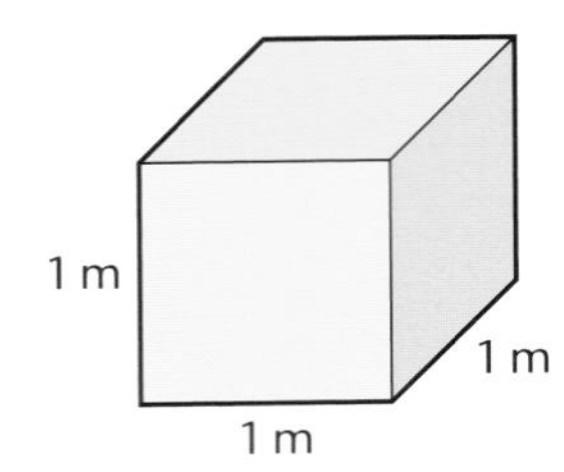

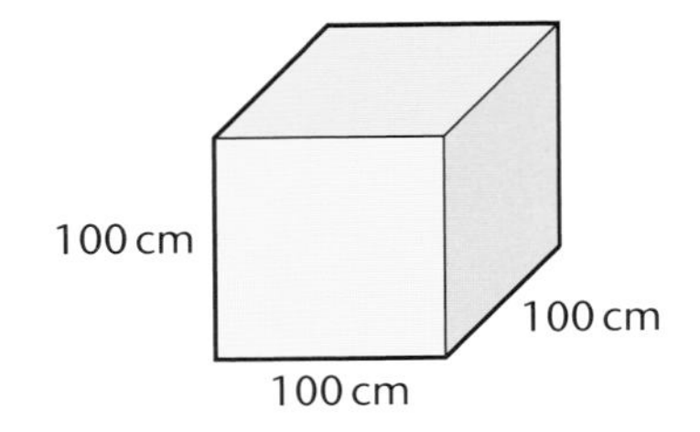

> 1 m = 100 cm.

> The **cubes** are the same size. They have the same volume.

Volume
$= 1\text{ m} \times 1\text{ m} \times 1\text{ m}$
$= 1\text{ m}^3$

Volume
$= 100\text{ cm} \times 100\text{ cm} \times 100\text{ cm}$
$= 1\,000\,000\text{ cm}^3$

> Volume is measured in **cubic** units cm^3 or m^3.

$1\text{ m}^3 = 1\,000\,000\text{ cm}^3$

> A common mistake is to think that $1\text{ m}^3 = 100\text{ cm}^3$. The correct answer gives one million cm^3 in 1m^3.

EXAMPLE 12

(a) Convert 3 m^3 to cm^3. (b) Convert 4000 cm^3 to m^3.

(a) $3\text{ m}^3 = 3 \times 1\,000\,000\text{ cm}^3 = 3\,000\,000\text{ cm}^3$

(b) $4000\text{ cm}^3 = 4000 \div 100\,000\text{ m}^2 = 0.004\text{ m}^3$

EXERCISE 8J

1 (a) Convert 4 m^2 to cm^2.
(b) Convert 1900 cm^2 to m^2.

2 (a) Convert 2.5 m^3 to cm^3.
(b) Convert $300\,000\text{ cm}^3$ to m^3.

3 Copy and complete these conversions.
(a) $3\text{ m}^2 =$ ____ cm^2
(b) $5\text{ m}^3 =$ ____ cm^3
(c) $70\,000\text{ cm}^2 =$ ____ m^2
(d) $\frac{1}{2}\text{ m}^3 =$ ____ cm^3
(e) $\frac{1}{2}\text{ m}^2 =$ ____ cm^2
(f) $3\,000\,000\text{ cm}^3 =$ ____ m^3

4 How many
(a) mm^2 in 1 cm^2 (b) cm^2 in 1 km^2?

> Draw squares to help you.

5 How many
(a) m^3 in 1 km^3 (b) mm^3 in 1 m^3?

6 A lake covers an area of 3.25 million square metres. What is its area in square kilometres?

7 A rectangular room measures 420 cm by 380 cm.

(a) What is the area of the floor in cm^2?

(b) What does the room measure in square metres?

8 A rectangular tank for a poisonous spider measures 85 cm by 65 cm by 45 cm.

(a) What is the volume of the container?

(b) This type of spider needs at least $\frac{1}{4}$ m^3 of space. Is this container suitable?

8.6 Accuracy of measurement

There are two kinds of measurement, discrete and continuous.

Discrete measure is for quantities that can be counted or only have certain values. For example, the number of spectators at a tennis final was 12 416.

Continuous measure can take any value in a range. The accuracy of the measure depends on the measuring instrument. For example, if you measure your height as 174 cm, it will not be exactly 174 cm.

You will meet these again in Chapter 10.

It is likely to be 174 cm **to the nearest** centimetre.

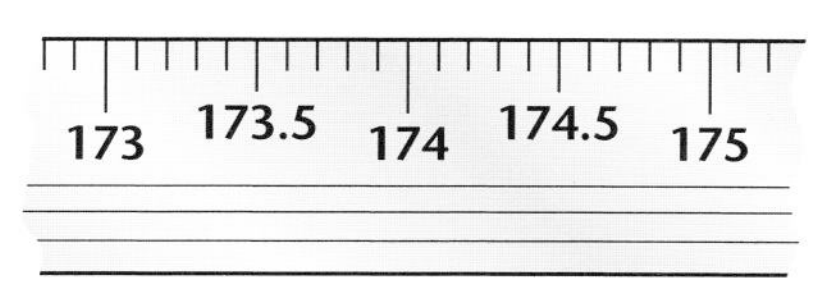

Think what this really means.

The true height lies between 173.5 cm and 174.5 cm.

This is written as 173.5 cm $\leqslant$ height $<$ 174.5 cm.

Notice that the $\leqslant$ sign is used for the 'smaller' value but the $<$ sign is used for the 'larger' value.

If the value was 174.5 cm, it would round up to 175 cm to the nearest cm.

In the same way, if your mass is 68 kg to the nearest kilogram, it must be nearer to 68 kg than to either 67 kg or 69 kg.

Your true mass is
67.5 kg $\leqslant$ mass $<$ 68.5 kg.

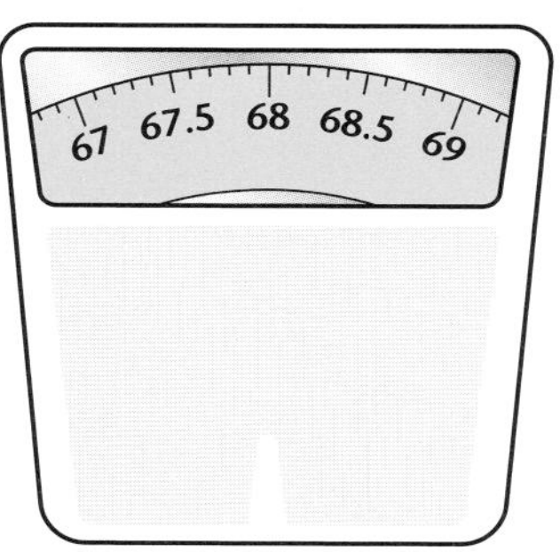

Any measurement given to the nearest whole unit may be inaccurate by up to one half in either direction.

EXAMPLE 13

A rectangle has sides of 8 cm and 5 cm, each measured to the nearest centimetre.

Write the range within which each measurement must lie.

8 cm means a length in the range $7.5 \text{ cm} \leqslant \text{length} < 8.5 \text{ cm}$
5 cm means a length in the range $4.5 \text{ cm} \leqslant \text{width} < 5.5 \text{ cm}$

EXERCISE 8K

1 These have been measured to the nearest whole unit. Write the range within which each measurement must lie.

(a) 9 cm (b) 12 kg (c) 65 cℓ (d) 10 seconds

For example, $4.5 \text{ cm} \leqslant 5 \text{ cm} < 5.5\text{cm}$.

2 These have been measured to the nearest millimetre. Write the range within which each measurement must lie.

(a) 14 mm (b) 1.4 cm (c) 3.8 cm (d) 7.5 cm

Change them all to mm.

3 These have been measured to the nearest centimetre. Write the range within which each measurement must lie.

(a) 125 cm (b) 1.25 m (c) 4.07 m (d) 6.20 m

What must you remember to do first?

4 For each question, state whether the measurement is discrete or continuous.

(a) The height of a door
(b) The number of sweets in a bag
(c) The mass of a cake
(d) The cost of pairs of shoes
(e) The capacity of a glass

5 A rectangular box is 17 cm long by 11 cm wide by 5 cm high to the nearest centimetre. Write down the range within which each length must lie.

8.7 Compound measures

Speed

You will need to know

- how to convert times between 'hours' and 'hours and minutes'

Speed is a measurement of how fast something is travelling.

It involves two other measures, **distance** and **time**. If you travel from Hamburg to Kiel, a distance of 100 kilometres, and it takes 2 hours, you have averaged 50 kilometres per hour.

Your speed would not be constant for the whole journey so the speed calculated is an *average* speed.

To calculate a speed you divide a distance by a time.

100 kilometres in 2 hours
$\frac{100}{2} = 50$ kilometres per hour.

There are three formulae connecting speed, distance and time.

This triangle will help you to remember the formulae.

$$\text{speed} = \frac{\text{distance}}{\text{time}} \qquad \text{time} = \frac{\text{distance}}{\text{speed}}$$

$$\text{distance} = \text{speed} \times \text{time}$$

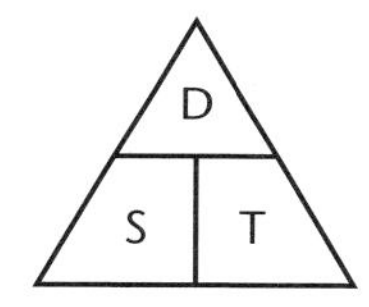

If you cover up the one you want to find you are left with the formula you need.

For example, cover up 'T', you are left with $\frac{D}{S}$.

which means that $T = \frac{D}{S}$ or $\text{time} = \frac{\text{distance}}{\text{speed}}$.

Speed can be measured in many different units.

The most common are

metres per second (m/s), kilometres per hour (km/h) and centimetres per second (cm/s)

The word 'per' means divide. Kilometres per hour (km/h) means you divide a distance in km by a time in hours.

When you do calculations involving speed, distance and time you must make sure that you are consistent with the units that you use.

For example,to calculate a speed in metres per second you need the distance in metres and the time in seconds.

EXAMPLE 14

Dalila went on her holidays and travelled a distance of 180 kilometres. The journey took $4\frac{1}{2}$ hours.
What was her average speed for the journey?

Distance in kilometres divided by time in hours gives speed in kilometres per hour.

$$\text{Average speed} = \frac{\text{total distance travelled}}{\text{total time taken}}$$

$$= \frac{180 \text{ kilometres}}{4.5 \text{ hours}}$$

$$= 40 \text{ km/h}$$

EXAMPLE 15

Simon averaged a speed of 80 km/h for a journey of 208 km.
How long did the journey take?

Distance in km divided by speed in km/h gives time in hours.

$$\text{Time} = \frac{\text{distance}}{\text{speed}} = \frac{208}{80} = 2.6$$

You must *not* read this as 2 hours 6 minutes.
It is 2.6 hours.

0.6 hours = 0.6 × 60 minutes = 36 minutes

Time taken for journey = 2 hours 36 minutes

To convert the decimal part to minutes you multiply by 60.
Look back at Example 5 to help you.

EXAMPLE 16

Ian drove for 2 hours 45 minutes at an average speed of 48 km/h. How far did he travel?

2 hours 45 minutes = $2\frac{3}{4}$ hours = 2.75 hours

2 hours 45 needs to be written in hours only.
45 minutes is $\frac{3}{4}$ or 0.75 of an hour, so the time is 2.75 hours.

Distance = speed × time
= 48 × 2.75
= 132

Distance = 132 kilometres

To change minutes to hours, divide by 60

$\frac{45}{60} = \frac{3}{4} = 0.75$

EXERCISE 8L

Remember to state the units in your answers.

1 I cycled 60 kilometres in 4 hours. What was my average speed?

2 Ferdinand ran the last 400 metres to school in 80 seconds. What was his average speed?

3 It took me 4 hours to travel to Cordoba at an average speed of 65 km/h. How far did I travel?

4 It is 14 kilometres along the Tree Top Walkway. I walk at an average speed of 4 km/h. How long will it take me to complete the walk?

5 I left home at 9 am to drive to Queenstown 150 kilometres away. I arrived there at 1 pm. What was my average speed?

6 A long-distance truck drove 320 kilometres in 5 hours 45 minutes. What was its average speed? Give your answer correct to 1 d.p.

7 How long would it take to travel 270 kilometres at a speed of 50 km/h?

8 I cycled for 3 hours at a speed of 12 km/h. I then walked for 2 hours at a speed of 3 km/h. How far did I travel altogether?

9 During a race Ali ran the first 300 metres in 2 minutes 10 seconds and the last 100 metres in 30 seconds. What was his average speed in m/s?

Look at Example 16 to help you.

10 The journey to a conference was in three parts.

- 50 kilometre drive to the airport took 1 hour 20 minutes.
- 1200 kilometre flight took 2 hours.
- 42 kilometre bus trip from the airport to the conference venue took 55 minutes.

(a) What was the total travelling time?

(b) What was the total distance travelled?

(c) What was the average speed for the whole journey?

Density

Density is defined as '**mass** per unit **volume**'.

It is the mass (usually given in grams) of one unit of volume of material (usually given in cm^3).

This means that density is calculated by dividing the mass of an object by its volume.

There are three formulae connecting density, mass and volume.

The word density is not in your syllabus. But mass per unit volume is included.

$$\text{density} = \frac{\text{mass}}{\text{volume}} \qquad \text{volume} = \frac{\text{mass}}{\text{density}}$$

$$\text{mass} = \text{density} \times \text{volume}$$

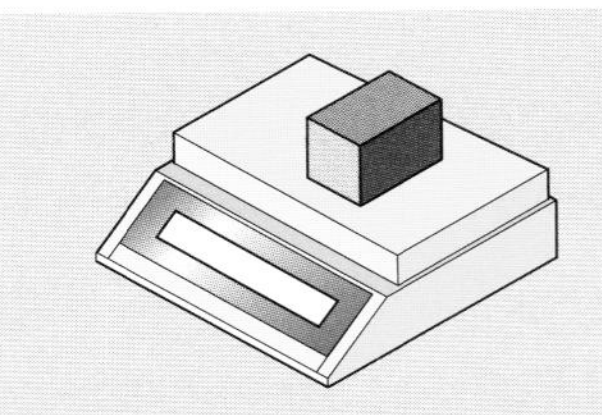

You can think of density as the mass of 1 cm^3.

This triangle will help you to remember the formulae. You use it like the one for speed, distance and time.

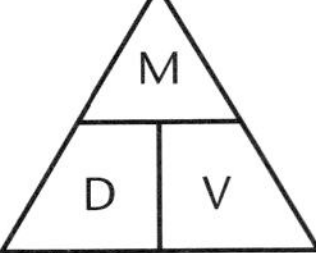

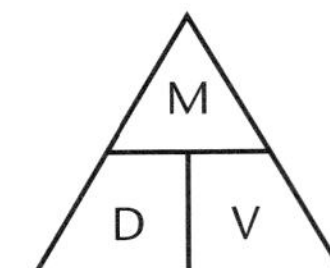

Cover the one you want to find. Read the formula you need. For example, to find density, cover D. Read $\frac{M}{V} = \frac{\text{mass}}{\text{volume}}$.

You need to be consistent with the units that you use.

- Density is usually in g/cm^3
- Volume is usually in cm^3
- Mass will then be calculated in g.

If mass is in kg and volume is in m^3, density will be in kg/m^3.

EXAMPLE 17

A piece of wood weighs 124 g and has a volume of 140 cm^3.
What is the density of the wood?

$$\text{Density} = \frac{\text{mass}}{\text{volume}} = \frac{124}{140} = 0.886 \text{ (3 s.f.)}$$

Density = 0.886 g/cm^3 (3 s.f.)

The terms 'mass' and 'weight' are not exactly the same but in this section we shall take them to be the same.

EXAMPLE 18

A measuring cylinder contains a liquid whose density is 1.18 g/cm^3. The volume of the liquid is 0.35 ℓ.
What is the mass of the liquid?

$0.35\ \ell = 0.35 \times 1000\ cm^3 = 350\ cm^3$

Mass = density × volume = $1.18 \times 350 = 413$

Mass = 413 g

Change the units of volume from litres to cm^3.
$1\ \ell = 1000\ cm^3$.

EXERCISE 8M

1 What is the density of a piece of material with a volume of 50 cm^3 and a mass of 225 g?

Remember to give your answer using the units g/cm^3.

2 Gold has a density of approximately 19 g/cm^3. What is the mass of a 250 cm^3 gold bar?

3 Flour has a density of 1.5 g/cm^3. What is the volume of a bag weighing 825 g?

4 Calculate the missing measurements in this table,

	Mass	Density	Volume
(a)	24 g	1.6 g/cm^3	
(b)		2.05 g/cm^3	40 cm^3
(c)	28 g		13.3 cm^3
(d)		3.2 g/cm^3	1.5 litres
(e)	5.7 kg	1.24 g/cm^3	

Change the units first in parts **(d)** and **(e)**.

5 (a) Which has the greater mass, a cubic metre of feathers or a cubic metre of steel?

(b) Which has the greater density, feathers or steel?

6 What is the volume of a concrete block whose mass is 50 kg and density is 1600 kg/m^3? Give your answer in m^3, correct to 2 decimal places.

7 A sculpture in a museum has a volume of about 3 m^3. The material from which it is made has a density of 11.4 g/cm^3.
What is the approximate mass of the sculpture?
Give your answer in kg.

EXAMINATION QUESTIONS

1 **(a)** A piece of rope is 70 m long, to the nearest metre.
Copy and complete the statement about the length of the rope.
Answer ....m $\leqslant$ length < ....m [2]
(b) Another piece of rope is 9 metres long , to the nearest 10 centimetres.
Write down the shortest possible length of this piece of rope. [1]

(CIE Paper 1, Jun 2000)

---

2 The distance between Zurich and New York is 6300 km.
(a) A plane takes $8\frac{1}{2}$ hours to fly from Zurich to New York.
(i) Calculate its average speed to the nearest 10 km. [2]
(ii) The plane leaves Zurich at 09.25.
What is the time in Zurich when it arrives in New York? [1]
(iii) The time in New York when the plane arrives is 11.55 on the same day.
What is the difference in time between Zurich and New York? [1]

(CIE Paper 3, Jun 2000)

---

3 The number of people watching a football match was 7200, correct to the nearest hundred.
Copy and complete the inequality.
Answer.....................$\leqslant$ number of people < .... [2]

(CIE Paper 2, Jun 2001)

4

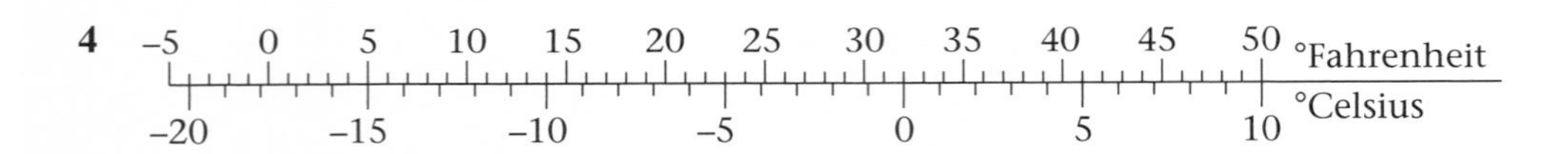

The diagram shows two temperature scales, Fahrenheit and Celsius, alongside each other.
(a) What temperature on the Celsius scale is equivalent to 0° on the Fahrenheit scale? [1]
(b) The temperature rises from $-15°$ Celsius to $10°$ Celsius.
How many degrees is this on the Celsius scale? [1]

(CIE Paper 1, Jun 2001)

5

OLIVER!

By LIONEL BART

Performances daily at 19.30.

FIRST PERFORMANCE – 25th APRIL
FINAL PERFORMANCE – 8th MAY

Tickets Adults $5.00
Children $3.50

Each performance ends at 22.05.

(a) How many performances were there? [1]
(b) How long was each performance? [2]

(CIE Paper 1, Jun 2001)

---

6 Elan's trouser length, L, is 76 cm, to the nearest centimetre.
Copy and complete the statement.
Answer ....cm $\leqslant L <$ ....cm

(CIE Paper 1, Nov 2001)

---

7 A family goes on holiday by car. They leave home at 18.30.
They travel 190 km during the first two hours.
They then travel 50 km during the next hour to reach their destination.
(a) At what time do they arrive at their destination? [1]
(b) Calculate the average speed for the whole journey. [2]

(CIE Paper 1, Nov 2001)

---

8 A spoon can hold 5 ml of medicine.
Write 5 ml in litres. [1]

(CIE Paper 1 Jun 2002)

---

9 The population, P, of a city is 280 000, to the nearest ten thousand.
Copy and complete the statement about P.
Answer .... $\leqslant P <$ .... [2]

(CIE Paper 1, Jun 2002)

---

10 Doreen cycles to her friend's home.
She leaves at 09.40 and arrives at 10.20.
(a) Write down the time taken
(i) in minutes, [1]
(ii) as a fraction of an hour in its lowest terms. [1]
(b) The distance Doreen cycles is 8.4 km.
Work out Doreen's average speed in km/h. [2]

(CIE Paper 1, Jun 2002)

---

11 A train leaves Johannesburg at 09.45 and arrives in Pretoria at 10.32.
How many minutes does the journey take? [1]

(CIE Paper 1, Nov 2002)

---

12 A bottle of mass 480 grams contains 75 centilitres of water.
(a) Write 75 centilitres in millilitres. [1]
(b) Write 75 centilitres in litres. [1]
(c) The mass of 480 grams is correct to the nearest 10 grams.
Copy and complete the statement
Answer ....g $\leqslant$ mass $<$ ....g [2]
(d) Write 480 grams in kilograms. [1]

(CIE Paper 3, Jun 2002)

---

13 On a journey a bus takes 35 minutes to travel the first 10 kilometres.
It then travels a further 20 kilometres in the next 40 minutes.
(a) The bus started the journey at 18.50
At what time did it complete the journey? [1]
(b) Calculate the average speed for the whole journey in
(i) kilometres /minute [2]
(ii) kilometres/hour. [1]

(CIE Paper 1, Jun 2003)

---

14 Jeff takes 10 minutes to walk 1 kilometre. Find his average walking speed in kilometres per hour. [2]

(CIE Paper 1, Nov 2003)

---

15 Write 0.4 kilograms in grams. [1]

(CIE Paper 1, Nov 2003)

---

16 The length of a road is 1300 metres, correct to the nearest 100 metres.
Copy and complete the following statement.
Answer ………………..m $\leqslant$ road length $<$ ……………………m [2]

(CIE Paper 1, Nov 2003)

---

17 Sergio's height is 142cm, to the nearest centimetre.
Copy and complete the statement about the limits of his height.
Answer ……………….. cm $\leqslant$ height $<$ ……………………cm [2]

(CIE Paper 1, Jun 2004)

---

18 The time in Dubai is 3 hours ahead of Birmingham.
(a) If it is 21.15 on Sunday in Birmingham, what time is it in Dubai? [1]
(b) An aircraft leaves Birmingham at 21.15 on Sunday and arrives in Dubai on Monday at 07.45 local time.
(i) How long did the journey take? [1]
(ii) The distance from Birmingham to Dubai is 5620 km.
Calculate the average speed of the aircraft. [3]

(CIE Paper 1, Jun 2004)

---

19 In New Zealand, a bus leaves New Plymouth at 8.10am and arrives in Wellington at 2.55pm.
(a) How long, in hours and minutes, does the journey take? [1]
(b) The distance from New Plymouth to Wellington is 355 km.
Calculate, in kilometres per hour, the average speed of the bus. [3]

(CIE Paper 1, Nov 2004)

---

20 The highest mountain in Argentina is Aconcagua.
Its height is 6960 metres, correct to the nearest twenty metres.
Write down the smallest possible height of Aconcagua. [1]

(CIE Paper 1, Nov 2005)

Chapter 9

Length, area and volume

This chapter will show you how to

✔ find perimeters
✔ find the areas of simple and compound shapes
✔ learn about the circle
✔ calculate the circumference and area of a circle
✔ calculate volumes

9.1 Perimeter

Compound shapes

The **perimeter** of a shape is the distance around the outside of it.

EXAMPLE 1

Calculate the perimeter of the shape drawn on the grid below.

The shape is drawn on a centimetre square grid.

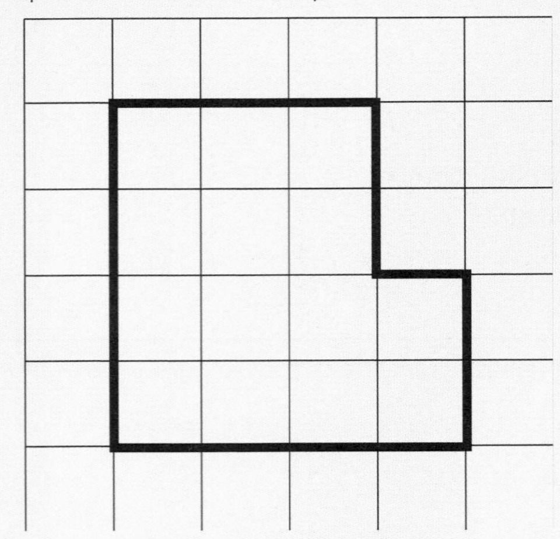

Start from the bottom left hand corner.

Perimeter = $4 + 3 + 2 + 1 + 2 + 4$ cm
$= 16$ cm

The shape has 6 sides so you need to add 6 lengths.

EXERCISE 9A

These shapes are drawn on a centimetre grid. Work out the perimeters of the shapes.

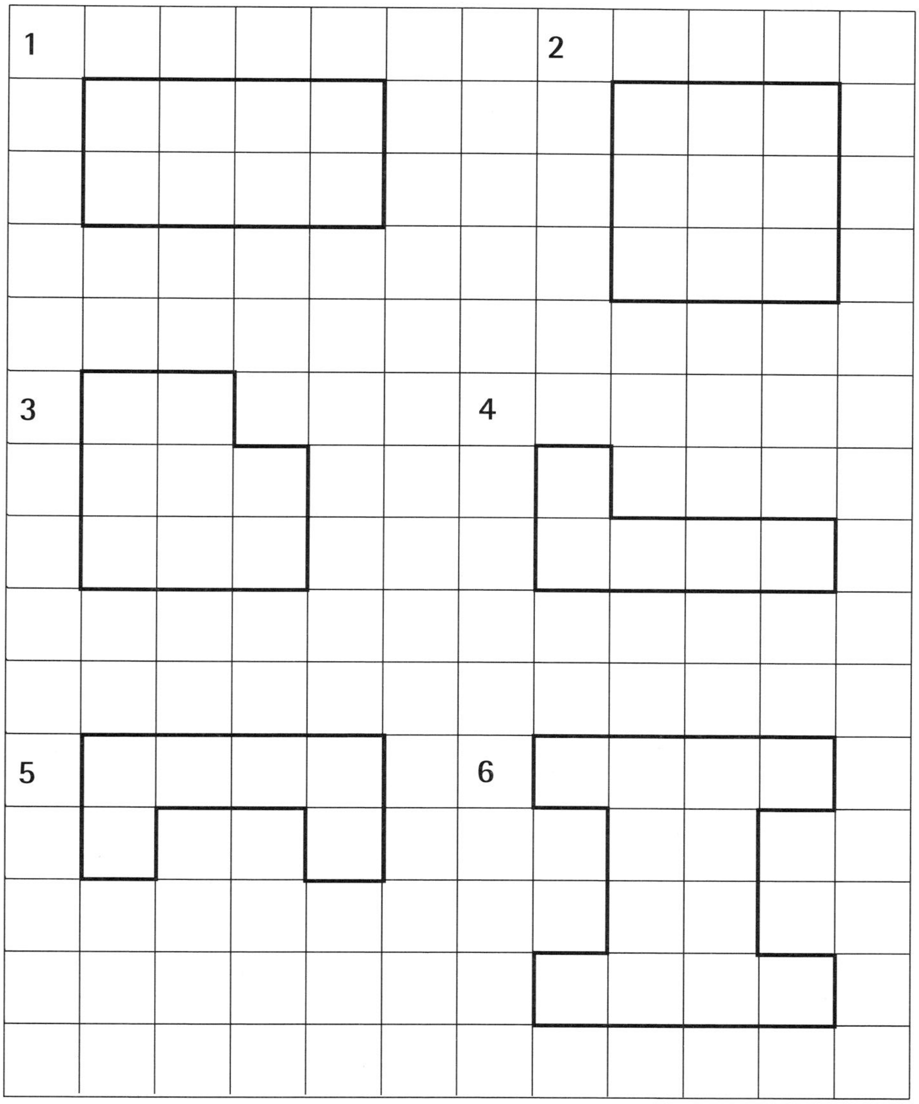

1

2

3

4

5

6

7 Work out the perimeters of these shapes.

(a) A square with side 5 cm.

(b) An equilateral triangle with side 6 cm.

(c) A rectangle with length 7 cm and width 5 cm.

(d) An isosceles triangle with two sides of length 7 cm and one side of length 5 cm.

EXAMPLE 2

Not all shapes are drawn on grids. You may have to work out some lengths for yourself.

Find the perimeter of these shapes

(a)

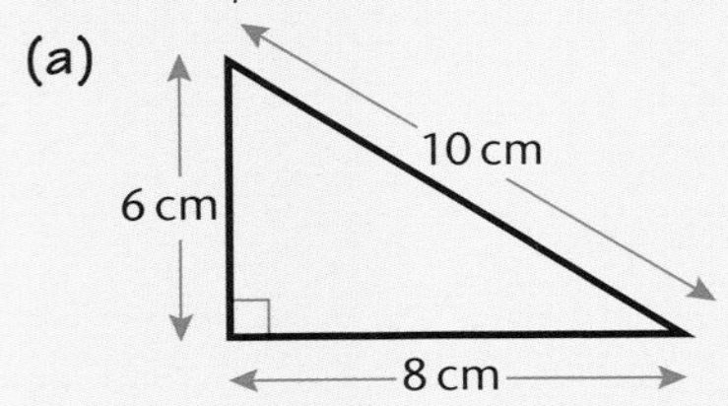

(b)

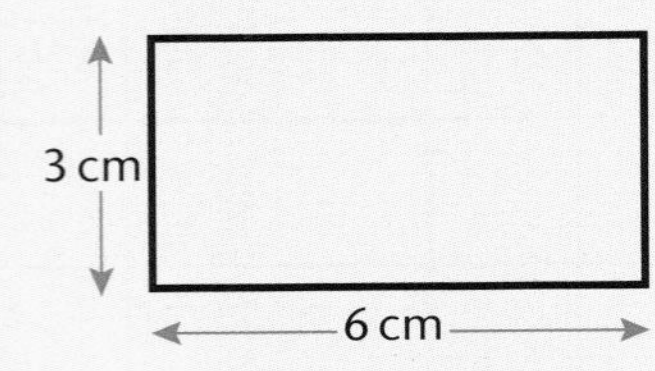

(c)

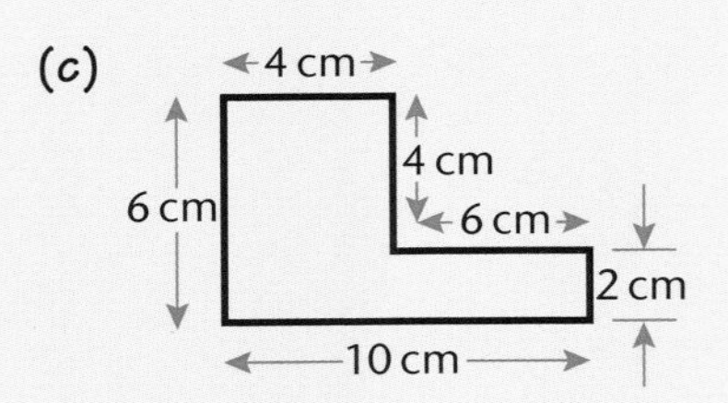

(a) Perimeter = 6 + 10 + 8 = 24 cm

(b) Opposite sides of a rectangle have the same length.

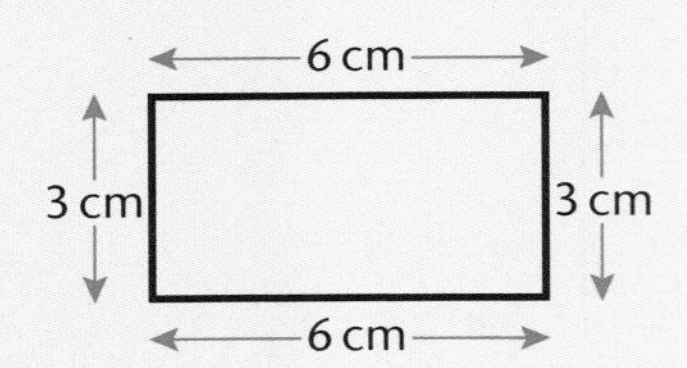

Perimeter = 3 + 6 + 3 + 6 = 18 cm

(c) Perimeter = 6 + 4 + 4 + 6 + 2 + 10 = 32 cm

EXERCISE 9B

Find the perimeters of these shapes. All lengths are in centimetres.

1

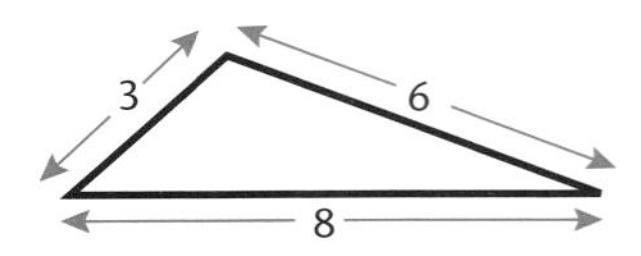

2

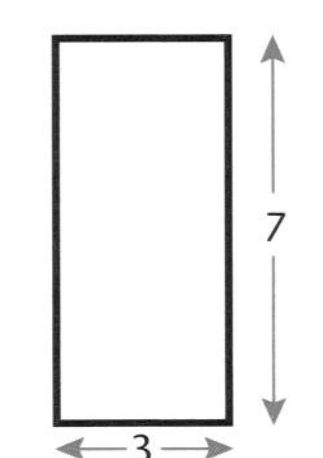

3

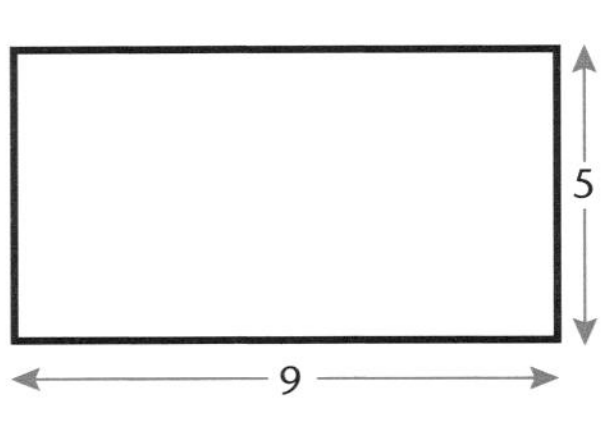

4

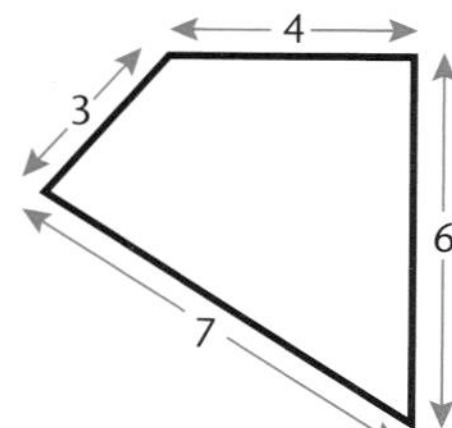

5

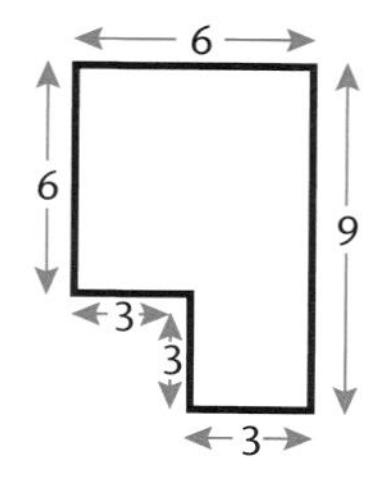

6

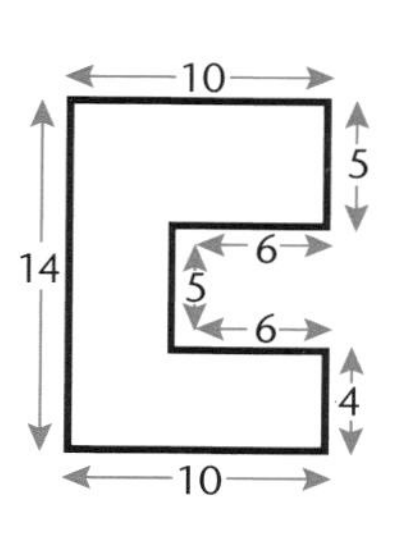

7

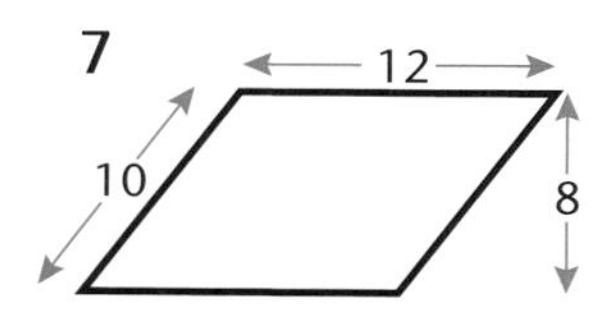

8

9

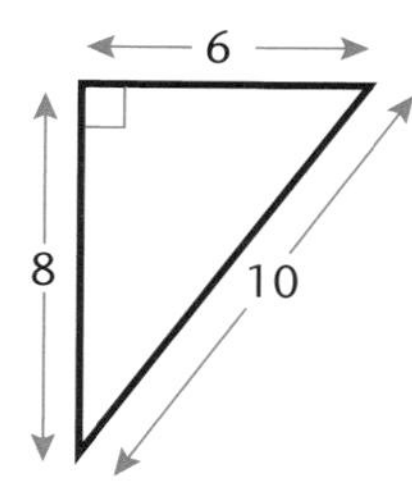

10

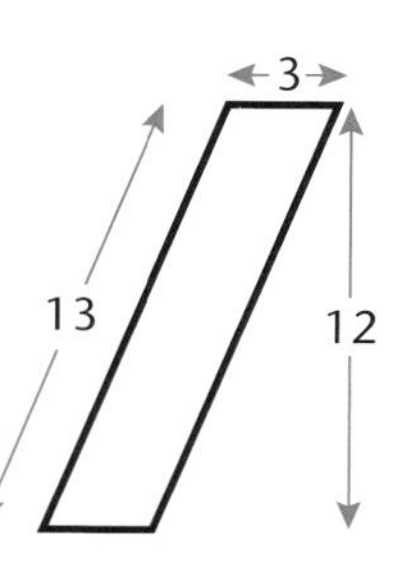

11

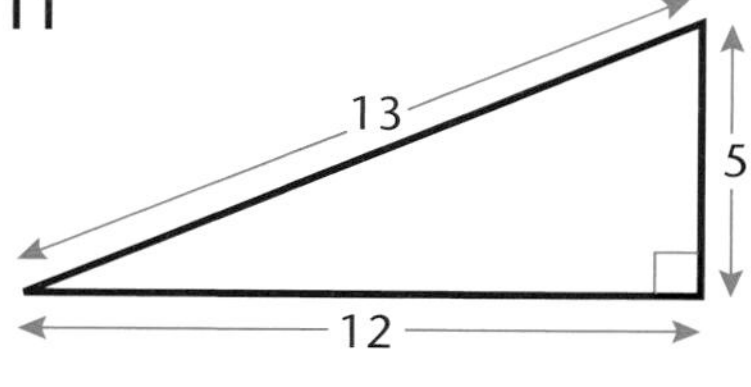

EXAMPLE 3

Find the perimeter of this compound shape.

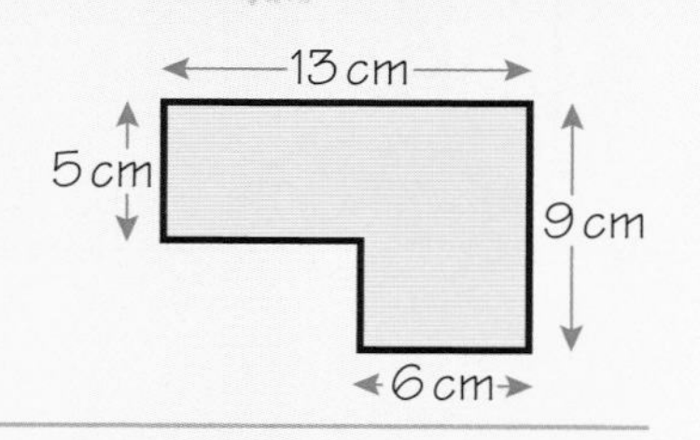

Look at the vertical lengths.
Look at the horizontal lengths.

Two of the lengths are not given on the diagram but they can be worked out.

Total width of shape = 13 cm

So $x + 6 = 13$

$x = 7$ cm

Total height of shape = 9 cm

So $y + 5 = 9$

$y = 4$ cm

Work out the missing lengths.

Perimeter = $5 + 13 + 9 + 6 + 4 + 7$

= 44 cm

$y = 4$ $x = 7$

Starting at the bottom left corner and work clockwise. Add 6 lengths.

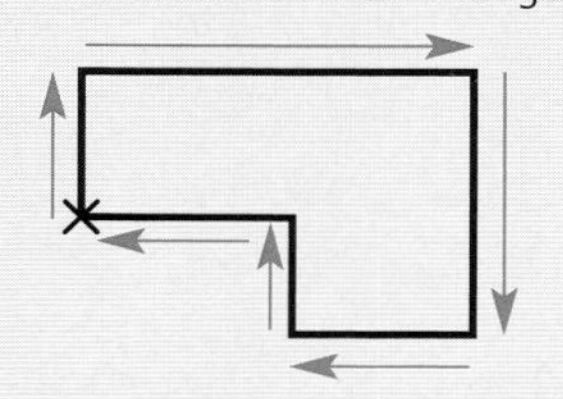

EXERCISE 9C

Find the perimeter of each shape. All lengths are in cm.

1

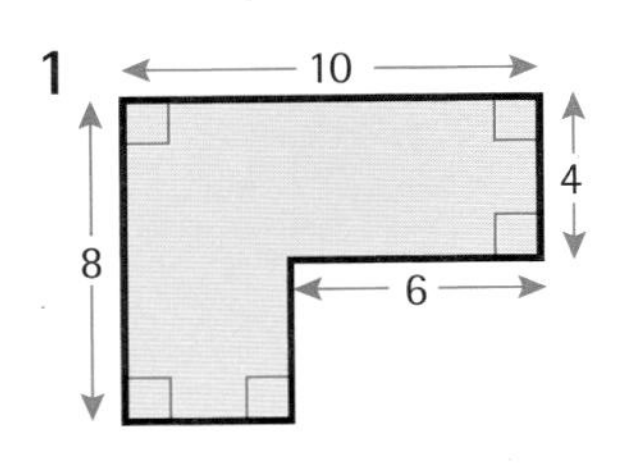

2

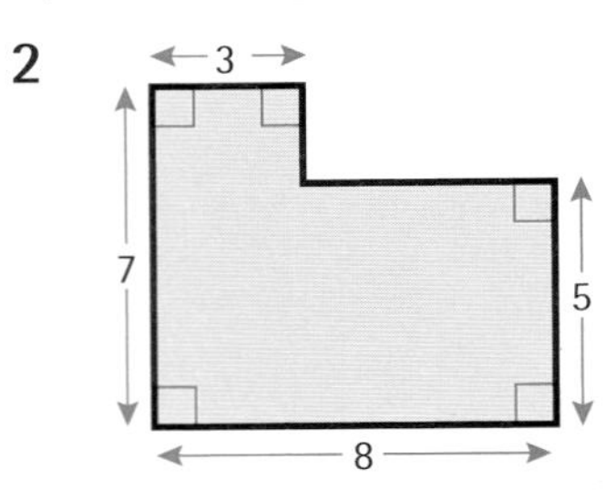

3

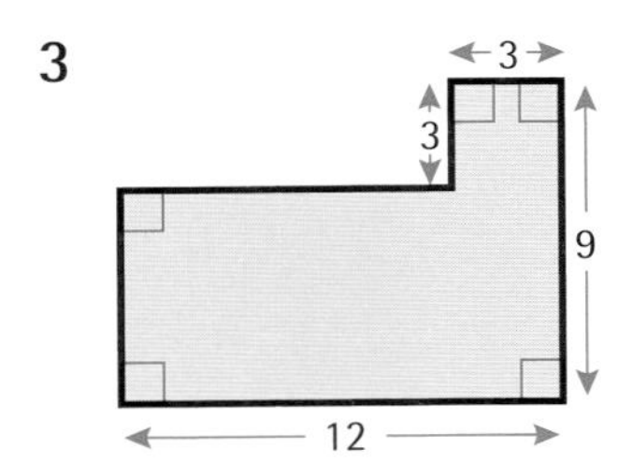

4

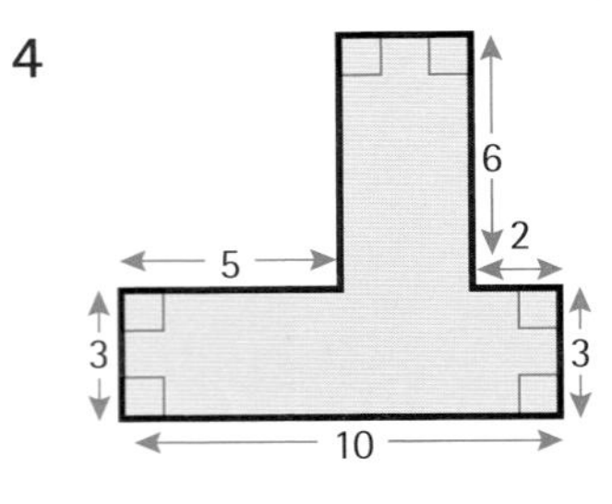

5

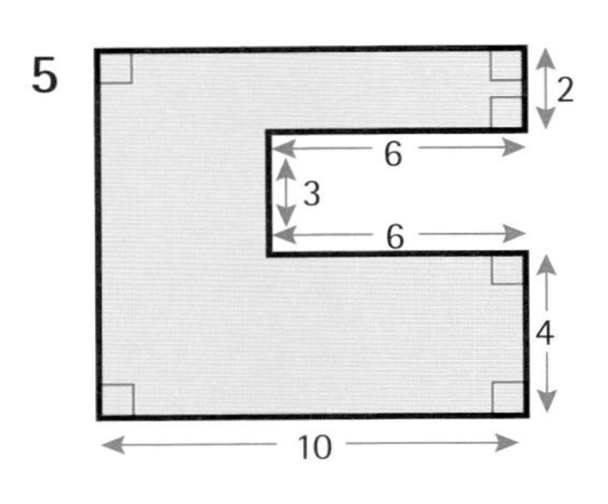

6

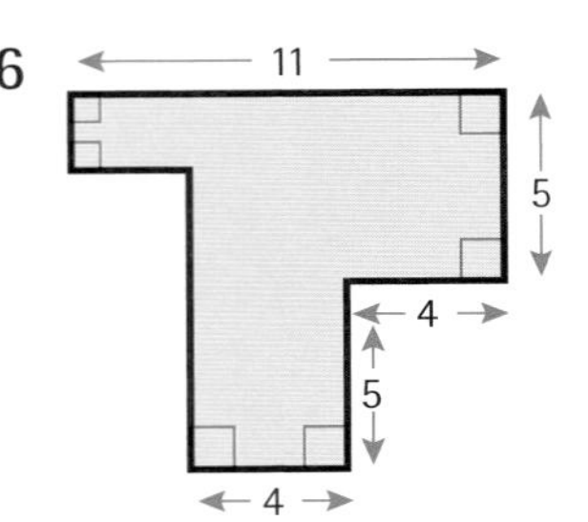

9.2 Area

The **area** of a shape is the amount of space inside it.

Rectangles

This **rectangle** has a length of 4 cm and a width of 3 cm.

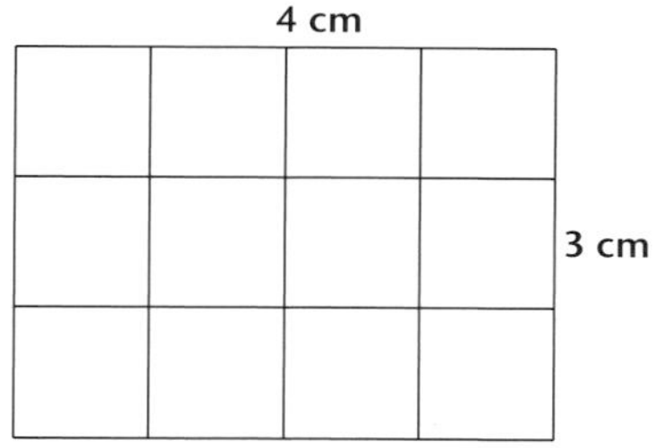

1 cm

1 cm

Area of 1 square centimetre

There are 12 squares in the diagram so the area is 12 square centimetres.

Find the area of the rectangles on this centimetre grid.

1

2

3

4

5

6

You may have noticed a quick way of working out the areas in the last exercise.

You can calculate the area using the formula

Area of a rectangle = length × width

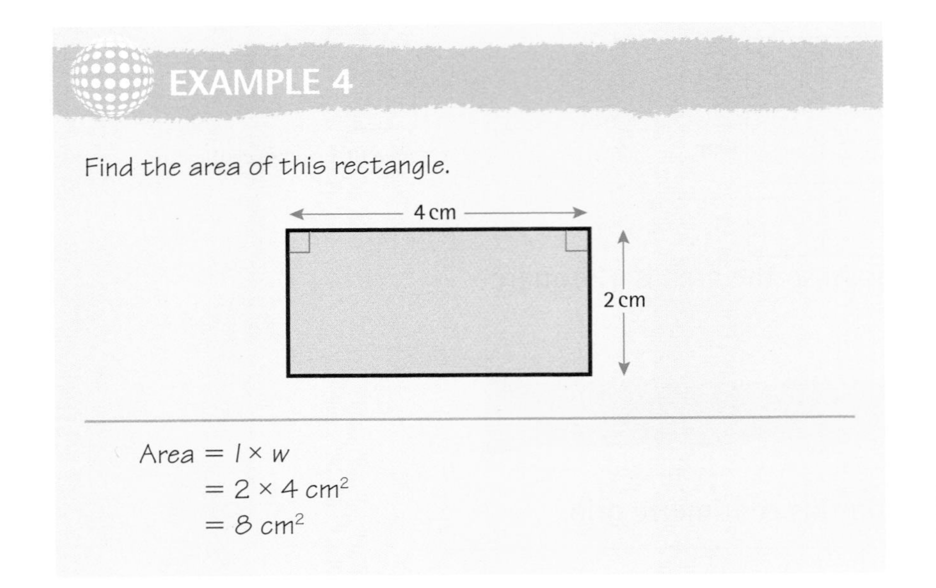

EXAMPLE 4

Find the area of this rectangle.

$$\text{Area} = l \times w$$
$$= 2 \times 4\ \text{cm}^2$$
$$= 8\ \text{cm}^2$$

Square units for area.

EXERCISE 9E

Find the area of these shapes.

All lengths are in centimetres.

1

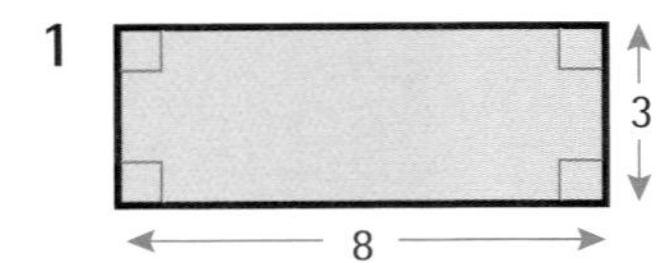

2

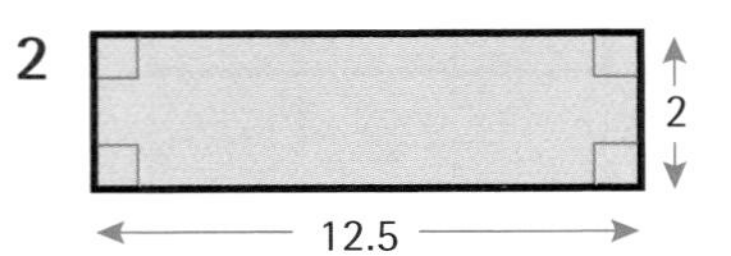

3

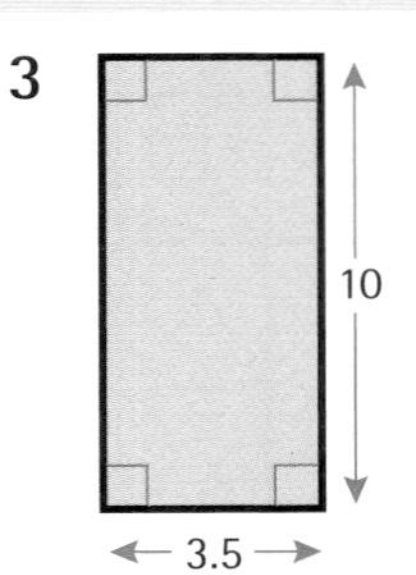

4

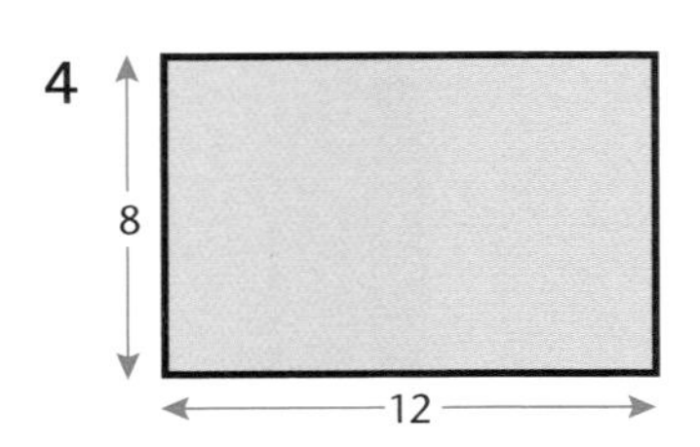

5

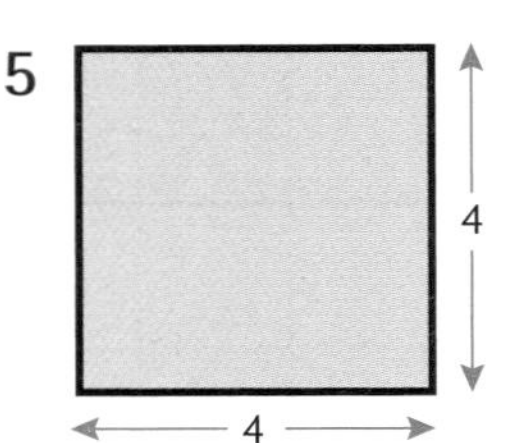

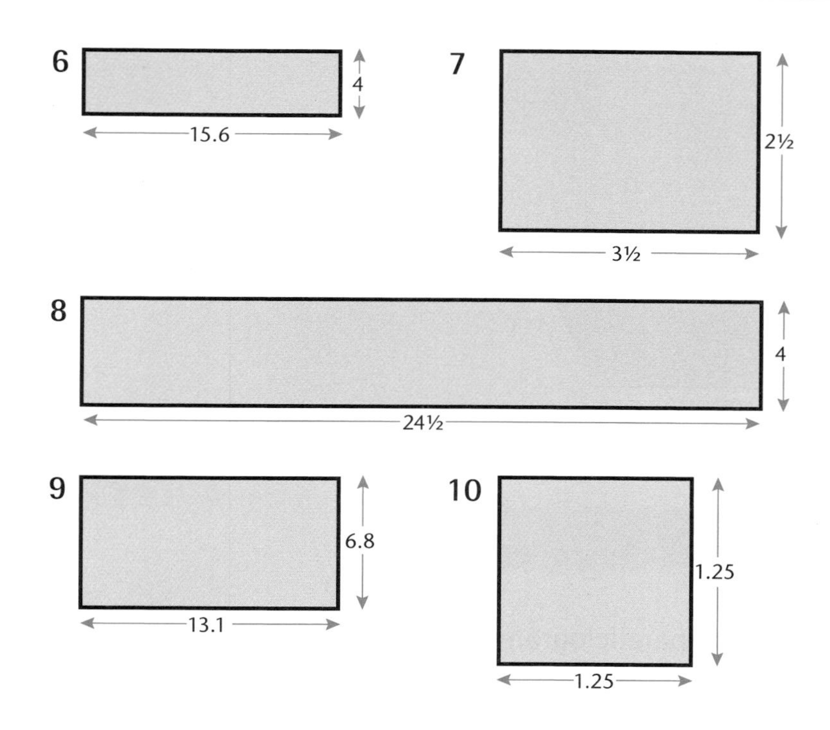

Parallelograms

A **parallelogram** has two pairs of parallel sides.

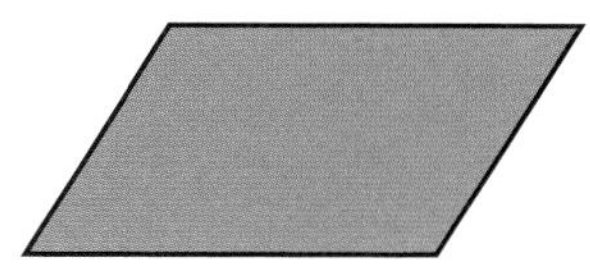

You can see in the diagrams below that cutting off the blue triangle from the left hand end of the parallelogram and adding it to the right hand end changes the parallelogram into a rectangle of the same area.

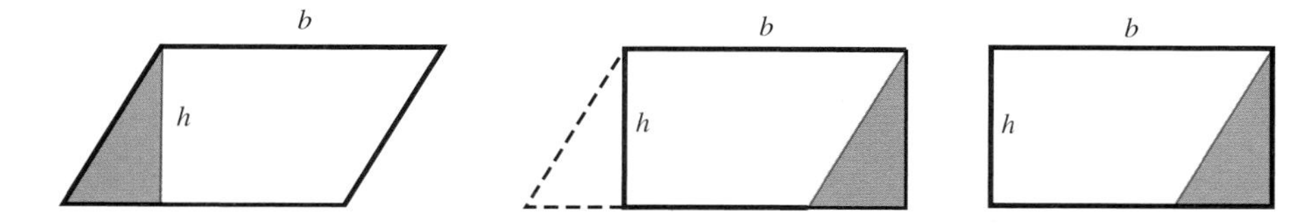

Area of a parallelogram = base × perpendicular height
$= b \times h$

EXAMPLE 5

Find the area of the parallelogram.

Area $= b \times h$

$= 7 \times 4 \text{ cm}^2$

$= 28 \text{ cm}^2$

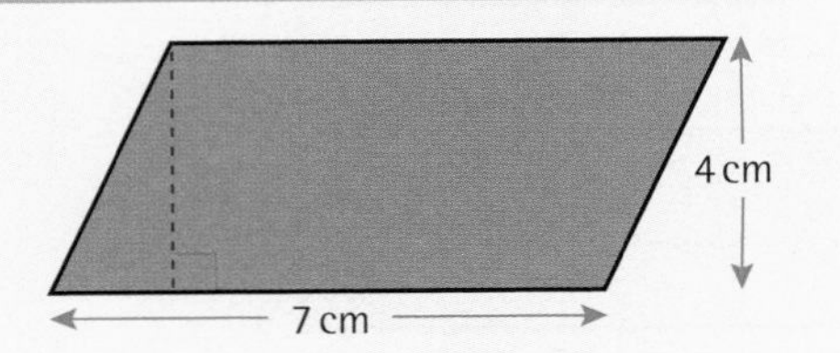

EXERCISE 9F

Calculate the area of each of the following parallelograms.
All lengths are in cm.

1

10

8

2

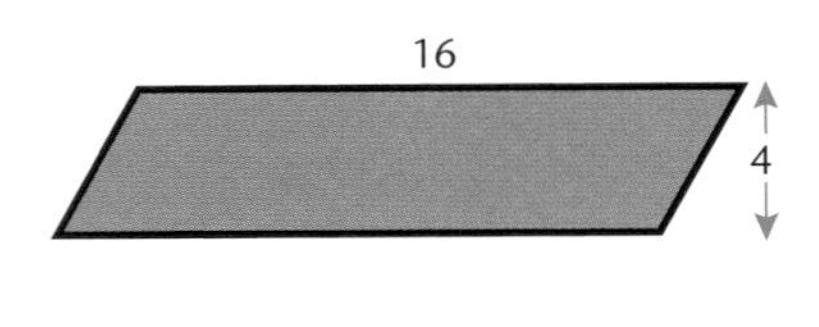

3

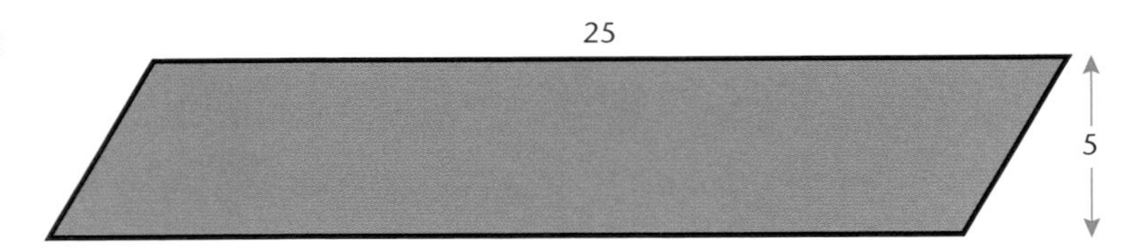

4

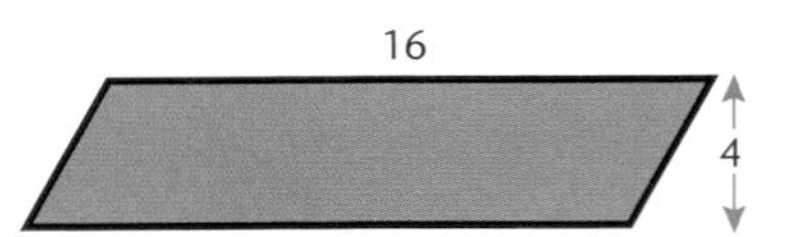

5

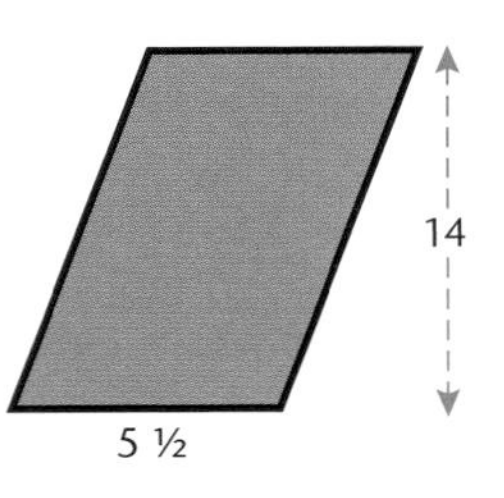

6

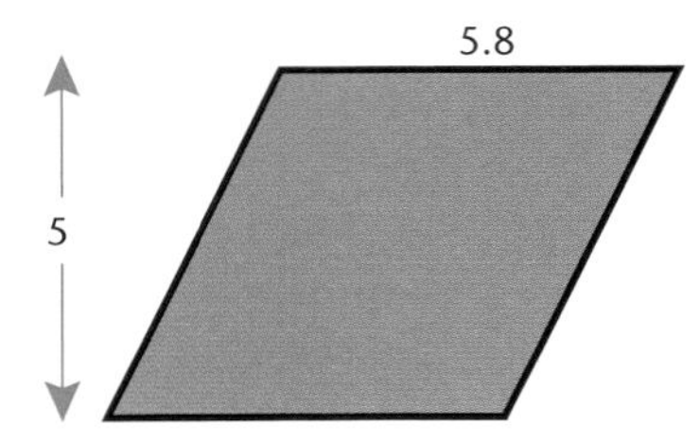

7

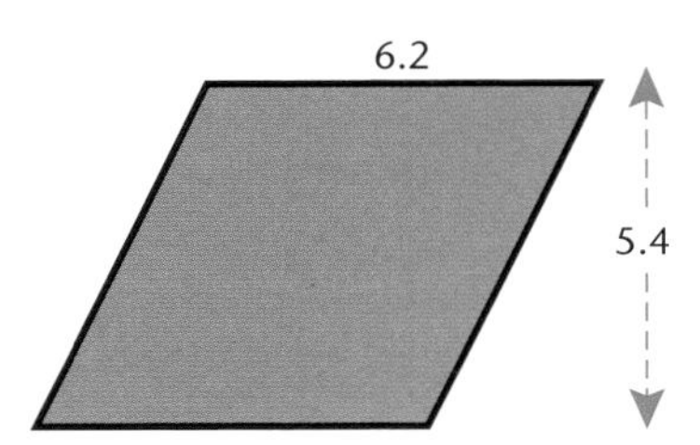

8

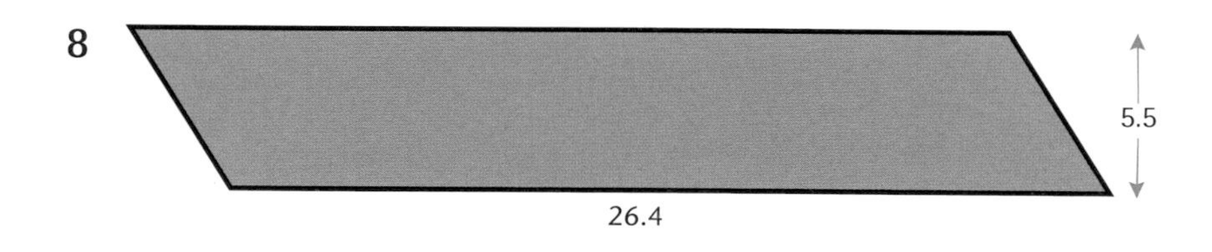

Triangles

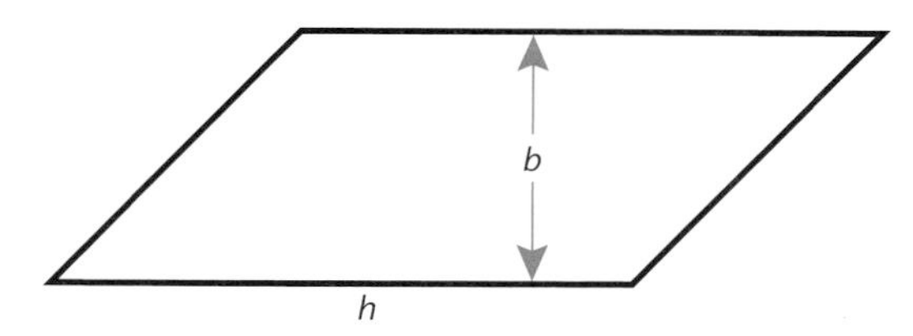

The diagonals in this parallelogram will split it into two equal triangles.

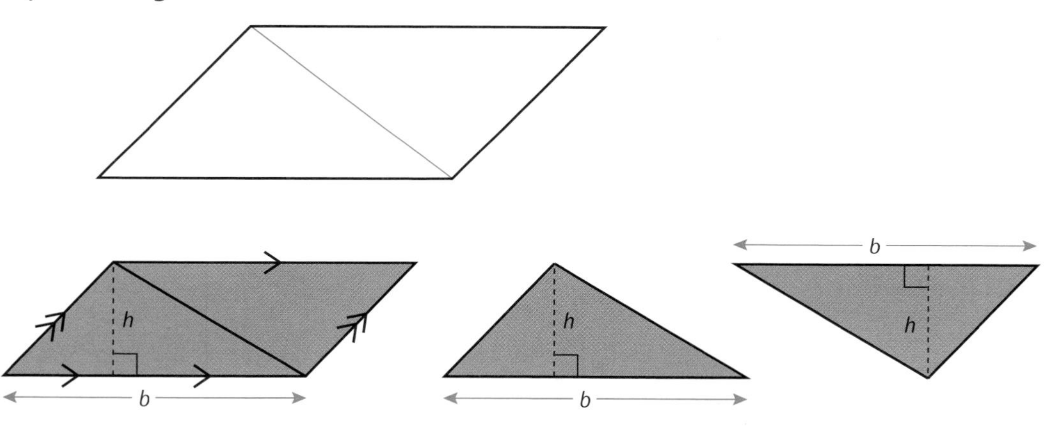

The two triangles have equal areas.
Each one is half the area of the parallelogram.

Area of a triangle $= \frac{1}{2} \times$ base $\times$ perpendicular height

$$= \frac{1}{2} \times b \times h$$

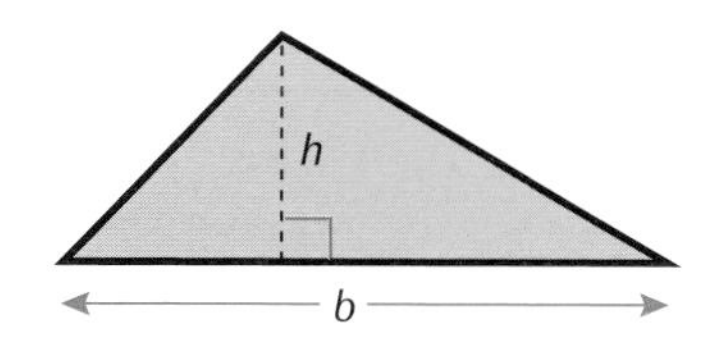

EXAMPLE 6

Find the area of each triangle.

(a)

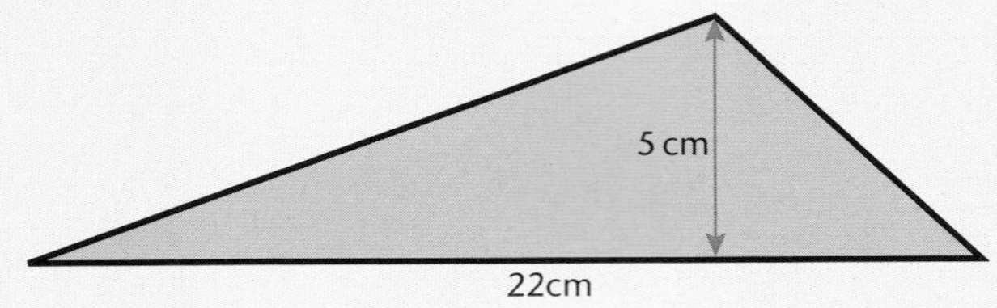

$$\text{Area} = \tfrac{1}{2} \times b \times h$$
$$= \tfrac{1}{2} \times 22 \times 5$$
$$= 55 \text{ cm}^2$$

(b) The area of a right-angled triangle can also be found this way.

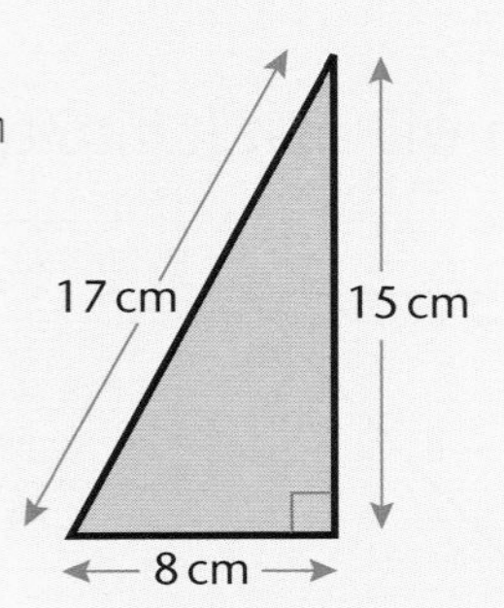

$$\text{Area} = \tfrac{1}{2} \times b \times h$$
$$= \tfrac{1}{2} \times 8 \times 15$$
$$= 60 \text{ cm}^2$$

EXERCISE 9G

Find the area of the following triangles. All lengths are in cm.

1

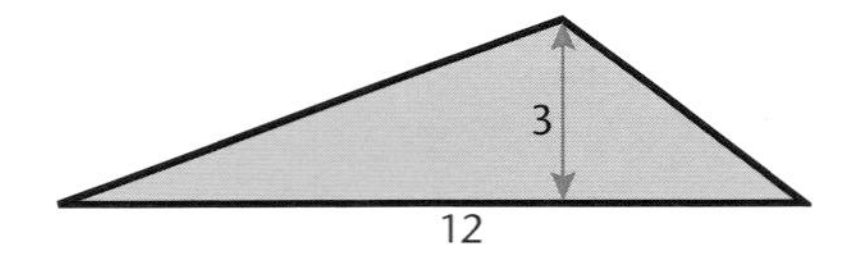

2

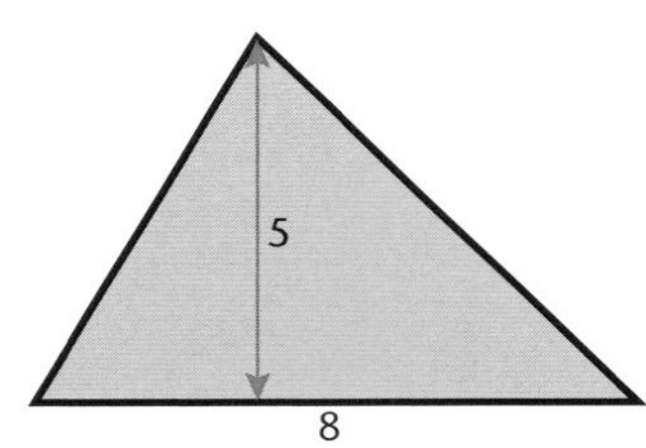

3

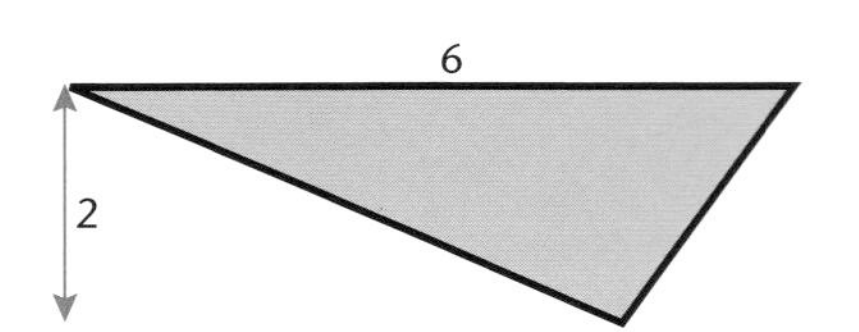

4

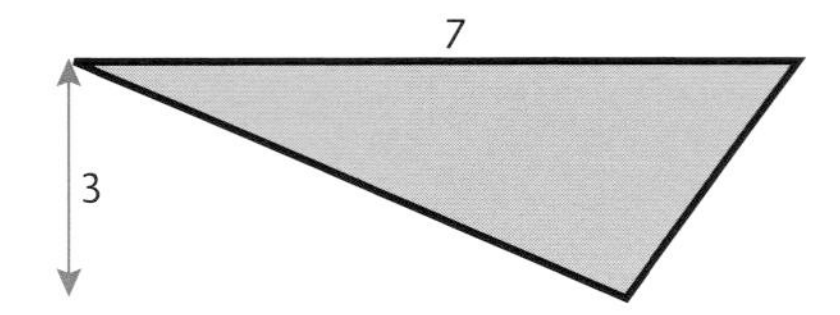

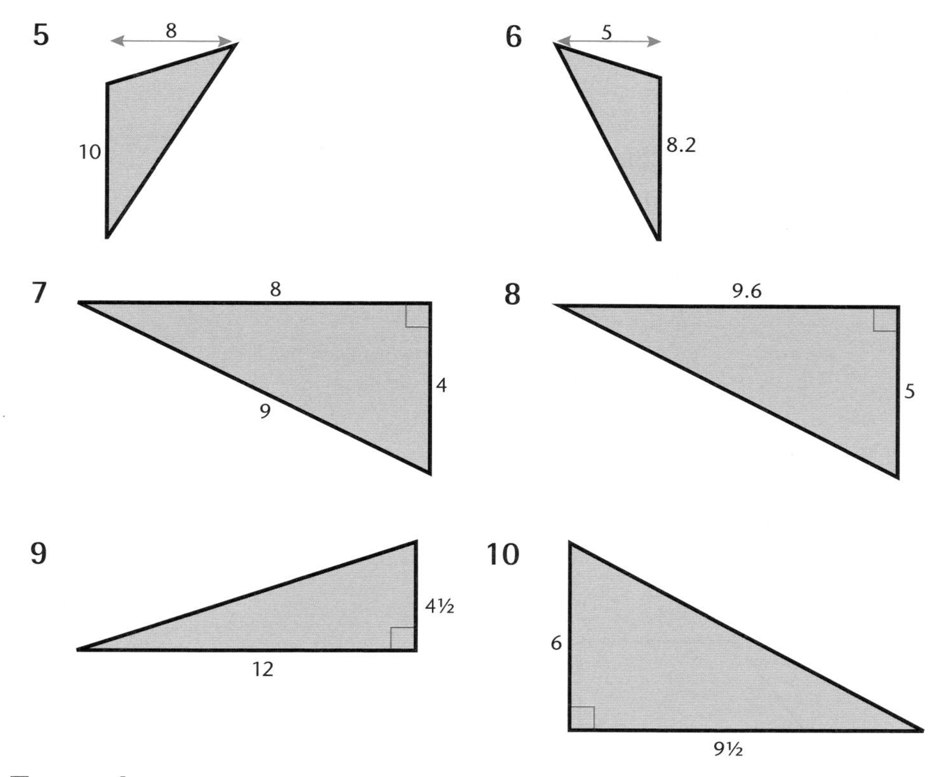

Trapezia

A **trapezium** is a quadrilateral with one pair of parallel sides.

This trapezium has

- parallel sides of length a and b
- perpendicular height h.

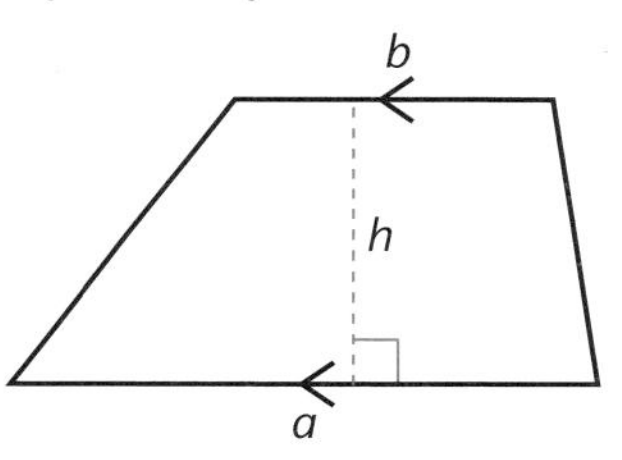

If you put an identical trapezium upside down next to it, you get a parallelogram

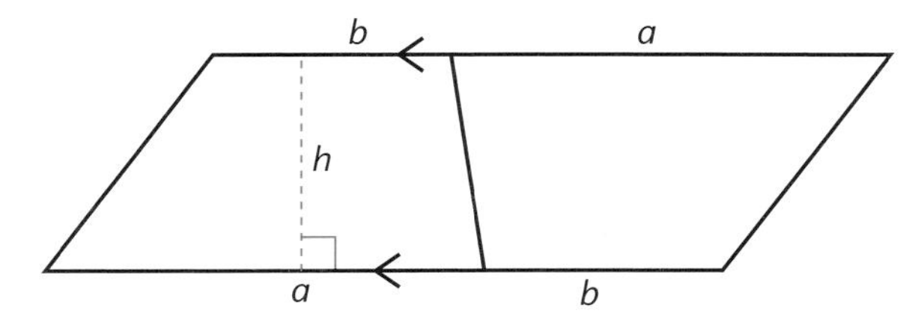

The parallelogram is made of two identical **trapezia**.

The base of this parallelogram is $(a + b)$ and its perpendicular height is h.

So the area of this parallelogram is base $\times$ perpendicular height $= (a + b) \times h$.

The area of the trapezium is half the area of this parallelogram.

Area of a trapezium $= \frac{1}{2} \times$ (sum of parallel sides)
$\times$ perpendicular height
$= \frac{1}{2} \times (a + b) \times h$

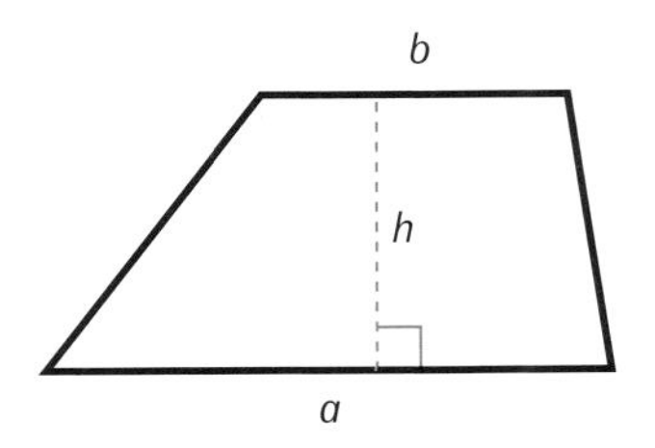

Another way of saying this is
$\frac{1}{2} \times$ sum of parallel sides $\times$ distance between them

EXAMPLE 7

Find the area of these shapes.

(a)

(b)

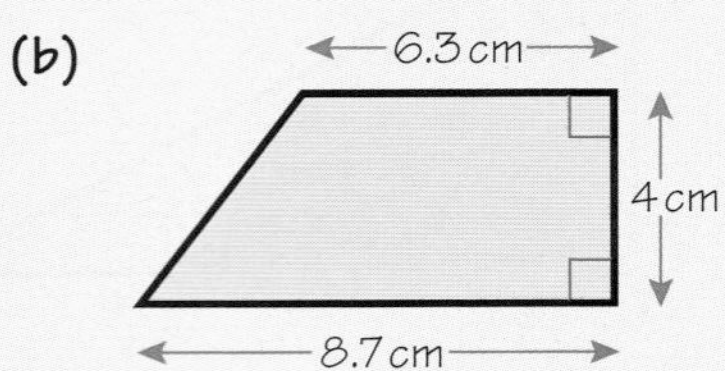

(a) Area $= \frac{1}{2}(a + b) \times h$
$= \frac{1}{2}(4 + 8) \times 3$
$= \frac{1}{2} \times 12 \times 3$
$= 6 \times 3$
$= 18\text{ cm}^2$

Order of operations brackets first.

Remember the units.

(b) Area $= \frac{1}{2}(a + b) \times h$
$= \frac{1}{2}(8.7 + 6.3) \times 4$
$= \frac{1}{2} \times 15 \times 4$
$= \frac{1}{2} \times 4 \times 15$
$= 2 \times 15$
$= 30\text{ cm}^2$

15×4 is the same as 4×15. It is easier to multiply 4 by $\frac{1}{2}$.

EXERCISE 9H

1 Find the area of these trapezia.

(a)

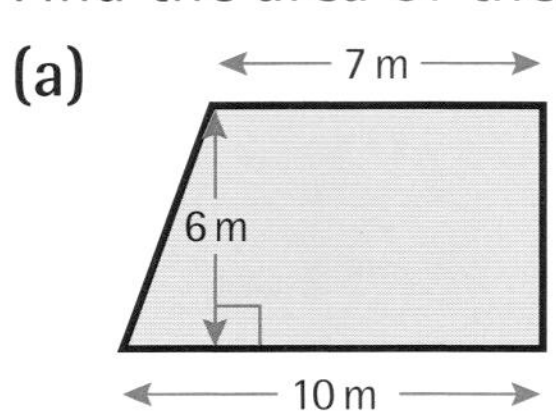

(b)

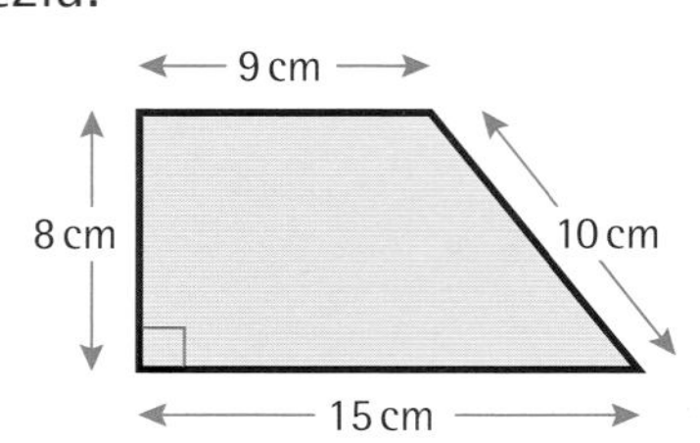

2 Work out the area of these trapezia.

All lengths are in cm.

(a)

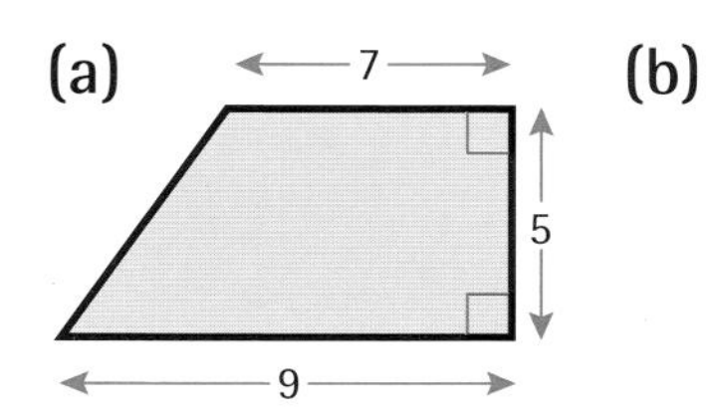

(b)

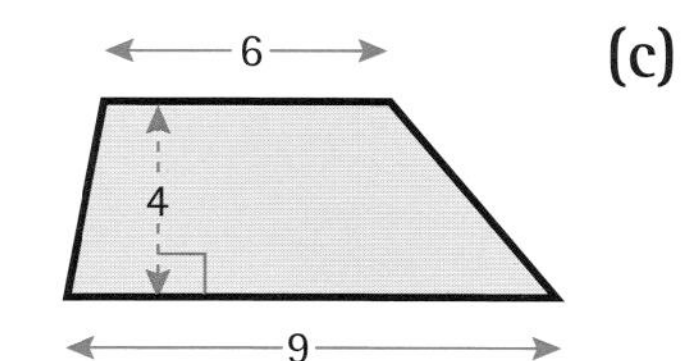

(c)

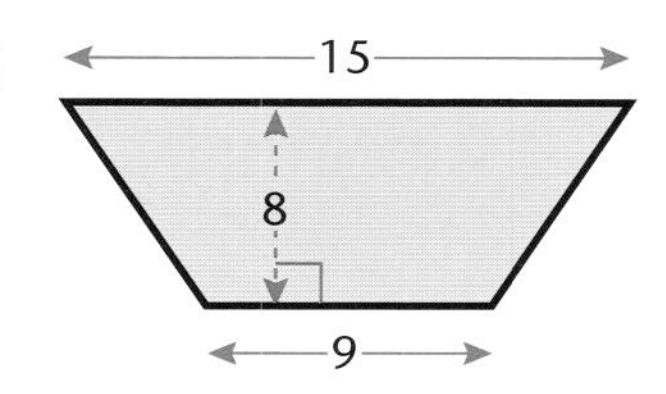

(d)

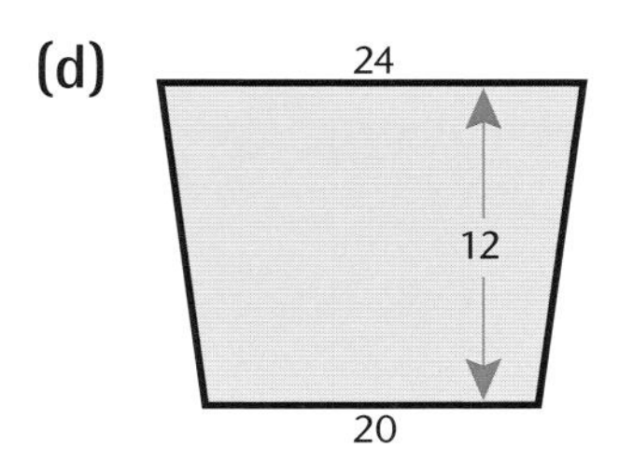

(e)

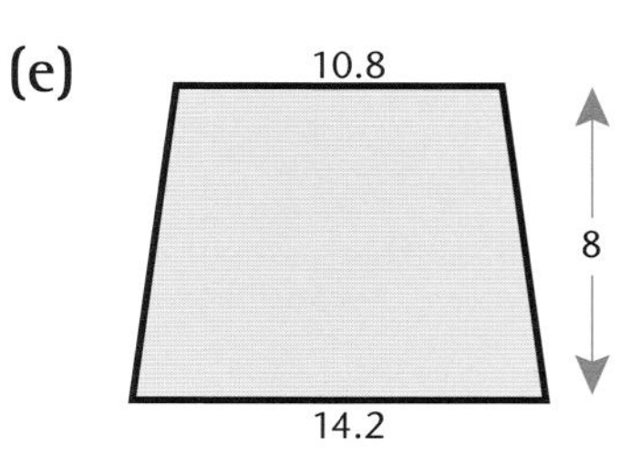

(f)

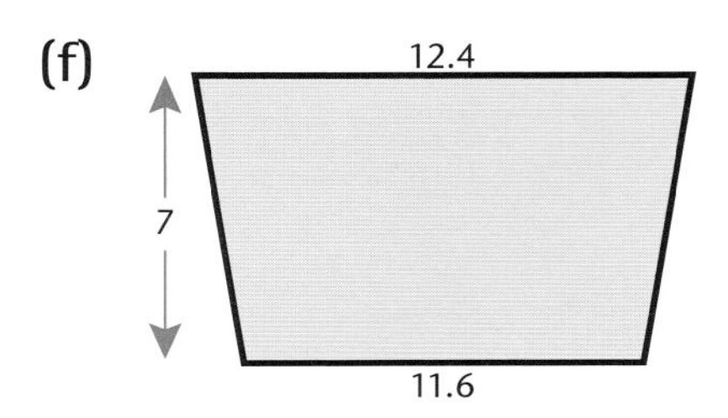

(g)

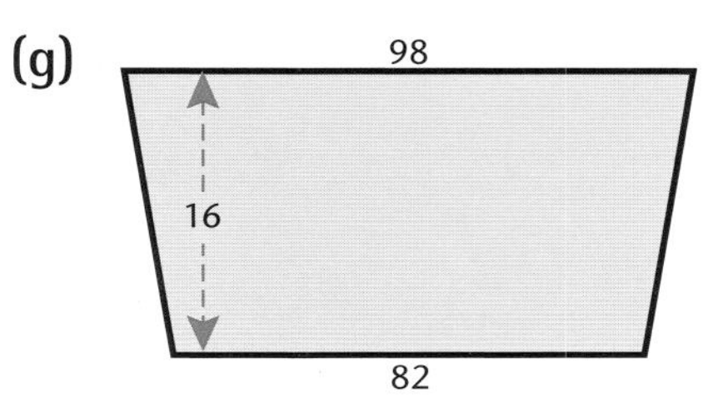

Areas of compound shapes

A **compound shape** is made from simple shapes.

To find the area of a compound shape, you split it into simple shapes.

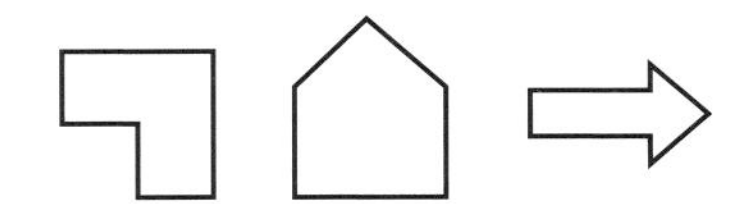

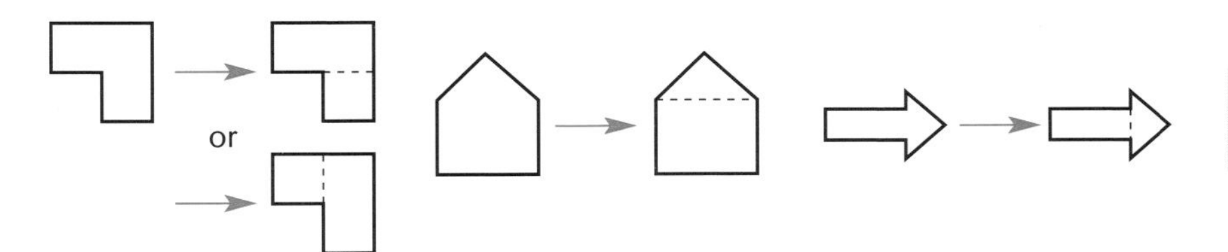

The simple shapes here are *rectangles* and *triangles*.

Use the formulae for areas of the simple shapes.

EXAMPLE 8

Find the area of this compound shape.

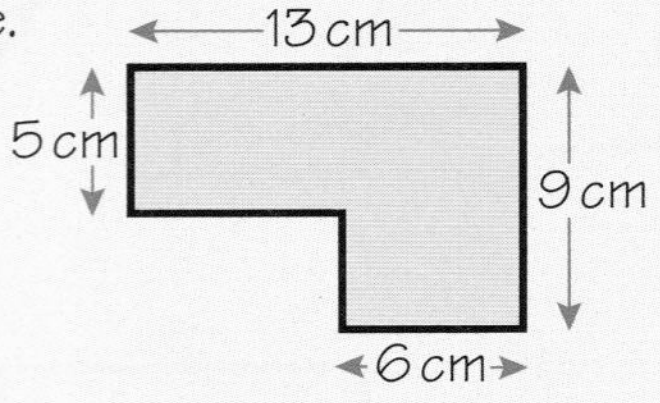

Look at the vertical lengths.
Look at the horizontal lengths.

Split the shape into two rectangles *A* and *B*.

You could split it like this instead

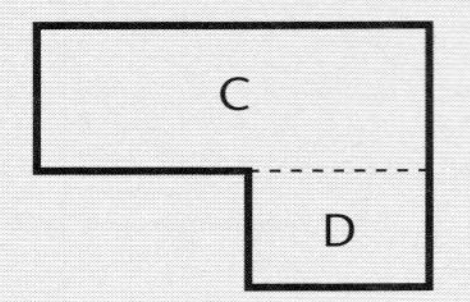

You would get the same answers.

Two of the lengths are not given on the diagram but they can be worked out.

Work out the missing lengths.

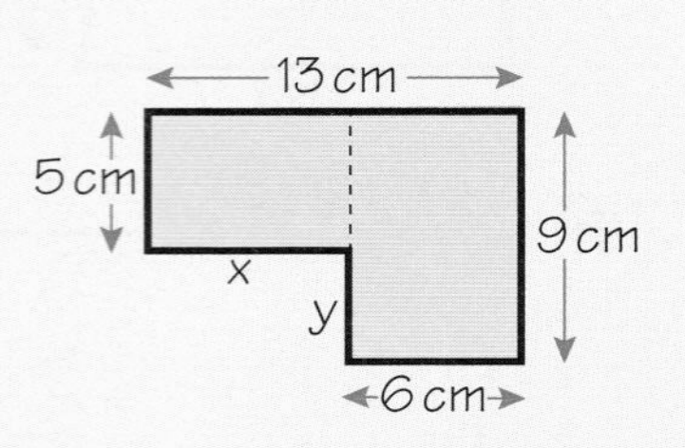

Total width of shape = 13 cm

So $x + 6 = 13$

$x = 7$ cm

Total height of shape = 9 cm

So $y + 5 = 9$

$y = 4$ cm

Area = area of *A* + area of *B*

$= (5 \times 7) + (9 \times 6)$ — Order of operations brackets first.

$= 35 + 54$

$= 89\ \text{cm}^2$

Another way to work this out is to add a small rectangle *E* to 'fill in' the missing part.

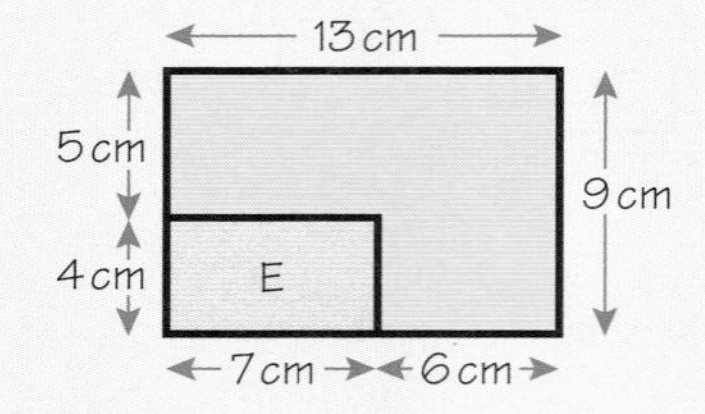

Area of large rectangle $= 13 \times 9 = 117\ \text{cm}^2$

Area of $E = 4 \times 7 = 28\ \text{cm}^2$

Area of shape = area of large rectangle − area of *E*

$= 117 - 28 = 89\ \text{cm}^2$

EXAMPLE 9

Split the shape into rectangle *A* and triangle *B*.

Find the area of this shape.

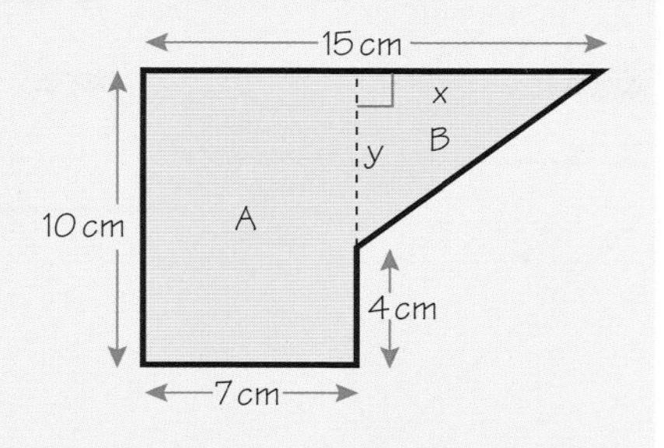

You need to find the lengths marked *x* and *y* before you can find the area of the triangle.

Total width of shape = 15

So $x + 7 = 15$

$x = 8$

Total height of shape = 10

So $y + 4 = 10$

$y = 6$

Area of triangle $= \frac{1}{2} \times h \times b$

Area = rectangle *A* + triangle *B*

$= (10 \times 7) + (\frac{1}{2} \times 6 \times 8) = 70 + 24 = 94 \text{ cm}^2$

EXERCISE 9I

1 Find the area of each shape.

All lengths are in cm.

(a)

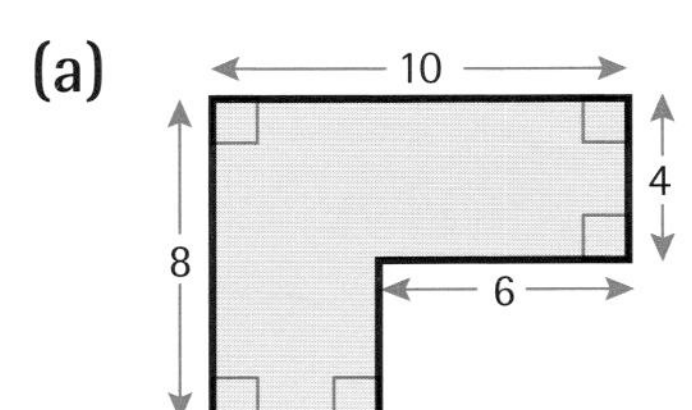

(b)

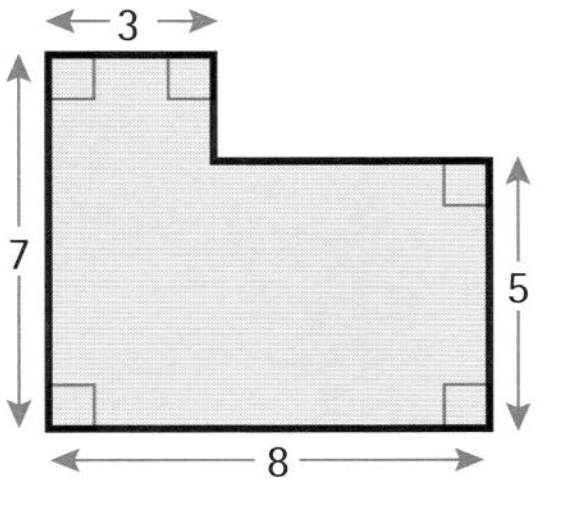

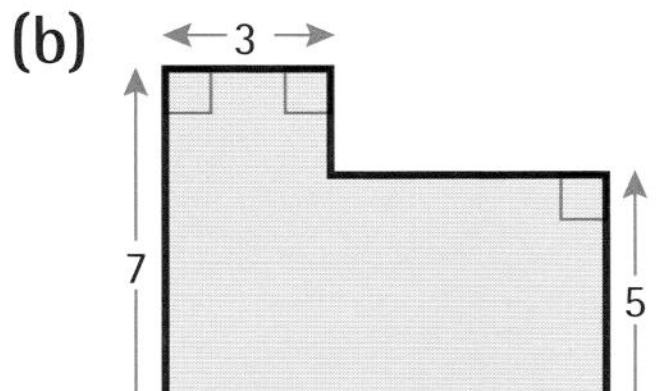

(c)

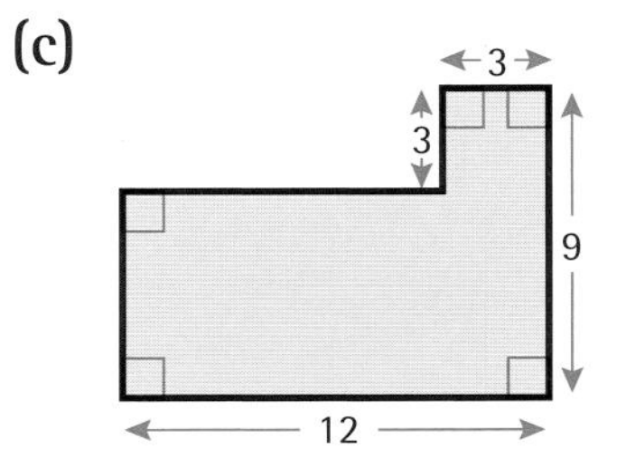

(d)

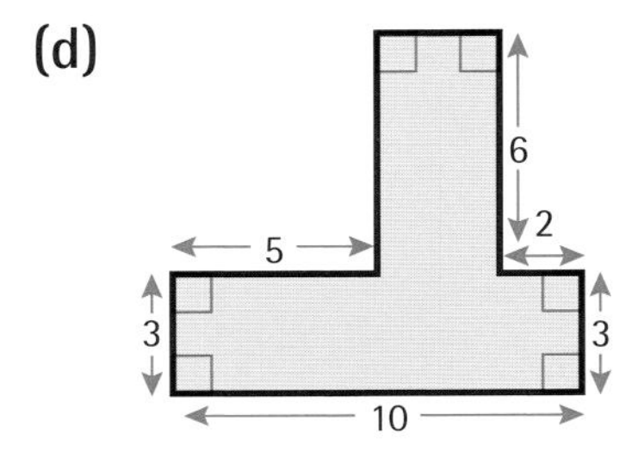

(e)

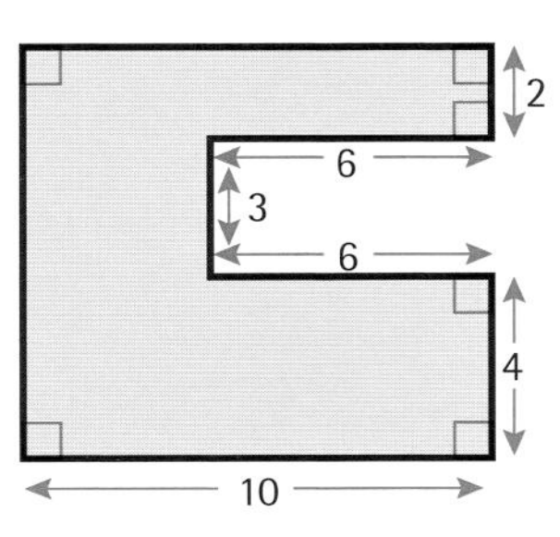

(f)

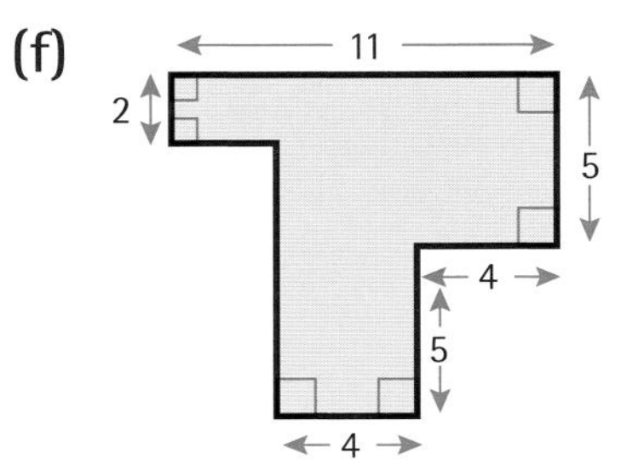

2 Work out the areas of these shapes.

All lengths are in cm.

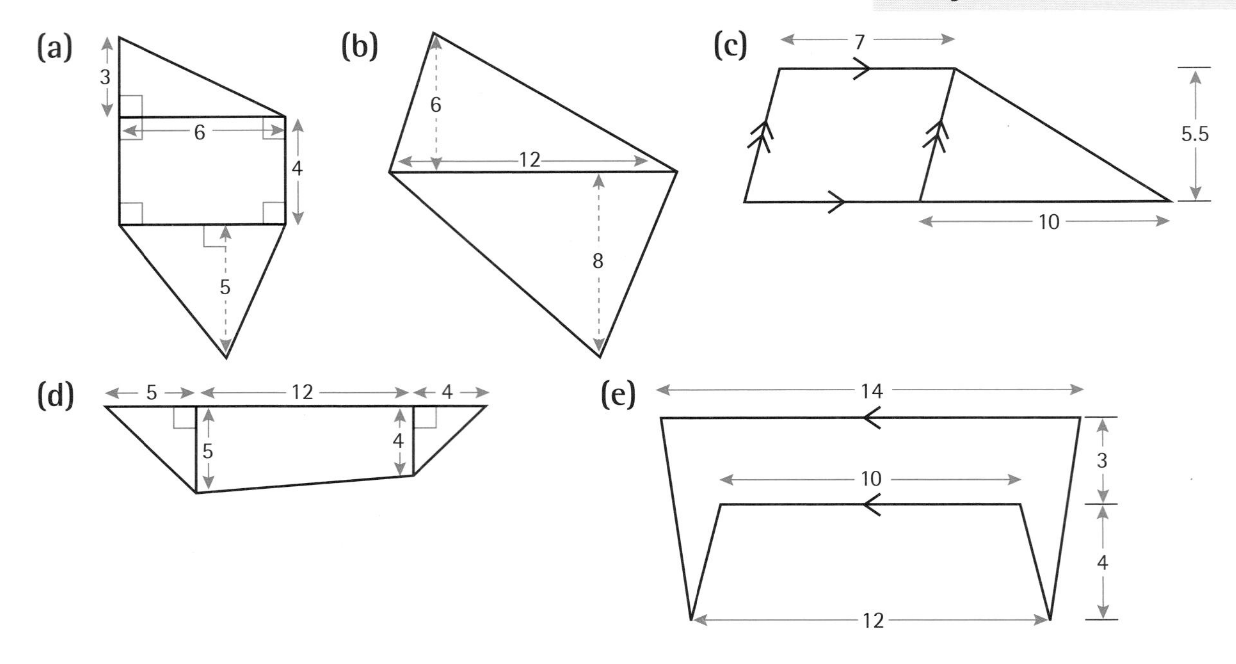

9.3 The circle

You need to know the following parts of the circle.

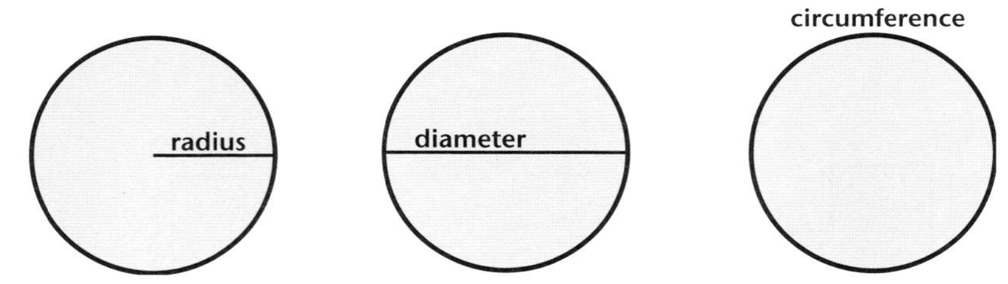

The **diameter** is twice the length of the **radius**.

The **circumference** is the distance around the outside.

Circumference of a circle

1 Take a piece of string and wrap it around the circumference of a cylinder. Make sure that you can tell how much of the string you have used.

2.Measure the length of the string used with a ruler.

3.Measure the diameter of the cylinder.

4.Work out circumference ÷ diameter.

5.Repeat the steps above for all the different cylinders that you can find.

You should find that your answer will be just over 3 every time. If you do it really carefully you should get 3.1, the value of this calculation is very special and has been given its own symbol π.

π (pi) is a letter of the Greek alphabet.

$$\frac{\text{circumference}}{\text{diameter}} = \pi$$

You can rearrange this formula as
Circumference $= \pi \times$ diameter

$C = \pi d$, where C is circumference and d is diameter.

Diameter $d = 2r$, where r is the radius, so $C = \pi \times 2r$ or

$C = 2\pi r$

The value of π is 3.141 592 654...
π is a non-recurring, non-terminating decimal.
Your calculator has a π key.
In calculations, you will use 3.142 or the π button on your calculator. Either value of π may be used in the examination. Make sure that you give your answer correct to 3 s.f.

EXAMPLE 10

A circle has diameter 36 cm.
Calculate its circumference.

$C = \pi d$
$= \pi \times 36$
$= 113$ cm (3 s.f.)

EXERCISE 9J

In all the following questions, take $\pi = 3.142$ or use the π button on your calculator.

1 Find the circumference of the circle when the diameter is

(a) 2 cm (b) 5 cm (c) 12 cm
(d) 18.4 cm (e) $8\frac{1}{2}$ cm

2 Find circumference of the circle when the **radius** is

(a) 6 cm (b) 15 cm (c) 24 cm
(d) 9.5 cm (e) $4\frac{1}{2}$ cm

3 The diameter of a coin is 20 mm. Calculate its circumference.

4 The diameter of a CD is 12 cm. Calculate its circumference.

5 The centre circle of a football pitch has a diameter of 6 m. Calculate the length of the white line which forms the circle.

6 A circular flower bed of diameter 4 m has a plastic strip around the edge of it. How long is this plastic strip?

7 A circular patch of grass has a radius of 7 m. Calculate the circumference.

8 The length of the minute hand of a clock is 2.5 m. Calculate the distance travelled by the tip of the hand in one hour.

9 A barrel of radius 26 cm has a metal band fixed around its top edge. How long is this strip of metal?

10 A tin of soup has a label around it. The radius of the can is 4 cm. How long is the label?

EXAMPLE 11

The distance around the edge of a circular pond is 10.5 metres. Calculate the radius of the pond.

Using the formula $C = \pi d$

$10.5 = \pi \times d$

$\frac{10.5}{\pi} = d$

which gives $d = 3.342\ldots$

and so $r = \frac{3.342\ldots}{2} = 1.67$ m (3 s.f)

Remember that $d = 2 \times r$, so divide d by 2 to get r.

EXERCISE 9K

In all the following questions, take $\pi = 3.142$ or use the π button on your calculator.

1. Find the diameter of the circle when the circumference is

(a) 12.56 cm (b) 25.681 cm (c) 1.2 cm

(d) 314.2 cm (e) 8.6 cm

2 Find the radius of the circle when the circumference is

(a) 25.12 cm (b) 15.662 cm (c) 24.77 cm

(d) 119.5 cm (e) 4.8 cm

3 The rim of a circular crater in Canada was found to be 10 900 m long. What is the diameter of the crater?

4 The circumference of the Earth is 40 000 km. Calculate the diameter of the Earth.

5 The distance around a tower is 274.4 m. Calculate the diameter of the tower.

6 A waterwheel has a circumference of 69 m. Calculate the radius of the waterwheel.

7 An circular athletics track is 400 m long. Calculate the radius of the track.

8 The label around a can of soup is 40.8 cm long. Calculate the radius of the can.

Rolling

When a wheel rolls, the distance that a point on the wheel travels in one revolution is the circumference of the wheel.

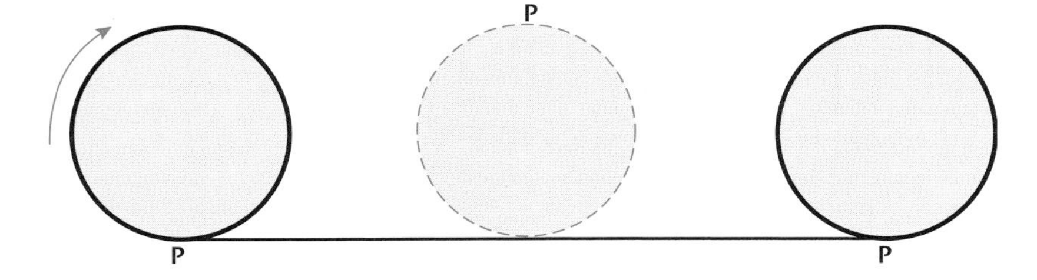

EXERCISE 9L

In all the following questions, take $\pi = 3.142$ or use the π button on your calculator.

1 A bicycle wheel has a diameter of 72 cm. How far does the bicycle travel in one revolution?

2 Surveyors use an instrument called an odometer to measure distance. It has a wheel of diameter 31.83 cm. How far does it travel in one revolution of the wheel?

3 A circular running track has a diameter of 127.3 m. How many laps of the track will an athlete have to run to complete an 800 m race?

4 The diameter of a car wheel is 60 cm. Find the distance travelled in one revolution of the wheel.

More challenging questions

EXAMPLE 12

This protractor is a semicircle with radius 5 cm.
Calculate the perimeter of the protractor.

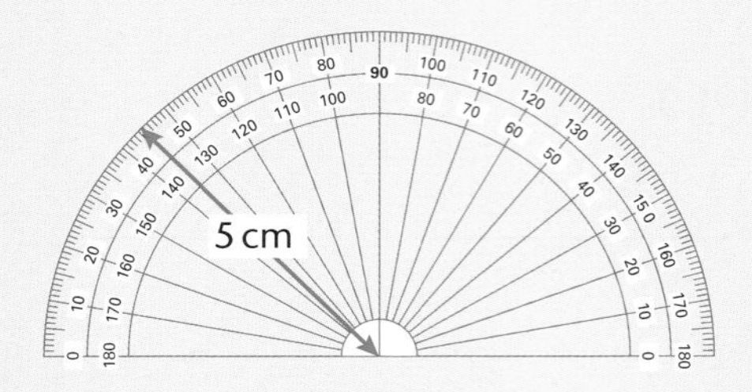

The circumference of a whole circle of radius 5 cm is $2 \times \pi \times 5$ cm.
So the distance around a semicircle of radius 5 cm is
$\pi \times 5 = 15.708\ldots$ cm

Leaving out the '2' means you calculate half of the circumference.

Perimeter = semicircular distance + diameter
= 15.708… + 10
= 25.7 cm (3 s.f.)

Don't forget to add on the straight edge.

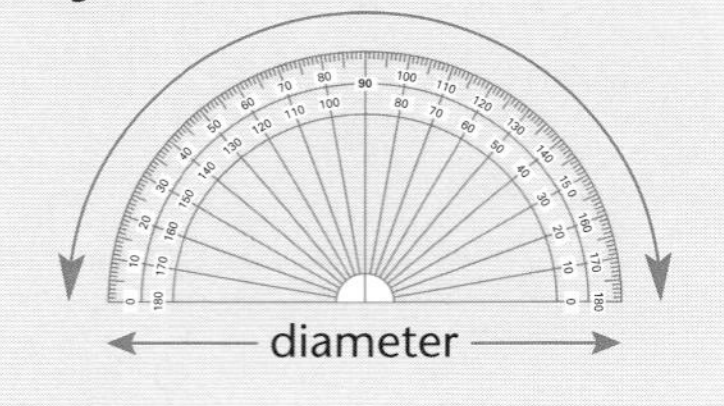

EXERCISE 9M

1 A 'quarter-light' window in the shape of a quarter of a circle of radius 52 cm. Calculate the length of the perimeter.

Use Example 12 to help you.

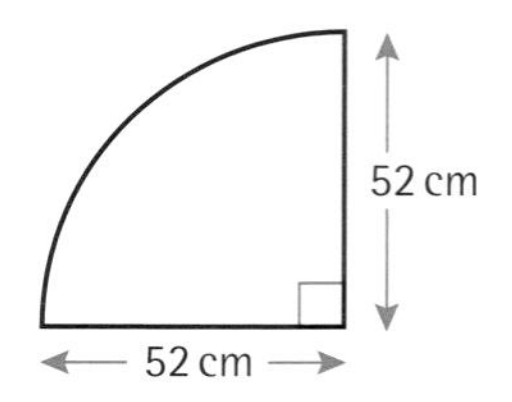

2 A door in the shape of a semicircle on top of a rectangle. Calculate the length of the perimeter.

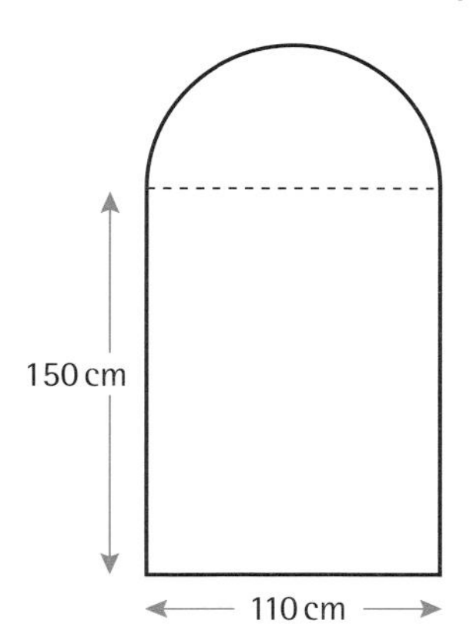

You need the diameter of the semicircle. Look at measurements on other parts of the diagram to find this.

3 A running track in the shape of two semicircular ends joined by two 'straights'. Calculate the length of the perimeter.

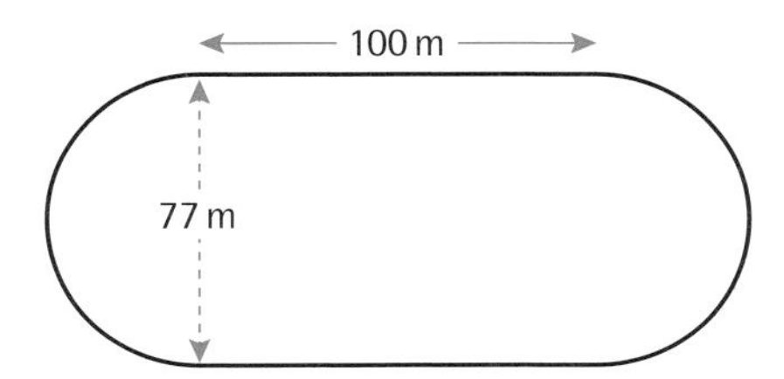

4 A frame has two circles of wire. The larger one has diameter 40 cm. The smaller one has diameter 20 cm. They are held together by four straight pieces, as shown in the diagram.

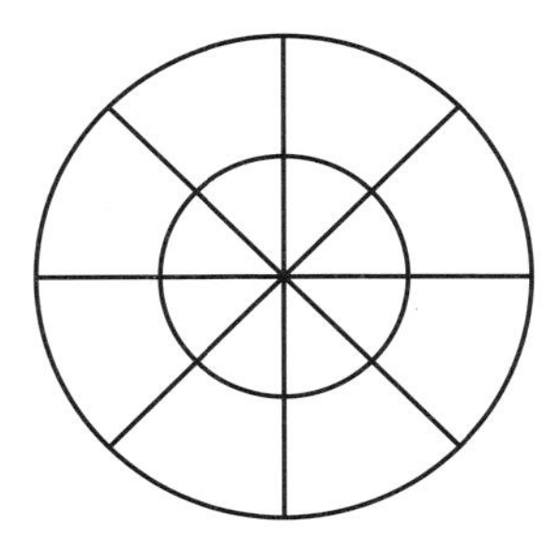

Calculate the total length of wire needed to make the frame.

Area of a circle

EXAMPLE 13

Calculate the area of these circles

(a) Radius = 10 cm

(b) Diameter = 18.6 cm

(a) $A = \pi r^2$
$= \pi \times 10^2$
$= \pi \times 100$
$= 314\text{ cm}^2$ (3 s.f.)

or $A = \pi r^2$
$= \pi \times 10^2 = 314.159\ldots$
$= 314\text{ cm}^2$ (3 s.f.)

Using the π button on the calculator.

(b) $r = d \div 2 = 18.6 \div 2 = 9.3$ cm

$A = \pi r^2$
$= \pi \times 9.3^2$
$= 272\text{ cm}^2$ (3 s.f.)

$A = \pi r^2$ uses the radius.

First find the radius.

Using the π button on the calculator.

EXERCISE 9N

1 Calculate the areas of these circles.

(a)

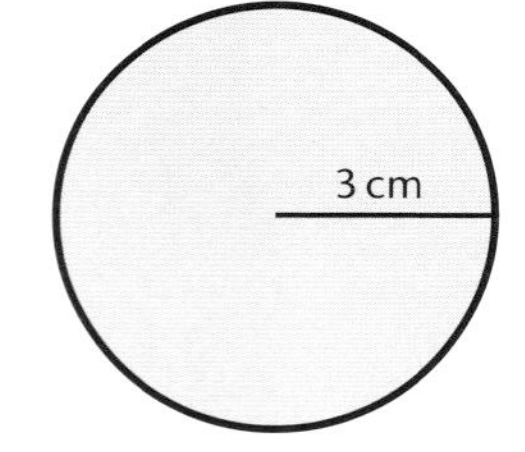

(b)

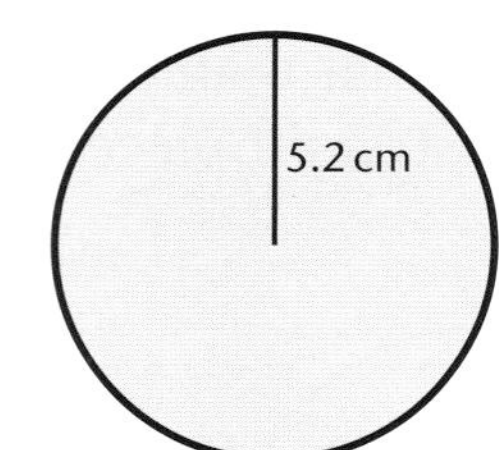

(c)

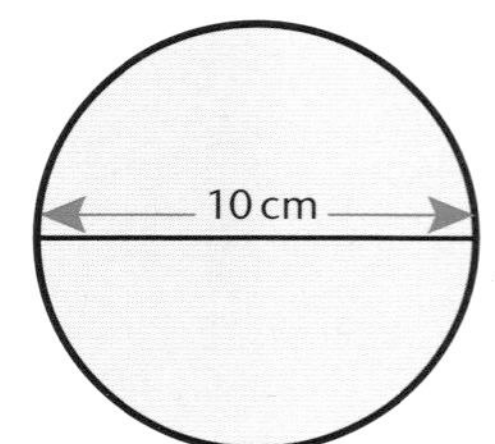

(d)

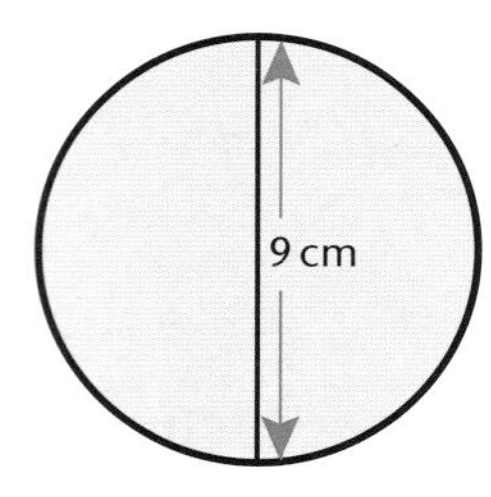

2 Calculate the areas of these circles. Use the π button on your calculator.

(a)

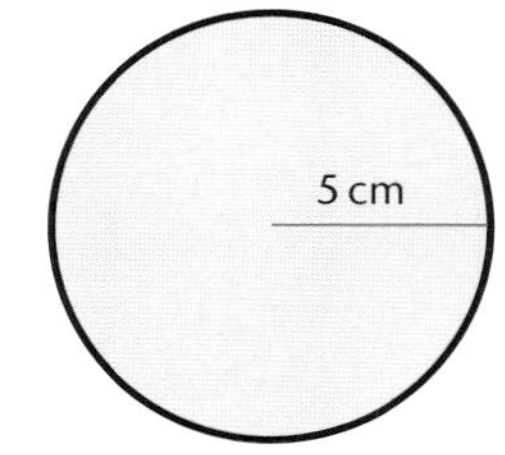

(b)

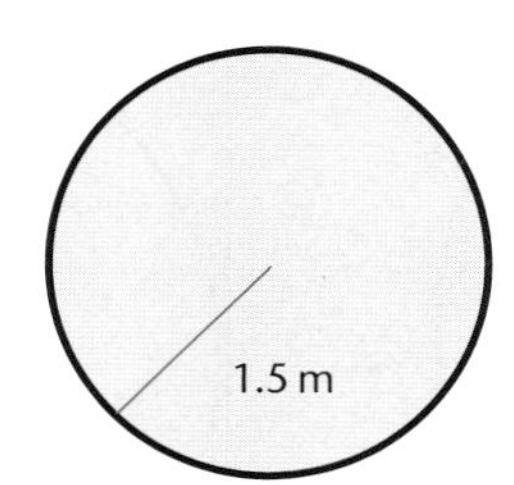

(c)

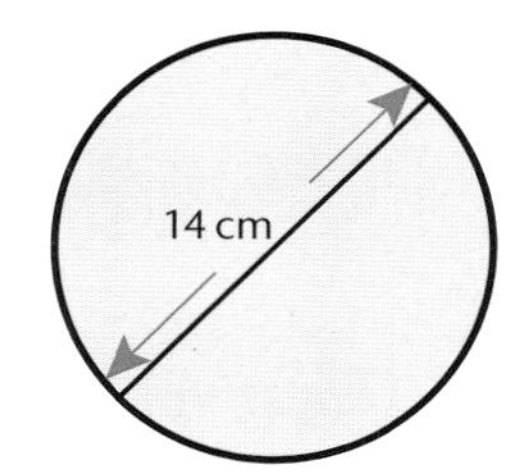

(d)

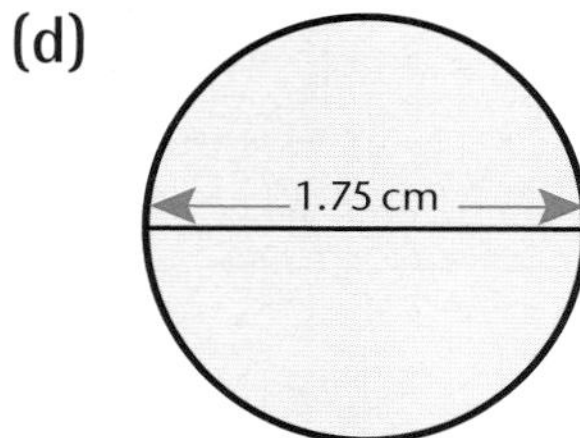

3 Calculate the area of these circles.

(a) Diameter = 12 cm **(b)** Diameter = 8 cm

(c) Diameter = 30 cm

4 Calculate the area of these circles.

(a) Radius = 23.5 cm **(b)** Radius = 6.7 cm

(c) Radius = 0.84 m

5 A circular table has a radius of 0.5 m. Calculate the area of the table.

6 A circular pond has a diameter of 4.8 m. Calculate the area of the pond.

7 The base of a vase is a circle with diameter 12 cm. Calculate the area of the base of the vase.

8 The cross-section of cylinder is a circle with diameter 24 cm. Calculate the area of the cross-section of the cylinder. Give your answer to the nearest whole number.

9 A circular paving slab has a circumference of 100 cm.

(a) Calculate the radius of the paving slab.

(b) Calculate the area of the paving slab.

10 Find the area shaded purple in each of these.

(a)

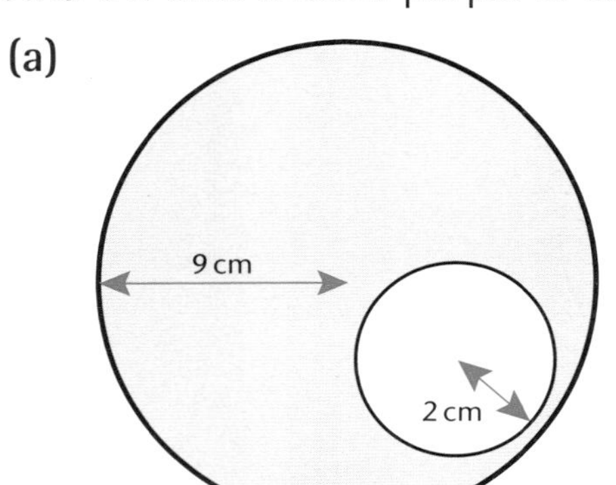

(b)

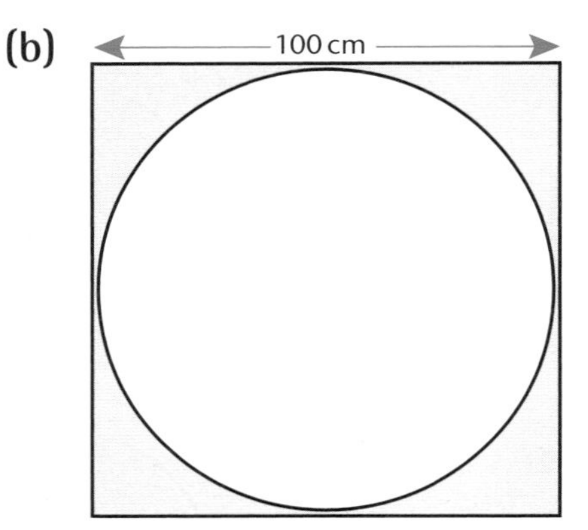

EXAMPLE 14

A circle has area 215 cm². Calculate its diameter.

Using the formula $A = \pi r^2$

$$215 = \pi \times r^2$$

$$\frac{215}{\pi} = r^2$$

$$r^2 = 68.436\ldots$$

so $r = \sqrt{68.436\ldots}$

$= 8.272$

$d = 2r = 2 \times 8.272\ldots = 16.5$ cm (3 s.f.)

$\sqrt{\ }$ means square root.

Keep the full calculator display at this stage.

Only round your answer at the end.

You can calculate the area of compound shapes involving circles or parts of circles.

EXERCISE 90

1 A circle has area 180 cm².
Calculate the radius of the circle.

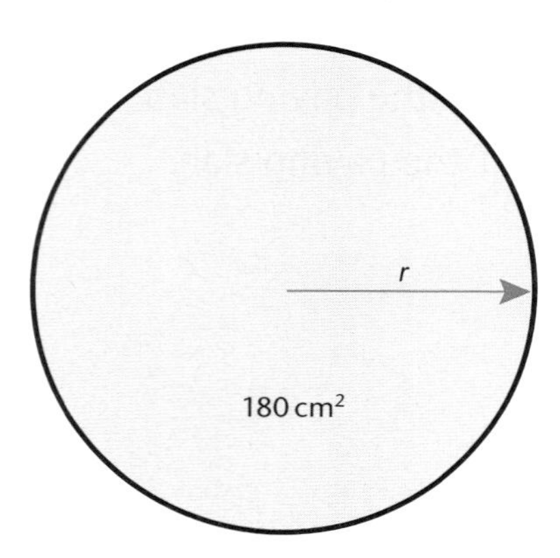

2 A circle has area 164 cm^2. Calculate its diameter, giving your answer to 3 s.f.

3 A circular plate has area 500 cm^2. Calculate the diameter of the plate. Give your answer to the nearest millimetre.

4 A circular garden has area 250 m^2.

(a) Calculate the radius of the garden.

(b) Calculate the circumference of the garden.

5 The area of the cross section of a water wheel is 45 m^2. Calculate the circumference of the water wheel. Give your answer to the nearest metre.

6 A square of side 14 cm and a circle have the same area.

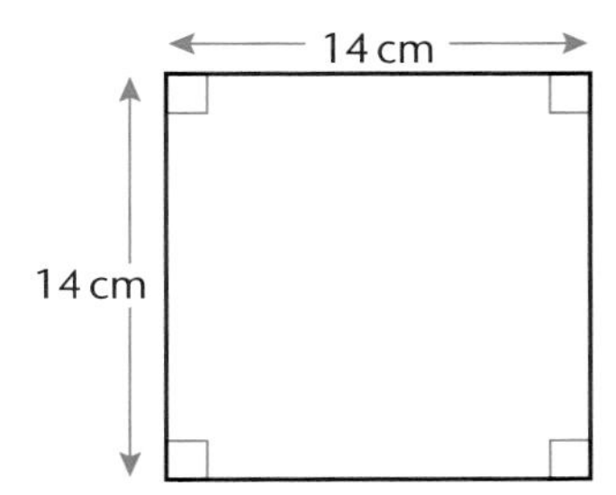

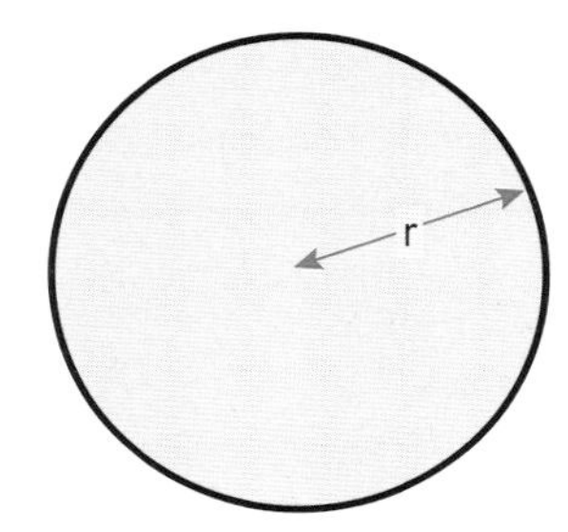

Find the radius of the circle.

7 In the diagram, the area of circle B is twice the area of circle A. Work out the radius of circle B.

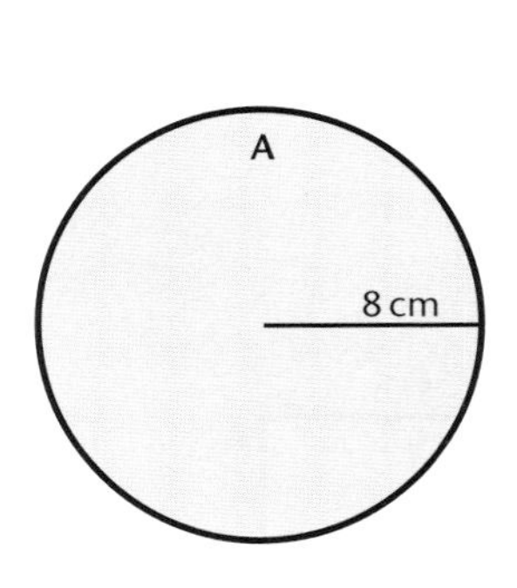

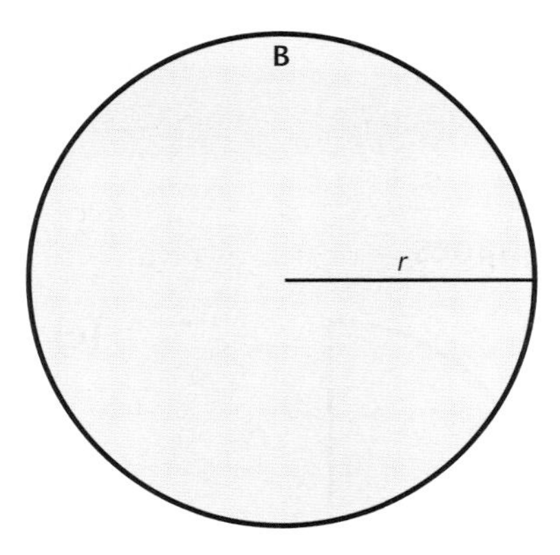

EXAMPLE 15

A garden is in the shape of a rectangle with two semi-circles, one on the length of the rectangle and one on the width, as shown in the diagram.

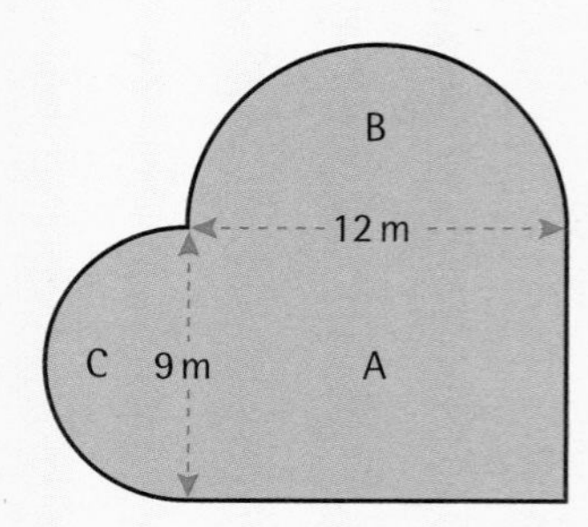

Calculate the area of the garden, giving your answer to the nearest square metre.

Area of rectangle $A = 12 \times 9 = 108\ \text{m}^2$

Area of semicircle $B = \frac{1}{2} \times \pi \times 6^2 = 56.548\ldots$

Area of semicircle $C = \frac{1}{2} \times \pi \times 4.5^2 = 31.808\ldots$

Total area $= 108 + 56.548\ldots + 31.808\ldots$

$= 196.35\ldots$

$= 196\ \text{m}^2$ (to nearest m^2)

Area of semi-circle $= \frac{1}{2}$ area of circle.
Diameter of $B = 12$ m, so radius $= 6$ m.
Diameter of $C = 9$ m, so radius $= 4.5$ m

If you round each of the three answers to the nearest whole number and then add them your answer would be $108 + 57 + 32 = 197\ \text{m}^2$ (which is incorrect).
You should round at the *end* of the calculation.

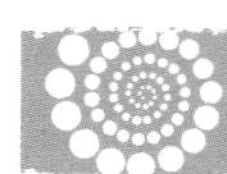

EXERCISE 9P

1 Find the area of each of these shapes.

(a)

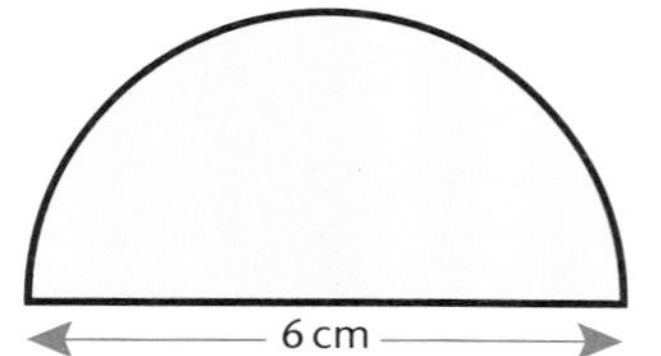

(b)

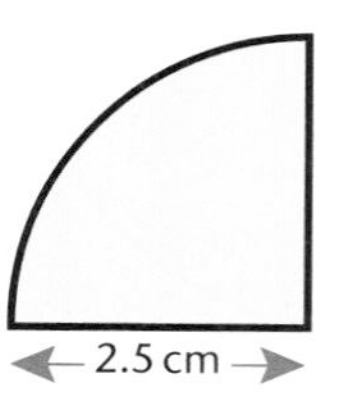

(c)

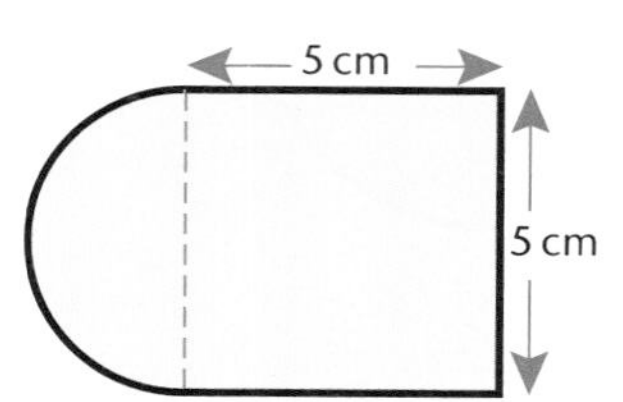

(d)

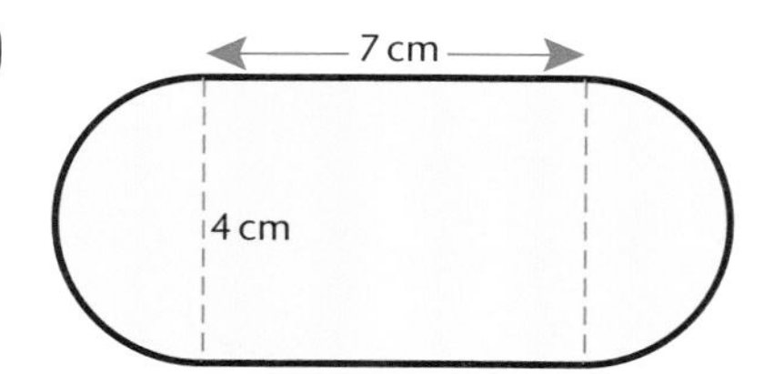

(e)

9.2 cm

6.4 cm

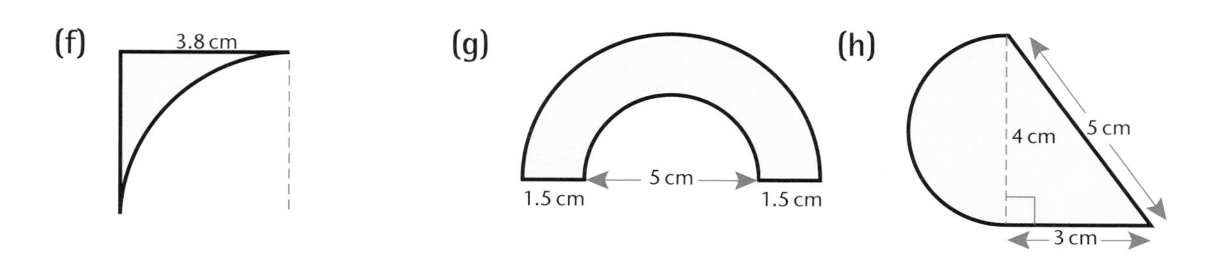

9.4 3-D objects

Nets of 3-D objects

Here is a closed box in the shape of a cuboid. It has been opened out so that you can see the shape of the card it is made from.

The 2-D shape that you see is called a **net**.

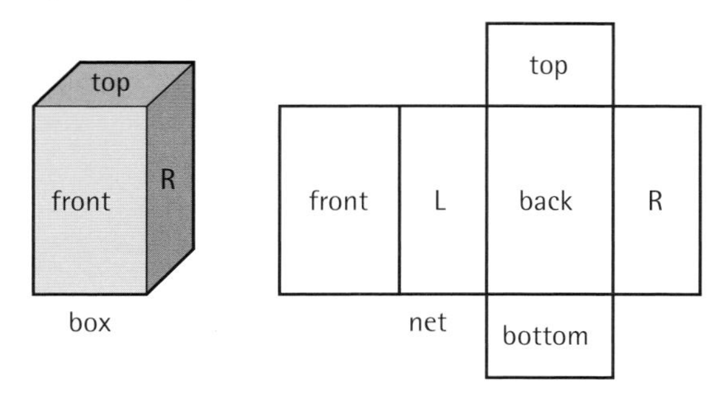

The net of a 3-D object is any 2-D shape that folds up to make the 3-D object.

EXAMPLE 16

Make an accurate drawing of the net of this 3-D object.

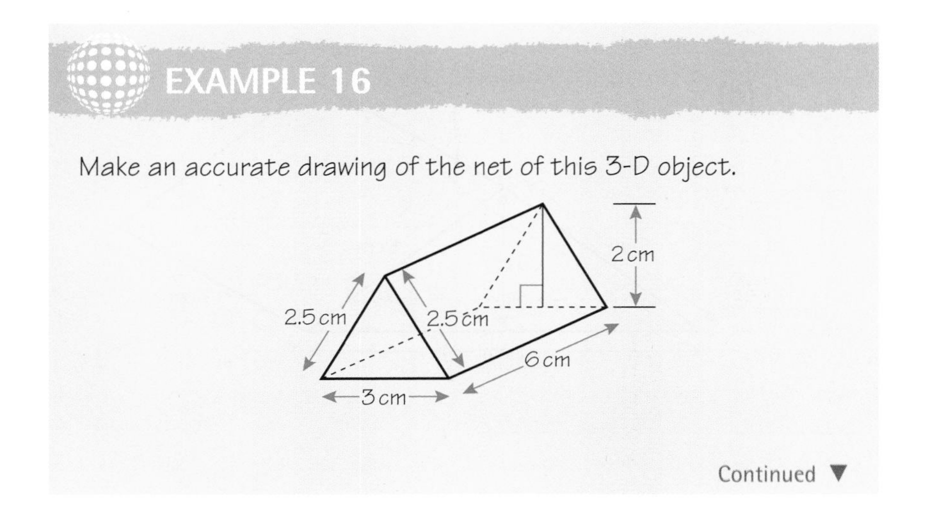

Continued ▼

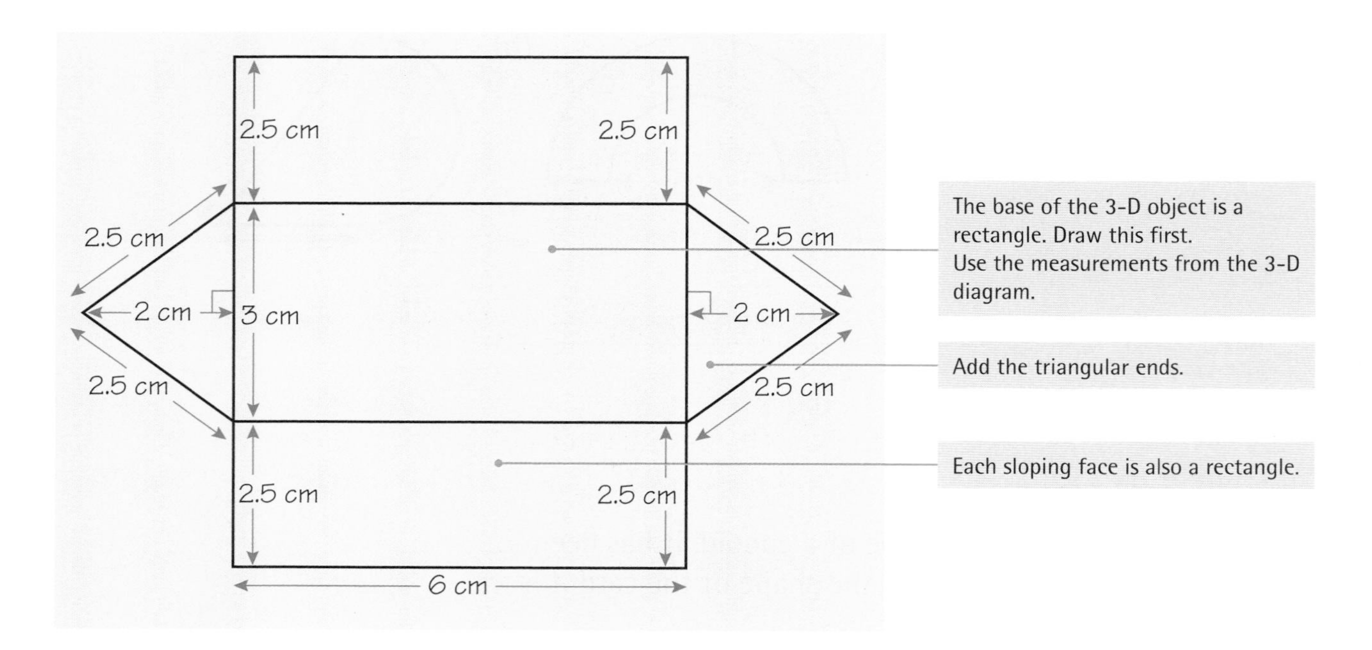

EXERCISE 9Q

1 Make accurate drawings of the nets of these 3-D objects.

(a)

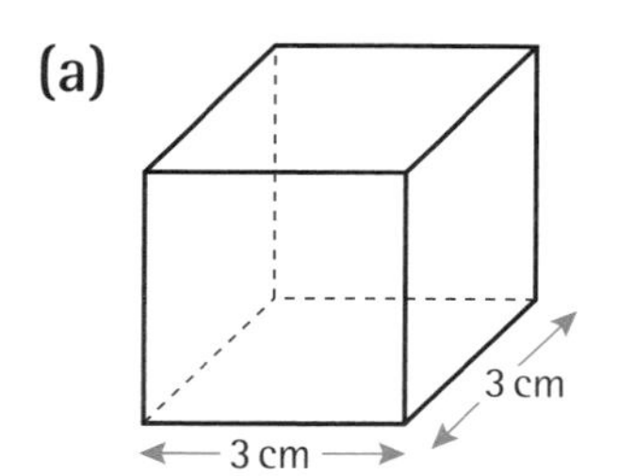

(b)

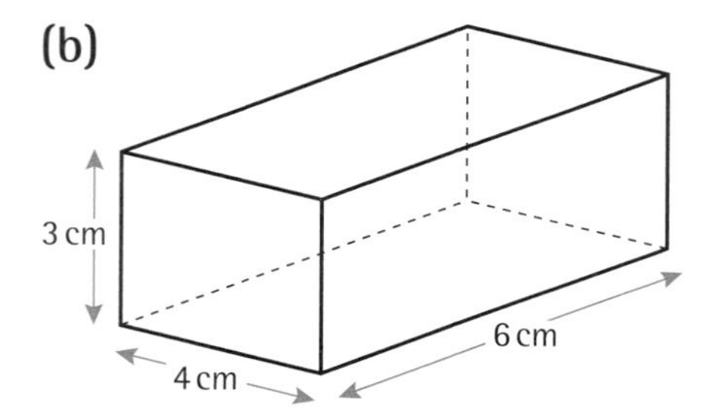

(c)

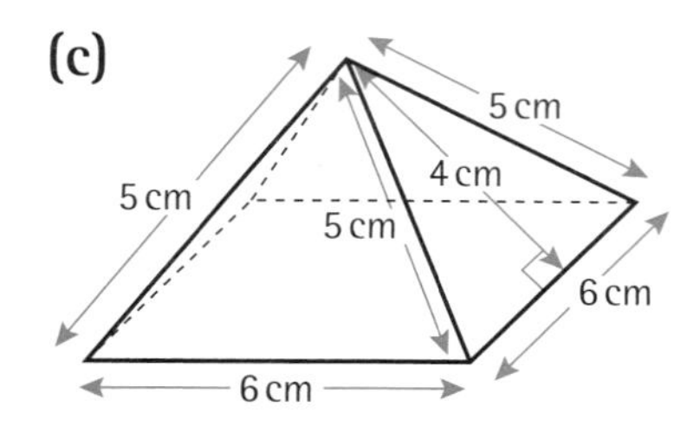

(d)

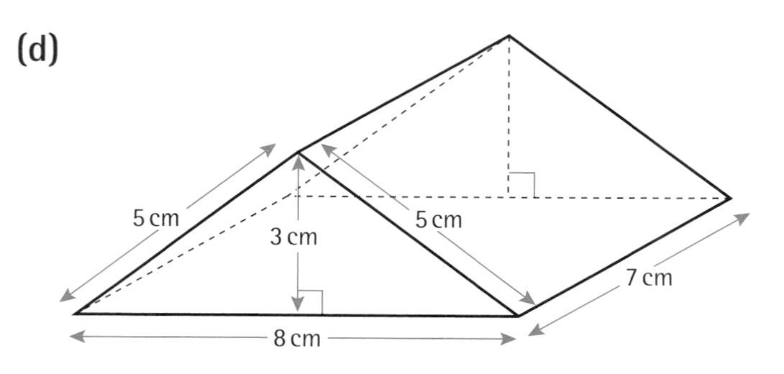

9.5 Prisms

You will need to know

- area of a rectangle = length × width
  area = $l \times w$
- area of a triangle = $\frac{1}{2}$ × base × height
  area = $\frac{1}{2} \times b \times h$
- area of a trapezium = $\frac{1}{2}$ × (sum of parallel sides) × height
  area = $\frac{1}{2} \times (a + b) \times h$

A **prism** is a 3-D object whose **cross-section** is the same all through its length.

In these prisms the cross-section is shaded.

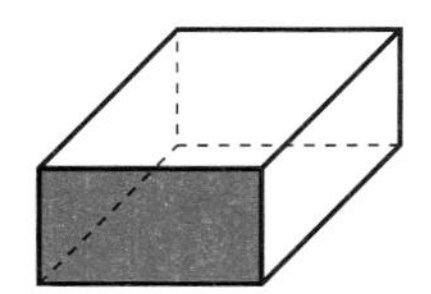

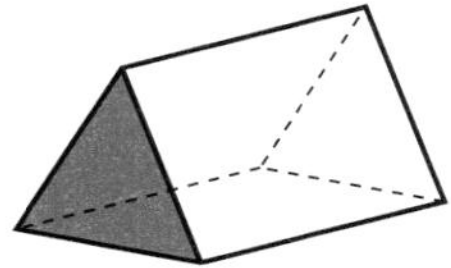

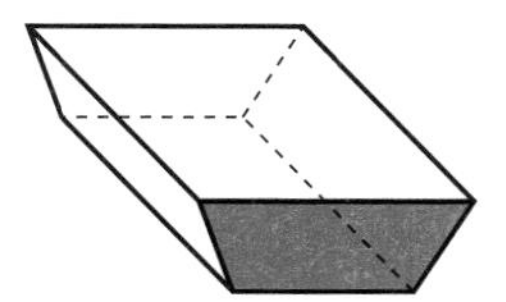

If you cut a 'slice' parallel to the end face, your slice will be the same shape as the end face.

Volume of a prism

A **cuboid** is a prism. Its cross-section is a rectangle.

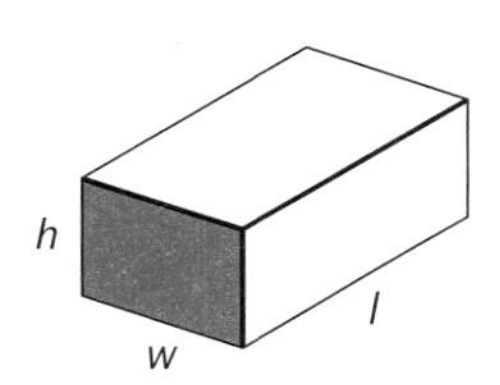

Imagine the cuboid made from cubes, for example,

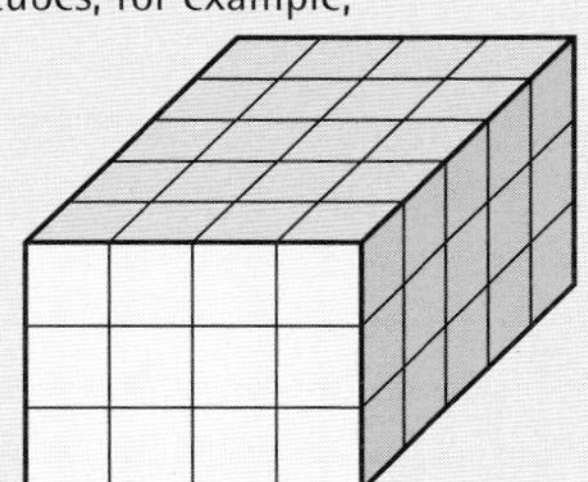

In the end face there are 3 × 4 cubes.
Along the length of the cuboid there are 5 'slices' of 3 × 4 cubes.
So there are 3 × 4 × 5 cubes in total.

To calculate its volume you use this formula

Volume of cuboid = height × width × length
$= h \times w \times l$

Another way of writing this formula is

Volume = area of end face × length

Area of end face = $h \times w$

You can use a similar formula to calculate the volume of *any* prism.

Volume of prism = area of cross-section × length

Area of cross-section
= area of end face

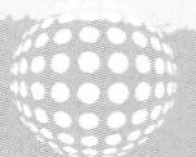

EXAMPLE 17

Find the volume of this cuboid.

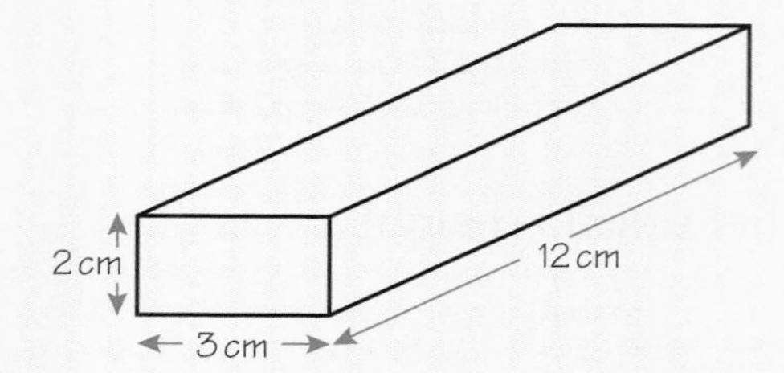

Volume $= h \times w \times l$

$= 2 \times 3 \times 12$

$= 72\ \text{cm}^3$

Cubic units for volume.

EXAMPLE 18

The cross-section of this prism is a trapezium.
Calculate the volume of the prism.

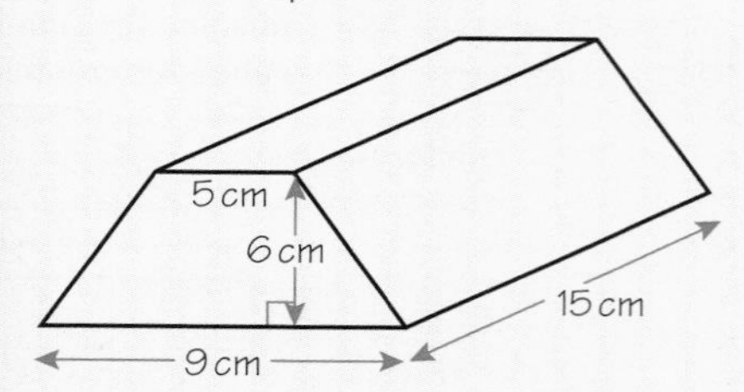

Volume = area of cross-section × length

Area of cross-section $= \frac{1}{2}(a + b) \times h$

$= \frac{1}{2}(5 + 9) \times 6$

$= \frac{1}{2} \times 14 \times 6$

$= 7 \times 6$

$= 42\ \text{cm}^2$

For this trapezium $a = 5$, $b = 9$, $h = 6$.

Order of operations brackets first.

Volume of prism $= 42 \times 15$

$= 630\ \text{cm}^3$

EXAMPLE 19

Calculate the volume of this prism.

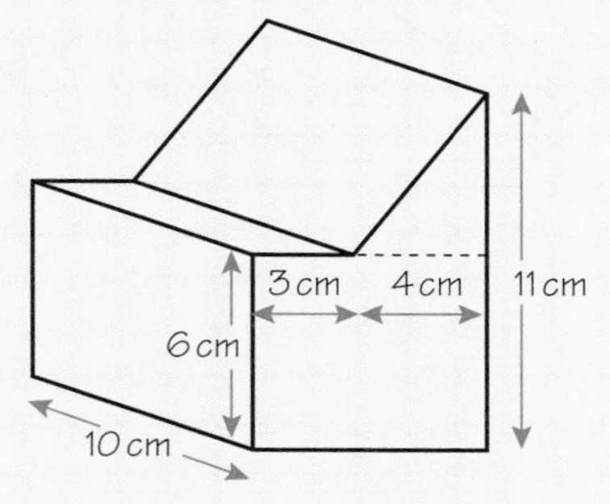

Area of rectangle $= 7 \times 6 = 42$ cm^2

Area of triangle $= \frac{1}{2} \times 4 \times 5 = 10$ cm^2

Area of cross-section $= 42 + 10 = 52$ cm^2

Volume of prism = area of cross-section × length

$= 52 \times 10 = 520$ cm^3

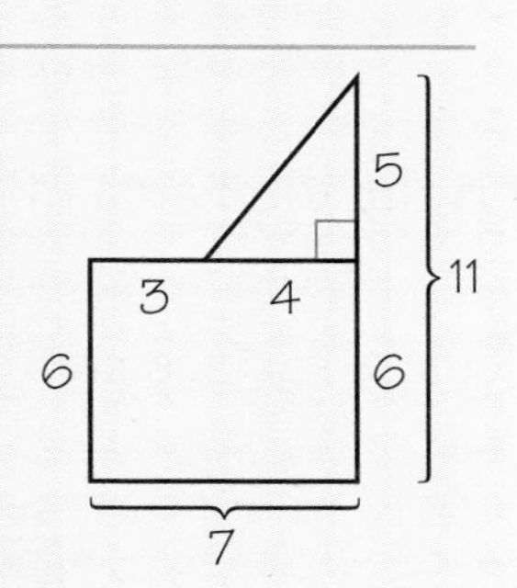

First find the area of the cross-section (end face). Split it into a rectangle and a triangle, as shown. Then add the two areas together.

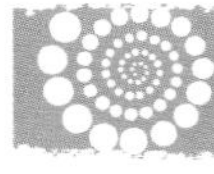

EXERCISE 9R

1 Find the volume of each prism.

(a)

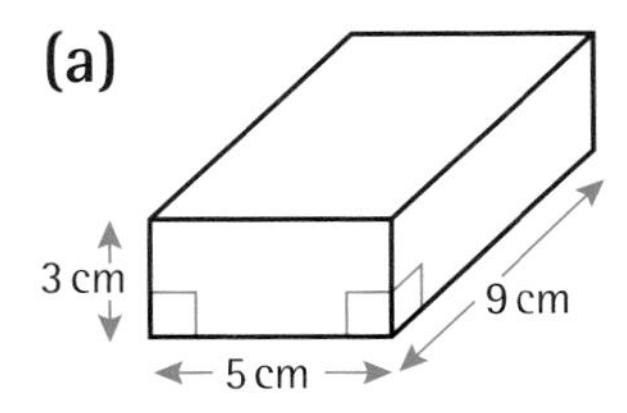

(b)

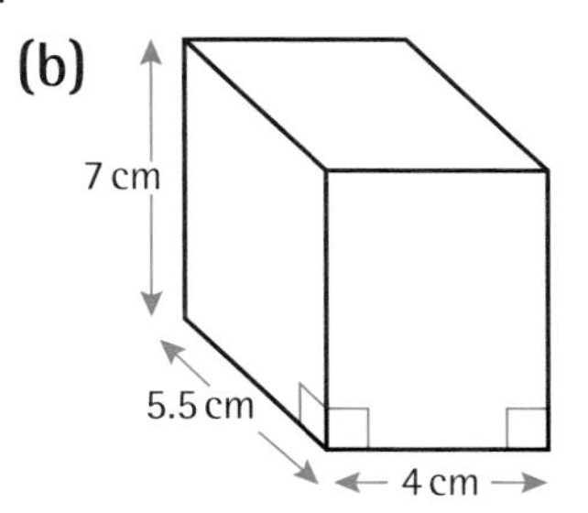

(c)

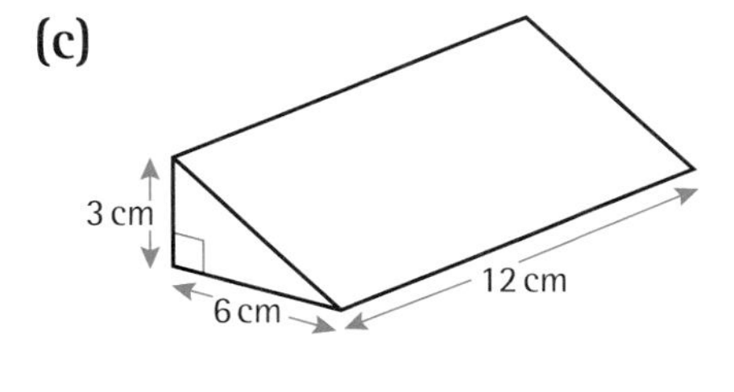

2 Find the volume of each prism.

(a)

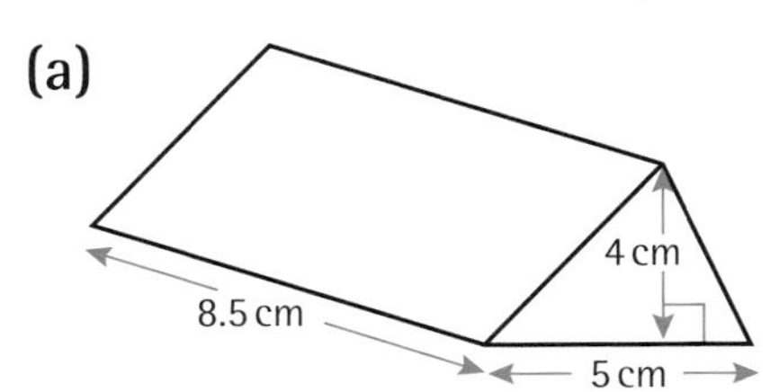

(b)

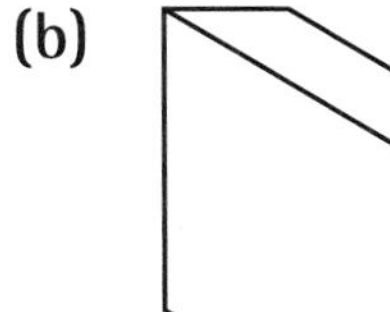

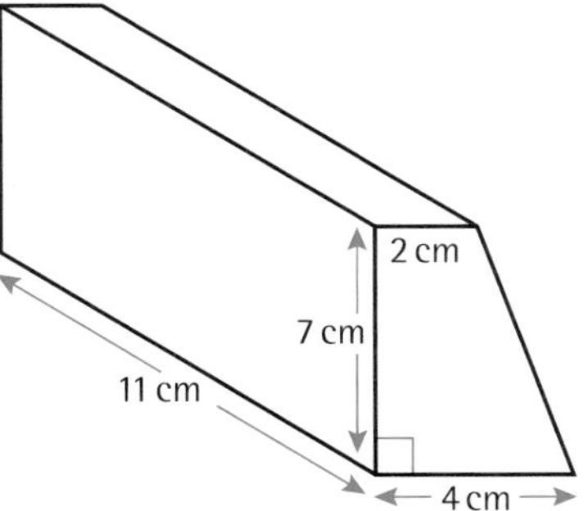

3 Calculate the volume of these prisms.

(a)

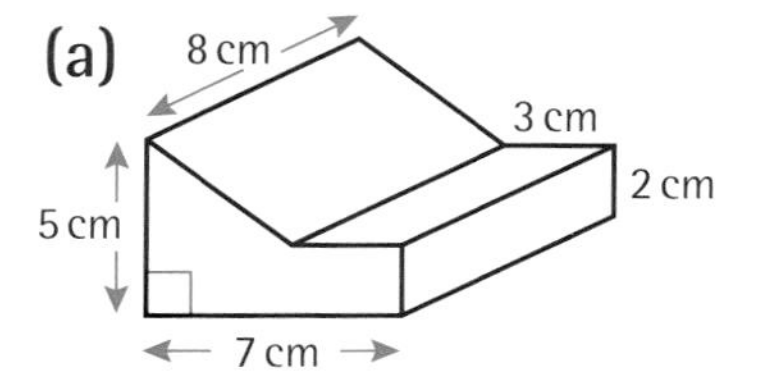

(b)

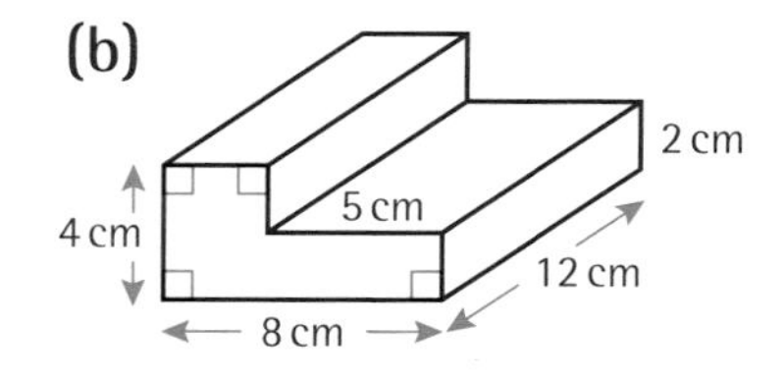

(c)

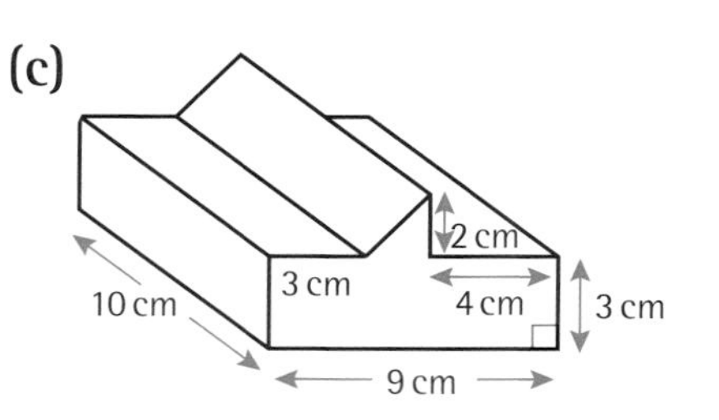

Split the end face into simpler shapes to calculate its area.

Surface area of a prism

The surface area of a 3-D object is the total area of all of its faces.

You can use nets to help you calculate surface area.
For the cuboid in Example 17

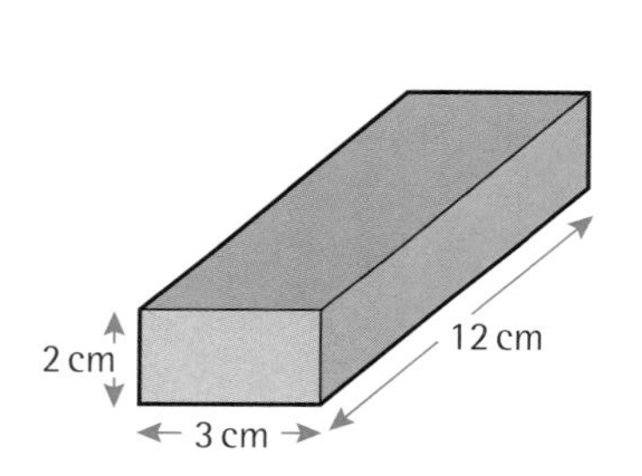

the net is

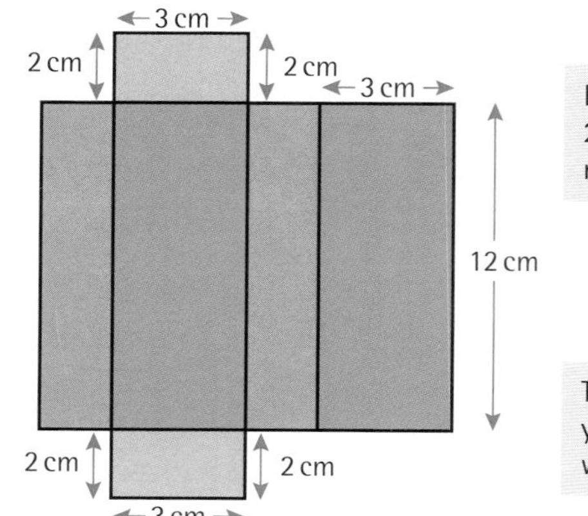

Look back at page 265 for help with nets.

The colour shows you which face is which.

The net is the 3-D object 'opened out', so you can see all the faces at once.

These are 2 green faces, area 3×12
2 orange faces, area 2×12
2 blue faces, area 3×2

$$\begin{aligned}\text{Total surface area} &= 2 \times (3 \times 12 + 2 \times 12 + 3 \times 2)\\ &= 2(36 + 24 + 6)\\ &= 2 \times 66\\ &= 132 \text{ cm}^2\end{aligned}$$

You can use this formula to calculate the surface area of a cuboid.

$$\text{Surface area} = 2(h \times w + h \times l + w \times l)$$
$$= 2(hw + hl + wl)$$

These are the areas of the faces in its net.

EXAMPLE 20

Calculate the surface area of this cuboid.

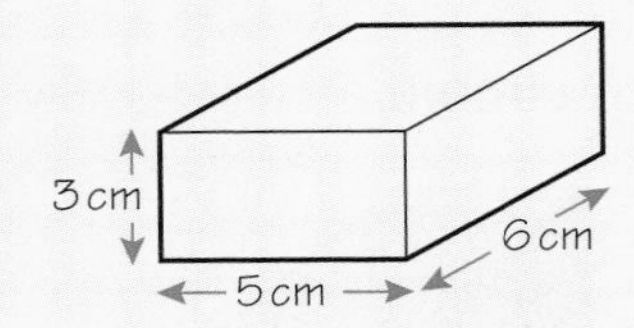

$h = 3$, $w = 5$, $l = 6$

$$\text{Surface area} = 2(hw + hl + wl)$$
$$= 2(3 \times 5 + 3 \times 6 + 5 \times 6)$$
$$= 2(15 + 18 + 30)$$
$$= 2 \times 63$$
$$= 126 \text{ cm}^2$$

EXAMPLE 21

Calculate the surface area of this prism.

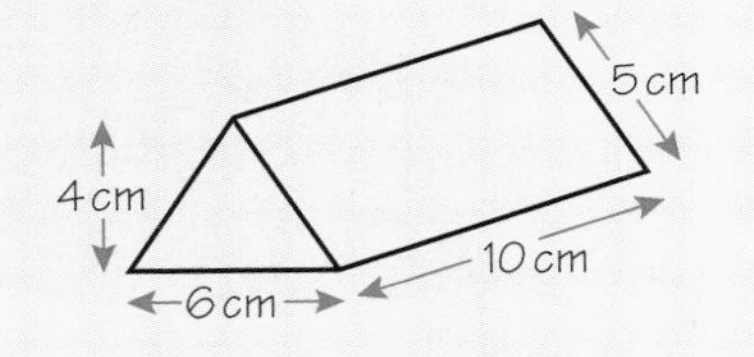

Sketch the net and label the measurements.

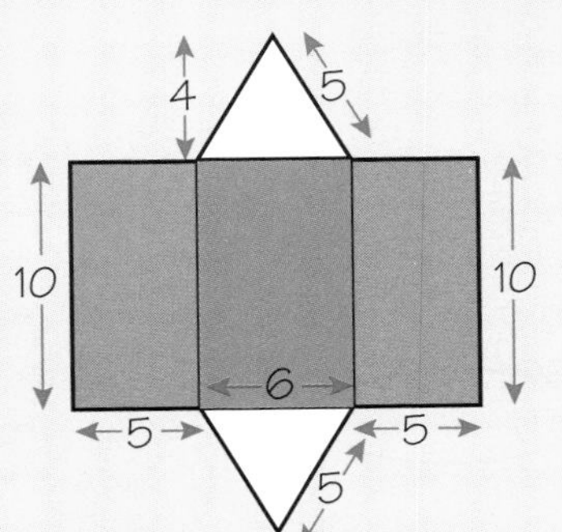

Area of each orange face $= 10 \times 5$
$= 50 \text{ cm}^2$

2 of these

Area of green face $= 6 \times 10$
$= 60 \text{ cm}^2$

Continued ▼

Area of triangle $= \frac{1}{2} \times h \times b$
$= \frac{1}{2} \times 4 \times 6$
$= 12\ \text{cm}^2$

2 of these

Surface area of prism $= 2 \times 50 + 60 + 2 \times 12$
$= 100 + 60 + 24$
$= 184\ \text{cm}^2$

EXERCISE 9S

1 Calculate the surface area of each cuboid.

(a)

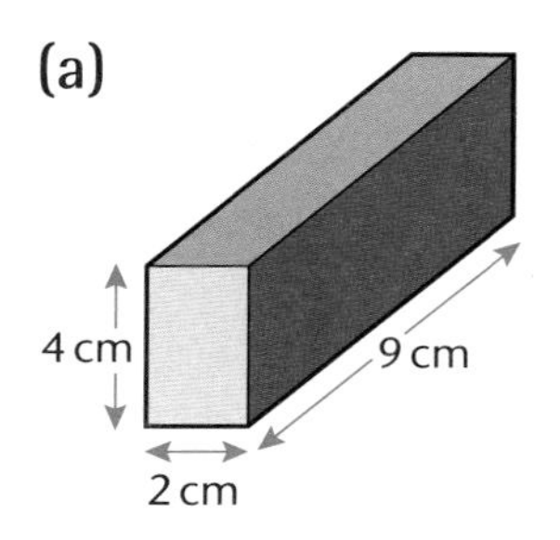

(b)

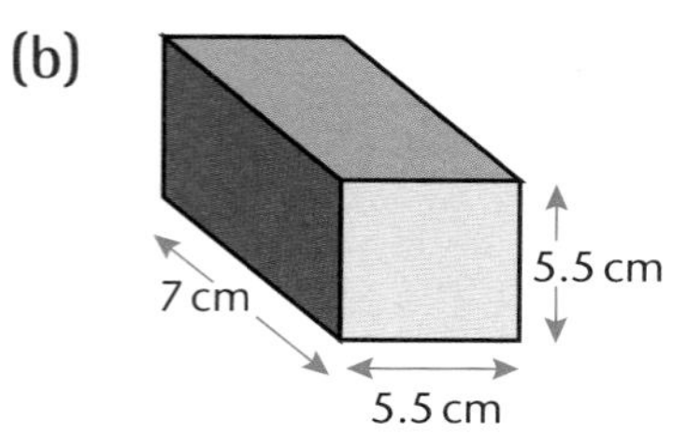

(c)

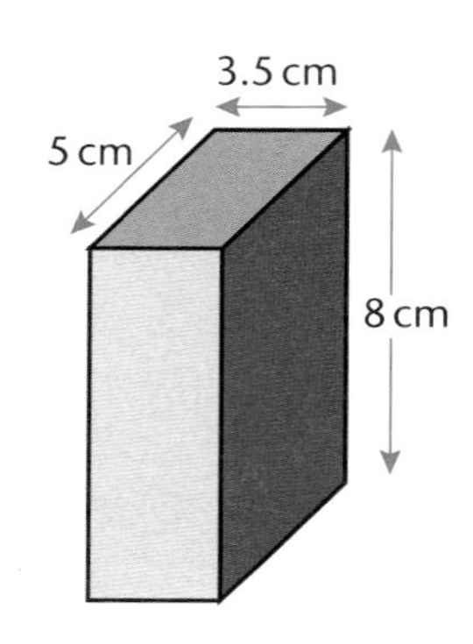

2 Calculate the surface area of each prism.

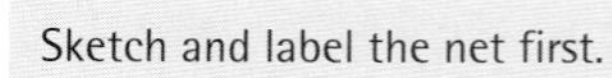
Sketch and label the net first.

(a)

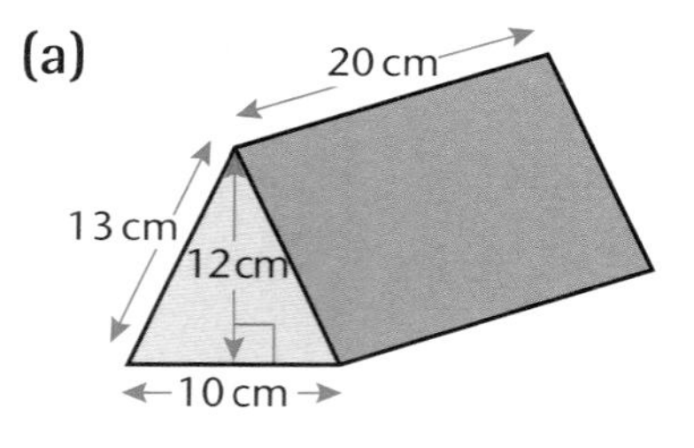

(b)

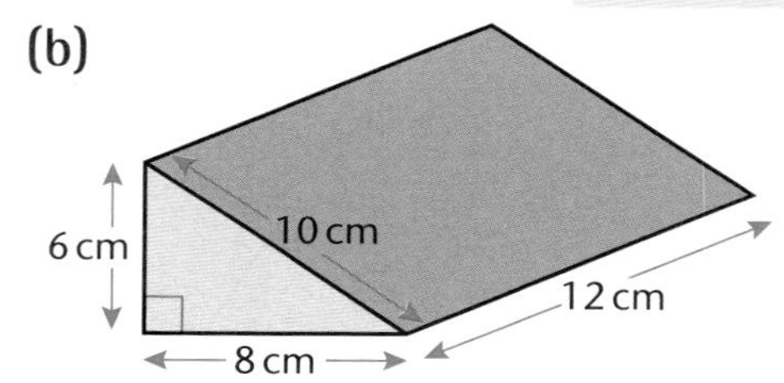

9.6 Cylinders

Volume of a cylinder

You will need to know

- area of a circle $= \pi r^2$

A cylinder is a prism whose cross-sectional area is a circle.

Volume of a cylinder = area of cross-section × height

$$V = \pi r^2 h$$

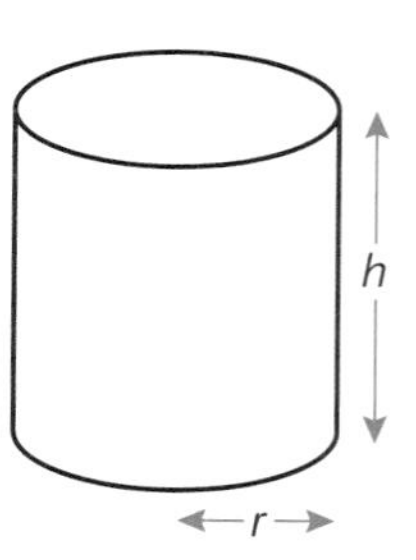

EXAMPLE 22

Find the volume of a cylinder of radius 9 cm and height 30 cm.

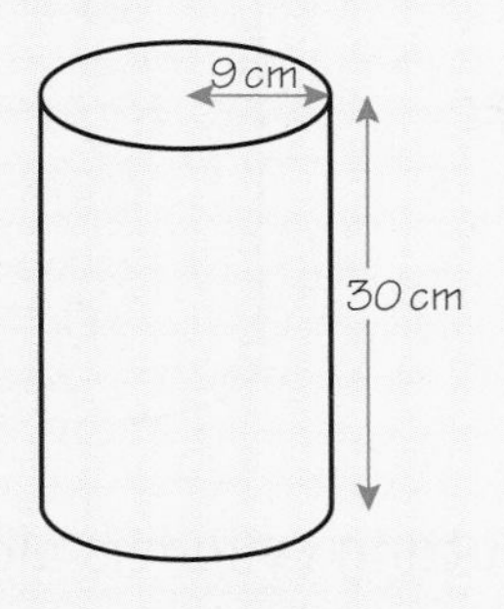

Volume $= \pi r^2 h$
$= \pi \times 9^2 \times 30$
$= \pi \times 81 \times 30$
$= 7630 \text{ cm}^3$ (3 s.f.)

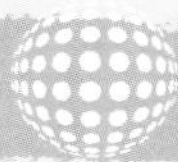

EXAMPLE 23

A cylindrical container holds 8.5 litres of liquid.
The height of the container is 24 cm.
Calculate the radius of the base.

8.5 litres $= 8.5 \times 1000 \text{ cm}^3 = 8500 \text{ cm}^3$

The volume is 8.5 ℓ. You must use 8500 cm³

Volume $= \pi r^2 h$

$8500 = \pi \times r^2 \times 24$

Substitute the values you know. $V = 8500$, $h = 24$

$\dfrac{8500}{\pi \times 24} = r^2$

$112.73\ldots = r^2$

Rearrange the formula to make r^2 the subject – divide by π and by 24.

$\sqrt{112.73\ldots} = r$

Use the √ key on your calculator.

radius $= 10.6$ cm (3 s.f.)

EXERCISE 9T

1 Find the volume of these cylinders.

(a) Base radius = 2 cm, height = 8 cm
(b) Base radius = 3 cm, height = 5 cm
(c) Base diameter = 14 cm, height = 20 cm
(d) Base radius = 4 cm, height = 2 m

2 Find the capacity, in litres, of an oil drum of base radius 30 cm and height 80 cm.

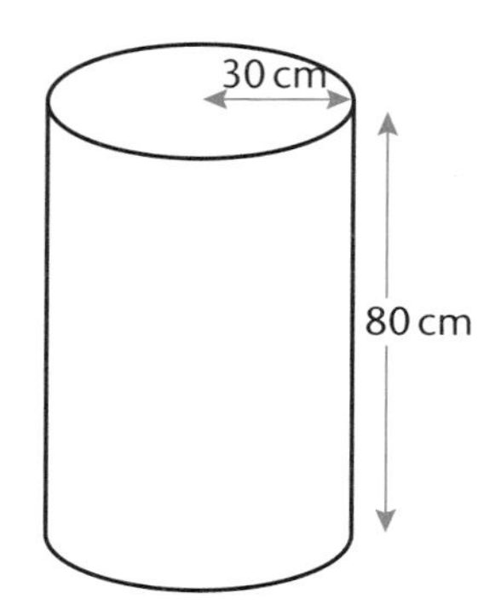

$1000 \text{ cm}^3 = 1\ell$

3 These objects were made by cutting cylinders along axes of symmetry. Find the volume of each object.

(a)

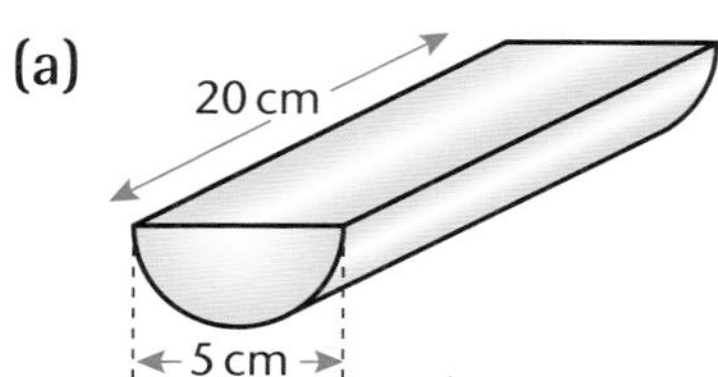

(b)

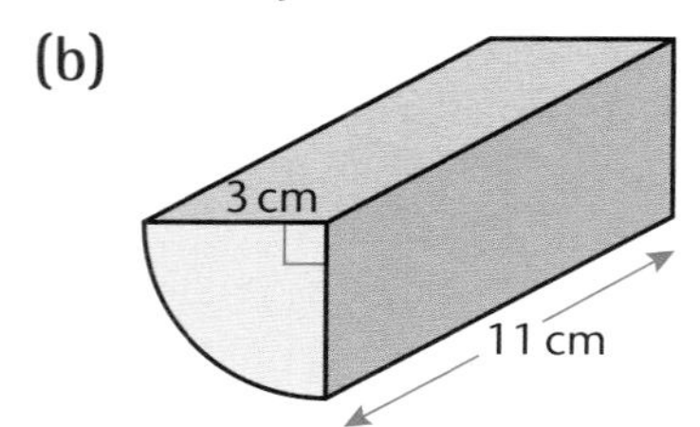

You are give the *diameter*

4 The diagrams show two cylindrical tins of cat food.

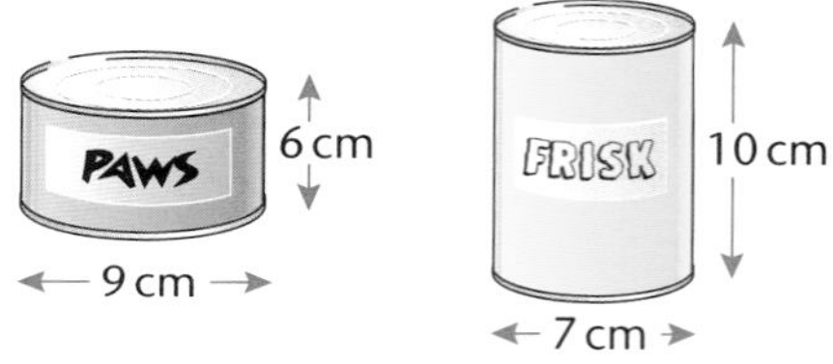

Which tin holds the most food?

5 A cylinder of height 27 cm holds 3.6 litres of liquid. What is the radius of the base of the cylinder?

Use Example 23 to help you.

6 A hollow metal tube, 1.5 m long, is in the form of a cylinder. The external radius is 7 cm and the internal radius is 6 cm.

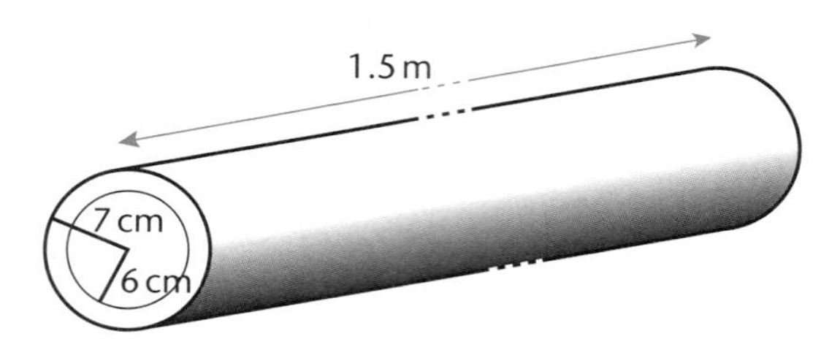

(a) Calculate the volume of metal in the tube.

(b) The mass of 1 cm^3 of the metal is 3.4 grams. Calculate the mass of the tube.

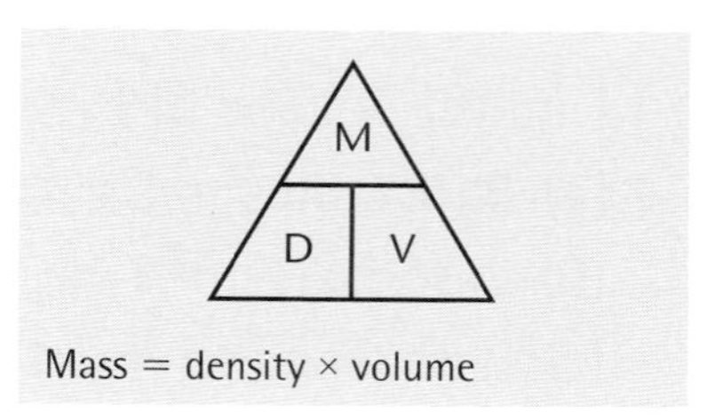

Mass = density × volume

7 A cylindrical metal rod 2 m long has a diameter of 4 cm.
The mass of the rod is 17 kg.
Calculate the density of the metal.

8 Cylindrical tins of diameter 10 cm and height 12 cm are packed *two deep* in a box in the shape of a cuboid 40 cm long, 20 cm wide and 24 cm high.

(a) How many tins will fit into the box?

(b) Calculate the volume of one tin.

(c) What is the total volume of all the tins?

(d) Calculate the volume of the box.

(e) What is the volume of unused space in the box when the tins are packed into it?

Surface area of a cylinder

You will need to know

- **circumference of a circle $= 2\pi r$**

The net of this cylinder is

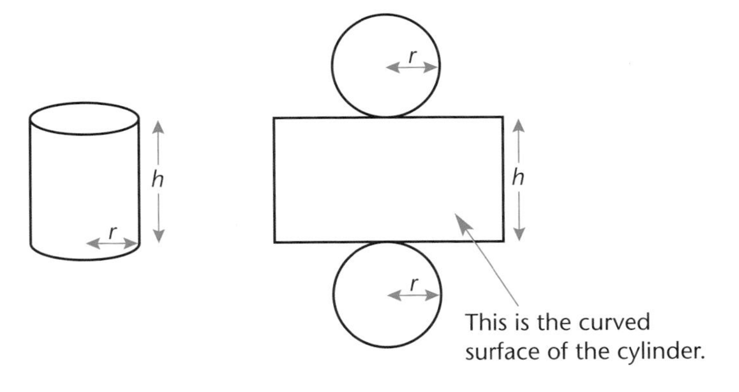

When you peel the label off a tin of beans, the label is rectangular.

The width of the rectangle is the same as the circumference of the circular top.

When you wrap the label round the tin, its top edge makes a circle.

Curved surface area $= 2\pi r \times h$
$= 2\pi rh$

The area of each circular end $= \pi r^2$

There are 2 of these.

Total surface area of a cylinder $= 2\pi rh + 2\pi r^2$

Area of curved surface + area of ends.

EXAMPLE 24

Calculate the surface area of a cylinder with base radius 12 cm and height 20 cm.

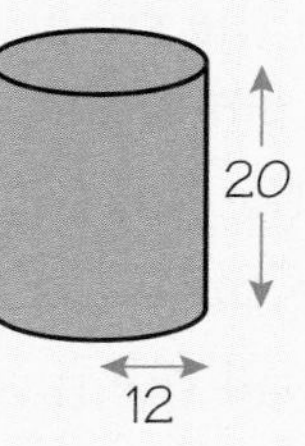

$$\text{TSA} = 2\pi rh + 2\pi r^2$$
$$= 2\pi \times 12 \times 20 + 2\pi \times 12^2$$
$$= 1507.96 + 904.78$$
$$= 2410 \text{ cm}^2 \text{ (3 s.f.)}$$

Numbers before letters.

Square units for area.

EXAMPLE 25

A thin cylindrical metal rod is 2 metres long and has diameter 4.6 cm.
Calculate its total surface area.

2 metres = 2 × 100 cm = 200 cm

Diameter = 4.6 cm, radius = $\frac{1}{2} \times 4.6 = 2.3$ cm

$$\text{TSA} = 2\pi rh + 2\pi r^2$$
$$= 2\pi \times 2.3 \times 200 + 2\pi \times 2.3^2$$
$$= 2890.265\ldots + 33.24\ldots$$
$$= 2923.503\ldots$$
$$= 2920 \text{ cm}^2 \text{ (3 s.f.)}$$

Mixed units. Change them all to centimetres.

Substitute $h = 200$, $r = 2.3$

Using the π key.

EXERCISE 9U

1 Find the total surface area of each cylinder in Exercise 9T question 1.

2 A cylindrical cushion is 30 cm long.
The radius of each end is 10 cm.
Calculate the area of fabric in the cover for this cushion.

3 A tin of beans has a diameter of 7.6 cm and a height of 11 cm.
The label on the tin overlaps by 1.5 cm.
The height of the label is 5 mm less than the height of the tin.

Beware – mixed units!

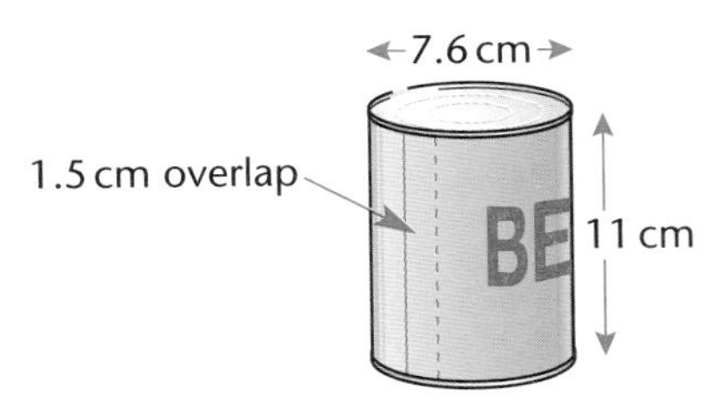

Calculate the surface area of the label around the tin.

4 A cylinder of height 10 cm has a curved surface area of 754 cm^2.
Calculate the volume of this cylinder.

5 Calculate the total surface area of each shape in Exercise 9T question 4.

Sketch the net.

EXAMINATION QUESTIONS

1

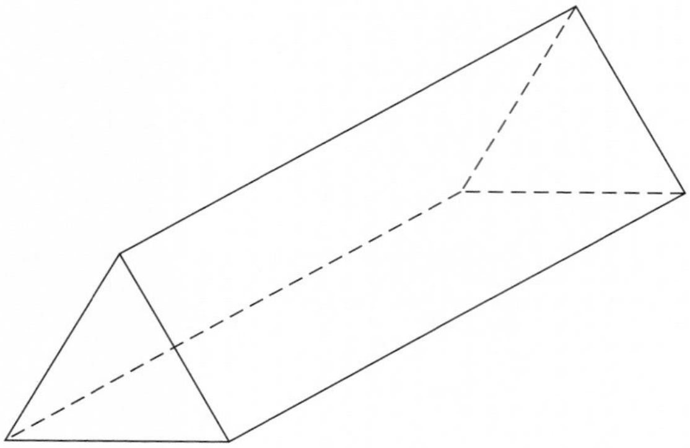

NOT TO SCALE

The diagram shows a triangular prism of length 4 cm.
The cross-section is an equilateral triangle of side 3 cm.
Draw an accurate net of the prism. [3]

(CIE Paper 1, Jun 2000)

---

2 A square has sides of length 9cm.

(a) Write down its area. [1]

(b) Write down its perimeter. [1]

(CIE Paper 3, Nov 2000)

---

3 A circular pool of water has a radius of 6 metres.

(a) (i) Work out the circumference of the pool. [2]

(ii) John walks round the circumference of the pool in 20 seconds
Work out his average speed in metres per second.
Give your answer correct to 1 decimal place. [2]

(b) Calculate the surface area of the pool, giving your answer

(i) in square metres, [2]

(ii) in square centimetres. [1]

(c) The water is 25 centimetres deep.
Work out the volume of water in the pool , giving your answer

(i) in cubic centimetres, [2]

(ii) in litres. [1 litre = 1000 cm^3] [1]

(CIE Paper 3, Nov 2000)

4

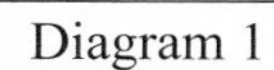

Diagram 1

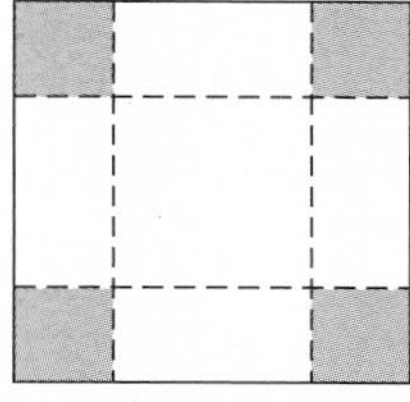

Diagram 2

NOT TO SCALE

A square card (Diagram 1) has sides of length 11 cm.

(a) Write down the area of the card. [1]

(b) Four equal squares, each with sides of length 2 cm, are cut from the corners of the square card, as shown in Diagram 2.
Work out the area of card remaining. [1]

(c) The card is now folded along the broken lines to make a box without a lid.
Work out the volume of this box. [2]

(CIE Paper 1, Jun 2001)

5

A W B

Z X

D C

NOT TO SCALE

ABCD is a rectangle. $AB = 140$ mm and $BC = 80$ mm.
W, *X*, *Y*, *Z* are the mid-points of *AB*, *BC*, *CD* and *DA* respectively.

(a) Work out the area of triangle *AWZ*. [1]

(b) Work out the area of the rhombus *WXYZ*. [2]

(CIE Paper 3, Jun 2001)

---

6 *ABCD* is a trapezium of height 8 cm. *AB* = 11 cm and DB = 15 cm.

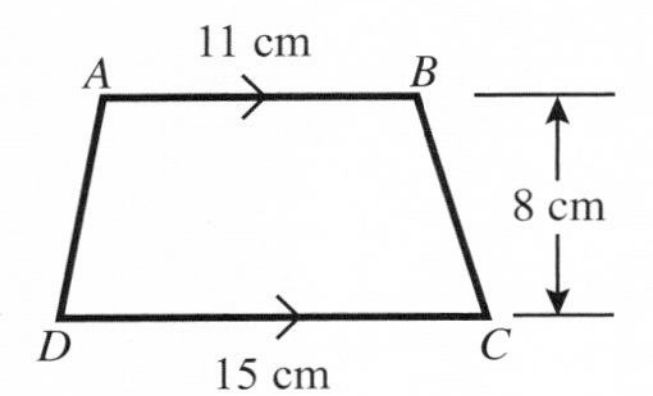

NOT TO SCALE

(a) Calculate the area of *ABCD*. [2]

ABCD is the cross section of a prism. The volume of the prism is 2600 cm^3.

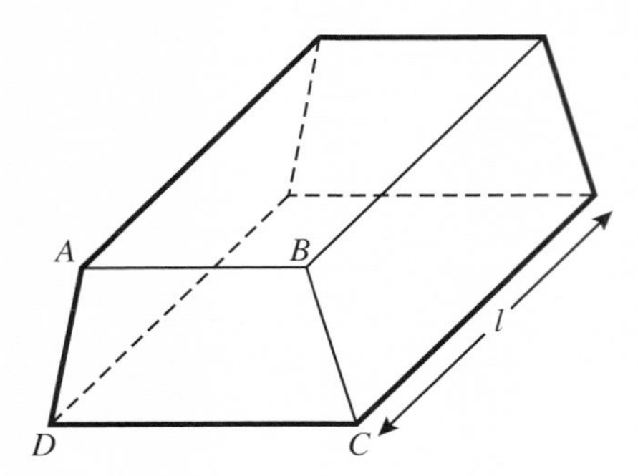

NOT TO SCALE

(b) Calculate the length *l*, of the prism. [?]

(CIE Paper 1, Nov 2001)

---

7 The diagram shows a triangular prism. *AB* = 5cm, *BC* = 6cm and the angle *ABC* is 90º. The prism has a length of 24 cm.

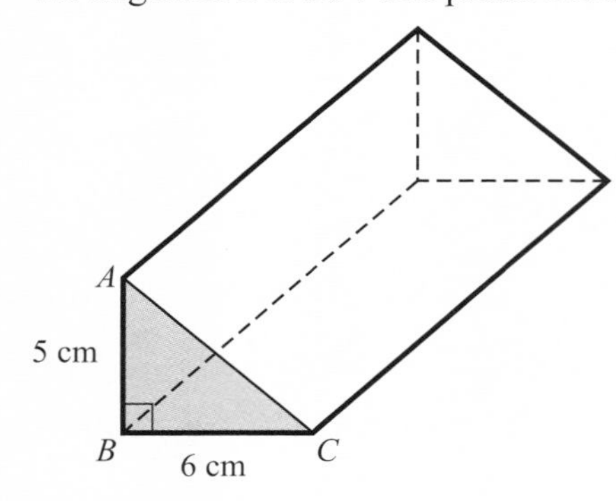

NOT TO SCALE

Calculate the volume of the prism. [2]

(CIE Paper 1, June 2002)

---

8 The diagram shows a square of side 8 cm and four congruent triangles

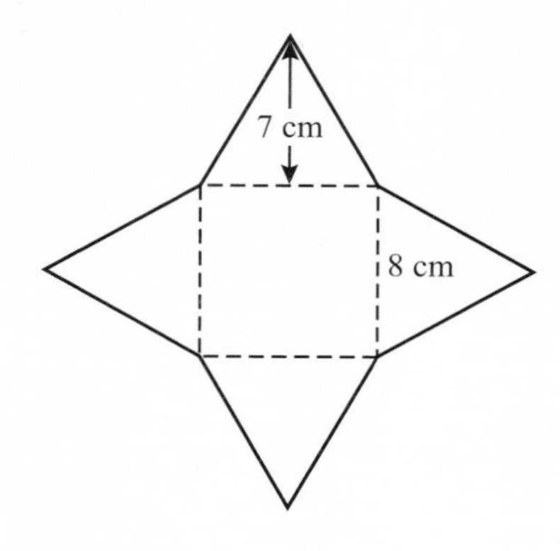

(a) Calculate
  (i) the area of one triangle, [2]
  (ii) the area of the whole shape. [2]

(b) The shape is the net of a solid.
  Write down the special name for this solid. [1]

(CIE Paper 1, June 2003)

---

9 The cuboid shown in the diagram has $EF = 4\text{cm}$, $FG = 6\text{cm}$ and $AE = 3\text{cm}$.

(a) Calculate
  (i) the volume of the cuboid, [2]
  (ii) the surface area of the cuboid. [2]

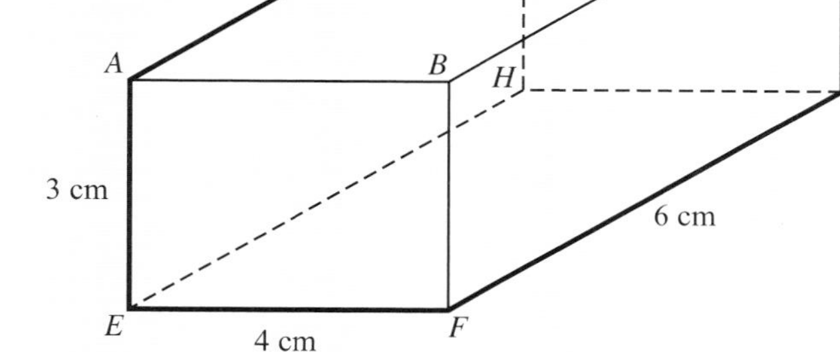

NOT TO SCALE

(b) The cuboid is divided into two equal triangular prisms.
One of them is shown in the diagram.

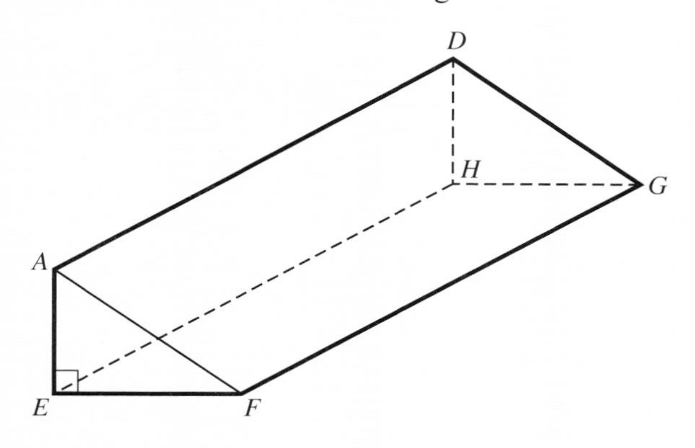

NOT TO SCALE

  (i) Write down the volume of the triangular prism. [2]
  (ii) Work out the area of the rectangle *AFGD*. [2]

(CIE Paper 3, Nov 2002)

Chapter 10

Handling data

This chapter will show you how to

- ✔ identify different types of data
- ✔ construct tally charts or frequency tables for discrete and grouped data
- ✔ design and use two-way tables
- ✔ draw pictograms, frequency diagrams, pie charts and line graphs
- ✔ construct scatter diagrams and identify correlation
- ✔ construct frequency polygons
- ✔ compare two data sets

10.1 Types of data

Primary data is information you collect directly yourself, for example, from questionnaires.

Secondary data is information that you get from existing records, for example, newspapers, magazines, the internet.

Qualitative data contains descriptive words, for example a colour (red, green), or an activity (climbing, sailing), or a location (London, Paris). It is sometimes called categorical data.

Quantitative data contains numbers, such as temperatures, masses, areas, lengths, time, number of TVs or cars.

There are two types of quantitative data.

1 **Discrete data** can only have particular values. Discrete data is 'countable'.
Discrete data examples

- Scores on a dice 4, 2, 6
- Goals scored in a match 0, 2, 3

You can't score $2\frac{1}{2}$ goals!

2 **Continuous data** can take any value in a particular range.
Continuous data examples

- Mass 72 kg, 15.3 g, 5 tonnes
- Temperature −4°C, 25.8°C, 100°C
- Length 800 m, 300 000 km, 2.6 mm

2.6 mm is measured to the nearest tenth of a millimetre.

Continuous data cannot be measured exactly. The accuracy depends on the measuring instrument, for example, a ruler, or thermometer.

EXERCISE 10A

1 State whether each set of data is quantitative or qualitative.

(a) Height (b) Age
(c) Eye colour (d) Place of birth
(e) Distance (f) Shoe size

2 State whether each set of data is discrete or continuous.

(a) Cost in pence
(b) Number of creatures in a rock pool
(c) Time
(d) Mass
(e) Area
(f) Score on a dartboard
(g) Hours worked

3 State whether each source will give primary or secondary data.

(a) Collecting data by observing traffic
(b) Downloading data from the internet
(c) Looking at data from the 2001 Census
(d) Using data found in a newspaper
(e) Giving people a questionnaire

4 (a) How could you collect data on the following
   (i) Car engine sizes and acceleration
   (ii) Pet ownership in the UK
   (iii) How people from one company travel to work
   (iv) Climate – local and national?
(b) For each type of data you describe, say whether it is primary or secondary, qualitative or quantitative.

10.2 Data collection

Putting data into a bar chart, pie chart or frequency diagram helps you to analyse it.

It is difficult to see if the vowels all occur the same number of times.

If you have a large amount of data, you can organise it in a table.

Here are the vowels in the first four sentences of a book.

i, a, o, e, a, i, i, e, i, i, o, a, i, a, a, e, a, a, i, e, e, a, i, o, e, i, i, o, a, i, a, a, i, e, o, i, a, i, a, a, i, i, i, i, e, a, e, u, o, o, e, i, a, u, i, e, e, o, o, e, e, u, e, e, a, a, i, a, u, i, e, a, u, i, e

You could put the results into a **tally chart**.

Vowel	Tally	Frequency
a	~~IIII~~ ~~IIII~~ ~~IIII~~ ~~IIII~~	20
e	~~IIII~~ ~~IIII~~ ~~IIII~~ III	18
i	~~IIII~~ ~~IIII~~ ~~IIII~~ ~~IIII~~ III	23
o	~~IIII~~ IIII	9
u	~~IIII~~	5

This is the total for each vowel.

Tallies are easier to count if you group them in 5s ~~IIII~~

You can now see that i occurred most often.

Work through the data, making a **tally mark** in the correct row for each vowel.

The complete table showing the tally marks and the **frequency** is called a **frequency table**.

When you have a large set of continuous data, you usually group the results together in equal sized groups. These are classes or **class intervals**.

EXAMPLE 1

The heights in centimetres of dancers in a musical production were

161, 168, 161, 165, 161, 160, 164, 167, 163, 162, 166, 161, 168

Put these heights into a tally chart.

There are 13 heights here.

Height (cm)	Tally	Frequency
160	|	1
161	||||	4
162	|	1
163	|	1
164	|	1
165	|	1
166	|	1
167	|	1
168	||	2
	Total	13

Add the tallies to find the frequencies.

Check the total in the frequency column is the same as the original number of values.

You can also write class intervals using $<$ and $\leqslant$ notation.

Instead of 140–149 you could write $140 \leqslant h < 150$.

This means the height h is 140 or more, but *less than* 150.

Using this notation, the table in Example 1 is

Height h (cm)	Tally	Frequency
$140 \leqslant h < 150$	|||	3
$150 \leqslant h < 160$	~~||||~~ |||	8
$160 \leqslant h < 170$	~~||||~~ ~~||||~~ |||	13
$170 \leqslant h < 180$	~~||||~~	5
$180 \leqslant h < 190$	||	2
	Total	31

This includes all heights *up to* 180 cm, but not 180 cm.

A height of 180 cm is recorded in this class.

Data-capture sheets

Before you collect data, you can prepare a tally chart to record it in. A pre-prepared tally chart to record data is called a **data-capture sheet**.

For example, to collect data on newspapers people buy, you could prepare a chart like this

Newspaper	Tally	Frequency
The Representative		
The Star		
Daily Dispatch		

EXERCISE 10B

1 The frequency table shows the type and number of pets treated one week.

Copy and complete the tally marks and the frequency table.

Pet	Tally	Frequency
Dog	𝍸 ||	
Cat		9
Bird	𝍸 |	
Other		
	Total	32

2 These are the trees in Thornicombe Wood.

oak	oak	birch	elm	chestnut
oak	chestnut	birch	chestnut	sycamore
elm	elm	birch	chestnut	birch
elm	birch	chestnut	elm	sycamore
sycamore	oak	chestnut	birch	chestnut
oak	elm	oak	sycamore	oak

Copy and complete the frequency table to show these trees.

Tree	Tally	Frequency
Oak		
Birch		
Elm		
Chestnut		
Sycamore		
	Total	

3 These are the cell phone networks used by 40 students.

O_5 Realtime Fonenet Fresh IQ-mobile
Realtime Fonenet Fonenet Realtime Fresh
IQ-mobile Fresh Realtime O_5 Realtime
IQ-mobile O_5 IQ-mobile Fresh O_5
O_5 Fresh IQ-mobile Fonenet Fonenet
Realtime Fonenet IQ-mobile IQ-mobile Fonenet
O_5 IQ-mobile Fonenet O_5 Realtime
Fresh Realtime Fonenet Fresh O_5

Copy and complete the frequency table for this information.

Tree	Tally	Frequency
O_5		
Realtime		
Fonenet		
Fresh		
IQ-mobile		
	Total	

4 These are the colours of 36 cars in a car park.

blue white red black silver blue
red green black green red black
green black silver red blue red
white blue green white red white
blue red black silver blue silver
red white blue silver blue green

(a) Design a data collection sheet for the colours of the cars in the car park.

(b) Use the information to complete your data collection sheet.

5 A café sells drinks. The drinks are tea, coffee, hot chocolate and lemonade.
The manager of the shop wants to find out about how many of these drinks she sells in a day. Design a suitable data-capture sheet for the manager to use.

6 When a coin is thrown, it can land on its head or on its tail.
Carl is going to throw a coin 50 times. Design a data collection sheet for Carl to use.

7 The list below shows the calls made by a salesperson over 29 weeks.

22 27 18 23 25 19 17 20 23 20
18 22 20 19 24 21 16 19 18 22
14 19 18 21 20 19 25 25 21

Copy and complete this frequency table for the data.

Calls per week	Tally	Frequency
1–10		
11–20		
21–30		
	Total	29

10.3 Two-way tables

Bus timetables and league tables are all two-way tables.

Two-way tables are similar to frequency tables. They show two or more types of information at the same time.

EXAMPLE 2

The table shows the type and outcome of matches played by a cricket team.

	Home matches	Away matches
Won	5	2
Drawn	4	6
Lost	4	5

The two types of information are
1 home or away
2 win, draw, lose.

(a) How many matches were played altogether?
(b) How many matches were lost altogether?
(c) How many matches in total were not drawn?

Continued ▼

Often the easiest way of answering these questions is to extend the two-way table to include the totals going across → and downwards ↓.

	Home matches	Away matches	Total
Won	5	2	7
Drawn	4	6	10
Lost	4	5	9
Total	13	13	26

The totals across (13 + 13 = 26) and down (7 + 10 + 9 = 26) should be the same.

(a) 26 matches played altogether.

(b) 9 matches lost altogether.

(c) 26 − 10 = 16 matches were not drawn.

You could also work out total lost + total won = 7 + 9 = 16.

EXAMPLE 3

In an office survey of 32 staff, 6 women said they walked to work, 10 men came by bus, and 4 men cycled. Of the remaining 11 women, only 1 cycled and the rest came by bus or walked.

(a) Draw a two-way table to show this information.

(b) Complete the table.

(c) How many women went by bus?

(d) How many people walked to work?

(e) What percentage of people went by bus?

(a)

	Walked	Cycled	Bus	Total
Men		4	10	
Women	6	1		17
Total				32

(b)

	Walked	Cycled	Bus	Total
Men	1	4	10	15
Women	6	1	10	17
Total	7	5	20	32

The total number of men is 32 − 17 = 15.

The number of men that walked is 15 − (4 + 10) = 1.

The number of women who came by bus is 17 − (6 + 1) = 10.

Once you have filled in all the values you can calculate the totals.

From the table

(c) 10 women went by bus.

(d) 7 people walked to work.

(e) 20 people out of 32 went by bus.

As a fraction this is $\frac{20}{32}$

As a percentage this is $\frac{20}{32} \times 100\% = 62.5\%$

For more on percentages see Section 3.5.

EXERCISE 10C

1 In a class of 30 people, 6 men and 8 women own a bicycle. There were 17 women in the survey. Copy and complete the two-way table to show this information.

	Bicycle	No bicycle	Total
Men	6		
Women	8		17
Total			30

2 In a school survey of 50 boys and 50 girls, 41 boys were right-handed and only 6 girls were left-handed. Copy and complete the two-way table.

	Left-handed	Right-handed	Total
Girls			
Boys			
Total			

Use the table to work out an estimate of the percentage of left-handed pupils in the school.

3 In a survey the male population of Poynton (Central) was 3522. The number of females in Poynton (West) was 3898. The population of Poynton (Central) was 6792. The total population of Poynton was 13 433.

(a) Construct a two-way table to show this information.

(b) Complete the table.

(c) What percentage of the population of Poynton is female? Give your answer to the nearest whole number.

4 The table gives the Science test results for a local school.

		Grade					
		3	4	5	6	7	8
Science	Boys	11	28	34	31	15	1
	Girls	4	20	36	43	22	5

(a) Copy the table and extend it to find the totals for each row and column.

(b) How many pupils took the test?

(c) What percentage of boys achieved a level 5 or higher?

(d) What percentage of girls achieved a level 7?

Two-way tables can show different types of information. Transport timetables, calendars, holiday brochure information, statistics from a census and currency conversion tables are all types of two-way table.

EXAMPLE 4

The table shows the monthly rainfall (in mm) and the maximum and minimum temperatures (in °C) for Paris.

The letters stand for the months.

	J	F	M	A	M	J	J	A	S	O	N	D
Rainfall	20	16	18	17	16	14	13	12	14	17	17	19
Max. temperature	6	7	11	14	18	21	24	24	21	15	9	7
Min. temperature	1	1	3	6	9	12	14	14	11	8	4	2

(a) Which month has the most rain?

(b) Which months have the smallest temperature range?

(c) Which months have the largest temperature range?

The range of temperature is maximum – minimum.

From the table

	J	F	M	A	M	J	J	A	S	O	N	D
Rainfall	20	16	18	17	16	14	13	12	14	17	17	19
Max. temperature	6	7	11	14	18	21	24	24	21	15	9	7
Min. temperature	1	1	3	6	9	12	14	14	11	8	4	2
Range °c	5	6	8	8	9	9	10	10	10	7	5	5

(a) January has the most rain (20 mm).

(b) January, November, December.

Each of these has temperature range 5 °C.

(c) July, August, September.

Each has temperature range 10°C.

EXAMPLE 5

The table below shows the cost of a holiday. The prices are per person, in dollars.

Group	Number of days						Extra night
	2	3	4	5	6	7	
5/6 adults sharing	170	178	185	190	193	196	25
4 adults sharing	173	184	190	197	199	205	25
3 adults sharing	179	192	202	213	220	227	25
2 adults sharing	179	192	202	213	220	227	25
Child	148	148	148	148	148	148	25

You find this type of table in holiday brochures.

Each price is *per person.*

(a) Find the cost of a 3-day holiday for 4 adults and 3 children.

(b) What is the cost of a holiday for 2 adults and 2 children for 10 days?

(a) A 3-day holiday for 4 adults costs 4 × $184 = $736.
For 3 children costs 3 × $148 = $444.
Total cost of the holiday = $736 + $444 = $1180.

Use the prices in the '3 days' column. Read the rows for 4 adults ($184 each) and child ($148 each).

(b) 2 adults for 7 days = 2 × $227 = $454.
2 adults for 3 extra nights = 2 × (3 × $25) = $150.
So 2 adults for 10 days = $454 + $150 = $604.
2 children for 7 days = 2 × $148 = $296.
2 children for 3 extra nights = 2 × (3 × $25) = $150.
So 2 children for 10 days = $296 + $150 = $446.
Total cost of holiday = $604 + $446
= $1050

Work out the cost for 7 days then add on **3** extra nights.

EXERCISE 10D

1 The table shows the distances in kilometres (km) between some major French cities.

Bordeaux						
870	Calais					
658	855	Grenoble				
649	1067	282	Marseille			
804	1222	334	188	Nice		
579	292	565	776	931	Paris	
244	996	536	405	560	706	Toulouse

The table shows that the distance between Calais and Paris is 292 km.
Find

(a) the distance between Bordeaux and Marseille

(b) the distance between Toulouse and Grenoble

(c) the total distance from Paris to Calais to Bordeaux and then back to Paris.

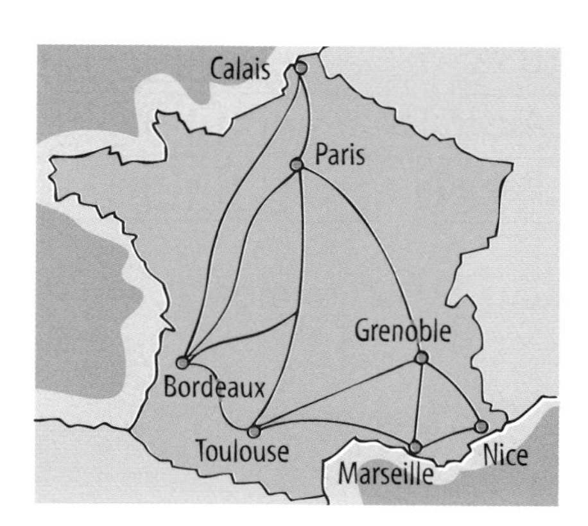

2 Use the two-way table in Example 5 to find the cost of

(a) a 5-day holiday for 3 adults and 4 children

(b) an 8-day holiday for 6 adults (no children).

10.4 Pictograms

You record survey data in a tally chart or frequency table. You can then display the data in a picture or diagram as well.

A diagram can help you to see patterns in the data.

In a **pictogram** a picture or symbol represents an item or number of items.

The table shows the amount of gold produced in tonnes each year, in the top-producing mines in four different countries.

Country	Gold produced (tonnes)
South Africa	625
USA	325
Australia	250
Canada	150

You can show this information in a pictogram.

South Africa	□□□□□□□□□□□□[
USA	□□□□□□[
Australia	□□□□□
Canada	□□□

Key □ = 50 tonnes

The **key** tells you what each symbol represents.

[represents 25 tonnes

EXERCISE 10E

1 The pictogram shows the number of students in different Mathematics classes who own cell phones.

Class	
A	
B	
C	
D	

Key represents 2 cell phones

10 students in Class B own cell phones.
7 students in Class D own cell phones.

(a) Complete the pictogram to show this information.

(b) How many students own cell phones in total?

2 Fiona asks her friends what their favourite sport is.
The results are shown below.

Sport	Frequency
Football	4
Netball	6
Riding	8
Other	6

Draw a pictogram to represent her results. Use the symbol ☺ to represent 4 friends.

3 This table shows the sales from a canteen drinks' machine.

Tea	40
Coffee	47
Chocolate	26
Soup	18
Fruit juice	14

Think how you will display numbers such as 47.

Draw a pictogram to represent these drinks sales.
The symbol you choose should be easy to split into 5 equal parts.

10.5 Frequency diagrams for discrete data

Bar charts

Bar charts can show patterns or trends in data. In a bar chart, the bars can be either vertical or horizontal. They must be of equal width.

Bar charts can be used for quantitative or qualitative data.

EXAMPLE 6

The table shows the frequency of vowels occurring in the first four lines of a book.

Vowel	a	e	i	o	u
Frequency	20	18	23	9	5

This is qualitative data.

Vowels are the letters a, e, i, o, u.

Draw a bar chart for this data.

Vowels in the first four lines

Frequency

25 20 15 10 5 0

a e i o u

Vowel

Choose a sensible scale.

Frequency on the Vertical axis.

The height of each bar represents the frequency.

Give your bar chart a title.

Leave gaps between the bars.

Label the axes and bars.

You could plot this bar chart with horizontal bars.

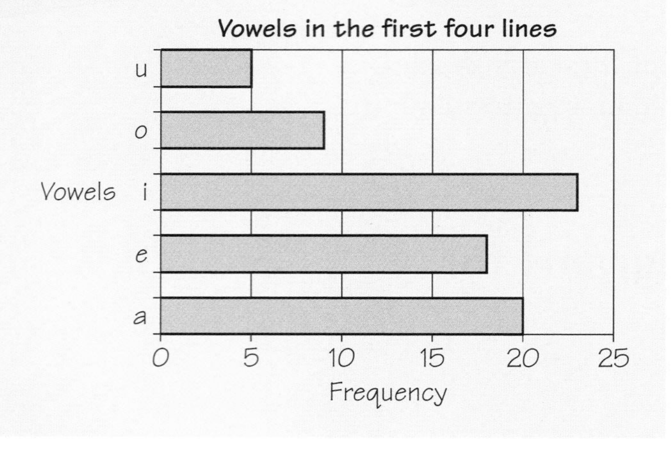

The length of each bar represents the frequency.

There is still a gap between the bars.

When you draw a bar chart make sure that you

- label the horizontal and vertical axes clearly
- give the chart a title
- use a sensible scale to show all the information clearly
- leave equal spaces between the bars.

EXERCISE 10F

1 The table shows the number of cars parked in three hospital car parks at 2 pm on one afternoon.

Car park	Number of cars
Staff (S)	40
Visitors (V)	70
Casualty (C)	65

(a) Draw a bar chart to show this information.

(b) Work out how many more cars were parked in the Visitors car park than in the Staff car park.

2 Emma asks her friends what type of TV programme they like best.

Draw a bar chart to show the results.

Type of TV programme	Frequency
Cartoons	4
Drama	2
Quizzes	1
Soaps	6

3 The following frequency table shows the results of a survey by a crisp manufacturer to find the most popular flavour among boys and girls.

Flavour	Frequency	
	(boys)	(girls)
Plain	6	6
Cheese and Onion	12	4
Ready Salted	21	12
Prawn Cocktail	4	9
Salt and Vinegar	5	16
Roast Chicken	2	3

(a) Draw two bar charts to display this.

(b) How many people took part in the survey?

4 This bar chart represents sales of cars at an auction.

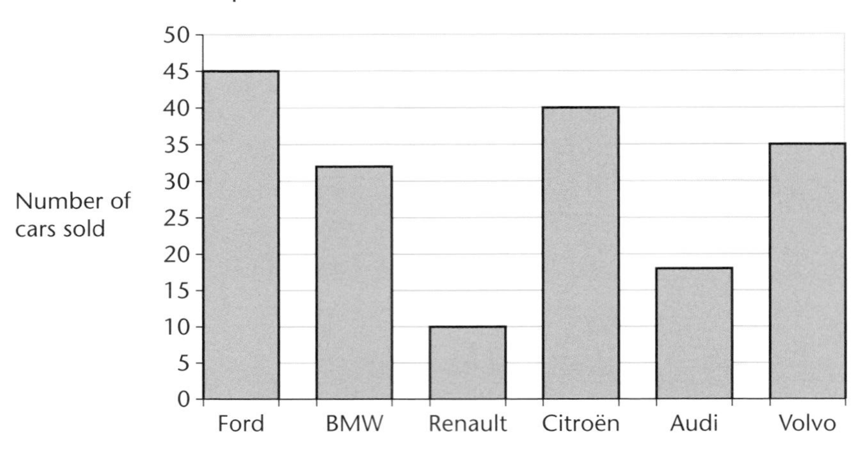

Use the bar chart to answer these questions.

(a) How many cars were sold altogether?

(b) Which make of car totalled exactly 10% of all the cars sold?

(c) Which make of car totalled exactly one quarter of all the cars sold?

(d) Which make of car totalled almost $\frac{1}{5}$ of all the cars sold?

(e) One make of car sold four times as many as another make of car. Which two makes of cars were these?

5 A survey of the most common birds in the UK gave these results.

Bird	Number (millions of pairs)
Blackbird	4.7
Blue tit	3.5
Chaffinch	5.8
Robin	4.5
Sparrow	3.8
Wood pigeon	2.4
Wren	7.6

Draw a bar chart to represent this information.

6 In a survey, families were asked how many holidays they had taken last year. Draw a bar chart to show this data.

Number of holidays	Number of families
0	2
1	14
2	17
3	8
4	1

10.6 Frequency diagrams for continuous data

If you collect data that is grouped, you will need to draw a frequency diagram similar to a bar chart, but with no gaps between the bars and a scale on both axes.

EXAMPLE 7

The heights of 31 sunflowers were measured.

Height (cm)	Frequency
$140 \leqslant h < 150$	3
$150 \leqslant h < 160$	8
$160 \leqslant h < 170$	13
$170 \leqslant h < 180$	5
$180 \leqslant h < 190$	2

Draw a frequency diagram to show this data.

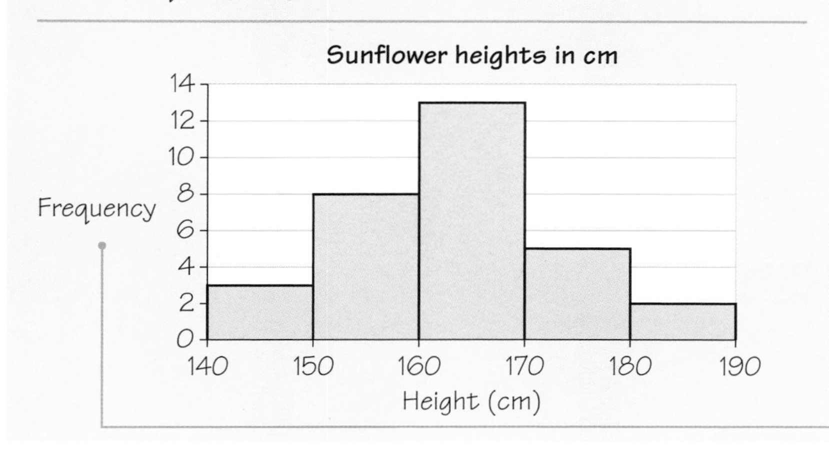

For continuous data there are no gaps between the bars.

The width of each bar is the same as the class interval.

Frequency on the vertical axis.

EXERCISE 10G

1 Draw a frequency diagram to show the following information.

Length, x (cm)	Frequency
$0 < x \leq 5$	6
$5 < x \leq 10$	11
$10 < x \leq 15$	8
$15 < x \leq 20$	5

2 This table gives the age range of the members of a local sports club.

Draw a frequency diagram to show the spread of ages.

Age	Frequency
$0 \leq$ age < 10	23
$10 \leq$ age < 20	45
$20 \leq$ age < 30	56
$30 \leq$ age < 40	36
$40 \leq$ age < 50	49
$50 \leq$ age < 60	32
$60 \leq$ age < 70	16

10.7 Pie charts

Pie charts show how data is shared or divided.

Interpreting pie charts

The whole pie chart represents the total number of items.

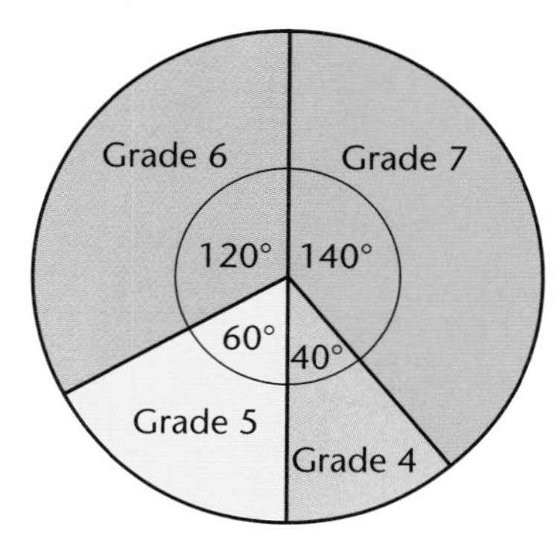

There are 360° in a circle.

The pie chart shows the results of a Science test for 180 pupils.

The chart shows that the whole 360° represents 180 pupils.

This means that 1° represents $\frac{180}{360} = 0.5$ pupils.

140° represents $140 \times 0.5 = 70$ pupils, so 70 pupils achieved grade 7.

60° represents $60 \times 0.5 = 30$ pupils, so 60 pupils achieved grade 5, and so on.

Pie charts are not particularly useful for reading off accurate values.

Pie charts are good for showing comparisons. From these two pie charts, it is easy to see the difference in spending patterns of the men and women surveyed.

Goods bought in the last month.

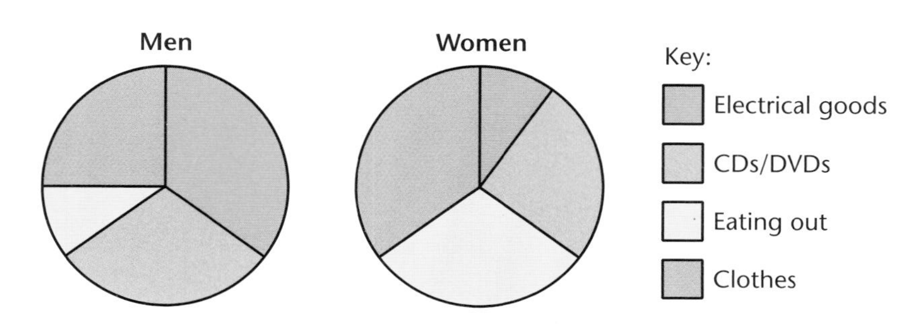

Pie charts should be labelled or accompanied by a **key** showing each category of item.

Drawing pie charts

The sections of the pie chart are called sectors.

To draw a pie chart, you first calculate the angle for each sector. You draw the angles using a protractor. Label each sector to show what it represents.

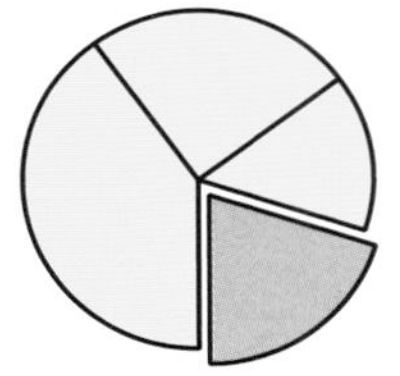

A sector is like a 'slice' of the pie.

For help using a protractor see Section 6.1.

EXAMPLE 8

In a pet shop survey, people were asked about the pets they owned.

Pet	Dog	Cat	Bird	Fish	Other
Frequency	20	37	15	32	16

(a) How many pets were recorded in total?
(b) Calculate the angle of the pie chart sector for each pet.
(c) Draw a pie chart to show this information.

(a) $20 + 37 + 15 + 32 + 16 = 120$
120 pets were recorded

Total frequency = total number of pets.

(b) 360° represents 120 pets
So $\frac{360°}{120} = 3°$ represents 1 pet

The angle for one item is always $\frac{360°}{\text{total number of items}}$.

Pet	Frequency	Sector angle calculation	Angle
Dog	20	$20 \times 3°$	60°
Cat	37	$37 \times 3°$	111°
Bird	15	$15 \times 3°$	45°
Fish	32	$32 \times 3°$	96°
Other	16	$16 \times 3°$	48°
Total	120	Total angle	360°

Check that the angles add up to 360°.

(c) Type of pets owned

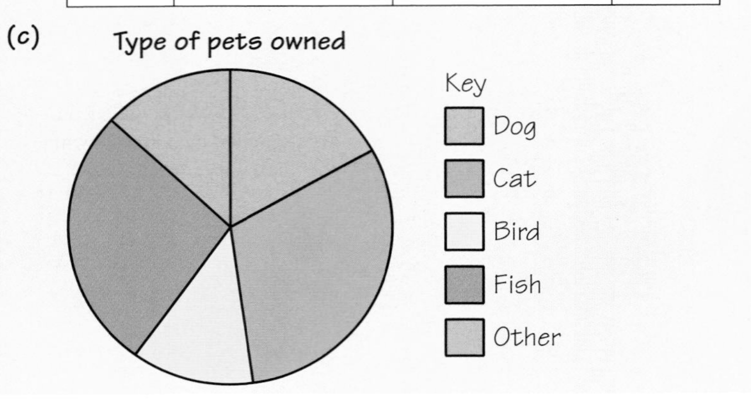

The key tells you what each sector represents.

EXAMPLE 9

Thirty people were asked how they travelled to work.
The results are shown in the frequency table.

Mode of travel	Frequency
Walk	7
Bus	9
Car	13
Cycle	1

Draw a pie chart to show this information.

360° represents 30 people

So $\frac{360°}{30} = 12°$ represents 1 person

Mode of travel	Frequency	Sector angle calculation	Sector angle
Walk	7	$7 \times 12°$	84°
Bus	9	$9 \times 12°$	108°
Car	13	$13 \times 12°$	156°
Cycle	1	$1 \times 12°$	12°
Total	30	Total angle	360°

Check that the total frequency equals the number of people and that the angles add up to 360°.

How people travel to work

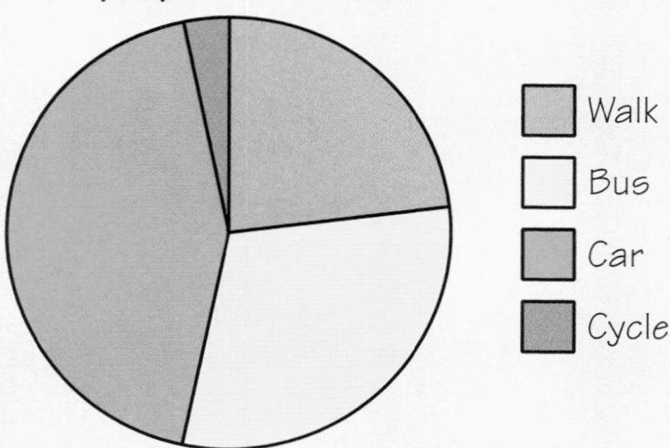

EXAMPLE 10

This pie chart shows how a family spends its money in a week.

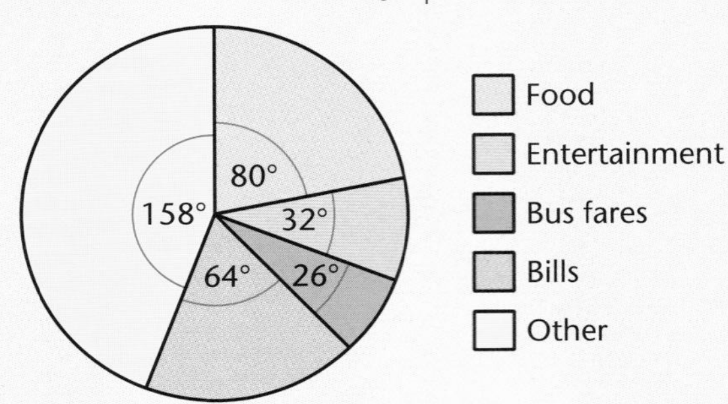

The amount spent on food is \$120.
How much do they spend on each of the other items?

Food has a sector of 80° and the amount for food is \$120.

80° represents \$120

1° represents $\frac{\$120}{80} = \$1.50$

Entertainment $= 32 \times \$1.50 = \$48$

Bus fares $= 26 \times \$1.50 = \$39$

Bills $= 64 \times \$1.50 = \$96$

Other $= 158 \times \$1.50 = \$237$

Check $\$120 + \$48 + \$39 + \$96 + \$237 = \$540$

and $360 \times \$1.50 = \$540$

Work out how much an angle of 1° represents.

If the sector angles were not labelled, you could measure them using a protractor.

The whole pie chart is 360° so the total of all the amounts of money must be 360 × \$1.50.

EXERCISE 10H

1 In a café, the number of people eating meals for lunch is shown below.

Fish pie	16
Sausages	10
Omelette	17
Salad	6
Chicken Pieces	19
Pizza	22

(a) How many people were in the café?

(b) What angle will represent 1 person?

(c) Draw a pie chart to show this information.

2 540 pupils were asked which was their favourite school subject.

The results are shown in this pie chart.

Work out the number who voted for each subject.

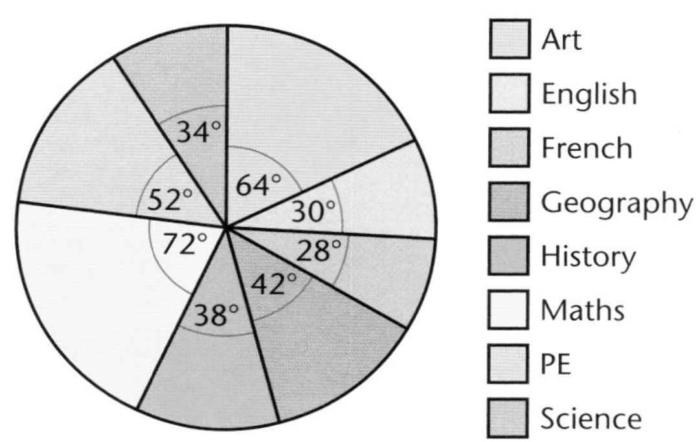

Use Example 5 to help you.

3 This pie chart shows what students did in the year after their exams.

135 pupils went to college.

Work out the number of pupils in each of the other categories.

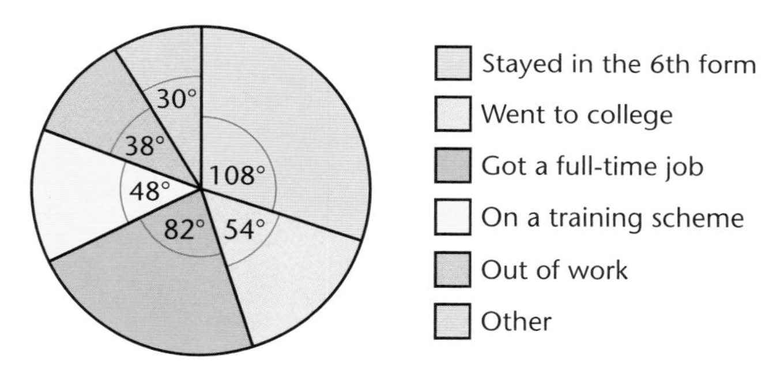

4 A packet of breakfast cereal showed the following nutritional information

Ingredient	Protein	Carbohydrate	Fat	Fibre
Amount per 100g of cereal	15 g	62.5 g	10 g	12.5 g

(a) If 360° represents 100 g of cereal, what angle represents 1 g?

(b) Calculate the angle of the sector for each ingredient.

(c) Draw a pie chart to show this information.

5 A different cereal brand has ingredients in these proportions.

400 g carbohydrate
150 g protein
120 g fibre
50 g fat

(a) Copy and complete the calculation and table below.

$$1 \text{ g of ingredient} = \frac{360°}{\text{total weight of cereal}} = \frac{360}{\square} \text{ g} = 0.\square°$$

Ingredient	Amount in g	Sector angle calculation	Angle
Carbohydrate	400 g	$400 \times 0.\square°$	
Protein	150 g	$150 \times 0.\square°$	
Fibre	120 g	$120 \times 0.\square°$	
Fat	50 g	$50 \times 0.\square°$	
Total		Total angle	360°

(b) Draw a pie chart to show this information.

(c) Compare your pie charts for questions 4 and 5.
Which brand could claim it has

(i) less carbohydrate

(ii) less fat?

10.8 Scatter diagrams

Scatter diagrams help you to compare two sets of data. They show if there is a connection or relationship called the **correlation**, between the two quantities plotted.

Sometimes scatter diagrams are called **scatter graphs** or **scattergrams**.

The following table shows the masses and heights of 10 men registered at a gym.

Height (cm)	Mass (kg)
166	65
169	73
172	67
161	62
177	75
171	72
168	66
165	67
170	70
176	75

For a scatter diagram, plot the height along the *x*-axis (horizontal) and the mass along the *y*-axis (vertical).

Scatter graph to show height–mass relationship

Mass (kg)
80
75
70
65
60
55
155 160 165 170 175 180 185
Height (cm)

You do not need to start the axes at zero. Find the smallest and largest values in each set of data to help you decide on the scale. The straight line shows the line of best fit.

The scatter diagram suggests that the taller you are the heavier you are.

A good way to show this is by drawing a straight line through, or as close to, as many points as possible. This line is called the **line of best fit**.

Try to have equal numbers of points above and below the line of best fit.

You draw it 'by eye', using a ruler.

Here the line of best fit slopes from bottom left to upper right. This is called a **positive correlation**.

Positive correlation means as one variable gets bigger, so does the other.

EXAMPLE 11

The following table shows how the fuel consumption (in litres per 100 km) changes as the speed of a car increases.

Speed (kph)	Fuel consumption (litres/100 km)
20	9.6
30	8.7
40	8.0
50	6.8
60	6.0
70	5.5
80	4.5

(a) Plot a scatter diagram for this data.

(b) Draw in a line of best fit.

(c) Comment on the correlation.

(a)(b)

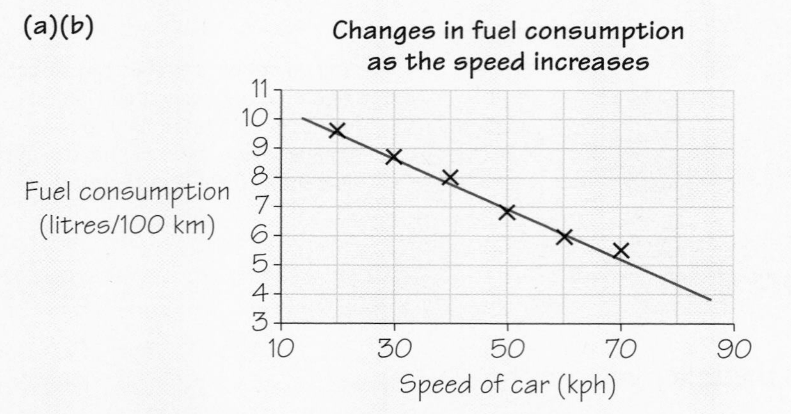

As the speed *increases* the fuel consumption *decreases*. A car uses more fuel when it first starts accelerating.

(c) The line of best fit shows **negative correlation**.

EXAMPLE 12

This table shows the results of a recent Geography test (as a percentage) and the hand span measurements (in cm) of students taking the test.

(a) Plot a scatter diagram for this data.

(b) Comment on the correlation shown.

Test result (%)	Hand span (cm)
32	15
85	19
54	21
47	16
41	23
36	14
29	18
57	17
67	20
60	21

(a)

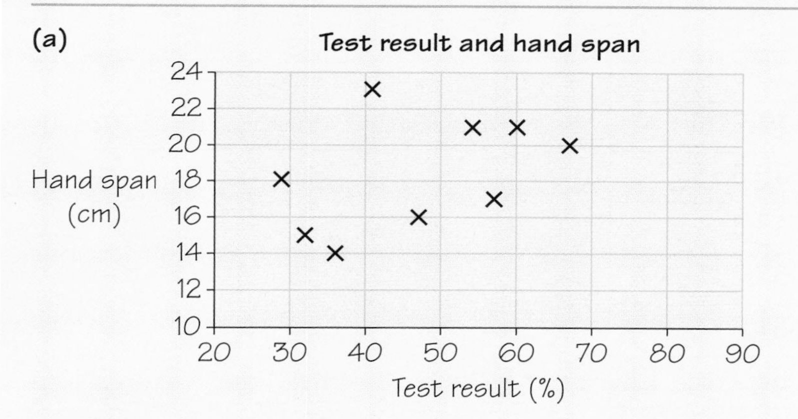

The points are randomly spread out, so you cannot draw a line of best fit.

There is no *linear* relationship because the points do not lie on or near a *line*.

(b) There is no linear connection between the test results and the hand span measurement. There is **no correlation**.

EXERCISE 10I

1 Here are four sketches of scatter diagrams.

1 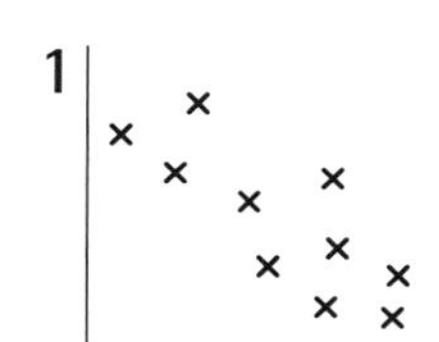

2 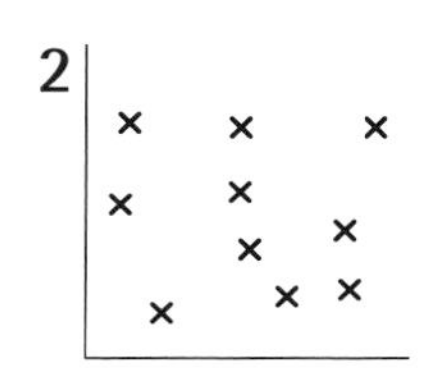

3 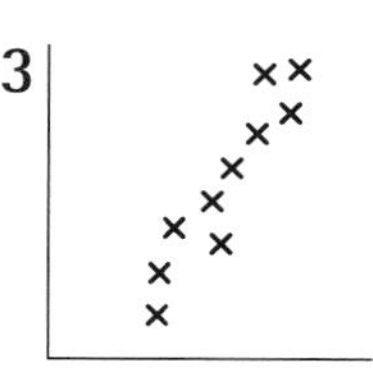

4 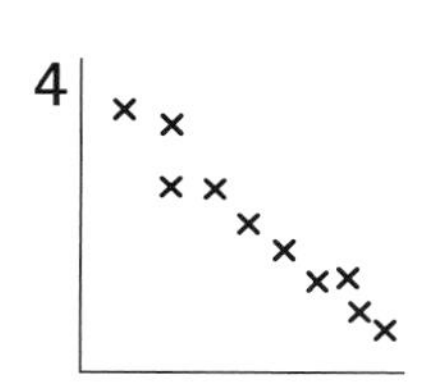

Which ones show

(a) positive correlation **(b)** negative correlation

(c) zero correlation?

2 What type of correlation would you expect if you drew a scatter graph of the following?

(a) The football league position (where top of the league = 1) against the number of goals conceded.

(b) How much a person earns against their height.

(c) The number of ice creams sold against the temperature during the day.

(d) The marks gained in a practice IGCSE examination against those gained in the actual IGCSE examination.

(e) The size of a car engine against the amount of fuel used by that engine.

3 In a science experiment, one end of a metal bar is heated. The results show the temperature at different points along the metal bar.

Position (cm)	Temperature (°C)
1	15.6
2	17.5
3	36.6
4	43.8
5	58.2
6	61.6
7	64.2
8	70.4
9	98.8

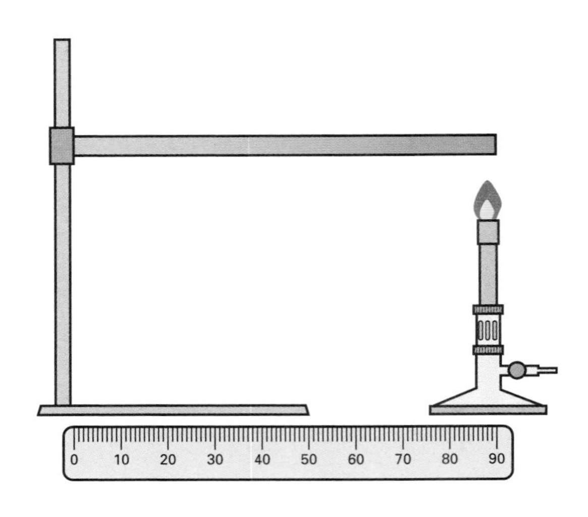

(a) Draw a scatter diagram to show this data. Let the x-axis represent the position between 0 and 10 cm, and the y-axis represent the temperature between 0 and 100°C.

(b) Draw the line of best fit.

(c) Comment on the correlation shown.

4 The table shows the number of female competitors taking part in each Olympic Games from 1948 to 1984.

Year	Female competitors
1948	385
1952	518
1956	384
1960	610
1964	683
1968	781
1972	1070
1976	1251
1980	1088
1984	1620

(a) Plot a scatter diagram for this data.

(b) Draw the line of best fit.

(c) State what type of correlation you see.

(d) Describe the trend or pattern.

Put 'Year' on the horizontal axis.

5 The table below shows the mean annual temperature (in °C) for 10 major cities and their latitude.

City	Mean annual temperature (°C)	Latitude (degrees)
Mumbai	32	19
Kolkata	26	22
Dublin	12	53
Hong Kong	26	22
Istanbul	18	41
London	12	51
Oslo	10	60
New Orleans	21	30
Paris	15	49
St Petersburg	7	60

Latitude describes position on the globe, as degrees North from the Equator.

(a) Plot the latitude along the x-axis (horizontal) from 0 to 70 degrees.
Plot the mean temperature along the y-axis (vertical) between 0 and 40°C.

(b) What type of correlation does the diagram show?

(c) What happens to the temperature as you move further north from the Equator?

6 The table below shows the examination results from twelve pupils in mathematics and science.

Mathematics mark	Science mark
42	38
83	58
29	23
34	17
45	30
47	35
55	47
74	55
61	36
59	50
53	37
77	63

(a) Draw a scatter diagram to show this information.

(b) Draw the line of best fit.

(c) What type of correlation is it?

(d) Estimate the science mark for a pupil who gained 38 marks in the mathematics examination.

Put the mathematics mark along the x-axis from 10 to 100.

EXAMINATION QUESTIONS

1 Arantaxa had the following scores on the eighteen holes on a golf course.
6, 2, 5, 4, 6, 5, 4, 6, 6, 4, 5, 6, 2, 6, 5, 4, 7, 6
Complete the frequency table below. [2]

Score	2	3	4	5	6	7
Frequency						

(CIE Paper 1, Nov 2000)

---

2 24 students estimated how many units of energy they had used during the afternoon. The results are shown in the table below.

Number of units of energy used	20	25	30	35	40
Number of students	4	6	9	3	2

Draw and label an accurate pie chart to show this information.
Show clearly how you calculated your angles. [5]

(CIE Paper 3, Jun 2002)

---

3 The results of the school's senior football team during a year are recorded, using W for a win, L for a loss and D for a draw. They are

L L W D L W L W
L L D L L W W L
W L L W D L L W

Copy and complete the table below to show these results.
Then display this information in a pie chart [6]

	Frequency	Pie chart angle
W		
L		
D		
TOTAL		360°

(CIE Paper 3, Nov 2002)

---

4 Fifty students take part in a quiz.
The table shows the results.

Number of correct answers	5	6	7	8	9	10	11	12
Number of students	4	7	8	7	10	6	5	3

(a) How many students had 6 correct answers? [1]
(b) How many students had less than 11 correct answers [1]
(c) A bar chart is drawn to show the results.
The height of the bar for the number of students who had 5 correct answers is 2 cm.
What is the height of the bar for the number of students who had 9 correct answers? [2]
(d) A pie chart is drawn to show the results.
What is the angle for the number of students who had 11 correct answers? [2]

(CIE Paper 3, Jun 2003)

---

5 A country has three political parties, the Reds, the Blues and the Greens.
The pie chart shows the proportions of the total vote that each party received in an election.

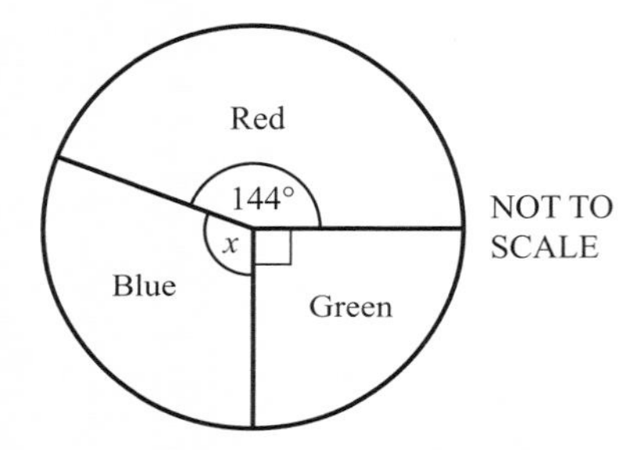

(a) Find the value of x. [1]
(b) What percentage of the votes did the Red party receive? [2]

(CIE Paper 1, Nov 2003)

---

6 In a school, the number of students taking part in various sports is shown in the table.

Sport	Number of students
Basketball	40
Soccer	55
Tennis	35
Volleyball	70

Draw a bar chart to show this data.
Show your scale on the vertical axis and label the bars. [4]

(CIE Paper 1, Jun 2004)

---

7 Grades were awarded for an examination.
The table below shows the number of students in the whole school getting each grade.

Grade	Number of students	Angle on pie chart
A	5	
B	15	
C	40	
D	20	
E	10	
Totals	90	

(a) Copy and complete the table above by calculating the angles required to draw the pie chart. [2]

(b) Draw an accurate pie chart to show the data in the table.
Label the sectors A,B,C,D and E.

(CIE Paper 3, Jun 2004)

---

8 Asif tests a six-sided spinner. The results are shown in the table below.

3	3	6	5	6	1	2	6	5	2
3	4	4	4	3	4	6	5	2	1
6	3	6	4	1	5	3	6	2	6
6	6	3	6	1	6	6	5	1	6
1	6	2	5	3	5	4	2	3	5
1	4	4	1	5	4	6	6	2	3

(a) Use these results to copy and complete the frequency table. [3]

Number	1	2	3	4	5	6
Frequency						

(b) Asif tests a different six-sided spinner. He draws a bar chart to show the results.

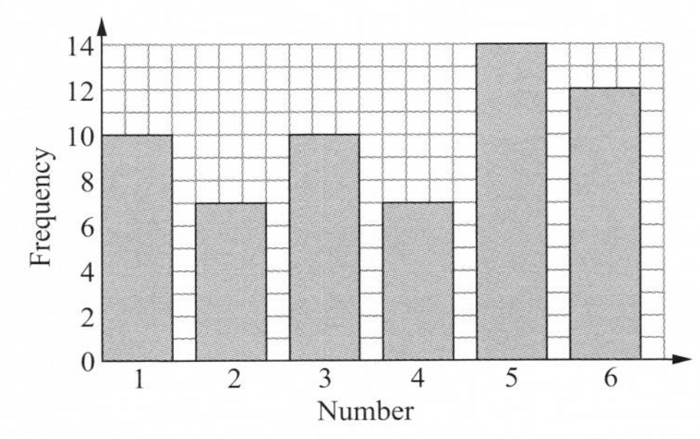

How many times did he spin the spinner? [2]

(CIE Paper 3, Jun 2005)

Chapter 11

Ratio and proportion

This chapter will show you how to

- ✔ write, use and simplify ratios
- ✔ write a ratio as a fraction and in the form 1 : n or n : 1
- ✔ divide quantities in a given ratio
- ✔ solve problems using direct and inverse proportion
- ✔ apply ratios to problems involving scales and maps

11.1 Ratio

Writing and using ratios

A **ratio** compares two or more quantities.

Ratios occur often in real life.
A scale model of a car is labelled 'Scale 1 : 18'.

This means that the length of any part of the real car is 18 times longer than the corresponding part on the model.

For example

Part	Length on model	Length on actual car
Length of car door	7 cm	$7 \times 18 = 126$ cm
Diameter of wheels	3 cm	$3 \times 18 = 54$ cm
Length of car	26 cm	$26 \times 18 = 468$ cm

You write ratio with a : between the numbers. For 1 : 18 you say '1 to 18'.

1 : 18 is an example of a ratio.

EXAMPLE 1

To make 12 small cupcakes you need 150 g of flour and 30 g of melted butter.

Continued ▼

Grace is making 36 cupcakes. How much of each ingredient does she need?

The multiplier is 3 so multiply the quantity of each ingredient by 3.

$36 \div 12 = 3$ so she will need three times as much of each of the ingredients.

Flour 3×150 g $= 450$ g   Melted butter 3×30 g $= 90$ g

EXAMPLE 2

Elderflower cordial needs to be diluted with water in the ratio 1 : 10. If you put 25 mℓ of cordial in a glass, how much elderflower drink will you make?

In this case, the multiplier is 10.

A ratio of 1 : 10 means that you need 10 times as much water as cordial.

Amount of water needed $= 10 \times 25$ mℓ
$= 250$ mℓ

Total amount = amount of cordial + amount of water.

Amount of elderflower drink $= 25$ mℓ $+ 250$ mℓ
$= 275$ mℓ

EXAMPLE 3

A recipe for a pudding for 6 people uses

180 g of butter
150 g of brown sugar
120 g of chopped walnuts

(a) How much sugar and chopped walnuts would you need if you used 270 g of butter?

(b) How many people would this pudding serve?

The multiplier is 1.5. Use it to calculate the quantities of the other ingredients.

(a) $9 \div 6 = 1.5$
Amount of sugar $= 1.5 \times 5 = 7.5$ g
Amount of chopped walnuts $= 1.5 \times 4 = 6$ g

(b) Number of people served $= 1.5 \times 6 = 9$ people

EXERCISE 11A

1 To make 12 fruit scones you need 200 g of flour, 100 g of margarine and 80 g of dried fruit.
How much of each ingredient will you need to make
(a) 24 scones **(b)** 6 scones **(c)** 15 scones?

2 A builder makes concrete with 7 bags of sand and 3 bags of cement.

(a) If he has 28 bags of sand, how many bags of cement will he need to make the concrete?

(b) The builder has 15 bags of cement. How many bags of sand should he order for the next day?

3 A Sunrise Smile drink uses cranberry juice, orange juice and tonic water in the ratio 5 : 3 : 1.

(a) If I have 25 mℓ of tonic water, how much of the other ingredients will I need?

(b) How much Sunrise Smile will I have altogether?

Find the total of all the ingredients.

4 The ratio of cats to dogs in a town is 4 : 3.

(a) How many dogs would you expect if there were 36 cats?

Make sure your answer is sensible!

(b) How many cats would you expect if there were 15 dogs?

5 A fruit stall sells 5 apples for every 3 oranges.

(a) Apples come in boxes of 60. The stall has 4 boxes of apples. How many oranges do they need?

(b) One day the stall sells 42 oranges. How many apples do they sell?

6 Cupro-nickel is used to make key rings. It is made from mixing copper and nickel in the ratio 5 : 2.

(a) How much copper would you need to mix with 3 kg of nickel?

(b) How much nickel would you need to mix with 2 kg of copper?

7 Potting compost is made by mixing 8 kg of peat with 3 kg of sand.

(a) How much sand would be needed to mix with 20 kg of peat?

(b) If I buy a 33 kg bag of compost, how much peat and sand will be in the bag?

8 A recipe for 475 mℓ of spaghetti sauce includes 150 g of minced meat, 250 g of tomatoes and 75 g of mushrooms.

(a) If I have only 200 g of tomatoes, how much of each other ingredient will I need?

(b) How many millilitres of sauce would this make?

Simplifying ratios

You will need to know

- how to simplify fractions

The method for simplifying a ratio is similar to simplifying fractions.

Simplifying fractions was covered in Section 3.1.

In Example 1 the amounts of flour and melted butter were 150 g and 30 g.

You could write this as a ratio 150 : 30
Dividing by 10 gives 15 : 3
Dividing by 3 gives 5 : 1

You can compare two quantities as long as they are in the same units.

You cannot divide this any further.

This means that you need 5 times as much flour as melted butter.

The three ratios 150 : 30, 15 : 3 and 5 : 1 are equivalent.

The **simplest form** is 5 : 1.

You could also say 'the ratio 5 : 1 is in its **lowest terms**'.

You can divide or multiply ratios to get them into their simplest form.

When a ratio is in its simplest form you cannot divide the numbers any further.

EXAMPLE 4

Write these ratios in their simplest form.

(a) 8 : 12 **(b)** 18 : 24 : 6 **(c)** 40 cm : 1 m
(d) 1 kg : 350 g **(e)** $\frac{3}{4} : \frac{1}{3}$ **(f)** 2.7 : 3.6

(a) 8 : 12 = 2 : 3 (dividing by 4)

Divide by a common factor.

(b) 18 : 24 : 6 = 3 : 4 : 1 (dividing by 6)

This ratio has 3 parts.

(c) 40 cm : 1 m = 40 cm : 100 cm Units not the same,
= 40 : 100 Change m to cm.
= 4 : 10 (dividing by 10)
= 2 : 5 (dividing by 2)

The units *must* be the same before you can simplify. This applies to (c) and (d).

Continued ▼

(d) 1 kg : 350 g = 1000 g : 350 g — Units not the same,

= 1000 : 350 — change kg to g.

= 100 : 35 — (dividing by 10)

= 20 : 7 — (dividing by 5)

(e) $\frac{3}{4} : \frac{1}{4} = (\frac{3}{4} \times 12) : (\frac{1}{3} \times 12)$ — (multiplying by 12)

= 9 : 4

(f) 2.7 : 3.6 = 27 : 36 — (multiplying by 10)

= 3 : 4 — (dividing by 9)

To be in its simplest form, all parts of the ratio must be whole numbers.

Find a number that both denominators divide into exactly. Multiply both fractions by this number.

Multiply by 10 to remove the decimal places.
Divide by 9 to simplify to its simplest form.

EXERCISE 11B

1 Write these ratios in their lowest terms.

(a) 5 : 10 (b) 12 : 8 (c) 6 : 30

(d) 9 : 15 (e) 40 : 25 (f) 48 : 36

(g) 120 : 70 (h) 28 : 49 (i) 1000 : 250

2 Write these ratios in their simplest form.

(a) \$1 : 20c (b) 6 mm : 3 m (c) 5ℓ : 250 mℓ

(d) 300 m : 2 km (e) 75c : \$5 (f) 350 g : 2 kg

(g) \$3.50 : \$1.25 (h) 45 cm : 2 m (i) 50 mm : 1 m

Change each part into the same units first.

3 Change these ratios into their simplest form.

(a) $\frac{1}{2} : \frac{1}{4}$ (b) $\frac{2}{3} : \frac{3}{4}$ (c) $\frac{5}{2} : \frac{7}{2}$

(d) 3.2 : 1.6 (e) 5.4 : 1.8 (f) 4.8 : 1.2

4 In a school there are 480 girls and 560 boys. Write the ratio of boys to girls in its lowest terms.

5 A drink was made with $1\frac{1}{2}$ cups of lemonade and $\frac{1}{4}$ cup of blackcurrant cordial. Write the ratio of lemonade to blackcurrant in its lowest terms.

6 The cost of materials for a stuffed toy was fabric \$3.20, stuffing \$1.20 and others 60c. Write the ratio of the cost of fabric to stuffing to others in its lowest terms.

Writing a ratio as a fraction

You can use fractions to help you solve ratio problems. When items are divided in a ratio 2 : 3, the fraction $\frac{2}{3}$ gives you another method of solving the problem. This is used in method B in Example 5.

EXAMPLE 5

In a school the ratio of boys to girls is 5 : 7.
There are 265 boys.

How many girls are there?

Method A (using multiplying)

Suppose there are x girls.

The ratio 5 : 7 has to be the same as 265 : x

$265 \div 5 = 53$ so multiply both sides of the ratio by 53

Boys: $5 \times 53 = 265$

Girls: $7 \times 53 = 371$

There are 371 girls in the school.

Method B (using fractions)

Suppose there are x girls.

The ratio 5 : 7 has to be the same as 265 : x

Using fractions, $\frac{5}{7} = \frac{265}{x}$

This can also be written as $\frac{7}{5} = \frac{x}{265}$

Multiplying both sides by 265, $\frac{7 \times 265}{5} = x$

This gives $x = 371$, so there are 371 girls in the school.

Multiplying a ratio by 2, 3, 4, 5 ... etc gives an equivalent ratio. Here the multiplier is 53.

When the unknown x is on the top, the equation is easier to solve.

See Section 3.2, Example 5 for how to solve this kind of equation.

EXERCISE 11C

1 On the lake the ratio of penguins to flamingos is 2 : 5. There are 235 flamingos. How many penguins are there?

Use the fraction method.

2 Antonio and Juanita share an inheritance from their aunt in the ratio of 4 : 3. If Antonio received \$480, how much did Juanita get?

3 At a football match the ratio of children to adults is 3 : 8. If there were 630 children, how many adults were there?

4 A piece of wood is cut into two pieces in the ratio 4 : 7.

  (a) If the shortest piece is 140 cm, how long is the longer piece?

  (b) How long was the original piece of wood?

5 A recipe for jam uses 125 g of fruit and 100 g of sugar.

  (a) What is the ratio of fruit to sugar in its lowest terms?

  (b) If I had 300 g of fruit how much sugar would I need?

  (c) What would be the total mass of the jam in part **(b)**?

6 In a box of strawberries the ratio of good to damaged fruit is 16 : 3.

  (a) If I found 18 damaged fruits, how many good ones did I have?

  (b) How many strawberries are in the box?

7 A magazine has pictures to text in the ratio of 3 : 2. If a magazine has 75 sections of pictures how many sections of text does it have?

8 The ratio of sand to cement to make mortar is 5 : 2.

  (a) If I use 3 buckets of cement, how many buckets of sand will I need?

  (b) A bucket of sand weighs 11 kg and a bucket of cement weighs 17 kg. How much will the mortar weigh?

Writing a ratio in the form 1 : n or n : 1

In Example 1, 150 : 30 simplified to 5 : 1.

Not all ratios simplify so that one value is 1.
For example, 12 : 8 simplifies to 3 : 2.

Dividing by 4.

If you are asked to write this ratio in the form n : 1 you cannot leave the answer as 3 : 2.

Look at which number has to be 1. Divide both sides of the ratio by that number.

$$12 : 8 = (12 \div 8) : (8 \div 8)$$
$$= 1.5 : 1$$

For $n : 1$ the number on the right-hand side has to be 1.

The '8' has to become a 1. Divide both numbers by 8.

You can write all ratios in the form $1 : n$ or $n : 1$.

For a ratio in the form $n : 1$ or $1 : n$, the n value does not have to be a whole number.

EXAMPLE 6

Write these ratios in the form $n : 1$.

(a) 27 : 4 **(b)** 6 cm : 25 mm **(c)** \$2.38 : 85c

(a) $27 : 4 = 6.75 : 1$

Divide both sides by 4.

(b) 6 cm : 25 mm = 60 mm : 25 mm
= 60 : 25
= 2.4 : 1

Divide both sides by 25.

Change of units needed in (b) and (c).

(c) \$2.38 : 85c = 238c : 85c
= 238 : 85
= 2.8 : 1

Divide both sides by 85.

EXAMPLE 7

Write these ratios in the form $1 : n$.

(a) 5 : 9 **(b)** 400 g : 1.3 kg **(c)** 40 minutes : $1\frac{1}{4}$ hours

---

(a) $5 : 9 = 1 : 1.8$

Divide both sides by 5.

(b) 400 g : 1.3 kg = 400 g : 1300 g
= 400 : 1300
= 1 : 3.25

Divide both sides by 400.

Change of units needed in (b) and (c).

(c) 40 mins : $1\frac{1}{4}$ hours = 40 mins : 75 mins
= 40 : 75
= 1 : 1.875

Divide both sides by 40.

EXERCISE 11D

1 Write the following ratios in the form $n : 1$.

(a) 15 : 3 (b) 7 : 2

(c) 9 : 4 (d) 12 : 5

(e) 3 kg : 200 g (f) 1 hour : 40 min

(g) 5 m : 40 cm (h) 1 day : 10 hours

(i) 2.5 ℓ : 500 mℓ

2 Write the following ratios in the form $1 : n$.

(a) 16 : 24 (b) 45 : 54

(c) 36 : 96 (d) 15 hours : 1 day

(e) 5 days : 4 weeks (f) 6 mm : 2.7 cm

(g) 18 ℓ : 4500 mℓ (h) 5 cm : 15 mm

Dividing quantities in a given ratio

Ivan and Stanislav share \$60 so that Ivan receives twice as much as Stanislav.

You can think of this as, Ivan : Stanislav = 2 : 1.

Ivan will get this much | Stanislav will get this much

Remember, a ratio compares quantities.

where each bag of money contains the same amount.

To work out each man's share, you need to divide the money into 3.

(2 + 1 = 3) equal **parts** (or **shares**).

1 part = \$60 ÷ 3 = \$20

Ian receives 2 × \$20 = \$40 2 parts

Simon receives 1 × \$20 = \$20 1 part

Check: \$20 + \$40 = \$60 ✓

To share quantities in a given ratio

1 Work out the total number of equal parts.

2 Work out the amount in 1 part.

3 Work out the value of each share.

You can use this method to share amounts between more than two people.

This fender guitar uses Alnico magnet pickups.

EXAMPLE 8

The alloy *Alnico* is made from five metals: cobalt, iron, nickel, aluminium and titanium in the ratio 6 : 5 : 3 : 1 : 1.
How many grams of each metal are there in 544 g of this alloy?

Total number of parts = 6 + 5 + 3 + 1 + 1 = 16 parts
1 part = 544 g ÷ 16 = 34 g

Cobalt	= 6 × 34 g = 204 g	6 parts
Iron	= 5 × 34 g = 170 g	5 parts
Nickel	= 3 × 34 g = 102 g	3 parts
Aluminium	= 1 × 34 g = 34 g	1 part
Titanium	= 1 × 34 g = 34 g	1 part
Check	204 g + 170 g + 102 g + 34 g + 34 g = 544 g ✓	

Check that your final answers add up to the original amount.

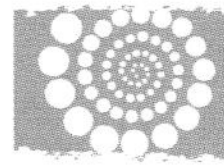

EXERCISE 11E

1 Divide these amounts in the ratio given.
  (a) 50 in the ratio 3 : 2 (b) 30 in the ratio 3 : 7
  (c) 45 in the ratio 5 : 4 (d) 64 in the ratio 3 : 5

2 Divide each of these amounts in the ratio given.
  (a) $200 in the ratio 9 : 1
  (b) 125 g in the ratio 2 : 3
  (c) 600 m in the ratio 7 : 5
  (d) 250 ℓ in the ratio 16 : 9

3 Divide $120 in these ratios.
  (a) 7 : 3 (b) 1 : 3 (c) 3 : 2 (d) 5 : 7

4 The ratio of pupils to teachers for a school trip is 13 : 2. If 135 people go on the trip, how many of them are teachers?

5 At a farm the ratio of black lambs to white lambs born is 2 : 9. If 132 lambs are born one year, how many of them are black?

6 A drink is made from orange juice, lemonade and lime juice in the ratio 5 : 6 : 1. If I make 960 mℓ of the drink, how much of each ingredient will I need?

7 One Saturday Jasmine decides to spend her revision time on Mathmatics, Geography and French in the ratio 7 : 5 : 3. If she spends a total of 4 hours revising, how much time will she spend on each subject?

Work in minutes.

8 Concrete is made from cement, sand and gravel in the ratio 1 : 3 : 4. If I want to make 2 tonnes of concrete, how much of each ingredient do I need?

9 Paul is 18 years old, John is 15 years old and Sarah is 12 years old. Their grandfather leaves them $255 to be divided between them in the ratio of their ages. How much will each of them receive?

Simplify the ratio first.

11.2 Proportion

Direct proportion

Two quantities are in **direct proportion** if their ratio stays the same as they increase or decrease.

Ratio method

For example, a mass of 5 kg attached to a spring stretches it by 40 cm.

A mass of 15 kg attached to the same spring will stretch it 120 cm.

The ratio mass : extension is

$$5 : 40 = 1 : 8 \text{ (dividing by 5)}$$
$$15 : 120 = 1 : 8 \text{ (dividing by 15)}$$

The ratio mass : extension is the same so the two quantities are in direct proportion.

You can use direct proportion to solve problems.

EXAMPLE 9

The cost of 5 apples is $1.40.
How much would 8 apples cost?

Suppose that 8 apples cost x cents.

The ratio 'number of apples' : 'cost' must stay the same.

So $5:140 = 8:x$

Using fractions, $\frac{140}{5} = \frac{x}{8}$

Multiplying both sides by 8, $\frac{140 \times 8}{5} = x$

$x = 224$

So 8 apples cost $2.24.

Working in cents for both costs.

Using the fractions method as in Example 5.

Unitary method

There is an easier way of solving Example 9.
First find the cost of 1 apple.

5 apples cost $1.40 or 140c

1 apple costs $\frac{140}{5} = 28$c

8 apples cost 28 × 8, = 224c or $2.24

This is easier than the method in Example 9. It only takes three lines.

This is called the **unitary method.**

In the unitary method, you find the value of *1 unit* of a quantity.

EXAMPLE 10

Susan has a part-time job.
She works for 15 hours a week and is paid $93.
Her employer wants her to work for 25 hours a week.
How much will Susan be paid now?

For 15 hours' work she is paid $93

For 1 hour's work she is paid $\frac{\$93}{15} = \$6.20$

For 25 hours' work she will get 25 × $6.20 = $155

Find how much she is paid for 1 hour.

To solve *direct* proportion problems you always do a *division* followed by a *multiplication.*

EXERCISE 11F

1 Find the cost of 1 unit for each of the following.
   (a) 5 bars of chocolate cost $2.25
   (b) 3 m of wood costs $3.36
   (c) 9 kg of carrots costs $4.05

2 If 8 pens cost $2.96, how much will 3 pens cost?

3 40 litres of petrol will take me 200 kilometres.
   How far can I go on 15 litres?

4 Five books weigh 450 g. How much would 18 books weigh?

5 Farah sews buttons onto trousers. She is paid $30 for sewing 200 buttons.
   How much would she get for sewing 35 buttons?

6 6 packets of soap powder cost $7.38.
   How much will 10 packets cost?

7 Jorina earns $67.20 for 12 hours' work. One week she works 17 hours. How much does she earn for that week?

8 A chocolate cake for 8 people uses 120 g of sugar.
   How much sugar would you need for a cake for 15 people?

9 Omar walks 500 m to school and it takes him 8 minutes. If he continues to walk at the same speed,
   (a) how far would he walk in 1 hour?
   (b) how long would it take him to walk 2.25 km?

10 My cat eats 2 tins of food every 3 days.
   How much does he eat in a week?

Inverse proportion

When two quantities are in direct proportion
- as one increases, so does the other
- as one decreases, so does the other.

See Section 8.8

$$\text{Time} = \frac{\text{distance}}{\text{speed}}$$
$$\frac{100}{50} = 2$$
$$\frac{100}{40} = 2\tfrac{1}{2}$$

Suppose you travel from Hamburg to Kiel, a distance of 100 kilometres.

If you travel at an average speed of 50 km/h it will take you 2 hours.

If you only average 40 km/h it will take you $2\frac{1}{2}$ hours.

If you *halve* the speed to 25 km/h you *double* the time to 4 hours.

The slower you travel, the more time it takes.

- As the speed decreases, the time increases.
- As the speed increases, the time decreases.

When two quantities are in **inverse proportion**, one quantity increases at the same rate as the other quantity decreases.

The best way to solve problems involving inverse proportion is to use the unitary method.

EXAMPLE 11

Two people take 6 hours to paint a fence.
How long will it take 3 people?

Read the problem first and decide whether it is *direct* or *inverse* proportion.
Use your common sense!
The more people painting, the less time it takes. This is inverse proportion.

2 people take 6 hours

1 person takes $6 \times 2 = 12$ hours

3 people will take $12 \div 3 = 4$ hours

- Work out how long it will take 1 person.
- 1 person takes twice as long as 2 people.

EXAMPLE 12

Tarik is repaying a loan from a friend.

He agrees to pay \$84 per month for 30 months.

If he can afford to pay \$120 per month, how many months will it take to repay?

Paying \$84 per month takes 30 months

Paying \$1 per month takes (84 × 30) months

Paying \$120 per month will take $\frac{84 \times 30}{120}$ months = 21 months

> Repaying *more* each month will take *less* time, so this is inverse proportion.

> 210 years !!!!!

> To solve *inverse* proportion questions you always do a *multiplication* followed by a *division.*

EXERCISE 11G

1 It takes 3 people 4 days to paint a shop.
How long would it take 1 person?

2 It takes 2 people 6 hours to make a suit.
How long would it take 3 people?

3 Eight horses need a trailer of hay to feed them for a week. How many horses could this feed for 4 days?

4 It takes 5 bricklayers 4 days to build a house.
How long would it take 2 bricklayers?

5 It takes 3 pumps 15 hours to fill a swimming pool.
How many pumps would be needed to fill the pool in 9 hours?

6 It takes 7 days for 6 people to dig a trench. How long would it take 14 people?

7 In a library 48 paperback books 20 mm wide fit on a shelf. How many books 24 mm wide would fit on the same shelf?

8 It takes 8 hours to fly to Chicago at a speed of 600 km/h.

(a) How long would it take to fly to Chicago at a speed of 800 km/h?

(b) If the journey took 12 hours, what speed was I travelling?

9 It takes 6 window cleaners 8 days to clean the windows of an office block.

(a) How long would it take 4 window cleaners?

(b) The windows need to be cleaned in 3 days for a special event. How many window cleaners are needed?

EXERCISE 11H

1 If 10 metres of material cost $23.50, what will 7 metres cost?

2 A man cuts a hedge in 45 minutes using a cutter with a blade 36 cm wide. How long would it take if the blade was only 15 cm wide?

3 It takes 200 tiles to tile a room when each tile covers 36 cm^2. If I use tiles which cover 25 cm^2, how many tiles would I need?

4 Alan works for 9 hours a week in a shop. He is paid $45. He increases his hours to 15. How much will he be paid?

5 It takes $2\frac{1}{2}$ hours to travel to Delhi by train at an average speed of 80 km/h. A new train travels at an average speed of 100 km/h. How long will the journey to Delhi take on the new train?

6 If 14 kg of potatoes cost $2.38, how much will 6 kg cost?

7 One load of feed lasts 40 sheep 9 days.

(a) How many sheep could this load feed for 5 days?

(b) For how many days could it feed 60 sheep?

8 In a hotel it takes a cleaner 24 minutes to clean 3 rooms.

(a) How long will it take her to clean 16 rooms?

(b) How many rooms can she clean in 4 hours?

11.3 Map scales

You will need to know

- how to convert between metric units of length

Map **scales** are written as ratios.

A scale of 1 : 50 000 means that 1 cm on the map represents 50 000 cm on the ground.

A scale of 1 : 25 000 means that 1 cm on the map represents 25 000 cm on the ground.

Always look carefully to see what scale is being used.

When you answer questions involving map scales you need to

- use the scale of the map
- convert between metric units of length so that your answer is in sensible units.

EXAMPLE 13

The scale of a map is 1 : 50 000. The distance between Kenton-on-Sea and Port Alfred on the map is 40 cm.
What is the actual distance between Kenton-on-Sea and Port Alfred? Give your answer in kilometres.

Distance on map = 40 cm
Distance on the ground = 40 × 50 000 cm
= 2 000 000 cm
= 2 000 000 ÷ 100 m
= 20 000 m
= 20 000 ÷ 1000 km
= 20 km

Work out the real distance in cm then convert to km.

Each 1 cm on the map is 50 000 on the ground.

1 m = 100 cm
1 km = 1000 m

EXAMPLE 14

The distance between two towns is 24 km. How far apart will they be on a map of scale 1 : 180 000?

Distance on the ground = 24 km
= 24 × 1000 m
= 24 000 m
= 24 000 × 100 cm
= 2 400 000 cm

Distance on map = $\frac{2\,400\,000}{180\,000}$ cm = 13.3 cm (to 3 s.f.)

Convert the real distance to cm before you divide by the scale of the map.

EXERCISE 11I

1 The scale of a map is 1 : 25 000. Find the actual distance represented by these measurements on the map.
 (a) 4 cm (b) 7 cm (c) 8 mm (d) 12.5 cm

2 A map has scale 1 : 200 000. What measurement on the map will represent
 (a) 4 km (b) 20 km (c) 15 km (d) 12.5 km?

3 This map has a scale of 1 : 3 000 000.
 (a) The distance on the map from Cair Faralel to Tashlan is 2.5 cm. How far is the actual distance?
 (b) Measure in a straight line the distance from Cair Faralel to Neruna. How far is the actual distance?

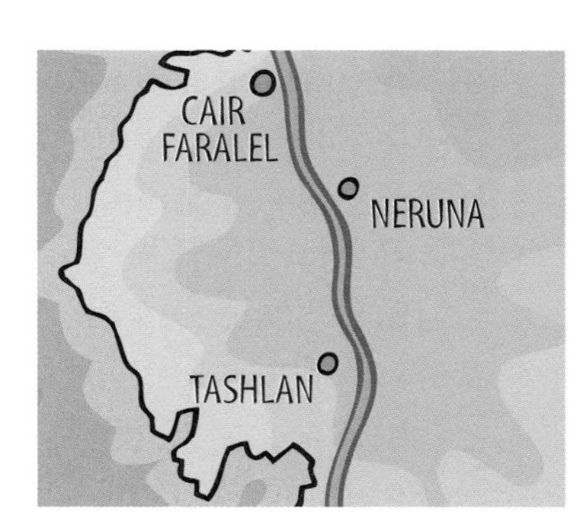

4 A map of Ireland has a scale of 1 : 500 000. What measurement on the map would represent these distances
 (a) Waterford to Dundalk, 400 km
 (b) Wicklow to Carrigart, 320 km
 (c) Dublin to Tralee, 160 km?

5 This map has a scale of 1 : 50 000. Use the map to work out the following distances.

(a) Between the ends of the 2 piers at Tynemouth.

(b) From Sharpness Point to Smuggler's Cave.

(c) The ferry crossing of the Tyne.

(d) The length of both piers.

(e) From the Coast Guard Station (CG Sta) to the Coast Guard Lookout (CG Lookout).

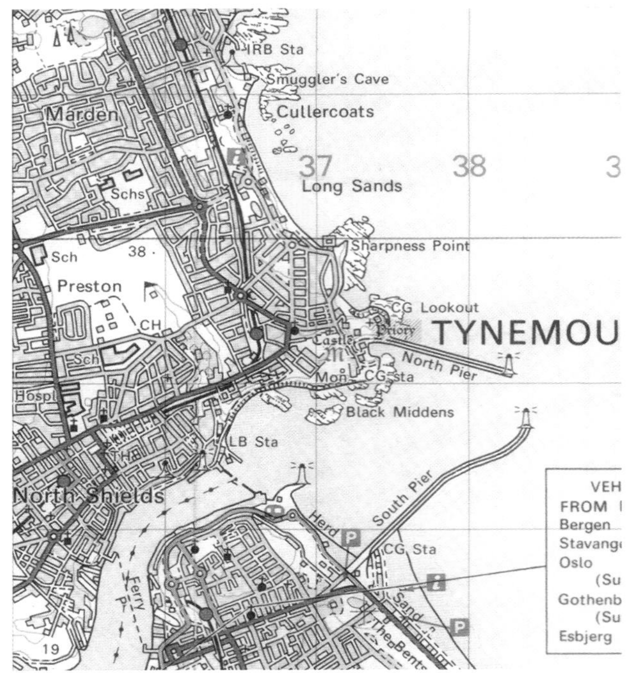

6 A model train is made to a scale of 1 : 40.

(a) The height of the model is 20 cm. How high is the real train?

(b) The real train is 7.4 m long. How long is the model?

7 A designer draws a plan of a garden using a scale of 1 : 50.

(a) The garden is 5.5 m long and 4.2 m wide. What are the measurements of the plan?

(b) She makes some scaled cut-outs of a shed, container and pond so she can try different layouts. Change these scaled measurements to find the real size of the items.

shed 4 cm × 2 cm
container 1.5 cm × 1 cm
pond 3 cm × 1 cm

EXAMINATION QUESTIONS

1 Cement, sand, aggregate and water are used to make concrete, in the ratio

Cement : Sand : Aggregate : Water = 2 : 5 : 8 : 1.

(a) Bobbie wants to make 1.2m^3 of concrete.
How much aggregate will he need? [1]

(b) Eddie wants to make concrete.
He uses 0.25 m^3 of cement.

(i) How much sand does he need? [1]

(ii) When water and aggregate have been added, how much concrete will he have? [1]

(CIE Paper 1, Jun 2001)

---

2 Claudia and Tania share a box of 30 biscuits in the ratio 2 : 3.
How many biscuits does Tania receive? [2]

(CIE Paper 1, Nov 2001)

---

3 Mahesh and Jayraj share $72 in the ratio 7 : 5.
How much does Mahesh receive? [2]

(CIE Paper 1, Jun 2003)

---

4 There are approximately 500 000 grains of wheat in a 2 kg bag.
Calculate the mass of one grain in grams. [2]

(CIE Paper 1, Jun 2003)

---

5 A model of a car has a scale of 1 : 25.
The model is 18 cm long.
Calculate, in metres, the actual length of the car. [2]

(CIE Paper 1, Jun 2004)

---

6 Antonia is making a cake.
She uses currants, raisins and sultanas in the ratio

currants : raisins : sultanas = 4 : 3 : 5.

The total mass of the three ingredients is 3.6 kilograms.
Calculate the mass of sultanas. [2]

(CIE Paper 1, Nov 2004)

---

7 Write, in its simplest form, the ratio 3.5 kilograms : 800 grams. [2]

(CIE Paper 1, Jun 2005)

---

8 Three friends, Cleopatra, Dalila and Ebony go shopping.
The money they each have is in the ratio

Cleopatra : Dalila : Ebony = 5 : 7 : 8.

Cleopatra has $17.

(a) How many dollars do they have in total? [2]

(b) Dalila spends $12 on a hat.
How many dollars does she have left? [1]

(CIE Paper 1, Nov 2005)

Chapter 12

Probability

This chapter will show you how to

- ✔ write a probability using words or values
- ✔ work out relative frequencies and estimate probabilities

12.1 Describing probability

Probability uses numbers to represent how likely it is that something (an **event**) will happen.

You can describe how likely it is that an event will happen using words such as **certain**, **likely**, **unlikely** and **impossible**.

You can use numbers to represent these based on the **probability scale**. All probability values lie between 0 and 1.

$0 \leqslant$ probability $\leqslant 1$.

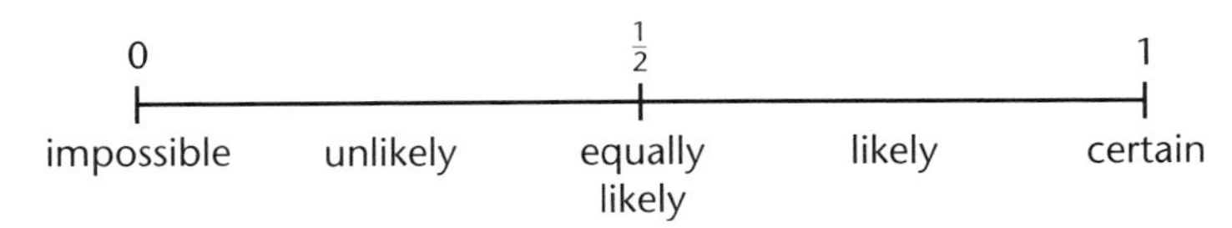

EXAMPLE 1

Describe the likelihood of these events using words.

(a) The sun will set today.

(b) You were born 30 years before your father.

(c) Getting a Head when you throw a coin.

(d) It will rain tomorrow.

(e) It will snow in summer.

(a) Certain
(b) Impossible
(c) Equally likely
(d) Possible
(e) Unlikely

You can write a probability as a fraction or a decimal.

probability	fraction	decimal
impossible	0	0
certain	1	1.0
equally likely	$\frac{1}{2}$	0.5
likely	e.g. $\frac{8}{10}$	0.8
unlikely	e.g. $\frac{1}{5}$	0.2

Converting between fractions and decimals was covered in Chapter 3.

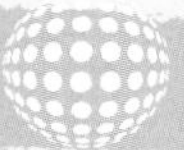

EXAMPLE 2

Write the probability of obtaining a Head when you toss a coin. Give the probability as a fraction.

There are two possible **outcomes** – a Head or a Tail. They are both equally likely.

It is equally likely to throw a Head or a Tail.
The probability $p(H) = \frac{1}{2}$.

p(H) is shorthand for 'probability of throwing a Head'.

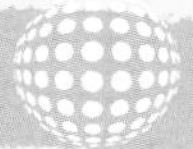

EXAMPLE 3

Write the probability of throwing a 6 on an ordinary six-sided dice. Write the probability as a decimal.

An ordinary six-sided dice has six faces with numbers 1, 2, 3, 4, 5 and 6. There are 6 possible outcomes.

The probability of throwing a 6 is 1 out of a possible 6 numbers.
The probability $p(6) = \frac{1}{6}$.

p(6) is shorthand for 'probability of throwing a six'.

In Examples 2 and 3 you are finding the probability of an event as the fraction.

$$\text{probability} = \frac{\text{number of successful outcomes}}{\text{total number of possible outcomes}}$$

Remember this, it is an important formula.

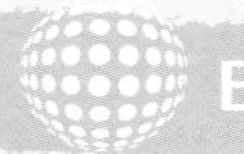

EXAMPLE 4

An eight-sided (octahedral) fair dice numbered 1 to 8 is thrown. Find

(a) the probability of obtaining a 3

(b) the probability of obtaining a prime number.

(a) The value 3 only occurs once on the dice, so there is only 1 successful outcome.

The total number of possible outcomes is 8.

$p(3) = \frac{1}{8}$

(b) The prime number values are 2, 3, 5 and 7, so there are 4 successful outcomes.

$p(\text{prime}) = \frac{4}{8} = \frac{1}{2}$

The possible outcomes are 1, 2, 3, 4, 5, 6, 7, 8.

$\frac{\text{number of successful outcomes}}{\text{number of possible outcomes}}$

The number of possible outcomes is still 8.

p(prime) is shorthand for 'probability that outcome is prime'.

EXERCISE 12A

1 Describe the likelihood of these events using words.

(a) The sun will rise tomorrow.

(b) England will win the next World Cup.

(c) You will eat five portions of fruit or vegetables today.

(d) You will see the President on your way home from school.

(e) You will throw an even number on an ordinary six-sided dice.

2 Copy the probability scale. Label each arrow with an event from the list below.

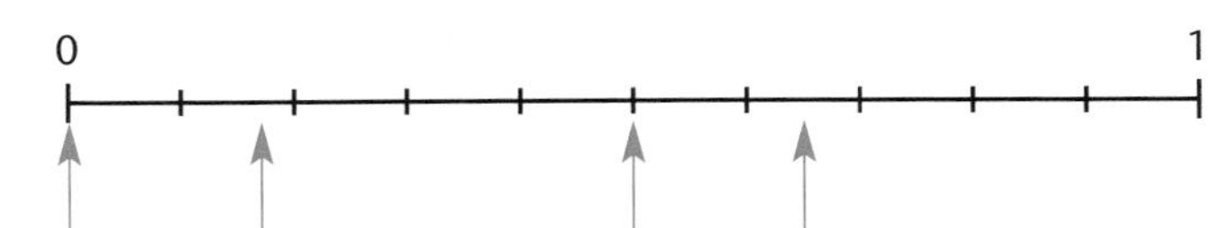

Using an ordinary six-sided dice and

(a) throwing a 1

(b) obtaining an even number

(c) getting a number >7

(d) obtaining a number less than or equal to 4.

3 Copy the probability scale from question 2. Draw and label arrows to show these probabilities.

(a) It will rain tomorrow.

(b) You throw a coin and get a 'Head'.

(c) A dropped drawing pin lands point-down.

(d) You throw a 7 on an ordinary six-sided dice.

(e) You buy a new music CD next week.

4 Write these probabilities using the notation p(event) = ☐ and in the form specified.

(a) The probability that you will be given homework is 0.4 (as a fraction).

(b) The probability that it will snow tomorrow is 0.01 (as a fraction).

(c) The probability of tossing a coin and getting a Tail is $\frac{1}{2}$ (as a decimal).

5 An ordinary six-sided dice is thrown. Find the probabilities of these events

(a) throwing a 2

Use the notation p(2).

(b) obtaining an even number

(c) throwing a number less than 4

(d) throwing a 3 or a 5

(e) not obtaining a 6.

6 In a raffle one hundred tickets are sold. What is the probability of winning if you buy

(a) 3 tickets (b) 11 tickets?

7 In a bag there are 11 blue sweets and 5 red sweets. What is the probability of picking out

(a) a blue sweet (b) a red sweet

(c) a green sweet (d) a blue or red sweet?

12.2 Relative frequency

The probability of obtaining a Head when a coin is tossed is p(H) = $\frac{1}{2}$.

In reality, if you toss a coin 10 times, you may not get 5 Heads.

The number of Heads you get (**successful trials**) will get closer to $\frac{1}{2}$ of the total number of throws the more times you toss the coin (total number of trials).

When you carry out a probability experiment, the experimental probability is called the **relative frequency**.

$$\text{Relative frequency} = \frac{\text{number of successful trials}}{\text{total number of trials}}$$

Remember this, it is an important formula.

This is also called experimental probability.

EXAMPLE 5

A coin is tossed 50 times. Estimate the probability of

(a) a Head **(b)** a Tail.

	Tally	Frequency
Head (H)	卌 卌 卌 卌 卌 III	28
Tail (T)	卌 卌 卌 卌 II	22
	Total	50

(a) $\text{Relative frequency} = \frac{\text{number of successful trials}}{\text{total number of trials}}$

$$p(H) = \frac{\text{number of heads}}{\text{total number of throws}}$$
$$= \frac{28}{50} \text{ or } 0.56 \text{ or } 56\%$$

(b) $p(T) = \frac{\text{number of tails}}{\text{total number of throws}}$
$$= \frac{22}{50} \text{ or } 0.44 \text{ or } 44\%$$

Check $0.56 + 0.44 = 1.00$ ✓

Check that the probabilities for all the events add up to 1.

$$\text{Frequency} = \text{total number of trials} \times \text{relative frequency}$$

Remember this, it is an important formula.

EXAMPLE 6

An ordinary six-sided dice is thrown 100 times.
The scores are recorded in a frequency table.

Score	1	2	3	4	5	6	Total
Frequency	13	18	19	14	15	21	100

(a) Find the relative frequency of each score.

(b) Compare the relative frequency of throwing a 5 with the theoretical probability of p(5).

(c) How could you get a better estimate for the probability?

(d) How many times would you expect to throw a 6 if you rolled the dice a total of 300 times?

(a)

Score	Frequency	Relative frequency
1	13	$\frac{13}{100} = 0.13$
2	18	$\frac{18}{100} = 0.18$
3	19	$\frac{19}{100} = 0.19$
4	14	$\frac{14}{100} = 0.14$
5	15	$\frac{15}{100} = 0.15$
6	21	$\frac{21}{100} = 0.21$
Total	100	$\frac{100}{100} = 1.00$

$\frac{\text{number of times 1 is thrown}}{\text{total number of throws}}$

(b) The theoretical probability p(5) is

$$\frac{\text{number of successful outcomes}}{\text{total number of outcomes}} = \frac{1}{6} = 0.167$$

The relative frequency for p(5) is 0.15, which is slightly lower.

(c) To improve the estimate, you could carry out more trials.

(d) Number of sixes $= 300 \times \frac{21}{100}$
$= 63$

As the number of trials increases, the relative frequency approaches the theoretical probability.

Frequency $=$ no. of trials $\times$ relative frequency

EXERCISE 12B

1 (a) Copy the tally chart shown.

	Tally	Frequency	Relative frequencies
Head (H)			
Tail (T)			
	Total	100	

(b) Toss a coin 100 times and record your results in the chart.

(c) Work out the relative frequency for Head and Tail from your results. Write them in the table.

(d) Compare your results with the theoretical probabilities for p(H) and p(T). How could you get a better estimate for the probability?

2 Copy the tally chart.

Type of number obtained	Tally	Frequency
An even number		
A square number		
A prime number		
A number greater than 4		
	Total	

Some values may appear in more than one column at the same time.

(a) Throw an ordinary six-sided dice 100 times and record your results in the chart.

(b) Calculate the relative frequency for each event.

(c) Compare your answers to the theoretical probabilities for p(even), p(square), p(prime) and p($>$4).

3 A bag contains 7 red counters and 3 blue counters. Pick a counter from the bag, record its colour, then replace it. Repeat 50 times.

(a) Draw a tally chart to record this information.

(b) Work out the relative frequencies for picking a red and blue counter.

(c) Compare your results with the theoretical values for p(R) and p(B).

For an event A, the probability of the event A *not* happening is given by

$$p(\text{not } A) = 1 - p(A)$$

Remember this, it is an important formula.

EXERCISE 12C

There are 26 letters in the English alphabet. Vowels are a, e, i, o, u. Consonants are all the others.

1 A letter of the English alphabet is chosen at random. What is the probability that it will be a vowel? What is the probability it will be a consonant?

2 One of the longest rivers in the world is called the MISSISSIPPI. If you choose one of these letters at random, what is the probability of choosing

(a) the letter I

(b) the letter S?

3 An ordinary six-sided dice is thrown. Work out

(a) p(3), the probability of throwing a 3

(b) p(1), p(2), p(4), p(5) and p(6)

(c) the probability of *not* throwing a 2.

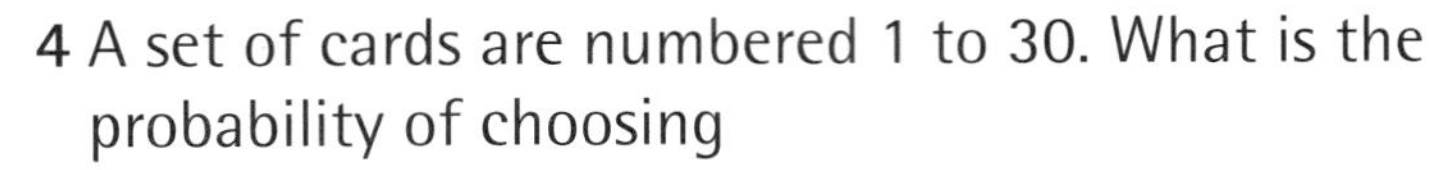

4 A set of cards are numbered 1 to 30. What is the probability of choosing

(a) a square number

(b) a prime number

(c) a number > 10

(d) a multiple of 3

(e) a factor of 24

(f) a card containing the number 2?

5 A bag contains 5 green counters, 4 blue counters, 2 red counters and 3 yellow counters.
What is the probability of picking

(a) a red counter

(b) a counter that is not yellow?

6 A bag contains 16 sweets of three different colours – red, green and yellow.

The probability of picking a green sweet is $p(G) = \frac{1}{4}$.

The probability of picking a red sweet is $p(R) = \frac{3}{8}$.

(a) Work out the probability of picking a yellow sweet.

(b) How many sweets of each colour are in the bag?

EXAMINATION QUESTIONS

1 A whole number is picked at random from the numbers 1 to 200, inclusive.

(a) What is the probability that it is **more than** 44?

Give your answer as

(i) a fraction in its lowest terms, [2]

(ii) a decimal. [1]

(b) What is the probability that the number is **at least** 180? [1]

(CIE Paper 1, Jun 2000)

---

2 A letter is chosen at random from the word MATHEMATICS.

What is the probability that it is M? [1]

(CIE Paper 1, Nov 2000)

---

3 Alex, Bernice, Divya, Elisa and Fernanda play a game.

They all have an equal chance of winning.

(a) What is the probability that Alex does not win? [1]

(b) Divya wins the first game and then drops out. The others continue playing.

What is the probability that Elisa wins the second game? [1]

(CIE Paper 1, Jun 2001)

---

4 Akaliza chooses a number from the list 99, 107, 211, 478 and 481.

Find the probability that the number is

(a) odd, [1]

(b) greater than 400, [1]

(c) a multiple of 5. [1]

(CIE Paper 1, Nov 2001)

---

5 Fifty students take part in a quiz.
The table shows the results.

Number of correct answers	5	6	7	8	9	10	11	12
Number of students	4	7	8	7	10	6	5	3

A student is chosen at random from the fifty students.
What is the probability that this student had

(a) exactly 10 correct answers, [1]
(b) at least 10 correct answers, [1]
(c) more than 1 correct answer? [1]

(CIE Paper 3, Jun 2003)

---

6 Aminata has a bag containing 35 beads.
The beads are either blue, yellow or red.
One bead is chosen at random.
The probability of choosing a blue bead is $\frac{2}{7}$ and the probability of choosing a yellow bead is $\frac{3}{5}$.
Calculate

(a) the number of blue beads in the bag. [2]
(b) the probability of choosing a red bead. [2]

(CIE Paper 1, Nov 2004)

---

7 A bag of 30 sweets contains 8 chocolates, 13 nougats and 9 toffees.
A sweet is selected at random.
What is the probability that it is a toffee? [1]

(CIE Paper 1, Jun 2005)

Chapter 13

Multiples, factors, powers and roots

This chapter will show you how to

- ✔ recognise odd, even and prime numbers
- ✔ write down multiples and factors of numbers
- ✔ recognise square and cube numbers and find square roots and cube roots
- ✔ revise and use the laws of indices
- ✔ find prime factors, the highest common factor (HCF) and the lowest common multiple (LCM)

13.1 Multiples, factors and prime numbers

All whole numbers are either odd or even.

Even numbers divide exactly by 2.

The even numbers are 2, 4, 6, 8, 10, ...

Odd numbers do not divide exactly by 2.

The odd numbers are 1, 3, 5, 7, 9, ...

A **multiple** of a number is a number in its 'times table'.

For example, the multiples of 4 are 4, 8, 12, 16, 20, ...

A **factor** of a number is a whole number that divides into it exactly. The factors of any number always include 1 and the number itself.

For example, the factors of 10 are 1, 2, 5 and 10.

The factors of a number come in pairs (except for square numbers).

For example, for 10 you can link the factors like this.

Each pair of factors multiplies to give 10.
$2 \times 5 = 10$
$1 \times 10 = 10$

A **prime** number has only *two* factors. The factors are always 1 and the number itself.

For example, 7 is prime. Its only factors are 1 and 7.

7 cannot be divided by any other whole numbers.

The only *even* prime number is 2.
1 is *not* a prime number.

1 only divides by itself so it has only *one* factor.

The first few prime numbers are, 2, 3, 5, 7, 11, ...

You need to remember these facts.

EXAMPLE 1

(a) Write down all the multiples of 4 which are less than 30.
(b) Write down all the multiples of 6 which are less than 30.
(c) Which multiples are common to both lists?

Do not include 30.

(a) 4, 8, 12, 16, 20, 24, 28
(b) 6, 12, 18, 24
(c) 12 and 24 are in both lists.

12 and 24 are multiples of both 4 and 6. They are called *common multiples*. You will meet these again in Section 13.5.

EXAMPLE 2

Here is a list of numbers.

1 4 5 7 12 17 20 21 28

Write down
(a) the odd numbers
(b) the multiples of 3
(c) the factors of 24
(d) the factors of 28
(e) the prime numbers.

(a) The odd numbers are 1, 5, 7, 17 and 21.
(b) The multiples of 3 are 12 and 21.
(c) The factors of 24 are 1, 4 and 12.
(d) The factors of 28 are 1, 4, 7 and 28.
(e) The prime numbers are 5, 7 and 17.

1 and 4 are factors of both 24 and 28. They are called *common factors*. You will meet these again in Section 13.5.

EXERCISE 13A

1 From this card

[1] [2] [5] [7] [8]

[9] [11] [12] [13] [27]

[33] [51] [69] [81] [86]

(a) list the even numbers.
(b) list the odd numbers.
(c) list the prime numbers.
(d) which is the largest prime number?
(e) which is the smallest odd number?
(f) which numbers are even and prime?
(g) which numbers are odd and prime?

2 Write down the first six multiples of these numbers.

(a) 3 (b) 9 (c) 11 (d) 15 (e) 21

3 Find all the factors of these numbers.

(a) 14 (b) 20 (c) 36 (d) 44

4 Which of these numbers are

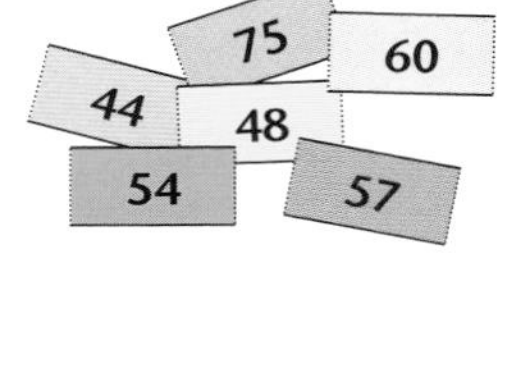

(a) multiples of 2 (b) multiples of 3
(c) multiples of 5 (d) multiples of 9
(e) multiples of both 2 and 3
(f) multiples of 2, 3 and 5?

5 Find the missing factors.

(a) The factors of 12 are 1, ☐, 3, 4, 6, ☐.
(b) The factors of 36 are 1, 2, 3, ☐, 6, 9, ☐, 18, 36.
(c) The factors of 18 are 1, 2, ☐, 6, ☐, 18.

Make factor pairs.

6 (a) Which of these numbers are multiples of

(i) 4 (ii) 7 (iii) 13 (iv) 23?

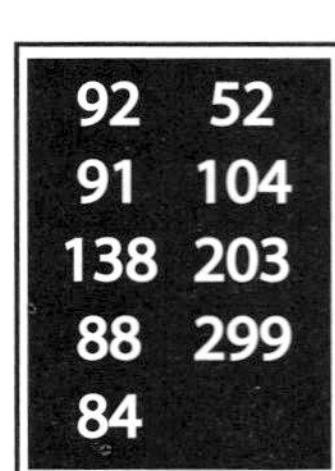

(b) Which of these numbers are multiples of 7 and 13?
(c) Which of these numbers are multiples of 4, 7 and 13?

7 (a) List all the factors of 10.
(b) List all the factors of 25.
(c) Which factors are common to both lists?

8 Find the smallest number greater than 50 that is
(a) a multiple of 2 (b) a multiple of 3
(c) a multiple of 5 (d) a multiple of 9.

9 (a) List all the factors of 24.
(b) List all the factors of 16.
(c) List all the factors of 30.
(d) Which numbers are factors of 24 and 16?
(e) Which numbers are factors of 16 and 30?
(f) Are there any numbers that are factors of 24, 16 and 30?

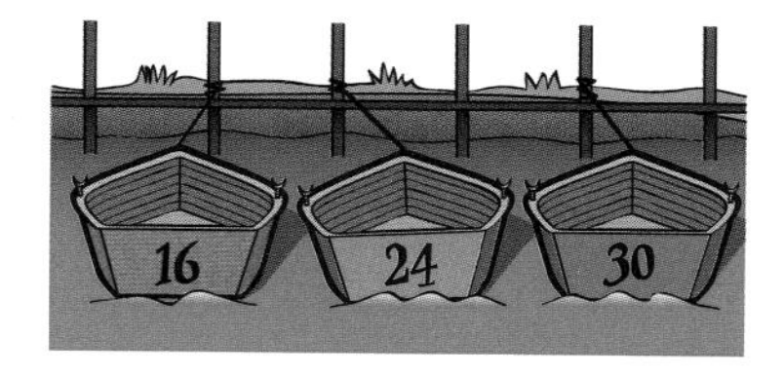

10 Find the biggest number less than 100 that is
(a) a multiple of 4 (b) a multiple of 6
(c) a multiple of 7 (d) a multiple of 8.

11 How many even prime numbers are there?

13.2 Squares, cubes and roots

Square numbers and cube numbers

A **square** number is what you get when you multiply a whole number by itself.

For example, $4 \times 4 = 16$, $7 \times 7 = 49$, $15 \times 15 = 225$

16, 49 and 225 are square numbers.

They are called square numbers because you can arrange them in a square pattern of dots.

Here is the pattern for $4 \times 4 = 16$.

You write 4×4 as 4^2.

The ^{2} is called a **power** (or **index**).

You say '4 squared' or 'the square of 4' or '4 to the power of 2'.

Square numbers have an odd number of factors.

For example,
the factors of 16 are

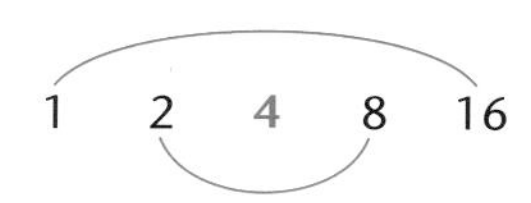

The factor pairs are 1 and 16, 2 and 8.
4 multiplies by itself ($4 \times 4 = 16$).

A **cube** number is what you get when you multiply a whole number by itself, then by itself again.

For example, $2 \times 2 \times 2 = 8$, $3 \times 3 \times 3 = 27$,
$10 \times 10 \times 10 = 1000$

8, 27 and 1000 are cube numbers.

They are called cube numbers because you can arrange them in a cube pattern.

Here is the pattern for $2 \times 2 \times 2 = 8$.

You write $2 \times 2 \times 2$ as 2^3.

The ^{3} is a power (or index).

You say '2 cubed' or 'the cube of 2' or '2 to the power of 3'.

You can cube or square any number. You may need to use a calculator.

You will need a calculator to square or cube numbers that are not whole numbers.

A calculator has a key for squaring numbers [x^2].

Some calculators have a key for cubing numbers [x^3].

Check how to square and cube numbers on your calculator.

On others you need to use the power key [y^x], [x^y] or [$x^■$]

To calculate 2^3 you input [2] [y^x] [3]. Answer is 8.

EXAMPLE 3

Work out these cubes and squares.

(a) 8^2
(b) 15^2
(c) 4^3
(d) 30^3
(e) $(8.3)^2$
(f) $(3.4)^3$

(a) $8^2 = 8 \times 8 = 64$

(b) $15^2 = 15 \times 15 = 225$

(c) $4^3 = 4 \times 4 \times 4 = 64$

(d) $30^3 = 30 \times 30 \times 30 = 27\,000$

(e) $(8.3)^2 = 68.89$ (using a calculator)

(f) $(3.4)^3 = 39.304$ (using a calculator)

You need to recognise all the squares of numbers up to $15^2 = 225$.

EXAMPLE 4

Write down

(a) all the square numbers between 40 and 90

(b) all the cube numbers less than 100.

(a) $6 \times 6 = 36$ (too small)

Start by trying any number.

$7 \times 7 = 49$

$8 \times 8 = 64$

$9 \times 9 = 81$

$10 \times 10 = 100$ (too big)

The square numbers between 40 and 90 are 49, 64 and 81.

Stop when the answer is $\geqslant 100$.

(b) $1 \times 1 \times 1 = 1$

$2 \times 2 \times 2 = 8$

$3 \times 3 \times 3 = 27$

$4 \times 4 \times 4 = 64$

$5 \times 5 \times 5 = 125$ (too big)

The cube numbers less than 100 are 1, 8, 27 and 64.

EXERCISE 13B

1 Work out these squares and cubes.

(a) 5^2 (b) 11^2 (c) 3^3 (d) 15^2

(e) 100^3 (f) 2^3 (g) 9^2 (h) 5^3

2 Work out these squares and cubes.

(a) 17^2 (b) 200^2 (c) 12^3 (d) 25^2

(e) $(4.9)^3$ (f) $(0.2)^2$ (g) $(2.5)^3$ (h) $(1.25)^2$

3 Work these out.

(a) $3^3 + 5^2$ (b) $6^2 - 4^2$ (c) $10^3 - 5^3$

4 Which is larger

(a) 4^3 or 5^2 (b) 2^3 or 3^2 (c) 6^3 or 7^2?

5 A square picture has a side of length 8.5 cm.
What is its area?

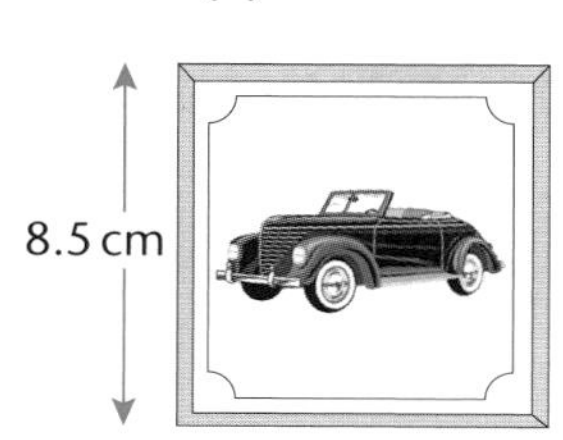

Square roots and cube roots

In the last section you saw that $4 \times 4 = 16$,
which you write $4^2 = 16$,
and that 16 is a square number.

The inverse of squaring is finding the **square root.**

The *inverse* is the reverse, or opposite, process. For example, the inverse of 'add' is 'subtract'.

The square root of 16 is 4, because $4^2 = 16$.
You write $\sqrt{16} = 4$.

Also $(-4) \times (-4) = 16$ or $(-4)^2 = 16$
So -4 is also a square root of 16 or $\sqrt{16} = -4$.

For more on negative numbers see Chapter 1.

negative × negative = positive
positive × positive = positive

All positive numbers have *two* square roots.

Squaring always gives a positive number, so (at this level) it is impossible to find the square root of a negative number.

For example $\sqrt{100} = 10$ and $\sqrt{100} = -10$
$\sqrt{36} = 6$ and $\sqrt{36} = -6$
You write $\sqrt{100} = \pm 10$ and $\sqrt{36} = \pm 6$

$\pm$ is shorthand for the positive and the negative value of a number.

The inverse of cubing is finding the **cube root.**

The cube root of 8 is 2, because $2^3 = 8$.
You write $\sqrt[3]{8} = 2$.

$2^3 = 2 \times 2 \times 2 = 8$
8 is a cube number.

When you cube a positive number the answer is positive.
For example $2 \times 2 \times 2 = 8$.

When you cube a negative number the answer is negative.
For example $-2 \times -2 \times -2 = -8$.

$(-2)^3 = (-2) \times (-2) \times (-2)$
$= (+4) \times (-2)$
$= (-8)$
positive × negative = negative

Positive numbers have a positive cube root.

Negative numbers have a negative cube root.

$\sqrt[3]{8} = 2 \quad \sqrt[3]{-8} = -2$

Any *positive* number has two square roots (for example $\sqrt{36} = \pm 6$).

The $\sqrt{\ }$ key only gives the positive square root.

You will not be penalised in the examination for including the negative root or for omitting it.

Any number (positive or negative) has a cube root.

Most calculators have a key for square roots $\sqrt{\ }$.

... and a key for cube roots $\sqrt[3]{\ }$.

On some calculators the cube root is a 'second function'.

Square roots and cube roots are not always whole numbers.
You may need to round the calculator answer.
Remember to round to the required degree of accuracy.

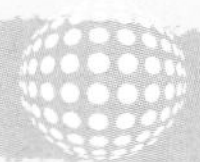

EXAMPLE 5

Without using a calculator, work these out.

(a) $\sqrt{49}$ (b) $\sqrt[3]{125}$

(a) $7 \times 7 = 49$, so $\sqrt{49} = 7$

(b) $3^3 = 3 \times 3 \times 3 = 27$ (too small)

$5^3 = 5 \times 5 \times 5 = 25 \times 5 = 125$

So $\sqrt[3]{125} = 5$

Try some values.

EXAMPLE 6

Use your calculator to work these out.

(a) $\sqrt{49}$ (b) $\sqrt{256}$ (c) $\sqrt{0.81}$

(d) $\sqrt{0.464}$ (e) $\sqrt{264\,196}$

(a) $\sqrt{49} = 7$ (b) $\sqrt{256} = 16$ (c) $\sqrt{0.81} = 0.9$

(d) $\sqrt{0.4624} = 0.68$ (e) $\sqrt{264\,196} = 514$

EXAMPLE 7

Use your calculator to work these out.

(a) $\sqrt{67}$ (b) $\sqrt{0.4755}$ (c) $\sqrt{218.75}$

(d) $\sqrt[3]{70}$ (e) $\sqrt[3]{2344}$ (f) $\sqrt[3]{0.038}$

Give your answers correct to 3 s.f.

Look back at Chapter 2, Section 2.1 for help with significant figures.

(a) $\sqrt{67} = 8.1853\ldots = 8.19$

(b) $\sqrt{0.4755} = 0.6895\ldots = 0.690$

(c) $\sqrt{218.75} = 14.790\ldots = 14.8$

(d) $\sqrt[3]{70} = 4.121\ldots = 4.12$

(e) $\sqrt[3]{2344} = 13.283\ldots = 13.3$

(f) $\sqrt[3]{0.038} = 0.3361\ldots = 0.336$

EXERCISE 13C

You will not be penalised in the examination for including the negative root or for omitting it.

See Example 5 if you need help.

1 Without using a calculator, work these out.

(a) $\sqrt{100}$ (b) $\sqrt{25}$ (c) $\sqrt{64}$ (d) $\sqrt{81}$

(e) $\sqrt[3]{8}$ (f) $\sqrt[3]{27}$ (g) $\sqrt[3]{64}$ (h) $\sqrt[3]{1000}$

2 Use your calculator to work these out.

(a) $\sqrt{8}$ (b) $\sqrt{15}$ (c) $\sqrt{84}$ (d) $\sqrt{3.5}$

(e) $\sqrt[3]{21}$ (f) $\sqrt[3]{285}$ (g) $\sqrt[3]{9.5}$ (h) $\sqrt[3]{15.7}$

Give your answers to 3 s.f.

3 Use your calculator to work these out.

(a) $\sqrt{2}$ (b) $\sqrt[3]{39}$ (c) $\sqrt{14}$ (d) $\sqrt{110}$

(e) $\sqrt[3]{500}$ (f) $\sqrt[3]{4008}$ (g) $\sqrt{8.3}$ (h) $\sqrt[3]{1.3}$

13.3 Calculating powers

You can calculate any number raised to any power.

For example, $3^4 = 3 \times 3 \times 3 \times 3 = 81$

$2^7 = 2 \times 2 \times 2 \times 2 \times 2 \times 2 \times 2 = 128$

In Chapter 7 you used index notation to simplify numerical or algebraic expressions. For example, $5 \times 5 \times 5 \times 5 = 5^4$ or $x \times x \times x = x^3$.

$2 \times 2 \times \ldots$ 7 times

Calculations like this can lead to very large answers. When they involve decimals a calculator is essential.

The power key on a calculator will work out powers of any number.

It usually looks like x^y or y^x or $x^{\square}$.

On some calculators it is a 'second function' key.

To work out $(2.3)^6$ you press

       or   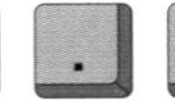  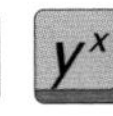  

which gives an answer of 148.035 889 = 148 (3 s.f.)

Check how to calculate powers on your calculator.

EXAMPLE 8

Use your calculator to work these out.

(a) 4^7 (b) 2^{12} (c) $(5.6)^4$ (d) $(0.8)^5$

(a) $4^7 = 16\,384$

(b) $2^{12} = 4096$

(c) $(5.6)^4 = 983.4496$

(d) $(0.8)^5 = 0.327\,68$

EXAMPLE 9

Find the value of x in each of these.

(a) $2^x = 64$ (b) $12^x = 20\,736$ (c) $3^x = 2187$

(a) $2^4 = 16$ (too small)

$2^5 = 32$ (too small)

$2^6 = 64$

$x = 6$

(b) If $12^x = 20\,736$ then $x = 4$

(c) If $3^x = 2187$ then $x = 7$

Use the power button and try a value of x.

EXERCISE 13D

1 Use your calculator to work these out.

(a) 2^5 (b) 5^7 (c) 15^4 (d) 9^6

(e) $(2.5)^4$ (f) $(0.2)^5$ (g) $(0.4)^4$ (h) $(0.1)^6$

2 Use your calculator to find the value of x.

(a) $2^x = 256$ (b) $7^x = 16\,807$ (c) $16^x = 4096$

(d) $9^x = 4\,782\,969$ (e) $2^x = 32\,768$

13.4 The laws of indices

You use the laws of **indices** (powers) to multiply or divide powers of the same number or variable.

A variable is a letter which stands for an unknown quantity.

The laws of indices

To multiply powers of the same number or variable you *add* the indices $x^m \times x^n = x^{m+n}$

To divide powers of the same number or variable you *subtract* the indices $x^m \div x^n = x^{m-n}$

You used the Laws of Indices for variables in Chapter 7.

These laws only work for powers of the *same* number (or variable).

EXAMPLE 10

Write these as a single power.

(a) 4×4^6 (b) $7^8 \times 7^5$ (c) $12^9 \div 12^3$ (d) $8^{14} \div 8^4$

(e) $\frac{5^6 \times 5^9}{5^7}$ (f) $x^6 \times x^5$ (g) $y^2 \div y^5$

Continued ▼

(a) $4 \times 4^6 = 4^1 \times 4^6 = 4^{1+6} = 4^7$

(b) $7^8 \times 7^5 = 7^{8+5} = 7^{13}$

(c) $12^9 \div 12^3 = 12^{9-3} = 12^6$

(d) $8^{14} \div 8^4 = 8^{14-4} = 8^{10}$

(e) $\dfrac{5^6 \times 5^9}{5^7} = \dfrac{5^{15}}{5^7} = 5^{15-7} = 5^8$

(f) $x^6 \times x^5 = x^{6+5} = x^{11}$

(g) $y^2 \div y^5 = y^{2-5} = y^{-3}$

$4 = 4^1$

From part (g) in the example, $y^2 \div y^5 = y^{-3}$.
Writing it out in full

$$y^2 \div y^5 = \frac{y \times y}{y \times y \times y \times y \times y}$$

$$= \frac{1}{y^3}$$

So $y^{-3} = \dfrac{1}{y^3}$

This rule works for all negative powers

$$x^{-n} = \frac{1}{x^n}$$

Any number (or variable) raised to the power of 1 is equal to the number (or variable) itself.

For example, $5^1 = 5$, $12^1 = 12$, $x^1 = x$, $y^1 = y$.

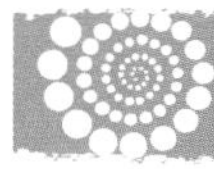

EXERCISE 13E

1 Write each of these as a single power.

(a) $3^6 \times 3^2$

(b) $5^4 \times 5^6$

(c) $4^2 \times 4^5 \times 4^2$

(d) $9^4 \times 9^8 \times 9^2$

(e) $6^4 \times 6^2 \times 6^6$

(f) $7^3 \times 7 \times 7^3 \times 5^4 \times 5^2$

2 Use the laws of indices to write each of these as a single power.

(a) $7^4 \div 7^2$

(b) $9^5 \div 9^2$

(c) $11^3 \div 11^2$

(d) $\dfrac{8^6}{8^1}$

(e) $\dfrac{5^9}{5^4}$

(f) $10^6 \div 10^3$

3 Use the laws of indices to write each of these as a single power.

(a) $8^2 \times 8^4$ (b) $10^6 \div 10^2$ (c) $y^2 \times y^3 \times y^2$

(d) $10^5 \times 10^2 \times 10^3$ (e) $3^5 \div 3^4$ (f) $m^3 \div m^2$

(g) $t^2 \times t^3 \times t^2$ (h) $3^4 \times 3^3 \div 3^6$ (i) $9^4 \div 9 \times 9^2$

4 Match each question from box A with an answer from box B.

A	B
$7^3 \times 7^5$	7^6
$3^4 \times 3^3$	7^3
$7^{10} \div 7^7$	3^8
$7^2 \times 7^3 \times 7^2$	3^6
$3^5 \div 3^3$	7^5
$7^{10} \div 7^5 \times 7$	7^8
$7^4 \times 7^2 \div 7^1$	7^9
$3^3 \times 3^4 \times 3$	3^2
$7^2 \times 7^3 \times 7^4$	3^7
$3^5 \times 3^3 \div 3^2$	7^7

5 Simplify the following by writing them as a single power.

(a) $\dfrac{4^3 \times 4^5}{4^6}$ (b) $\dfrac{7^6}{7^2 \times 7^3}$

(c) $\dfrac{2^5 \times 2^2 \times 2^4}{2^3 \times 2^2}$ (d) $\dfrac{8^7 \times 8^{10}}{8^{15}}$

(e) $\dfrac{6^5 \times 6^3 \times 6}{6^2 \times 6^4}$ (f) $\dfrac{2^5 \times 2^3}{2^4}$

(g) $\dfrac{3^3 \times 3^6 \times 3^2}{3^5}$ (h) $\dfrac{5^3}{5^4 \times 5^2}$

Calculate the value of your answers.

13.5 Using prime factors

Writing a number as the product of its prime factors

A factor of a number is a whole number that divides into it exactly.

See Section 13.1.

If a factor is also a prime number, it is called a **prime factor**.

For example, the factors of 10 are 1, 2, 5 and 10.
2 and 5 are also prime numbers.
2 and 5 are prime factors of 10.

You can write any number as the product of its prime factors.

For example $10 = 2 \times 5$.

For larger numbers you need to work systematically to find the prime factors.

Product means multiply.

Write 84 as the product of its prime factors.

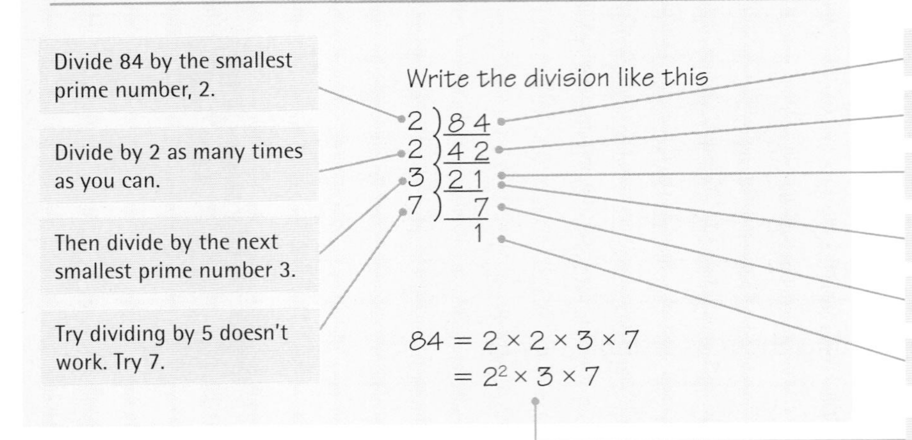

$84 = 2 \times 2 \times 3 \times 7$
$= 2^2 \times 3 \times 7$

The product of the numbers you divided by in **index form**.

This is the general method for writing a number as a product of its prime factors.

1 Start with the smallest prime.
   Divide as many times as you can.
2 Then try the next smallest prime.
   Divide as many times as you can.
3 Continue until you get to 1.
4 The answer is the product of the primes you divided by.

Prime numbers 2, 3, 5, 7, 11, 13, 17, 19, 23, ...

If you cannot divide by a particular prime, try the next one.

You can also write a number as the product of its prime factors by using 'factor trees'. This is how it works.

1 Divide your number by *any* factor. Write the factor pair on 'branches' coming from the original number.
2 Keep dividing the factors until you only have prime numbers at the ends of the branches.
3 The answer is the product of the primes on the branches.

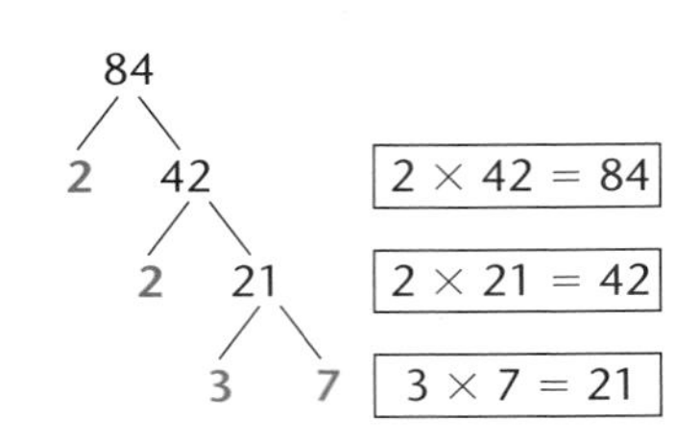

EXAMPLE 12

Write 1980 as the product of its prime factors, giving your answer in index form.

Method A

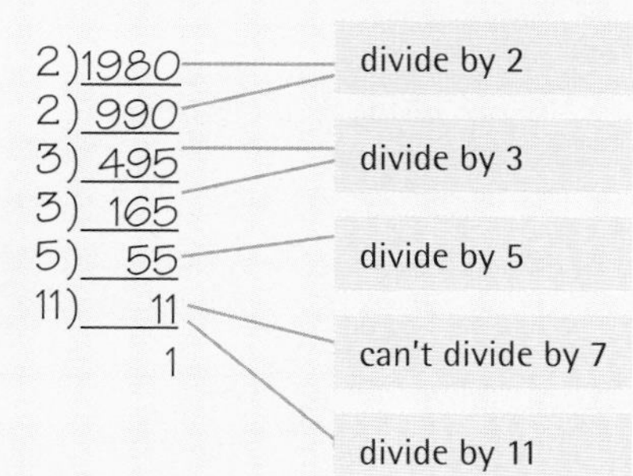

Method B

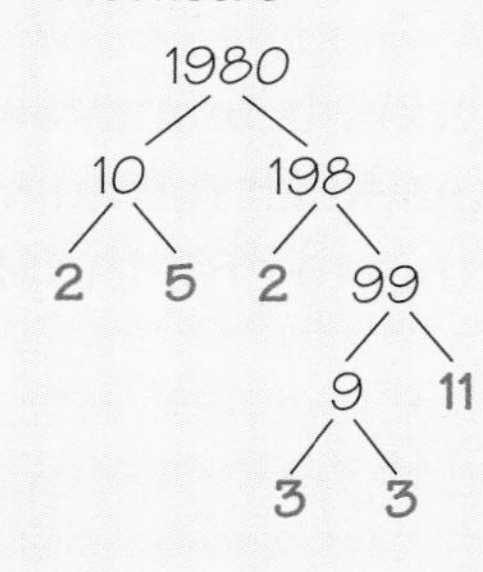

$1980 = 2 \times 2 \times 3 \times 3 \times 5 \times 11$
$= 2^2 \times 3^2 \times 5 \times 11$

$1980 = 2 \times 5 \times 2 \times 3 \times 3 \times 11$
$= 2 \times 2 \times 3 \times 3 \times 5 \times 11$
$= 2^2 \times 3^2 \times 5 \times 11$

EXERCISE 13F

1 Write these numbers as the product of their prime factors.
Give your answers in index form.

(a) 20	(b) 12	(c) 45	(d) 24
(e) 100	(f) 72	(g) 210	(h) 108

Use the method you prefer.

2 Which of these numbers are *prime* factors of 64?

2 3 4 5 7 8 11 12

3 (a) Write 48 as the product of its prime factors.
(b) Write 66 as the product of its prime factors.
(c) Which prime factors are common to both 48 and 66?

Which prime factors are in both lists?

4 (a) Write 60, 72 and 120 as the product of their prime factors.
(b) What are their common prime factors?

Highest common factor (HCF)

The **highest common factor** (HCF) of two or more numbers is the largest number that divides into all of them exactly.

The HCF of two numbers is the highest factor they have in common.

To find the HCF

1 Write each number as the product of its prime factors.
2 Pick out the prime factors common to all the numbers.
3 Multiply the common prime factors to give the HCF.

Use one of the methods from the previous section.

EXAMPLE 13

Find the highest common factor of 54, 72 and 90.

You could use the factor tree method instead.

2)54	2)72	2)90
3)27	2)36	3)45
3)9	2)18	3)15
3)3	3)9	5)5
1	3)3	1
	1	

Write each number as a product of its prime factors.

$54 \div 18 = 3$,
$72 \div 18 = 4$,
$90 \div 18 = 5$.

You can tell this from the factors left once you have taken out the common factors.

$54 = 2 \times 3 \times 3 \times 3$ $\quad$ $72 = 2 \times 2 \times 2 \times 3 \times 3$
$90 = 2 \times 3 \times 3 \times 5$

The factors common to all three are 2, 3, 3.

Pick out the factors common to all three.

So the HCF of 54, 72 and 90 is $2 \times 3 \times 3 = 18$.

Multiply them to get the HCF.

EXERCISE 13G

1 (a) Write 21 and 28 as the product of their prime factors.
(b) Find the highest common factor (HCF) of 21 and 28.

2 Find the HCF of these numbers.
(a) 15 and 25 (b) 32 and 56 (c) 27 and 45
(d) 16 and 20 (e) 25 and 45 (f) 24 and 36.

3 Find the HCF of these numbers.
(a) 4, 6 and 12 (b) 10, 30 and 35
(c) 12, 16 and 18 (d) 27, 36 and 72
(e) 40, 48 and 64 (f) 75, 100 and 150.

4 What is the highest number that will divide exactly into 60 *and* 140?

5 Find two numbers less than 60 that have a common factor of

(a) 5 (b) 16 (c) 28.

6 The quickest way to simplify a fraction is to divide both numbers by their HCF.

For example, $\frac{48}{60} \overset{\div 12}{=} \frac{4}{5}$. 12 is the HCF of 48 and 60.

By finding the HCF of the two numbers, simplify

(a) $\frac{20}{70}$ (b) $\frac{57}{76}$.

Lowest common multiple (LCM)

The **lowest common multiple** (LCM) of two or more numbers is the smallest number that is a multiple of all of them.

To find the LCM

1. Write each number as the product of its prime factors.
2. Pick out the highest power of each of the prime factors in the lists.
3. Multiply them to give the LCM.

The quickest way to add fractions is to find the LCM of the denominators and use this as a common denominator.
For example, $\frac{3}{12} + \frac{1}{16}$
LCM of 12 and 16 is 48

$\frac{3}{12} + \frac{1}{16} = \frac{3 \times 4}{48} + \frac{1 \times 3}{48}$

$= \frac{12}{48} + \frac{3}{48} = \frac{15}{48} = \frac{5}{16}$

EXAMPLE 14

Find the lowest common multiple of 54, 72 and 90.

$54 = 2 \times 3 \times 3 \times 3 \quad = 2 \times 3^3$

$72 = 2 \times 2 \times 2 \times 3 \times 3 \quad = 2^3 \times 3^2$

$90 = 2 \times 3 \times 3 \times 5 \quad = 2 \times 3^2 \times 5$

The highest power of each of the prime factors is shown in **red**.

The LCM of 54, 72 and 90 is $2^3 \times 3^3 \times 5 = 1080$.

Using the prime factors of 54, 72 and 90 from Example 13.

1080 = 20 × 54
1080 = 15 × 72
1080 = 12 × 90

1080 is the smallest number in all of the 54, 72 and 90 'times tables'.

EXERCISE 13H

1 Find the LCM (lowest common multiple) of these numbers.

(a) 10 and 8 (b) 8 and 9 (c) 18 and 12

(d) 12 and 16 (e) 20 and 90 (f) 45 and 60

(g) 45 and 54 (h) 12 and 66 (i) 33 and 88

2 Find the LCM of these numbers.

(a) 12, 16 and 24 (b) 15, 30 and 40

(c) 15, 25 and 50 (d) 72, 48 and 36

(e) 24, 40 and 120 (f) 30, 36 and 40

(g) 18, 30 and 75 (h) 45, 90 and 105

(i) 42, 56 and 90

3 Toby the dog barks every 4 seconds.
Mini the cat mews every 6 seconds.
Jake starts the timer when they bark and mew at the same time. How many seconds does the timer show when they next bark and mew at the same time?

4 Dara is waiting for the Paddington bus and Rajesh is waiting for the Euston bus.
They decide to wait until both buses come together before they get on.
The Paddington bus runs every 10 minutes.
The Euston bus runs every 8 minutes.
The buses start running at the same time in the morning.
What is the longest time they will have to wait?

5 In a group of handbell ringers
Anharay rings a bell every 6 seconds, Beth rings a bell every 9 seconds, Claudine rings a bell every 10 seconds. They start by all ringing their bells at the same time. How long is it before they next ring their bells at the same time?

EXAMINATION QUESTIONS

1 (a) The factors of 18 are 1, 2, 3, 6, 9, 18.
List all the factors of 48. [2]

(b) When written as a product of its **prime** factors, $18 = 2 \times 3^2$.
Write, as a product of its **prime** factors
(i) 48, [1]
(ii) 504. [1]

(c) (i) Find the highest common factor of 48 and 504. [2]
(ii) Find the lowest common multiple of 48 and 504. [2]
(iii) Find the lowest common multiple of 4800 and 50400. [1]

(CIE Paper 3, Jun 2000)

---

2 91 162 239 357 469

Which of the numbers above are
(a) multiples of 3 [1]
(b) multiples of 7 [1]
(c) multiples of 21? [1]

(CIE Paper 1, Nov 2000)

---

3 Work out $\dfrac{37^3 + 13^3}{37 + 13}$. [2]

(CIE Paper 1, Nov 2002)

---

4 Work out $\sqrt{(7.13^3 + 2.9^3)}$, giving
(a) your full calculator display, [1]
(b) your answer correct to 2 decimal places. [1]

(CIE Paper 1, Jun 2003)

---

5 20 21 22 23 24 25 26 27 28 29 30

From the set of numbers above, write down
(a) a multiple of 8, [1]
(b) a square, [1]
(c) a cube, [1]
(d) two prime numbers, [2]
(e) a factor of 156, [1]
(f) the square root of 784, [1]
(g) two numbers whose product is 567. [1]

(CIE Paper 3, Nov 2003)

---

6 **(a)** List all the factors of 30. [2]
(b) Write down the prime factors of 30.
(1 is not prime) [1]

(CIE Paper 1, Nov 2004)

---

7 2 3 5 9 12 15

From the set of numbers above, write down
(a) a multiple of 6, [1]
(b) a prime factor of 27. [1]

(CIE Paper 1, Jun 2005)

---

8 **(a)** Write down a number other than 1, which is a **factor** of both 14 and 35. [1]
(b) Write down a number which is a **multiple** of both 14 and 35. [1]

(CIE Paper 1, Jun 2006)

Chapter 14

Percentages and money

This chapter will show you how to

- ✔ use percentages in calculations
- ✔ calculate percentage increase and decrease and find percentage change
- ✔ write one quantity as a percentage of another
- ✔ investigate money problems
- ✔ work out simple and compound interest and use repeated proportional change

14.1 Percentages

Using percentages in calculations

You will need to know

- how to convert between fractions, decimals and percentages

Look back at Chapter 3 for help with percentages, decimals and fractions.

It helps to remember these percentages and fractions

$100\% = 1$ $50\% = \frac{1}{2}$ $25\% = \frac{1}{4}$ $75\% = \frac{3}{4}$

$10\% = \frac{1}{10}$ $33\frac{1}{3}\% = \frac{1}{3}$ $66\frac{2}{3}\% = \frac{2}{3}$

Percentage of a quantity

EXAMPLE 1

Work out 5% of \$60.

'of' means multiply

Write 5% as a fraction.

$5\% = \frac{5}{100}$

5% of \$60 $= \frac{5}{100} \times 60$

$= \frac{300}{100}$

$= \$3$

EXAMPLE 2

In a village $37\frac{1}{2}\%$ of the population are under 17 years of age. The village has a population of 1200. How many of them are under 17?

Number of people under 17 $= 37\frac{1}{2}\%$ of 1200

$= \frac{37.5}{100} \times 1200$

$= 0.375 \times 1200$

$= 450$

You do not really need this step, you can use your calculator on the line above.

To work out the percentage of a quantity, write the percentage as a fraction or decimal and multiply by the quantity.

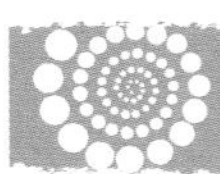

EXERCISE 14A

1 Work these out.
   (a) 10% of \$350 (b) 5% of 200 g
   (c) 25% of 64 litres (d) 1% of \$16.

2 Work these out.
   (a) 17% of 150 km (b) 45% of 630 kg
   (c) 88% of \$1280 (d) 11% of 32 mℓ.

3 Work these out.
   (a) $2\frac{1}{2}\%$ of \$160 (b) 5.5% of 125 kg
   (c) 23.5% of 5800 g (d) $7\frac{1}{4}\%$ of \$3560.

4 14% of the employees of a company are left-handed. There are 800 employees in the company. How many are left-handed?

5 In a town, 4% of the houses need repairs. There are 4625 houses in the town. How many need repairs?

6 12% of children at a football match received a free ticket for the following game. If there were 1875 children, how many received free tickets?

7 Juan got 80% in his Science test. The test was out of 75. How many marks did he score?

8 Syeda earns $43 500 a year. She pays 24% of this in income tax.
How much tax does she pay?

9 A survey in a village of 425 households found that 28% of them had satellite TV. How many households did not have satellite TV?

10 A journey to Munich is 350 kilometres. 68% of this is on motorways. How many kilometres is on other roads?

Percentage increase and decrease

You sometimes need to increase (or decrease) a quantity by a given percentage. There are two ways to do this.

Method A

Percentage increase

1 Work out the actual increase.
2 Add it to the original amount.

Percentage decrease

1 Work out the actual decrease.
2 Subtract it from the original amount.

Method B

Percentage increase

1 Add the % increase to 100%.
2 Convert this % to a decimal.
3 Multiply the original amount by this decimal.

Percentage decrease

1 Subtract the % decrease from 100%.
2 Convert this % to a decimal.
3 Multiply the original amount by this decimal.

Examples 3 and 4 use both methods.
You can choose the method you prefer.

EXAMPLE 3

Liz earns $150 per week at the moment.
Next week she is due to receive a pay rise of 6%.
What will her new weekly pay be?

Percentage increase.

Method A

Pay rise = 6% of $150

$= \frac{6}{100} \times 150$

$= \$9$

Liz's new weekly pay

$= \$150 + \$9$

$= \$159$

Method B

Increase = 6%

New % = 100% + 6%

= 106%

= 1.06

Divide by 100 to convert a % to a decimal.

Liz's new weekly pay

$= 1.06 \times \$150$

$= \$159$

EXAMPLE 4

A CD costs $14.50. With her student card, Alix gets 10% discount. What does she pay for the CD?

Percentage decrease.

A discount is a decrease in price.

Method A

Discount = 10% of $14.50

$= \frac{10}{100} \times \$14.50$

$= \$1.45$

Alix pays $14.50 − $1.45

$= \$13.05$

Method B

Discount = 10%

New price = 100% − 10%

= 90%

= 0.9

Divide by 100 to convert a % to a decimal.

Alix pays $0.9 \times \$14.50$

$= \$13.05$

EXERCISE 14B

1 Increase these by 14%.
   (a) $420 (b) 600 g (c) 84 litres

2 Decrease these by 4%.
   (a) 500 kilometres (b) $64 (c) 4250 kg

3 A pair of jeans cost $45. Christof gets a staff discount of 8%. What does Christof pay for the jeans?

4 Work out these percentage increases and decreases.

(a) Increase \$30 by 5%.

(b) Increase 58 by 12%.

(c) Decrease 3400 $m\ell$ by 20%.

(d) Decrease 480 by 32%.

(e) Increase \$135 by $7\frac{1}{2}$%.

(f) Decrease 890 $m\ell$ by 8.4%.

5 Miners were given a 4% pay rise this year. Before the pay rise their average wage was \$420 per week. What is their average wage now?

6 A café wants to increase some of its prices by 5%. Work out the new cost of these items. Give your answers to the nearest cent.

Pasta	\$2.60	Coffee	90c
Burger	\$1.80	Tea	80c
Salad	95c		

7 There has been a 6% decrease in the number of reported thefts in Palton this year. There were 250 reported thefts last year. How many have there been this year?

8 A rail company announces that all rail fares will increase by $7\frac{1}{2}$%.
If I pay \$28 for my ticket now, how much will it cost after the increase?

9 A new car costs \$8450. After 2 years the value of the car will have decreased by 43%. How much will the car be worth then?

10 During one month the number of members of a gym increased by 65%. If there were 60 members at the beginning, how many were there at the end?

11 In a shop, sales of DVDs decreased by 11% this week.
If they sold 430 last week, how many did they sell this week?

Writing one quantity as a percentage of another

Test marks are usually written as a fraction.

For example, in a Maths test, a score of 17 out of 25 is written as $\frac{17}{25}$.

You can write this score as a percentage.

$$\frac{17}{25} = \frac{17}{25} \times 100\% = 68\%$$

Here you are writing 17 as a percentage of 25.

To write one quantity as a percentage of another
1 write the two quantities as a fraction
2 multiply by 100 to convert to a percentage.

EXERCISE 14C

1 Write these scores as percentages.

(a) 6 out of 10 (b) 17 out of 20
(c) 24 out of 40 (d) 30 out of 35.

2 Work these out.

(a) 3 as a percentage of 12
(b) 8 as a percentage of 40
(c) 22 as a percentage of 60
(d) 65 as a percentage of 70.

3 In a Mathematics test Jasmine scored 16 out of 20. In her English test she scored 73%. In which subject did she do better?

Change both to percentages to compare them..

4 A sports centre has 1200 members, in the following groups.

Type	Men	Women	Girls	Boys
Number	602	392	118	88

What percentage of the members are

(a) men (b) girls
(c) children (d) female?

Give your answers to 1 s.f.

Finding percentage change

'Change' means increase *or* decrease.

You use this formula.

$$\text{percentage change} = \frac{\text{change}}{\text{original amount}} \times 100\%$$

The formula calculates the change as a percentage of the original amount.

EXAMPLE 5

Examples 5 and 6 are the type of question you can expect on your IGCSE papers.

The original cost of a DVD player is \$225.
In a sale it is reduced to \$180.
What is the percentage decrease in the price?

Always work out the change first. Don't forget – the *original* amount goes on the bottom of the fraction.

Decrease in price = \$225 − \$180
= \$45

$$\text{Percentage decrease} = \frac{\text{decrease}}{\text{original amount}} \times 100\%$$

Change is decrease here.

$$= \frac{45}{225} \times 100\%$$

$$= 20\%$$

EXAMPLE 6

The average attendance at Manchester United home games last season was 55 000.
This year the average attendance is 58 575.
What percentage increase is this?

Increase in attendance = 58 575 − 55 000
= 3575

$$\text{Percentage increase} = \frac{\text{actual increase}}{\text{original amount}} \times 100\%$$

Change is increase here.

$$= \frac{3575}{55\,000} \times 100\%$$

$$= 6.5\%$$

EXERCISE 14D

1 In a sale the following price changes were made. What was the percentage reduction for each item?

(a) a coat was \$120, now \$80

(b) a shirt was \$24, now \$18

(c) shoes were \$40, now \$20

(d) trousers were \$30, now \$18.

2 A car was for sale at \$3600. The price was raised to \$3780.
What was the percentage increase?

3 Ali had a pay rise from \$240 per week to \$255 per week.
What was his percentage increase in pay?

4 The number of tigers in an area of India has decreased from 108 to 96 in two years. What is the percentage decrease in tigers in this area?

5 The cost of a holiday will rise from \$480 this year to \$528 next year.
What is the percentage increase?

6 The rent that Peter paid was increased from \$280 per month to \$315 per month.
What was the percentage increase?

7 A hospital reports that the number of sport injuries in people over 60 years old is increasing. In 1995 there were 682 sport injuries but in 2004 there were 1088 in this age group.
What was the percentage increase?

8 There were 350 members of the Matis tribe living in the Amazon Basin.
The number fell to 198 one year later.
What percentage decrease is this?

9 In a garage 15 cars were priced under \$10 000. After a price rise the number under \$10 000 had dropped to 9.
What percentage reduction is this?

10 The population of a port in Chile increased from 183 000 people to 300 000 in just 20 years.
What percentage increase is this?

11 The number of sea lions seen at Bogoslof in Alaska had fallen from 5000 in 1973 to 120 in 2002.
What is the percentage reduction in sea lions seen in this area?

12 In 1970 survival suits gave 3 hours of protection in the Bering Sea. Modern survival suits will protect you for 24 hours. What is the percentage increase in time?

14.2 Working with money

Percentage profit and loss

A shopkeeper buys items from a wholesaler at **cost price**.
She sells the items at the **selling price**.

When you make money on the sale of an item you make a **profit**.
When you lose money on the sale of an item you make a **loss**.

If a garage buys a car for $1000 and sells it for $750, the garage makes a loss.

$$\text{percentage profit or loss} = \frac{\text{actual profit (or loss)}}{\text{cost price}} \times 100\%$$

The cost price is the *original* price of the item. It goes on the bottom of the fraction.

where the actual profit (or loss) is the difference between the cost price and the selling price.

EXAMPLE 7

A used car salesman buys a car for $2500 and then sells it for $3200.
What is his percentage profit?

Cost price = $2500.
Selling price = $3200.

$$\text{Profit} = \$3200 - \$2500$$
$$= \$700$$

$$\text{Percentage profit} = \frac{\text{profit}}{\text{cost price}} \times 100\%$$
$$= \frac{700}{2500} \times 100\%$$
$$= 28\%$$

EXAMPLE 8

Peter bought a new motorbike for $5450 but traded it in for another one a year later.
He got a trade-in value of $4578 for his bike.
What percentage loss is this?

Loss = $5450 − $4578
= $872

$$\text{Percentage loss} = \frac{\text{loss}}{\text{cost price}} \times 100\%$$

$$= \frac{872}{5450} \times 100\%$$

$$= 16\%$$

EXERCISE 14E

1 A collector bought a stamp for $8 and sold it for $10.
   (a) How much profit did he make?
   (b) What was the percentage profit?

2 Harvey bought a computer game for $25 in December.
   He sold it for $15 in the summer.
   (a) How much money did he lose?
   (b) What was the percentage loss?

3 For each item find the percentage profit or loss.

	Item *a*	Item *b*	Item *c*	Item *d*	Item *e*	Item *f*
Cost price	$50	$34	$6	$73.50	$125	$3500
Selling price	$65	$24	$7.50	$69.50	$84	$3635

4 A bicycle shop paid $84 for a bike.
   They sold it in a sale for $79.
   What was the percentage loss?

5 Ahmed sold his MP3 player to Jane for $30. He had bought it for $35.
   What was his percentage loss?

6 Sarah restores books. She bought one for \$8 and sold it for \$15. What was her percentage profit?

7 Pet food normally costs \$15.50 per sack.
If you buy two sacks the shop charges \$13 per sack.
(a) What would two sacks cost without the discount?
(b) What percentage discount is this?

Discount is an amount taken off the price. You work out % discount in the same way as % loss.

8 A shop buys a mobile phone for \$22 and sells it for \$35. What is the percentage profit?

9 Floyd has a ticket for a concert. It cost him \$32.50 and he sells it for \$45. What will be his percentage profit?

10 A DVD was sold for \$14.99. Its cost price was \$9. What was the percentage profit?

GST

GST stands for General Sales Tax and is known by different names in different countries.

GST, for example at 15%, is added to the cost of many items and services.

Usually the price of an article includes GST, but sometimes it doesn't. You need to look carefully to see how much you will have to pay.

This means that you pay an extra percentage on top of the cost of the item you buy.

You can use the percentage increase methods to calculate GST.

EXAMPLE 9

A digital camera is advertised for sale at \$240 (excluding GST). How much will you have to pay?

'Excluding' is the opposite of 'including'. You need to add GST on to the advertised price.

Method A

GST = 15% of \$240

$= \frac{15}{100} \times 240$

= \$36

Cost of digital camera
= \$240 + \$36
= \$276

Method B

Increase = 15%
New % = 100% + 15%
= 115%
= 1.15

$115 \div 100 = 1.15$
This is your multiplier to work out the % increase.

Cost of digital camera
= 1.15 × \$240
= \$276

The GST makes quite a difference to the price!

EXERCISE 14F

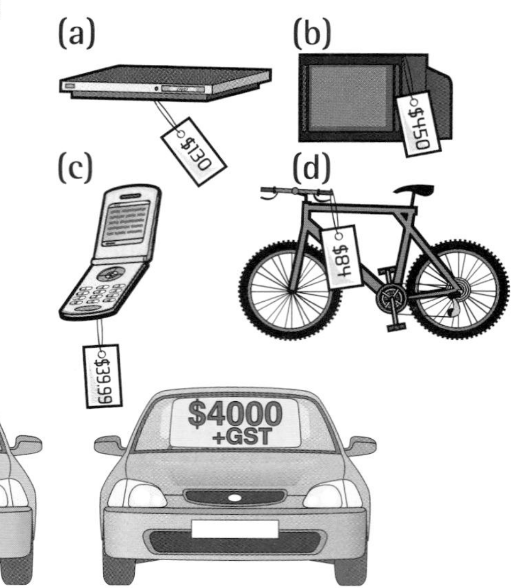

Take the GST rate as 17.5% in this exercise.

1 Work out the GST to be added to these prices.
Give your answers to the nearest cent.

2 Work out the total cost of these.
(a) a bag $8 + GST (b) a chair $74 + GST
(c) a CD $12 + GST (d) a car $2480 + GST.

3 Two cars in a garage had these prices marked.
Which one is the most expensive?

4 A computer costs $399 + GST.
What is the total cost?

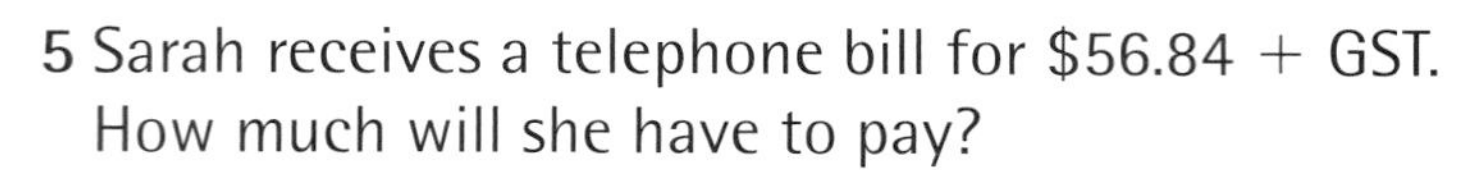

5 Sarah receives a telephone bill for $56.84 + GST.
How much will she have to pay?

6 A meal for four cost $73.58 plus GST.
(a) How much GST will be added to the bill?
(b) What is the total cost?
(c) The four people decide to share the bill equally.
How much will each pay?

7 A meal for 3 people cost $55.50 + GST. They decide to pay equal shares. How much will each person pay?

8 A car needed some brake fluid, some antifreeze and an adjustment to the fan-belt.
Labour costs were $60.50 and the cost of the parts used was $30.23.
GST was added to the total of these amounts.
How much was the total bill? Give your answer to the nearest cent.

Credit

Many items that you want to buy are expensive. You may not have enough money to pay the full cost.

When you buy **on credit**, you pay a **deposit** followed by a number of **regular payments**, usually each month.

The cost of buying an item on credit (the **credit price**) is usually more than paying cash for the full cost (the **cash price**).

EXAMPLE 10

Mr Hanif wants to buy a new washing machine. There are two ways he can pay.

Option A Pay the cash price of $470.

Option B Pay a 10% deposit, and 24 monthly payments of $19.30.

Mr Hanif chooses Option B.

(a) How much deposit does he pay?
(b) What is the total of his monthly payments?
(c) What is the total credit price?
(d) What is the extra cost of buying on credit rather than paying the cash price?

(a) Deposit = 10% of \$470 = $\frac{10}{100} \times 470$ = \$47

The deposit is 10% of the cash price.

(b) Total of monthly payments = 24 × \$19.30
= \$463.20

(c) Total credit price = deposit + monthly payments
= \$47 + \$463.20
= \$510.20

(d) Extra cost of buying on credit = \$510.20 − \$470
= \$40.20

Find the difference between the credit price and the cash price.

EXERCISE 14G

1 Salim is buying a new motorbike. He decides to buy it on credit. He pays a 10% deposit and 12 monthly payments of $275.

(a) How much is the deposit?
(b) How much does he pay in monthly instalments?
(c) How much does Salim pay in total?
(d) What is the difference between the cash price and the credit price of the motorbike?

2 You can buy a car in two ways.

Option A $6800 cash price.

Option B $2500 deposit and 36 monthly payments of $210.

(a) What is the total cost of paying on credit (Option B)?

(b) How much more does it cost to buy on credit?

3 A cell phone costs $199 cash. The phone company also allows you to pay for the phone with a 20% deposit and $17 added to your monthly bill for a year.
Which method is cheaper?
How much do you save using the cheaper method?

4 A sofa costs $384. The credit terms are 10% deposit and 15 monthly payments of $28.
What is the difference between the cash price and the credit price?

5 To pay for a surf board (cash price $124.50), James takes out a credit agreement. He pays a 15% deposit and 15 weekly payments of $9.
How much more is the credit price than the cash price?

6 A house costs $89 000. A company offers a mortgage of 5% deposit and monthly payments of $400 for 25 years.

(a) How much is the deposit?

(b) What is the total cost of the house?

Wages

If you work, you are sometimes paid by the hour. You get a **basic rate** of pay, which is a fixed amount per hour.

If you work more hours than your normal **working week** you get an **overtime** rate. This is more than your basic rate.

Basic pay = number of hours in normal working week × basic rate of pay

Overtime pay = number of hours of overtime × overtime rate of pay

Common rates of overtime pay are
'Time and a quarter'
= 1.25 × basic rate of pay
'Time and a half'
= 1.5 × basic rate of pay.
'Double time'
= 2 × basic rate of pay.

EXAMPLE 11

Vijay is paid \$6 per hour for a 38-hour working week.
His overtime is paid at 'time and a half'.
How much does he earn in a week when he works 50 hours?

'Time and a half' means $1.5 \times$ the basic rate of pay.

Vijay works for 50 hours = 38 hours + **12** hours.

So he works his normal working week + **12** hours of overtime.

Basic pay $= 38 \times \$6 = \$228$

Overtime rate of pay $= 1.5 \times \$6 = \$9$

So overtime pay $= 12 \times \$9 = \$108$

Vijay's total wage for the week $= \$228 + \$108 = \$336$

EXAMPLE 12

Helen is paid a basic rate of \$7 per hour for a 40-hour working week.
Overtime is paid at the rate of 'time and a half'.
In a particular week she earns \$353.50.
How many hours of overtime did she work?

Helen's basic pay $= 40 \times \$7 = \$280$

Amount earned by working overtime
$= \$353.50 - \$280$
$= \$73.50$

Work out her basic pay. Then you will know how much of the \$353.50 came from overtime.

Rate of pay for overtime $= 1.5 \times \$7$
$= \$10.50$

Work out the rate of pay for overtime.

Number of hours overtime $= \$73.50 \div \$10.50$
$= 7$

Divide the overtime pay by the rate of pay for overtime.

Helen worked 7 hours of overtime.

EXERCISE 14H

1 What is the basic pay for a 38-hour week if the hourly rate is

(a) \$6 (b) \$8.45 (c) \$9.30?

2 The hourly rate at McHill Builders is $6.60.
What is the hourly rate for overtime if they pay
(a) double time
(b) time and a half
(c) time and a quarter?

3 Jane is paid $7.20 per hour for a 40-hour week and then time and a half for any overtime. One week she works 47 hours. How much is she paid?

4 Three workers at 'Johnson Builders' work 45 hours one week. Their normal working week is 38 hours. Overtime is paid at time and a half. How much will they each get if their hourly rate is
Scott, a carpenter – $9.70
Ali, a bricklayer – $8.30
Liam, a labourer – $7.50?

5 At a call centre the pay is $6.80 per hour, Monday to Friday.
The overtime rate of pay on Saturday is 'time and a quarter'.
The overtime rate of pay on a Sunday is 'time and a half'.

Here is a timesheet for one of the workers.

Day	Mon	Tue	Wed	Thu	Fri	Sat	Sun
Hours	8	8	4	8	5	6	4

A timesheet shows the hours worked in a week.

Work out the total pay for the week.

6 Ali is paid $8 per hour for a 40-hour week and double time for any overtime. One week he was paid $368. How many hours of overtime did he work?

7 A worker in a call centre is paid $323 for a 38-hour week. She is paid time and a half for working on a Sunday. What is she paid per hour on a Sunday?

8 A postman is paid $420 for a 40-hour week. When he works overtime he is paid time and a half.
How many hours' overtime has he worked in a week when he is paid $514.50?

Best buys

Cans of cola come in two sizes, a 330 mℓ can for 39c or a 500 mℓ can for 63c.

You can work out which is the best value for money in two ways.

The 'best buy' is the best value for money.

- Work out the cost of 1 mℓ of cola for each can.
- Work out how many mℓ you get for 1c for each can.

You are looking for the *least* cost for 'best buy'.

You are looking for the *most* mℓ for 'best buy'.

EXAMPLE 13

Which is the best buy
a 330 mℓ can for 39c or a 500 mℓ can for 63c?

Method A

330 mℓ can 330 mℓ costs 39c

1 mℓ costs $\frac{39}{330}$ c = 0.118 … c

500 mℓ can 500 mℓ costs 63c

1 mℓ costs $\frac{63}{500}$ c = 0.126c

1 mℓ costs less in the 330 mℓ can so it is the best buy.

Find the cost of 1 mℓ.

You must remember which method you are using so that you know whether to pick the biggest answer or the smallest one!

Method B

330 mℓ can You get 330 mℓ for 39c

so you get $\frac{330}{39}$ mℓ for 1c

= 8.46 … mℓ for 1c

500 mℓ can You get 500 mℓ for 63c

so you get $\frac{500}{63}$ mℓ for 1c

= 7.93 … mℓ for 1c

You get more mℓ for 1c in the 330 mℓ can so it is the best buy.

Find how many mℓ you get for 1c.

EXERCISE 14I

1 Find the number of grams for 1c.

(a) 10c for 40 g (b) 18c for 324 g (c) 500 g for 15c

2 Find the cost of 1 gram.

(a) 25 g for $2 (b) 300 g for 75c

(c) $2.50 for 500g

Change the costs to cents.

3 A large box of cereal costs $2.78 for 500g.
A small box costs $1.45 for 250 g.

(a) How many grams do I get for 1c in each box?

(b) Which one is the best buy?

Change the costs to cents.

4 A cell phone company has different tariffs.

Tariff	A	B	C
No. of calls	50	100	500
Total cost	$2.60	$5.15	$25.80

(a) Find the cost of 1 call for each tariff.

(b) Which tariff is the best buy?

5 I can buy a 545 g jar of jam for $1.35, or a 350 g jar for 90c. Which is the best buy?

6 A large block of chocolate costs $1.84 for 250 g. A small block costs 80c for 60 g. Which is the best buy?

Bills and services

Many people pay household bills and services (such as electricity and gas) **quarterly**. This means they pay four times a year.

4 times a year = every 3 months.

Sometimes bills are made up of two parts.

- a fixed amount of money (a **standing charge**),
- an amount of money depending on the number of units of gas or electricity used.

You pay the standing charge even if you haven't used any gas.

Total amount = standing charge + cost of units used

The number of units of gas or electricity you have used is worked out from the gas or electricity meter reading.

EXAMPLE 14

Mr Choi buys his gas from Cosygas.
The quarterly charges are shown on the right.
In March his meter reading was 12 027 units.
In June his meter reading was 14 967 units.

(a) How many units of gas did Mr Choi use during the quarter from March to June?

(b) Calculate the cost **(i)** of the first 1200 units **(ii)** of the remaining units.

(c) How much does Mr Choi pay for his gas for this quarter?

COSYGAS
Standing charge = \$15.36
2c per unit for first 1200 units used
1.5c per unit for the remainder

You can use this method for any bill calculation where there is a standing charge.

(a) Units used = 14 967 − 12 027 = 2940

(b) (i) 2c per unit for 1200 units = 1200 × 2c = 2400c
= \$24

(ii) Number of units used at 1.5c per unit = 2940 − 1200
= 1740

1.5c per unit for 1740 units = 1740 × 1.5c = 2610c
= \$26.10

Convert these answers to \$.

(c) Gas bill = \$15.36 + \$24 + \$26.10
= \$65.46

Standing charge + cost for 1200 units + cost for remainder.

EXERCISE 14J

1 Mrs Tan buys her electricity from **Top Power**.
Their charges are

TOP POWER
10.5c per unit for the first 100 units
6.5c per unit for the remainder

Mrs Tan uses 670 units.

(a) How much do the first 100 units cost?

(b) How much does the remainder cost?

(c) What is the total cost for 670 units?

2 Mr Patel also buys electricity from **Top Power**. His meter reading in December was 17 228. In January it was 18 003.

(a) How many units of electricity did he use?

(b) What is the total cost of his electricity for this month?

3 Work out the electricity bills for these houses. There is no standing charge.

House	Previous reading	Present reading	Unit price for first 100 units	Unit price for remaining units
(a) 32 Front St.	10 325	19 436	2.7c	1.5c
(b) 34 Front St.	34 219	35 106	3.4c	2.8c
(c) 36 Front St.	61 754	68 127	2.3c	1.75c

4 Morag's quarterly charge for gas from **Ngas** is

- 3.2c per unit for the first 100 units
- 2.4c per unit for the remainder
- standing charge $12.75.

Her previous meter reading was 43 249. The new reading is 56 310.

(a) How many units of gas has she used?

(b) What is the cost for the gas used?

(c) What is her total bill for gas for this quarter?

5 Morag's next bill from **Ngas** was $183.95.

(a) What was the cost for the gas (*without* the standing charge)?

(b) What was the cost of the first 100 units used?

(c) How much did the remaining units cost?

(d) How many units did she use altogether?

Work out how many units she used at the rate of 2.4c per unit then add on 100 units.

6 The quarterly charges for the Talkalot telephone company are

Standing charge = $18.50
3c for cheap rate calls
7.5c for normal rate calls

(a) Salma made 168 cheap rate calls. How much do these cost?

(b) She made 98 normal rate calls. How much do these cost?

(c) What is the total for the calls and standing charge?

(d) 17.5% GST is added to the bill. What is the total telephone bill?

Foreign currency

When you go on holiday or if you buy goods from abroad you will need to use a **currency** other than your own.

This means that you will need to convert from one currency to another.

To do this you need to know the **rate of exchange**.

EXAMPLE 15

Mr and Mrs Lee go on holiday from Singapore to China.

They change 500 Singapore dollars (S$) into Chinese Yuan (Y).
The rate of exchange was S$1 = 4.5665Y.

(a) How many Yuan did they receive?

(b) At the end of their holiday they had 40 Yuan left.
They change this back into S$ at the same exchange rate.
How many dollars did they receive?

Answers to 2 d.p.

(a) S$1 = 4.5665Y

S$500 = 500 × 4.5665Y

= 2283.25Y

Continued ▼

(b) 4.5665Y = S$1

1Y = S$1 ÷ 4.5665

1Y = S$0.21898...

40Y = 40 × S$0.21898...

= S$8.76 (2 d.p.)

Notice that the rate of exchange is written the other way round.

This is like the unitary method (see p335).

EXERCISE 14K

1 A train ticket from Singapore to Malaysia can be paid for in either Singapore dollars (S$) or Malaysian Ringgits (MR). If the ticket cost S$30 and the exchange rate was S$1 = 2.407MR, how much was the ticket in MR?

2 A cup of coffee costs €2.50 in France. Copy and complete the table to show how much you would you have to pay in each of the currencies in the table.

Currency	Rate of exchange for 1€	Cost of a cup of coffee
Swedish Krona (K)	10.470	K
Russian Rouble(R)	43.728	R
Turkish Lira (L)	2.1505	L
Algerian Dinar (D)	95.507	D

3 Mr Ndele travels from South Africa to Botswana on business. He changes 4000 Rand (R) into Pula ready for his trip. The exchange rate is R1 = P0.01242.

(a) How many Pulas did he receive?

(b) In Botswana he sold goods to the value of P5000. How many Rands did he receive for the goods if the exchange rate remained the same?

4 (a) Mrs Dickinson changed 400 Australian dollars into New Zealand dollars when the exchange rate was \$1 = NZ\$1.2542. How many NZ dollars did she receive?

(b) She went shopping in Auckland and spent NZ\$358. She changed the remainder of her money back into Australian dollars but the exchange rate had changed to A\$1 = NZ\$1.2275. How many Australian dollars did she receive?

Savings

If you put money into a bank savings account or a building society, the bank or building society pays you **interest** (extra money). The interest is worked out as a percentage of the money you **invest** (save).

If you borrow money from a bank or a building society, you have to pay interest on the money you borrow.

You will look at compound interest in Section 14.3.

There are two ways of calculating interest. These are simple interest and compound interest.

Simple interest

The **rate of interest** is fixed.

The interest is a percentage of the **principal** (the money you invest).

You receive the same amount of interest each year.

You calculate

$$\text{Simple interest} = \left(\frac{\text{rate}}{100} \times \text{principal}\right) \times \text{time}$$

For 5 years | Interest for 1 year | 5 years

$$\left(\frac{R}{100} \times P\right) \times T = \frac{R}{100} \times P \times T = \frac{PRT}{100}$$

(put the letters in alphabetical order)

This gives you a formula for working out simple interest.

$$I = \frac{P \times R \times T}{100} = \frac{PRT}{100}$$

I = simple interest $\quad$ P = principal

R = rate of interest (% *p.a.*) $\quad$ T = time (in years)

EXAMPLE 16

Carla invests \$600 at a rate of 5.5% for 36 months.
How much simple interest will she receive?

$$I = \frac{P \times R \times T}{100} = \frac{600 \times 5.5 \times 3}{100} = \$99$$

Carla receives \$99 interest.

36 months = 3 years.
T must be in years.

You can use the formula to calculate *P* or *R* or *T* if you are given the value of *I*.

EXAMPLE 17

How much money do I need to invest over 6 years at a rate of interest of 4% to earn simple interest of \$210?

$T = 6$, $R = 4$ and $I = 210$ P is unknown

$$I = \frac{P \times R \times T}{100}$$

$$210 = \frac{P \times 4 \times 6}{100}$$

$$210 = \frac{P \times 24}{100}$$

$$\frac{210 \times 100}{24} = P$$

$$P = 875$$

You need to invest \$875.

Multiply both sides by 100 and divide by 24.
For help in solving equations like this, see Chapter 7.

EXERCISE 14L

1 Find the simple interest on
   (a) \$400 invested for 2 years at a rate of 10% p.a.
   (b) \$250 invested for 5 years at a rate of 4% p.a.
   (c) \$875 invested for 48 months at a rate of 5.5% p.a.
   (d) \$4350 invested for $2\frac{1}{2}$ years at a rate of 6.25% p.a.

per annum (p.a.) means 'per year' or 'each year'.

2 Kim's grandmother gives her $150. Kim invests it at a rate of 8% for 3 years until she starts college.
   (a) How much simple interest will she earn on her money?
   (b) How much will she have in total at the end of the 3 years?

Add the interest earned to the principal ($150).

3 Ronaldo has $2300 to invest. The Tiger Bank gives 7% interest for the first year and 5% for the following years. The Panda Bank gives 6% p.a.
He wants to invest his money for 4 years. Which bank will pay him the most interest?

4 How much would I have to invest to receive $48 interest after 5 years at a rate of 8%?

5 How long will it take to make $96 interest if I invest $800 at a rate of 4% p.a?

6 I invested $300 for 5 years and received $52.50 simple interest. What was the rate of interest?

14.3 Repeated proportional change

Generally, when you invest money the interest you earn is calculated using **compound interest**.

If you borrow money, the interest you pay is calculated using compound interest.

- The interest you receive in year 1 is added to the principal.
- In year 2 your interest is calculated as a percentage of your principal *plus* your interest from year 1.
- In year 3 your interest is calculated as a percentage of your principal *plus* your interest from years 1 and 2.
- and so on ...

The rate of interest is fixed.

For compound interest
- The interest you receive each year is *not* the same.
- The interest is calculated on the amount invested in the first place *plus* any interest already received.

EXAMPLE 18

Venetia puts $800 into a building society account.
The rate of interest is 5% p.a. compound interest.
How much does she have in her account after 2 years?

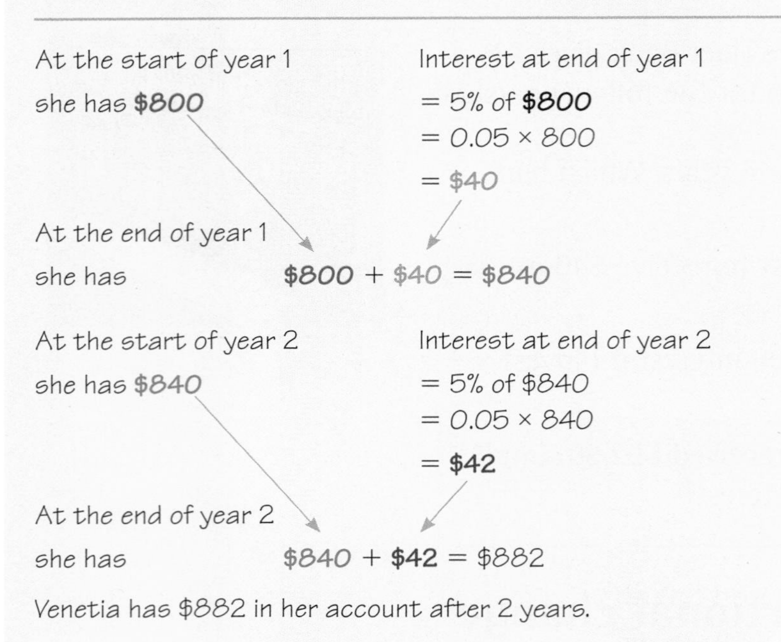

5% = 0.05 as a decimal.

Add the interest to the principal.

To calculate the amount of money you will have if you invest it at compound interest it is quickest to

1 Add the rate of interest on to 100%.
2 Convert this percentage to a decimal.
3 Multiply the original amount of money by this decimal as many times as the number of years the money is invested.

Using this method for example-
Venetia invested $800 at a rate of interest of 5% pa. for 2 years.

- 100% + 5% = 105%
- 105% = 1.05
- $800 × 1.05 × 1.05 = $882

This method gives you the final amount of money, not the interest.

You can use a multiplier where anything is increasing or decreasing. This is called repeated proportional change.

The 1.05 above is called the multiplier.

EXAMPLE 19

There are 5000 whales of a certain species but scientists think that their numbers are reducing by 8% each year.
Estimate how many of these whales there will be in 3 years' time.

The whales are reducing by 8% each year.

$100\% - 8\% = 92\%$

$92\% = 0.92$ as a decimal

Number of whales in 3 years' time

$= 5000 \times 0.92 \times 0.92 \times 0.92$

$= 3893.44$

Estimated number of whales = 3890 (3 s.f.)

The number of whales reduces by 8% of the number *at the start of each year.* Each % calculation is done on a *different* number so you use the same method as for compound interest problems.

The whales are *reducing* in number so you *subtract* the % from 100%. 0.92 is the multiplier.

3 years, so multiply by 0.92 3 times.

EXERCISE 14M

1 Find the compound interest on
   (a) $200 invested for 2 years at a rate of 10%
   (b) $750 invested for 2 years at a rate of 5%
   (c) $4500 invested for 2 years at a rate of 2%.

2 Grace invests $120 for 3 years at a rate of 4% compound interest.
How much will she have in total at the end of the 3 years?

3 How much will Zarah have at the end of 3 years if she invests $350 at a rate of 6% compound interest?

4 A new car cost $12 000. Each year the car depreciates by 8% of its value at the start of the year. What will the car be worth at the end of 3 years?

'Depreciate' means it reduces in price.

5 Tristan and Isobel inherit $200 each. Isobel invests her money for 2 years at a rate of 6% compound interest. Tristan invests his money at a rate of 5% compound interest for 3 years.
Who will have the most money at the end of their investment?

6 The population of a rare breed of owl is on the increase in parts of Europe. It is estimated that the present population of 2000 pairs is increasing at the rate of 4% each year. How many pairs will there be at the end of 3 years?

7 The seal population in Scotland is estimated to be declining at the rate of 15% each year. In 2008 there were 3000 seals. How many seals would you expect there to be in 2011?

8 Perry Biscuits give staff an annual pay increase of 5% for every year they stay with the firm.

(a) Kim earns $12 500 a year. How much will she earn in 2 years' time?

(b) Lucas earns $11 800 a year. How much will he earn in 3 years' time?

9 A baby octopus increases its body mass by 5% each day for the first week of its life. An octopus was born weighing 10 kg.

(a) How much did it weigh at the end of the first day?

(b) How much did it weigh at the end of the third day?

(c) How much did it weigh at the end of one week?

10 Amina earns $21 000 a year. This increases by 4% each year. After how many years will she earn over $23 000 a year?

EXERCISE 14N

1 Find

(a) 15% of $250 (b) $4\frac{1}{2}$% of 800 g.

2 Brooktown has a population of approximately 230 000 people. 7% of them are over 80 years of age. How many people are over 80?

3 An advertisement for "Chewy Gums" says you now get 15% more for the same price. If the original packs were 250 g, how much will the new packs contain?

4 A car manufacturer employed 1280 people. Due to a reduction in car sales they reduced their workforce by 40%. How many people work there now?

5 Carl scored 21 out of 25 in his French test and 16 out of 20 in his Geography test. Which was his better result?

6 What percentage of 300 g is 125 g?

7 In a survey in an nature reserve in Kenya the number of Oxpecker birds had fallen from 18 000 in 2004 to 12 000 in 2008. What is the percentage decrease?

8 A school population increased from 650 in 2005 to 1010 in 2009. What was the percentage increase?

9 A bike was bought for $125 and sold for $145. What was the percentage profit?

10 A car bought for $12 500 new was sold for $8750. What was the percentage loss?

11 A TV was for sale at $220 plus GST ($17\frac{1}{2}$%). What was the total cost of the TV?

12 Five friends went out for a meal which cost $52 plus GST ($17\frac{1}{2}$%).

(a) Work out the full cost of the meal.

(b) How much did each of them pay if they shared the cost equally?

13 Liam is buying a washing machine (cash price $320) on credit. He pays 15% deposit and 12 monthly payments of $25.

(a) How much deposit does he pay?

(b) What is the total cost of the machine on credit?

14 To buy a car costing $6800 on credit you have to pay 20% deposit and 36 monthly payments of $210. How much more does the car cost if you buy it on credit?

15 Mamet is paid $7.20 per hour for a 38-hour week. When he works overtime he is paid at the rate of 'time and a half'.
One week he works 45 hours.
How much will he be paid?

16 Lee works 16 hours a week and earns \$97.60. When he works overtime he is paid 'double time'. One week he earns \$146.40.
How many hours of overtime did he work?

17 A 375 g box of icing sugar costs \$1.25. A 750 g box costs \$3.
Which size is the best buy?

18 A 50 mℓ tube of toothpaste costs 78c. A larger 120 mℓ tube costs \$1.90.
Which one is the best buy?

19 Gas from *Hotburn* costs 3.5c per unit for the first 100 units and then 2.2c per unit for the rest. There is also a standing charge of \$23.50.
How much is my total bill if I use 475 units of gas?

20 An electricity meter reading was 14 358 in April and 15 842 in May. The cost of electricity is 5.3c per unit for the first 150 units and 3.4c per unit for the rest. What is the total bill?

21 What is the simple interest earned on \$375 invested for 6 years at a rate of 5.5%?

22 How many years would it take to earn \$720 simple interest if I invested \$2000 at a rate of 9%?

23 I invest \$400 for 3 years at a rate of 4% compound interest. How much will I have in total at the end of this time?

24 The amount of water in a lake in Africa is estimated at 3 million litres. Due to drought, the amount of water is reducing by 15% each year. How much water will there be in the lake at the end of 3 years?

EXAMINATION QUESTIONS

1 Tiago's father buys a car for $18 500.
During the first year its value falls by 20%.
Calculate its value at the end of the first year. [2]

(CIE Paper 1, Jun 2000)

---

2 Alexa invests $380 for 6 years at 5% per year simple interest.
How much interest does she receive? [2]

(CIE Paper 1, Jun 2000)

---

3 In April 1998, one US dollar was worth 128.65 Japanese Yen.
Tomoaki changed 20 000 Yen into dollars.
How much did he receive? Give your answer to 2 decimal places. [2]

(CIE Paper 1, Jun 2000)

---

4 A farmer employs 40 workers. Their wages are shown in the table below.

Weekly wage ($)	80	100	120	150
Number of workers	25	8	5	2

(a) Calculate the total amount paid in wages in a week. [2]
(b) After a good harvest, the farmer increases the wages of all of his workers by 5%.
(i) Work out the new weekly wage of each of the 25 lowest paid workers. [2]
(ii) Work out how much more he now pays in wages each week. [2]

(CIE Paper 1, Nov 2000)

---

5 Carla imported 2000 boxes of pencils. She paid $10 000 for them.

(a) $1 = 1.04 euros. Change $10 000 into euros. [2]

(b) She had to pay an import tax of 15% of the amount she paid for them. Work out, in euros, the exact amount of tax she paid. [2]

(c) Show that the total cost of the pencils to Carla was 11 960 euros. [1]

(d) Using the value 11 960, work out the cost of one box of pencils. [1]

(e) The boxes of pencils were sold in Carla's shops for 8 euros each. Work out her percentage profit to the nearest whole number. [2]

(CIE Paper 3, Jun 2001)

---

6 Tareq spends 85% of his pocket money.
How much does he spend when his pocket money is $12? [1]

(CIE Paper 1, Nov 2001)

---

7 Hassan picks 24kg of fruit.
He finds that 8% of the fruit is rotten.
Work out the mass of fruit which is rotten. [2]

(CIE Paper 1, Jun 2002)

---

8 In June 2000, one euro(€) was worth 0.59 British pounds (£).
Work out the value, in pounds, of a car which costs €12 800.
Give your answer to the nearest hundred pounds. [3]

(CIE Paper 1, Jun 2002)

---

9 When Carla started work she was paid $80 each week.
After 3 months her pay was increased by 15%.
After the increase how much was she paid each week? [2]

(CIE Paper 1, Nov 2002)

---

10 The population of a city is 550 000.
It is expected that this population will increase by 42% by the year 2008.
Calculate the expected population in 2008. [2]

(CIE Paper 1, Jun 2003)

---

11 Fifty students take part in a quiz.
The table shows the results.

Number of correct answers	5	6	7	8	9	10	11	12
Number of students	4	7	8	7	10	6	5	3

Work out the percentage of students who had **less than** 7 correct answers. [2]

(CIE Paper 3, Jun 2003)

---

12 Areeg goes to a bank to change $100 into riyals.
The bank takes $2.40 and then changes the rest of the money at a rate of $1 = 3.75 riyals.
How much does Areeg receive in riyals? [2]

(CIE Paper 1, Jun 2003)

---

13 The price of a book is $18. Sara is given a discount of 15%.
Work out this discount. [2]

(CIE Paper 1, Nov 2003)

---

14 Carlos buys a box of 50 oranges for $8.
He sells all the oranges in the market for 25 cents each.
(a) Calculate the profit he makes. [2]
(b) Calculate the percentage profit he makes on the cost price. [2]

(CIE Paper 1, Jun 2004)

---

15 Alix changed a traveller's cheque for 200 euros(€) when she visited the USA.
The exchange rate was 1 dollar = 1.05 euros.
How many dollars did she receive? [2]

(CIE Paper 1, Jun 2004)

---

16 On a certain day the conversion rate between dollars($) and Indian rupees was

$1 = 45 rupees

How many rupees were equivalent to $10 [1]

(CIE Paper 3, Nov 2004)

---

17 Ferdinand's electricity meter is read every three months.
The reading on the 1st April was 70683 units and on 1st July it was 71701 units.
(a) How many units of electricity did he use in those three months? [1]
(b) Electricity costs 8.78 cents per unit.
Calculate his bill for those three months.
Give your answer in dollars, correct to the nearest cent. [2]

(CIE Paper 1, Nov 2004)

---

18 Juana is travelling by plane from Spain to England.
She changes 150 euros (€) into pounds (£).
The exchange rate is €1 = £0.71.
Calculate how much she receives. [1]

(CIE Paper 3, Jun 2005)

---

19 Yasmeen is setting up a business.
She borrows $5000 from a loan company.
The loan company charges 6% per year simple interest.
How much interest will Yasmeen pay after 3 years. [2]

(CIE Paper 1, Jun 2005)

20

SALE
All items
35% Reduction

Abdul bought a spade in this sale. Its **original** price was $16.
How much did Abdul save? [2]

(CIE Paper 1, Jun 2005)

---

21 Lorenzo saves money for a motorbike.
The marked price of the motorbike is $900.
He pays a deposit of 35% of the marked price.
(a) Calculate his deposit. [2]
(b) He then makes 12 monthly payments of $60 each.
How much more than the $900 marked price does he pay altogether? [3]

(CIE Paper 1, Nov 2005)

Chapter 15

Sequences

This chapter will show you how to

- ✔ describe how a sequence continues
- ✔ find the next term in a sequence of numbers or diagrams
- ✔ find and use rules for the nth term of a sequence

15.1 Number patterns

A number pattern or number **sequence** is a list of numbers. There is often a connection between the numbers in the list.

Each number in a number sequence is called a **term**.

EXAMPLE 1

Here is a number sequence 3, 5, 7, 9, 11, …

(a) Write down the 1st term.
(b) Write down the 4th term.
(c) Describe the rule for continuing the sequence.
(d) Write down the next three terms.

(a) 3, 5, 7, 9, 11, … The 1st term is 3.
(b) 3, 5, 7, 9, 11, … The 4th term is 9.
(c) The rule for this sequence is add 2 to find the next term.
(d) 13, 15, 17.

You add 2 to 3 to get 5, then add 2 to 5 to get 7, and so on.

Because $11 + 2 = 13$, $13 + 2 = 15$, $15 + 2 = 17$.

EXAMPLE 2

For the sequence 24, 20, 16, 12, …

(a) Describe the rule for continuing the sequence.
(b) Write down the 5th and 6th terms.

(a) Subtract 4 each time.
(b) 5th term = 8 6th term = 4.

$24 - 4 = 20$, $20 - 4 = 16$, $16 - 4 = 12$.

Because $12 - 4 = 8$ and $8 - 4 = 4$.

EXERCISE 15A

1 For this number sequence 4, 6, 8, 10, 12, ...
  (a) Write down the 1st term.
  (b) Write down the 4th term.
  (c) Describe the rule for continuing the sequence.
  (d) Write down the next three terms.

2 For this number sequence 7, 11, 15, 19, 23, ...
  (a) Write down the 3rd term.
  (b) Write down the 5th term.
  (c) Describe the rule for continuing the sequence.
  (d) Write down the next three terms.

3 For each sequence
- describe the rule for continuing the sequence
- write down the 5th and 6th terms.

  (a) 6, 9, 12, 15, ... (b) 4, 9, 14, 19, ...
  (c) 22, 32, 42, 52, ... (d) 6, 13, 20, 27, ...
  (e) −4, −2, 0, 2, ... (f) 10, 7, 4, 1, ...

4 For this number sequence 25, 21, 17, 13, ...
  (a) Write down the 2nd term.
  (b) Write down the 4th term.
  (c) Describe the rule for continuing the sequence.
  (d) Write down the next three terms.

5 For this number sequence 40, 34, 28, 22, ...
  (a) Write down the 1st term.
  (b) Write down the 3rd term.
  (c) Describe the rule for continuing the sequence.
  (d) Write down the next three terms.

6 For each sequence
- describe the rule for continuing the sequence
- write down the 5th and 6th terms.

  (a) 76, 66, 56, 46, ... (b) 60, 54, 48, 42, ...
  (c) 32, 28, 24, 20, ... (d) 87, 78, 69, 60, ...
  (e) 105, 90, 75, 60, ... (f) 36, 29, 22, 15, ...

7 Write down the next two numbers in each sequence.

(a) 20, 15, 10, 5, ... (b) 13, 10, 7, 4, ...

(c) 35, 25, 15, 5, ... (d) 26, 19, 12, 5, ...

15.2 Using differences

Terms next to each other are called **consecutive** terms.

In the sequence

4, 10, 16, 22, ...

4 and 10 are consecutive terms,
10 and 16 are consecutive terms.

To write the next terms in a sequence it helps to look at the **differences** between consecutive terms.

sequence	4,		10,		16,		22, ...
differences		+6		+6		+6	

The difference between 4 and 10 is $10 - 4 = 6$.
or to get from 4 to 10 you add 6.

In this sequence the differences are all the same +6.

In some sequences the differences are not all the same.

EXAMPLE 3

For the sequence

3, 5, 8, 12, ...

(a) Find the differences between consecutive terms.

(b) Describe the pattern of the differences.

(c) Write down the next two terms.

(a)

sequence	3,		5,		8,		12, ...
difference		+2		+3		+4	

(b) The difference goes up by 1 each time.

(c) $12 + 5 = 17$
$17 + 6 = 23$

The next two differences will be +5 and +6.

EXERCISE 15B

1 For the sequence

2, 4, 7, 11, ...

(a) Find the differences between consecutive terms.

(b) Describe the pattern of the differences.

(c) Write down the next two terms.

2 For the sequence

6, 9, 13, 18, ...

(a) Find the differences between consecutive terms.

(b) Describe the pattern of the differences.

(c) Write down the 5th and 6th terms.

3 For each sequence find the next two terms. Use the differences between consecutive terms to help you.

(a) 10, 11, 13, 16, ... (b) 4, 5, 8, 13, ...

(c) 20, 25, 35, 50, ... (d) 1, 4, 9, 16, ...

(e) 1, 1, 2, 3, 5, 8, ... (f) 4, 8, 16, 28, ...

(g) 49, 36, 25, 16, ... (h) 17, 12, 8, 5, ...

(i) 55, 45, 36, 28, ... (j) 13, 9, 6, 4, ...

4 For these sequences, find

(i) the differences between consecutive terms

(ii) the next two terms

(a) 3, 6, 12, 24, ... (b) 1, 3, 9, 27, ...

(c) 40, 20, 10, ... (d) 10 000, 1000, 100, ...

15.3 Rules for sequences

You can use the **general rule** for a sequence to work out any term in a sequence. You need to know the position of the term.

The **term number** is the position of the term.

Sequence	1,	4,	7,	10, ...
	1st term	2nd term	3rd term	4th term
Term number	1	2	3	4

EXAMPLE 4

The general rule of a sequence is

$3 \times$ *the term number then add 1*

(a) Use the rule to find the first three terms.

(b) Use the rule to find the 15th term.

(a) 1st term $3 \times 1 + 1 = 3 + 1 = 4$

2nd term $3 \times 2 + 1 = 6 + 1 = 7$

3rd term $3 \times 3 + 1 = 9 + 1 = 10$

(b) 15th term $3 \times 15 + 1 = 45 + 1 = 46$

The term number of the 15th term is 15.

The general rule is also called the **general term** or the ***n*th term.**

The letter n is used to stand for the term number.

EXAMPLE 5

The nth term of a sequence is $2n + 5$.

(a) Write down the first four terms.

(b) What is the ratio between consecutive terms?

(c) Which term has a value of 45?

(d) Explain why 36 cannot be a term in this sequence.

(a) 1st term $2 \times 1 + 5 = 2 + 5 = 7$

2nd term $2 \times 2 + 5 = 4 + 5 = 9$

3rd term $2 \times 3 + 5 = 6 + 5 = 11$

4th term $2 \times 4 + 5 = 8 + 5 = 13$

Continued ▼

Substitute the term numbers for n in the formula.
For the 1st term $n = 1$
For the 2nd term $n = 2$
... and so on.

(b) sequence 7, 9, 11, 13, …

difference +2 +2 +2

The difference between consecutive terms is +2.

(c) $2n + 5 = 45$

$2n = 40$

$n = 20$

The 20th term is 45.

Every term follows the rule $2n + 5$. Find which value of n satisfies the equation $2n + 5 = 45$.

(d) If 36 is in the sequence then there is a whole number n for which $2n + 5 = 36$.

If $2n + 5 = 36$

then $2n = 31$

and $n = 15\frac{1}{2}$

but $15\frac{1}{2}$ is not a whole number so 36 is not in the sequence.

Part **(d)** is an extension of the method used in part **(c)**.

EXAMPLE 6

The nth term of a sequence is $n^2 - 1$.

(a) Write down the first four terms of the sequence.

(b) Write down the 12th term.

(a) 1st term $1^2 - 1 = 1 - 1 = 0$

2nd term $2^2 - 1 = 4 - 1 = 3$

3rd term $3^2 - 1 = 9 - 1 = 8$

4th term $4^2 - 1 = 16 - 1 = 15$

(b) 12th term $12^2 - 1 = 144 - 1 = 143$

Remember $n^2 = n \times n$.
So $2^2 = 2 \times 2 = 4$.

EXERCISE 15C

1 The general rule of a sequence is
3 × the term number then add 2

(a) Use the rule to find the first three terms.

(b) Use the rule to find the 15th term.

You can write this rule as $3n + 2$.

2 The general rule of a sequence is

2 × the term number then add 7

(a) Use the rule to find the first three terms.

(b) Use the rule to find the 10th term.

3 The general rule of a sequence is

4 × the term number then subtract 1

(a) Use the rule to find the first three terms.

(b) Use the rule to find the 50th term.

4 The nth term of a sequence is $2n + 3$.

(a) Write down the first four terms.

(b) What is the difference between consecutive terms?

(c) Write down the 20th term.

5 For each of the following sequences

- find the first four terms
- write down the difference between consecutive terms
- find the 30th term.

(a) nth term $3n + 5$ **(b)** nth term is $2n - 1$

(c) nth term $4n - 3$ **(d)** nth term is $3n + 7$

(e) nth term $5n + 1$ **(f)** nth term is $6n - 2$

6 Look at your answers to question 5. What do you notice about the rule for the nth term and the difference between consecutive terms for each sequence?

7 The nth term of a sequence is $2n + 4$.

(a) Work out the value of the 8th term.

(b) Which term has a value of 46?

(c) Explain why 35 is not a term in this sequence.

Use Example 5 to help you.

8 The nth term of a sequence is $3n - 1$.

(a) Calculate the value of the 6th term.

(b) Which term has a value of 59?

(c) Explain why 90 is not a term in this sequence.

9 The nth term of a sequence is $5n + 7$.

(a) Calculate the value of the 10th term.

(b) Which term has a value of 82?

(c) Explain why 110 is not a term in this sequence.

10 The nth term of a sequence is $n^2 + 1$.

(a) Write down the first four terms of the sequence.

(b) Write down the 12th term.

11 The nth term of a sequence is $n^2 + 4$.

(a) Write down the first four terms of the sequence.

(b) Write down the 9th term.

12 The nth term of a sequence is $n^2 - 3$.

(a) Write down the first four terms of the sequence.

(b) Write down the 13th term.

13 The nth term of a sequence is $n^2 + 3$.

(a) Which term has a value of 28?

(b) Explain why 50 is not a term in this sequence.

15.4 Finding the nth term

If the difference between consecutive terms is the same you can use it to find the rule for the nth term.

In the sequence 7, 11, 15, 19, ...

difference +4 +4 +4

the terms go up in fours.

The 4 × table also goes up in fours.
This tells you that the nth term includes $4n$.

$4n = 4 \times n$

To find the rest of the rule, compare the sequence to the 4 × table.

4 × table	4	8	12	16	...
Sequence	7	11	15	19	...

You have to add 3 to each number in the 4 × table to get the numbers in the sequence.

$4 + 3 = 7$, $8 + 3 = 11$, ...

So the rule for the nth term is $4n + 3$.

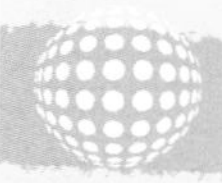

EXAMPLE 7

Find the nth term and the 50th term of the sequence

2, 5, 8, 11, ...

Sequence 2, 5, 8, 11, ...

Difference +3 +3 +3

The difference is 3 so the nth term includes $3n$.

3 × table	3	6	9	12	...
Sequence	2	5	8	11	...

You have to subtract 1 from each number in the 3 times table to get the numbers in the sequence.

So the rule for the nth term is $3n - 1$.

50th term $= 3 \times 50 - 1 = 150 - 1 = 149$.

Substitute $n = 50$ into the nth term.

EXERCISE 15D

Find the nth term and the 50th term of each of these sequences.

1 4, 7, 10, 13, ...

2 5, 7, 9, 11, ...

3 7, 11, 15, 19, ...

4 3, 8, 13, 18, ...

5 6, 9, 12, 15, ...

6 6, 13, 20, 27, ...

7 2, 12, 22, 32, ...

8 16, 25, 34, 43, ...

9 5, 6, 7, 8, ...

10 −4, −1, 2, 5, ...

For some sequences you can find the nth term by comparing your sequence with another one.

EXAMPLE 8

Find the nth term of the sequence

2 5 10 17 26 ...

Continued ▼

Try adding, subtracting, multiplying or dividing by a fixed number for each term.

In this case subtracting 1 gives

1 4 9 16 25 and we know the nth term of this sequence is n^2. Our sequence is 1 more than this so it has an nth term of $n^2 + 1$.

You need to be able to recognise important sequences like this.

EXERCISE 15E

1 Find the nth term of the following sequences.

(a) 5 8 13 20 29... (b) −1 2 7 14 23...

(c) 2 8 18 32 50... (d) 2 9 28 65 126...

(e) 4 11 30 67 128...

2 The sequence 1 3 6 10 15... has a nth term of $\frac{1}{2}n(n + 1)$. Find the nth term of the sequences

(a) 5 7 10 14 19... (b) 2 6 12 20 30...

(c) 3 7 13 21 31 (d) 1 5 11 19 29

3 The nth term of the sequence 2 6 12 20 30 is $n^2 + n$. Find the nth term of the sequences

(a) 5 9 15 23 33... (b) 6 12 20 30...

(c) 3 7 13 21 31... (d) 4 12 24 40 60...

4 Find the nth term of the sequences

(a) 2 5 10 17 26... (b) 3 5 8 12 17...

(c) 0.5 2 4.5 8 12.5...

15.5 Sequences of patterns

Sequences of diagrams can lead to number sequences.

The numbers of dots in these patterns make a number sequence.

EXAMPLE 9

This sequence of patterns is made from square tiles.

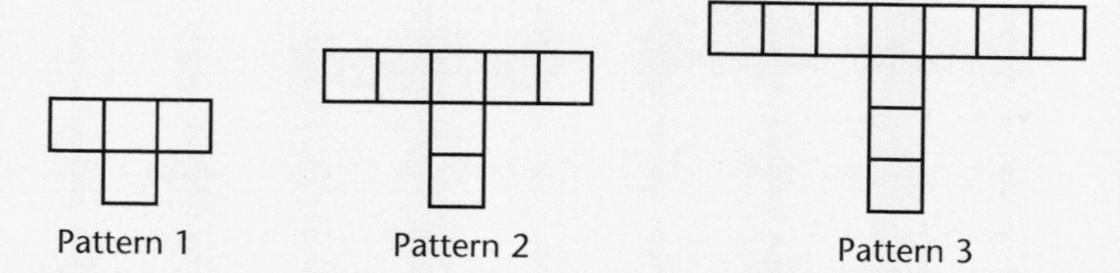

(a) Draw Pattern 4.
(b) How many tiles are needed for pattern 5?
(c) Describe the rule for continuing the sequence.
(d) How many tiles are needed for Pattern n?
(e) Describe how the patterns relate to the nth term.
(f) Which Pattern needs 37 tiles?

(a)

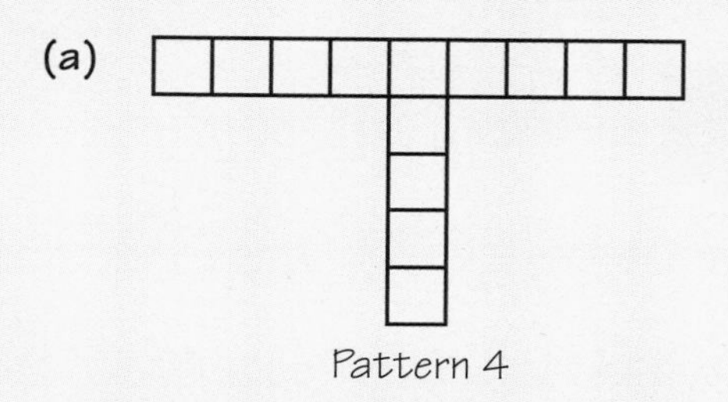

You could draw Pattern 5 to check, or you could work out that you add 1 tile to each arm, adding 3 in total. The new tiles go on the end of each 'arm' of the pattern.

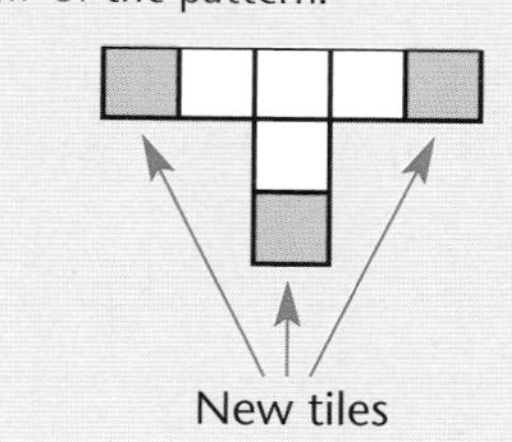

(b) 16 tiles for Pattern 5
(c) Add 3 tiles each time.
(d) The number sequence for the Patterns

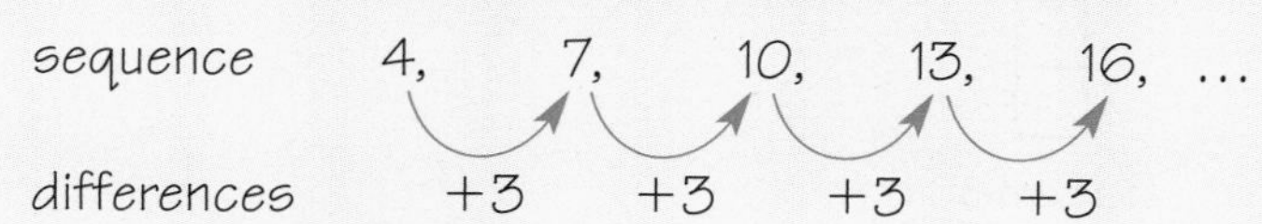

Pattern n is the nth term in the pattern sequence.

The difference is 3 so the rule includes $3n$.

Compare with the $3 \times$ table.

3 × table	3	6	9	12	…
Sequence	4	7	10	13	…

You have to add 1 to each number in the $3 \times$ table to get the numbers in the sequence.

The rule for the number of tiles needed in Pattern n is $3n + 1$.

Continued ▼

(e) 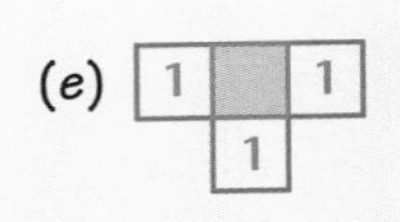

Pattern 1 has 1 lot of 3 tiles + 1 $= 3 \times 1 + 1$

Pattern 2 has 2 lots of 3 tiles + 1 $= 3 \times 2 + 1$

Pattern 3 has 3 lots of 3 tiles + 1 $= 3 \times 3 + 1$

etc.

Pattern n has n lots of 3 tiles + 1 $= 3n + 1$

(f) $3n + 1 = 37$

$3n = 36$

$n = 12$

Pattern 12 needs 37 tiles.

EXERCISE 15F

1 Square tiles are used to make a sequence of patterns.

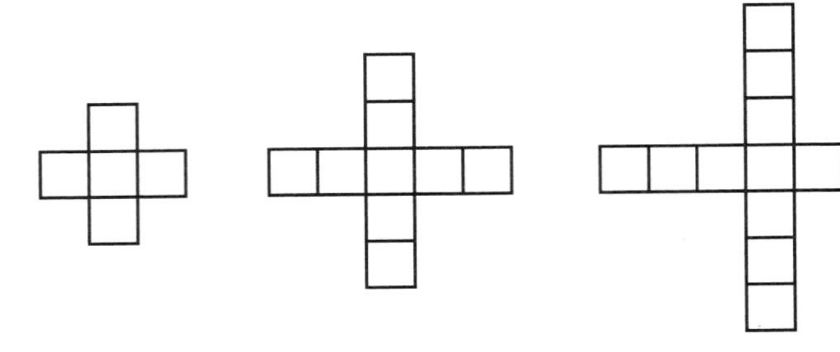

Pattern 1 Pattern 2 Pattern 3

(a) Draw Pattern 4.

(b) How many tiles are needed for Pattern 5?

(c) Describe the rule for continuing the sequence of the number of tiles needed.

(d) How many tiles are needed for Pattern n?

(e) Which pattern needs 45 tiles?

2 Matchsticks are used to make triangles.

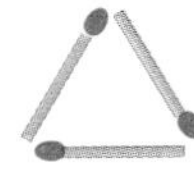

1 triangle

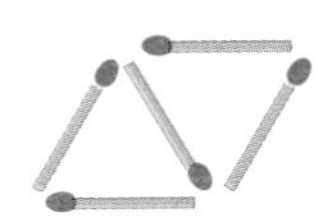

2 triangles

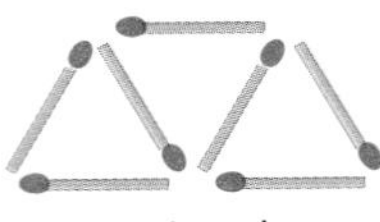

3 triangles

(a) Draw the pattern for 4 triangles.
(b) How many matchsticks are needed for 5 triangles?
(c) Describe the rule for continuing the sequence of the number of matchsticks needed.
(d) How many matchsticks are needed for n triangles?
(e) How many triangles can you make with 51 matchsticks?

3 Matchsticks are used to make pentagons.

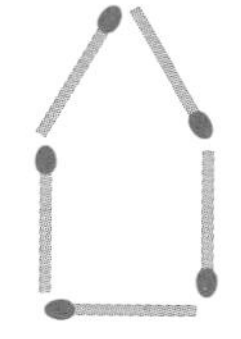
1 pentagon

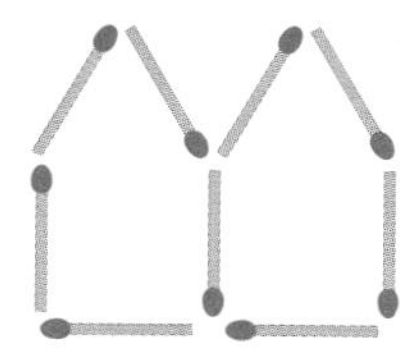
2 pentagons

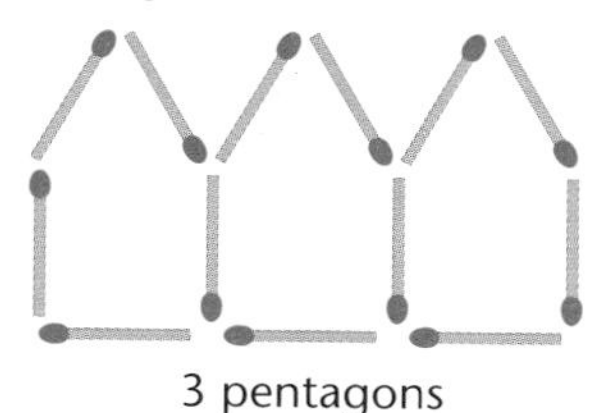
3 pentagons

(a) Draw the pattern for 4 pentagons.
(b) How many matchsticks are needed for 5 pentagons?
(c) Describe the rule for continuing the sequence of the number of matchsticks needed.
(d) How many matchsticks are needed for n pentagons?
(e) Describe how the patterns relate to the nth term.
(f) How many pentagons can you make with 85 matchsticks?

4 Matchsticks are used to make rectangles.

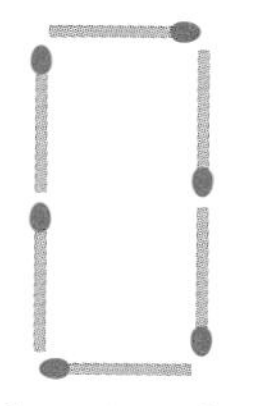
1 rectangle

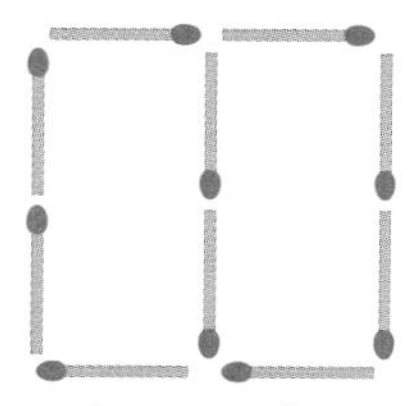
2 rectangles

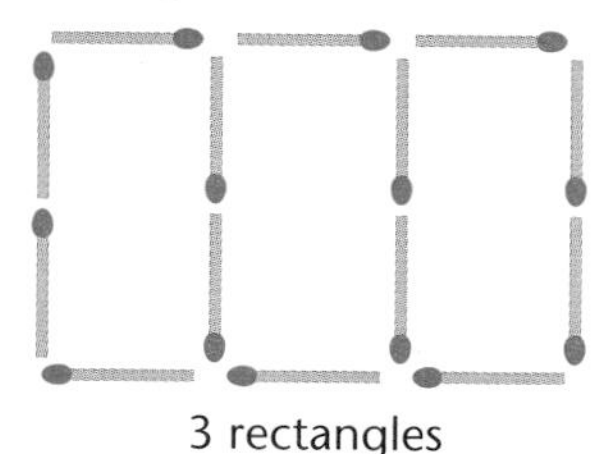
3 rectangles

(a) Draw the pattern for 4 rectangles.
(b) How many matchsticks are needed for 5 rectangles?
(c) How many matchsticks are needed for n rectangles?
(d) How many rectangles can you make with 66 matchsticks?

5 In a restaurant, tables are put together in a line to seat different numbers of people.

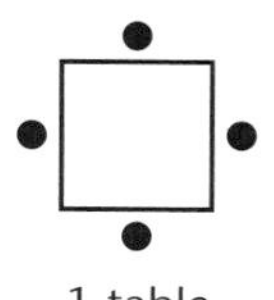
1 table
4 people

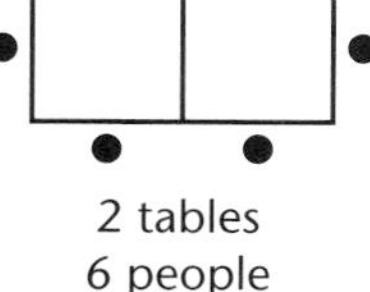
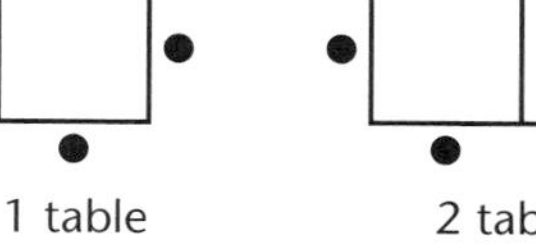
2 tables
6 people

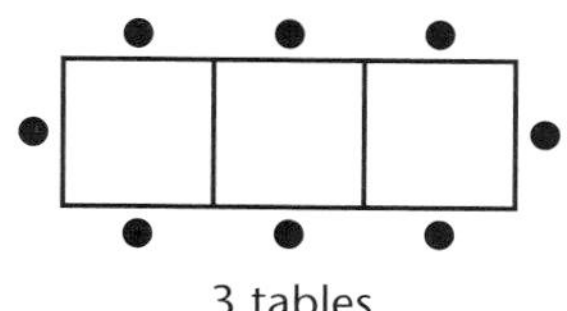

3 tables
8 people

(a) Draw the pattern for 4 tables.

(b) How many people can sit at 5 tables in a line?

(c) How many people can sit at n tables in a line?

(d) How many tables do you need to seat 24 people?

6 Dots are used to make a sequence of patterns.

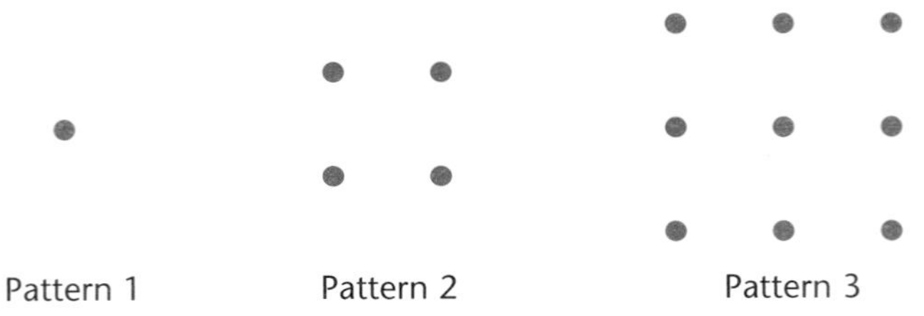

Pattern 1 Pattern 2 Pattern 3

(a) Draw Pattern 4.

(b) How many dots are needed for Pattern 5?

(c) Describe the rule for continuing the sequence of the number of dots.

(d) What is the name for the numbers in this sequence?

(e) How many dots are needed for Pattern n?

(f) Which pattern will have 144 dots?

7 Dots are used to make a sequence of triangles.

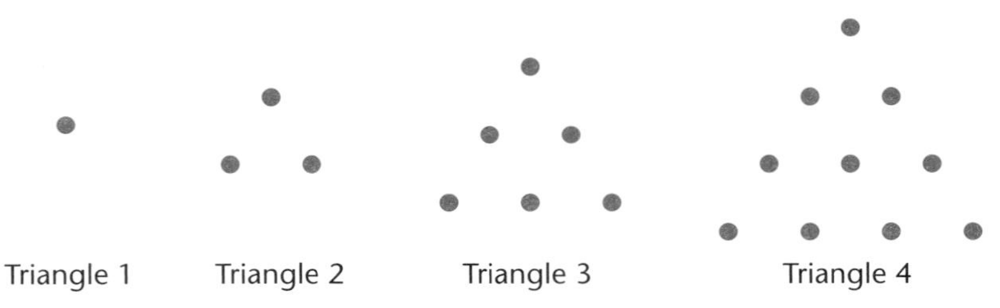

Triangle 1 Triangle 2 Triangle 3 Triangle 4

(a) Draw Triangle 5.

(b) How many dots are needed for Triangle 6?

(c) Describe the rule for continuing the sequence of triangles.

(d) The numbers in this sequence are called Triangular numbers.
One rule for finding the nth triangular number is $\frac{1}{2}n(n + 1)$.
What is the 10th triangular number?

EXAMINATION QUESTIONS

1

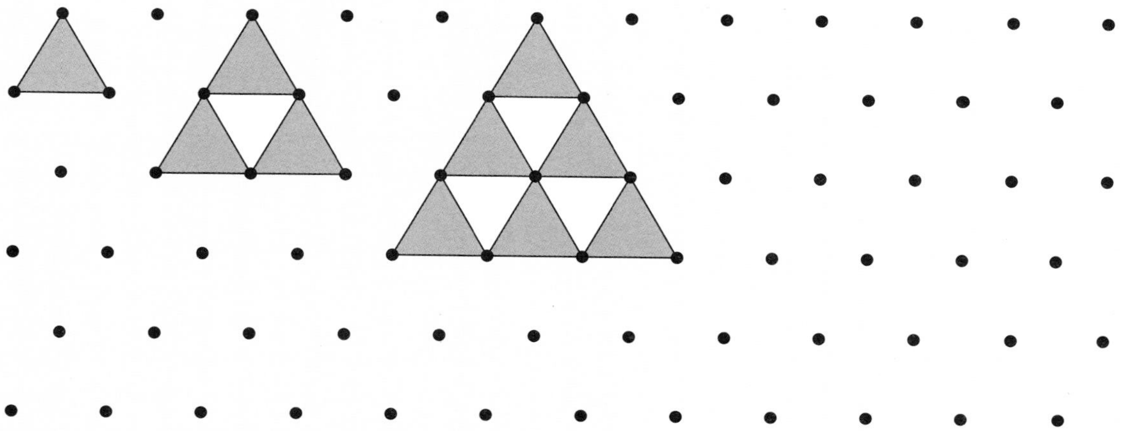

The diagram shows the first three triangles in a sequence of equilateral triangles of increasing size. Each is made from triangular tiles of side 1 cm which are either black or white.

(a) Draw the fourth equilateral triangle in the sequence above, shading in the black tiles. [2]

(b) Copy and complete the following table for equilateral triangles.

Length of base (cm)	1	2	3	4	5	6
Number of white tiles	0		3			
Number of black tiles	1		6			
Total number of tiles	1		9			

[6]

(c) Write down the special name to the given to the numbers in the total number of tiles row. [1]

(d) How many white tiles would there be in the equilateral triangle with base 10 cm? [3]

(CIE Paper 3, Nov 2000)

2 Rectangle 1 Rectangle 2 Rectangle 3

Look at the black and white dots in the rectangles above.
The rectangles form a pattern.
Copy and complete the table below.

Rectangle	Number of rows	Number of columns	Total number of dots	Number of black dots	Number of white dots
1	2	4	8	5	3
2	3	7	21	12	9
3	4	10	40	22	
4	5	13			
5					

[9]

(CIE Paper 3, Nov 2001)

3 Diagram 1 Diagram 2 Diagram 3 Diagram 4

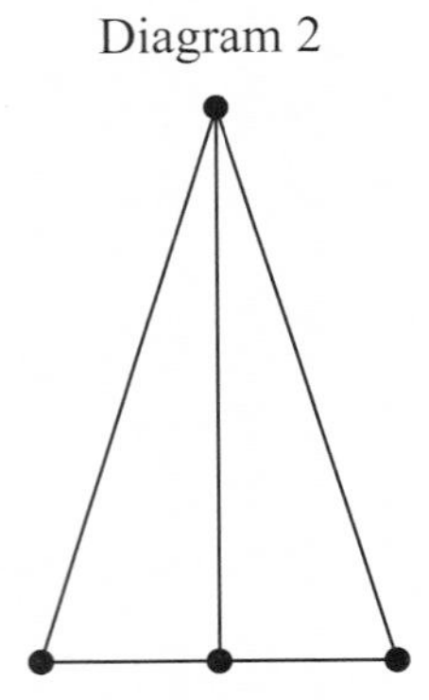

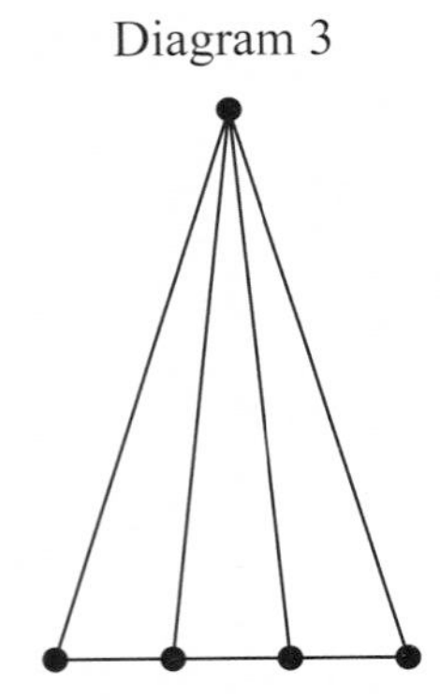

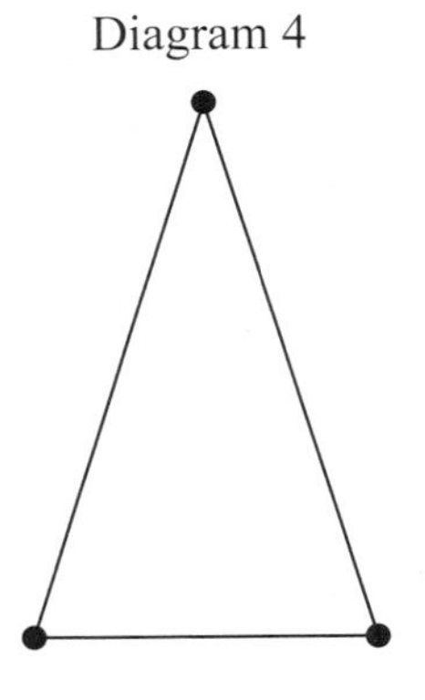

3 dots 1 triangle 4 dots 3 triangles 5 dots 6 triangles

Look at the diagrams above.

(a) Copy and complete Diagram 4 to continue the pattern. [2]

(b) Copy and complete the table below.

Diagram		1	2	3	4	5		n
Number of dots		3	4	5				

[3]

(c) Complete the table below.

Diagram		1	2	3	4	5	6	10
Number of triangles		1	3	6	10			

(d) A line is now drawn inside each of the diagrams as shown below.

Diagram 1 Diagram 2 Diagram 3

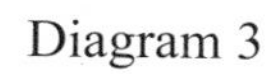

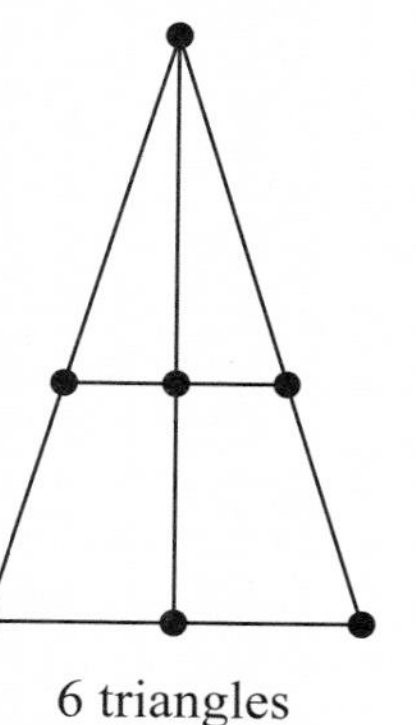

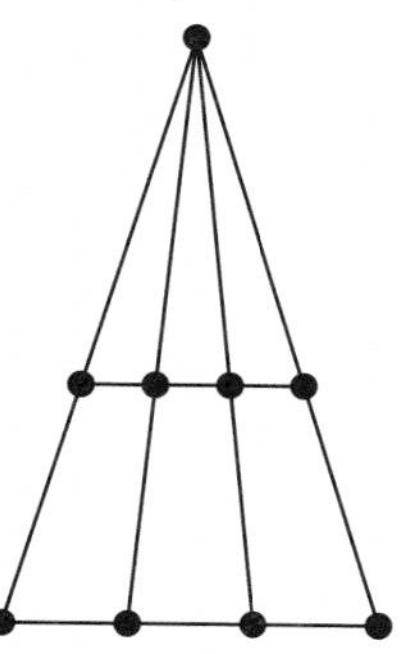

2 triangles 6 triangles

How many triangles are there in Diagram 3? [2]

(CIE Paper 3, Jun 2003)

---

4 (a) Look at the sequence of dots and squares below.

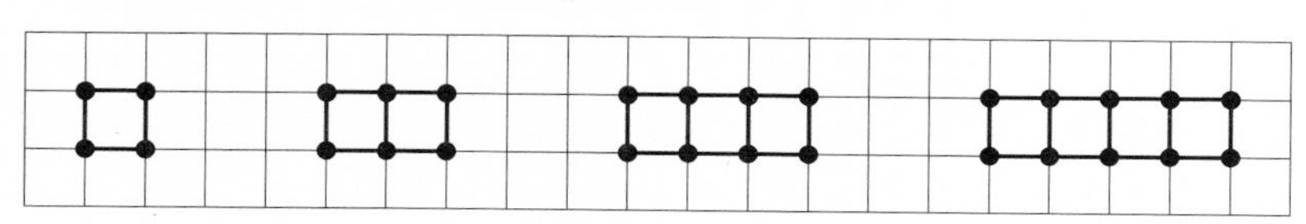

Number of dots	4	6	8	10
Number of squares	1	2	3	4

Find the number of dots when there are

(i) 5 squares, [1]

(ii) 9 squares, [1]

(iii) n squares. [2]

(b) Another sequence of dots and squares is shown below.

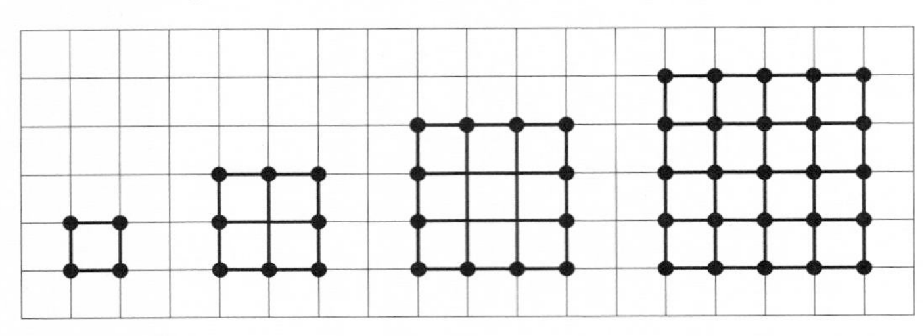

Diagram	1	2	3	4
Number of dots	4	8	12	16
Number of squares	1	4	9	16

(i) For Diagram 5, find
(a) the number of dots, [1]
(b) the number of squares. [1]

(ii) Find the number of dots in the diagram that has 144 squares. [2]

(iii) Find the number of squares in the diagram that has 40 dots. [2]

(CIE Paper 3, Jun 2004)

---

5 **(a)** The table below shows a pattern of numbers.
Copy and complete the two empty boxes. [2]

1	2	3	4	5		n
3	5		9			$2n + 1$

(b) The new table shows another pattern of numbers.
Copy and complete the two empty boxes. [2]

1	2	3	4	5		n
5	8	11	14			

(c) By looking at the patterns, copy and fill in the eight empty boxes. [5]

1	2	3	4	5	6		n
1	4	9	16				n^2
0	3	8	15				
4	9	16	25				

(CIE Paper 3, Jun 2002)

6 A pattern of numbers is shown below.

Row									
1					1				
2				2	3	4			
3			5	6	7	8	9		
4		10	11	12	13	14	15	16	
5	17	18	19	20	21	22	23	24	25
6									

(a) Write down row 6. [1]

(b) The last numbers in each row form a sequence.

1, 4, 9, 16, 25, ...

(i) What is the special name given to these numbers? [1]

(ii) Write down the last number in the 10th row. [1]

(iii) Write down an expression for the last number in the nth row. [1]

(c) The numbers in the middle column of the pattern form a sequence.

(i) Write down the next number in this sequence. [1]

(ii) The expression for the nth number in this sequence is $n^2 - n + 1$.

Work out the 30th number. [2]

(CIE Paper 3, Nov 2004)

---

7 Look at the sequence of numbers

7, 11, 15, 19, ...

(a) Write down the next number in the sequence. [1]

(b) Find the 10th number in the sequence. [1]

(c) Write an expression, in terms of n, for the nth number in the sequence. [1]

(CIE Paper 1, Jun 2005)

Chapter 16

Constructions and loci

This chapter will show you how to

- ✔ use a straight edge and compasses to construct
  - a triangle given all three sides
  - the perpendicular bisector of a line segment
  - the bisector of an angle
- ✔ construct angles of 60°
- ✔ understand, interpret and solve problems using simple loci (including scale drawing and bearings)

16.1 Constructions

You will need to know

- **how to use a pair of compasses**
- **how to measure accurately**

Standard constructions use only a straight edge (ruler) and a pair of compasses to draw accurate diagrams.

Your drawings must be accurate.

When you use compasses you must leave the construction **arcs** as evidence that you have used the correct method.

Arcs are parts of a circle drawn with compasses.

16.2 Construct a triangle given all three sides

Example 1 shows the method for constructing a triangle given the lengths of all three sides.

It helps to draw and label a sketch first.

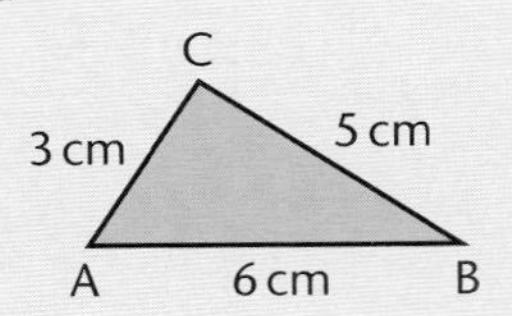

Use your ruler to draw the longest side 6 cm long. Label it *AB*.

EXAMPLE 1

Construct a triangle *ABC* where $AB = 6$ cm, $AC = 3$ cm and $BC = 5$ cm.

A ———— B

From the question, $AB = 6$ cm.

Open your compasses to a radius of 3 cm. Put the point on *A* and draw an arc in the space above line *AB*.

A ———— B

$AC = 3$ cm.

$BC = 5$ cm.

C

A ———— B

Open your compasses to a radius of 5 cm. Put the point on *B* and draw an arc to **intersect** the first arc. Where the arcs intersect is point *C*.

Point *C* is 3 cm from *A* and 5 cm from *B*.

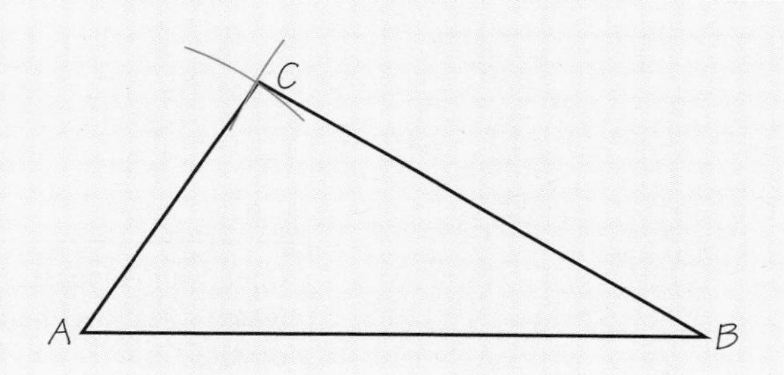

Join *C* to *A* and *B* to complete the triangle.

Remember, don't rub out the arcs!

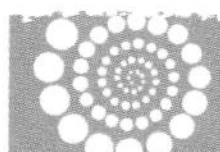

EXERCISE 16A

Construct triangles *ABC* with these measurements.

1 $AB = 8$ cm, $AC = 6$ cm, $BC = 5$ cm.

2 $AB = 9$ cm, $AC = 8.5$ cm, $BC = 4$ cm.

3 $AB = 7.5$ cm, $AC = 10$ cm, $BC = 4.5$ cm.

4 $AB = 6$ cm, $AC = 8$ cm, $BC = 10$ cm.

5 Equilateral triangle *ABC* with sides 6 cm.

6 Equilateral triangle *ABC* with sides 9.5 cm.

In an equilateral triangle all sides are the same length.

16.3 Constructing perpendiculars

Construct the perpendicular bisector of a line segment

To **bisect** means to cut in half.

A straight line has infinite length so you will only be looking at a part of it – a **line segment**.

line segment

A perpendicular bisector

- cuts a line segment in half
- is perpendicular (at 90°) to the line segment.

EXAMPLE 2

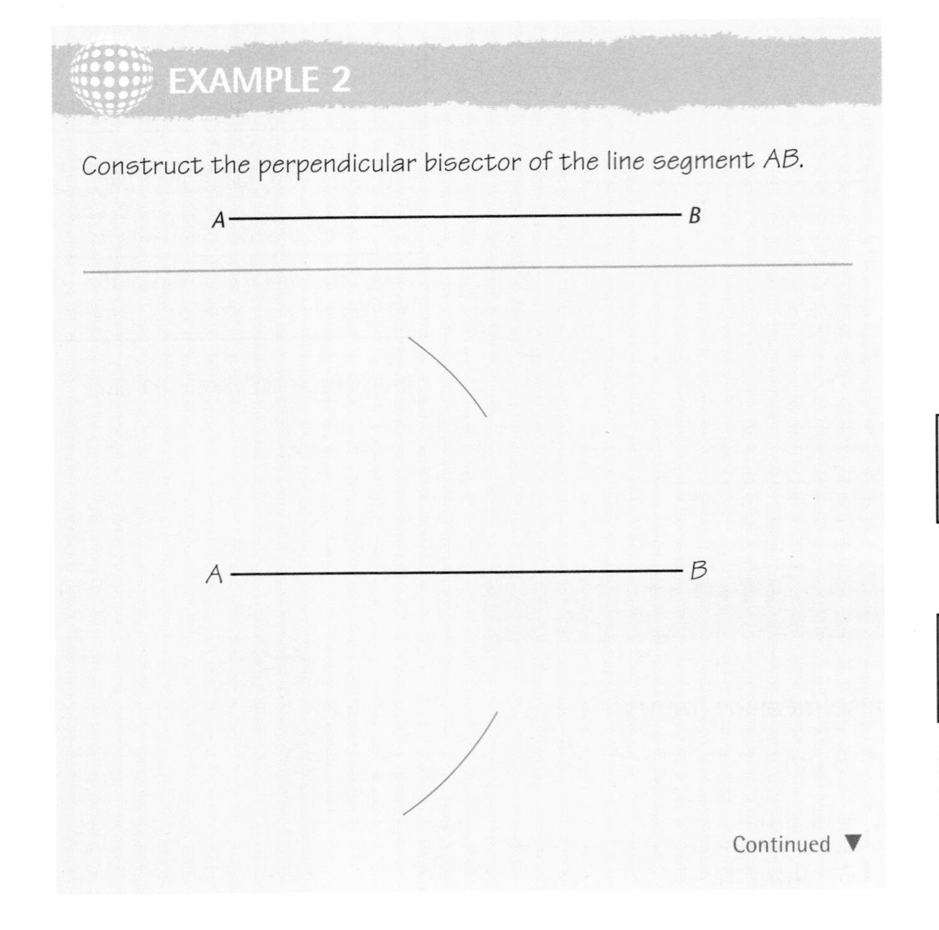

Open your compasses to a radius which is just over half the length of *AB*.

Put the compass point on *A* and draw one arc above *AB* and one below.

Keep the radius the same for both arcs.

Continued ▼

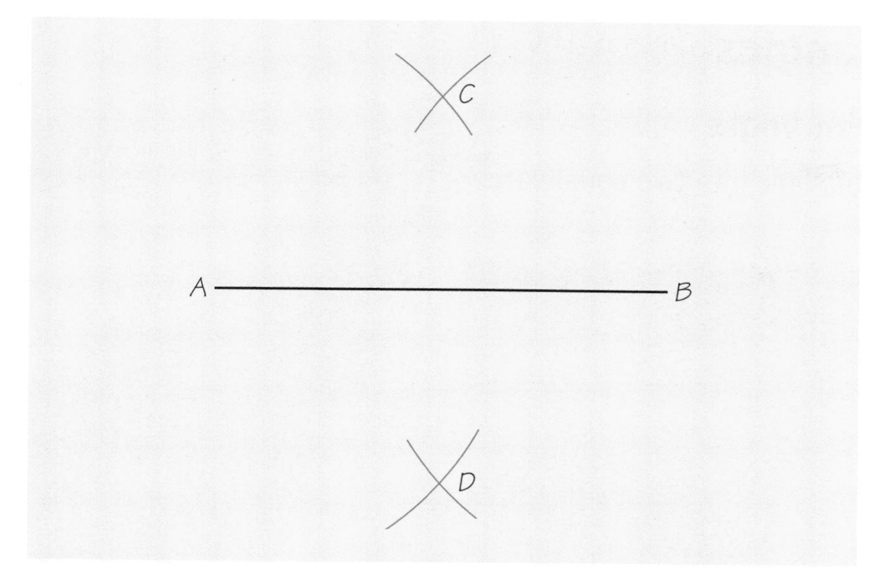

Keeping the radius the same, put your compass point on *B* and draw two more arcs to intersect the first pair.
Label these points *C* and *D*.

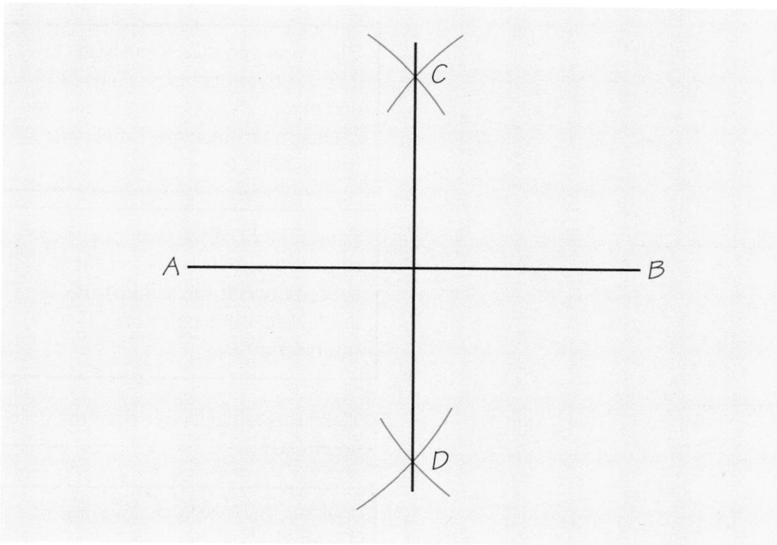

Join *CD*. This line is the perpendicular bisector of *AB*.

CD crosses *AB* at the mid-point of *AB*.

CD makes an angle of 90° with *AB*.

EXERCISE 16B

Draw six line segments of different lengths. Label the ends *A* and *B*. Construct the perpendicular bisector for each line segment.

Check by measuring that your perpendicular bisector passes through the **mid-point** of the line segment.

Check, using a protractor, that the angle between the two lines is 90°.

16.4 Constructing angles

Construct the bisector of an angle

The bisector of an angle divides an angle into two equal parts.

Sometimes this is called an *angle bisector.*

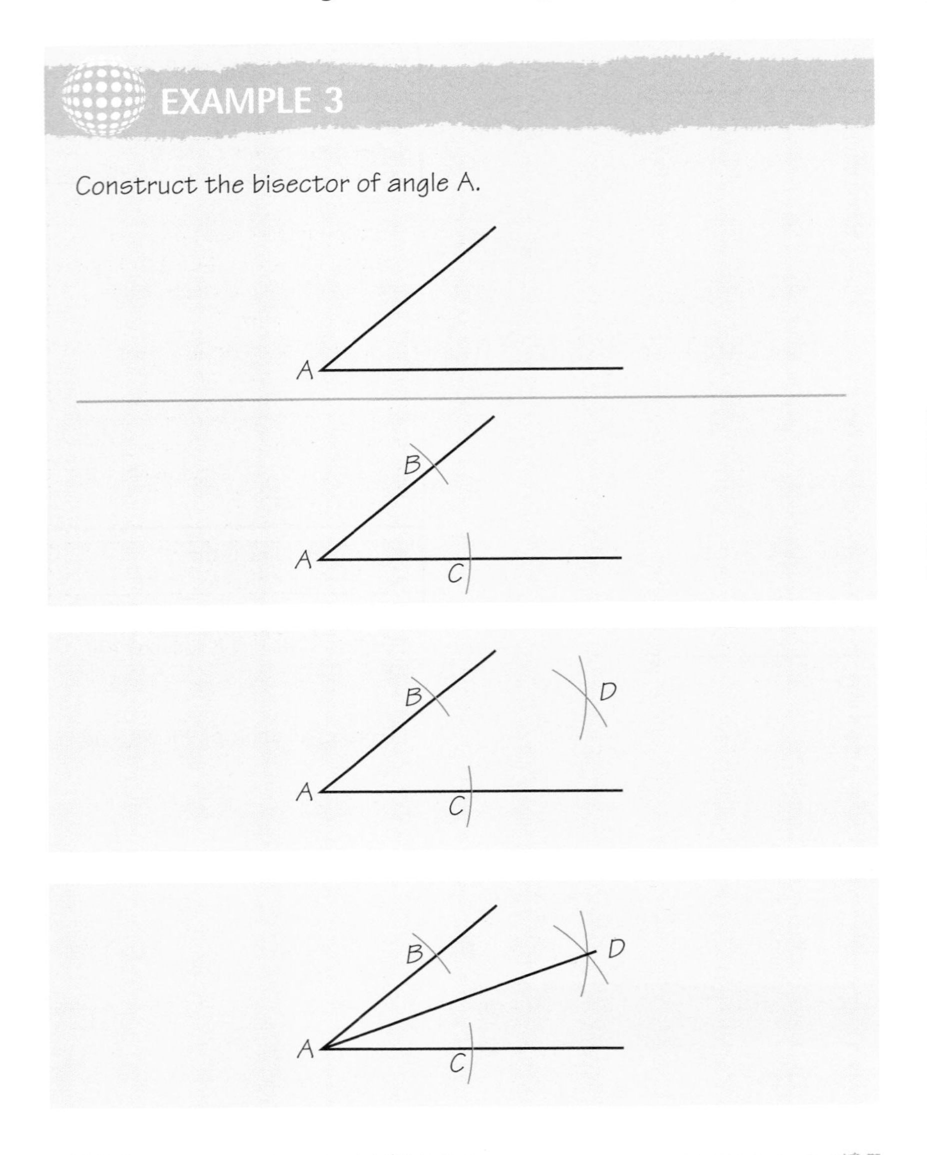

Open your compasses to about 2 cm. Put your compass point on *A* and mark off two small arcs, one on each arm of the angle. Label these points *B* and *C*.

From *B* and *C* draw two arcs of the same radius to intersect in the space between the arms of the angle. Label this point of intersection *D*.

Make this radius larger than the previous one.

Join *AD*. This line is the bisector of angle *A*.

AD divides angle *A* into two equal parts.

EXERCISE 16C

Draw six angles of different sizes. Construct the angle bisector for each of them.

Check the accuracy of your constructions by measuring the angles with a protractor.

Construct an angle of 60°

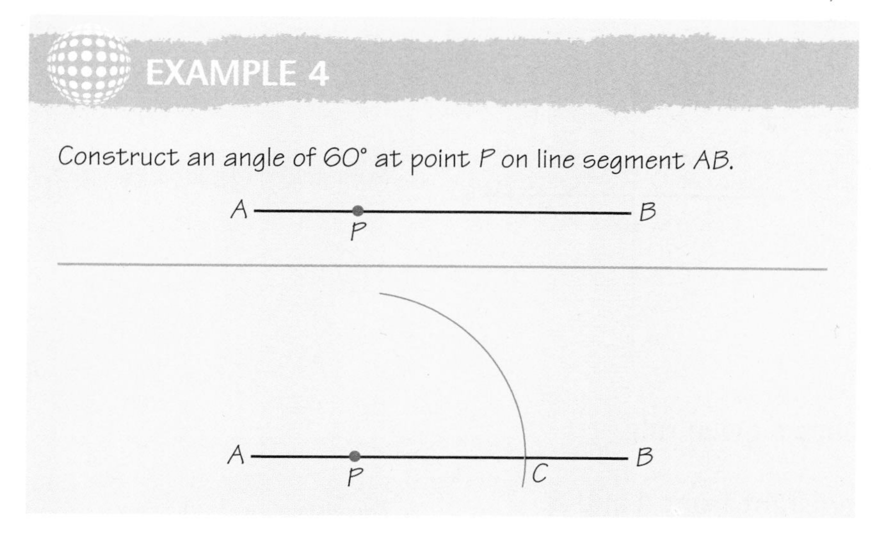

Open your compasses to a radius of about 3 cm.
Put the compass point on P and draw an arc which starts just below the line AB on one side of P and ends almost above P.
Label point C, where this arc intersects the line AB.

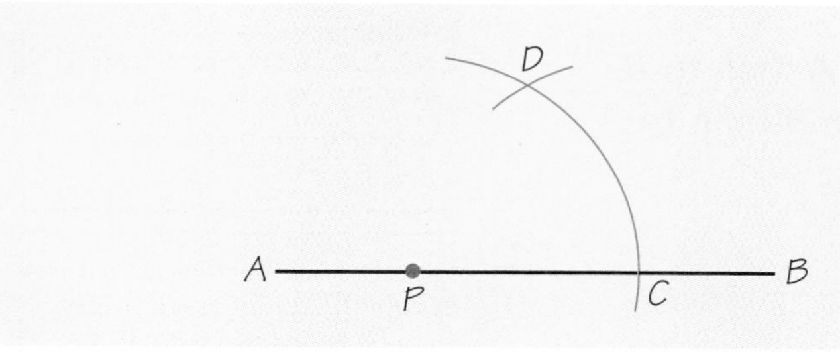

Keeping the radius the same, from point C draw an arc to intersect the first arc at point D.

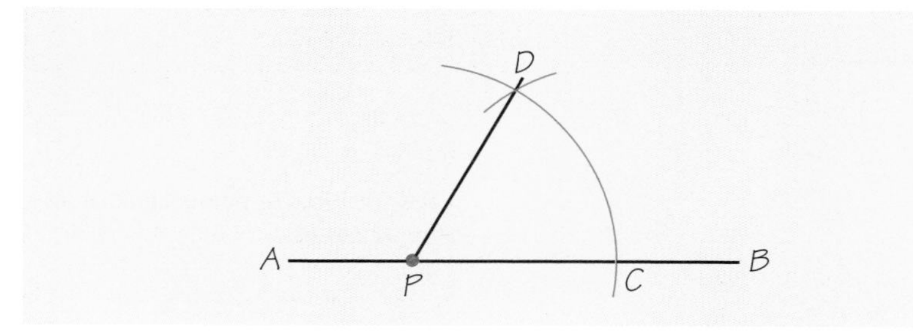

Join DP. Angle DPC is 60°.

Can you see that $\triangle DPC$ is equilateral?

EXERCISE 16D

Draw six line segments AB.

Mark point P on each line segment.

Construct a 60° angle at P for each line.

16.5 Understanding and using loci

The line *CD* is the perpendicular bisector of the line *AB*.

All the points on the line *CD* are exactly the same distance from *A* as they are from *B*. Some of them are shown on the diagram.

CD is an example of a **locus** (plural **loci**).

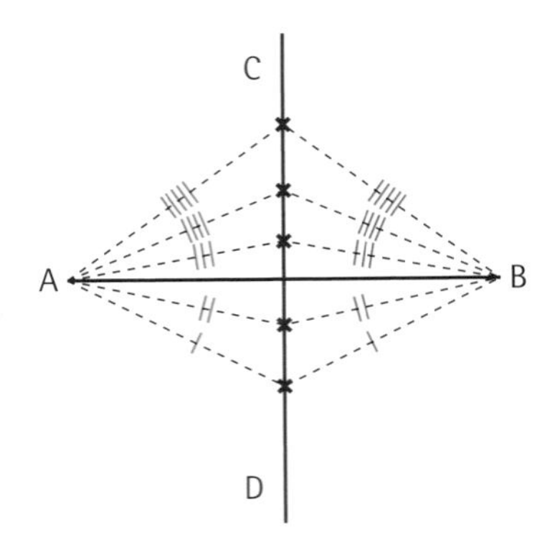

To construct the locus, see Example 2.

A locus is a set of points that obey a given rule.

For *CD* the rule is 'all points equidistant from *A* and *B*'.
All the points on line *CD* obey this rule.

Equidistant means 'the same distance from'. You will often see it in locus questions.

All the points to the left of *CD* are nearer to *A* than to *B*.
All the points to the right of *CD* are nearer to *B* than to *A*.

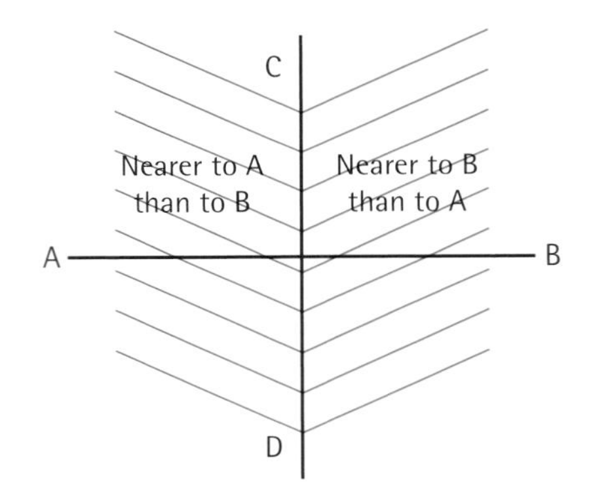

CD is the locus of points equidistant from *A* and *B*.

The locus of points equidistant from two fixed points is the perpendicular bisector of the two fixed points

EXAMPLE 5

What is the locus of points which are always 2 cm from a fixed point *P*?

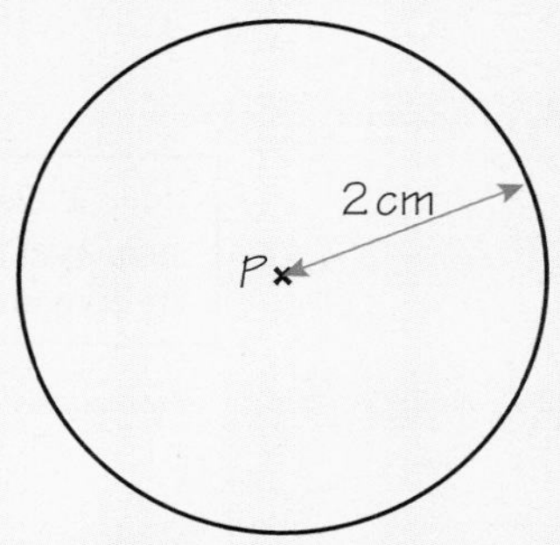

Circle, centre *P*, radius 2 cm.

All points on this circle obey this rule.

All points *inside* the circle are less than 2 cm from *P*. All points *outside* the circle are more than 2 cm from *P*.

The locus of points which are a fixed distance from a fixed point is a circle.

EXAMPLE 6

What is the locus of points which are equidistant from the two lines *AB* and *AC*?

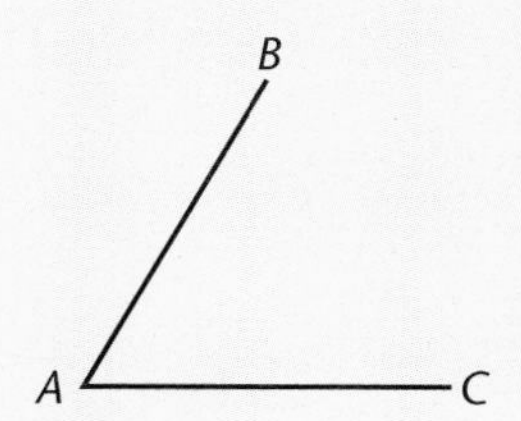

The locus of points which are equidistant from two fixed lines is the angle bisector of the two fixed lines.

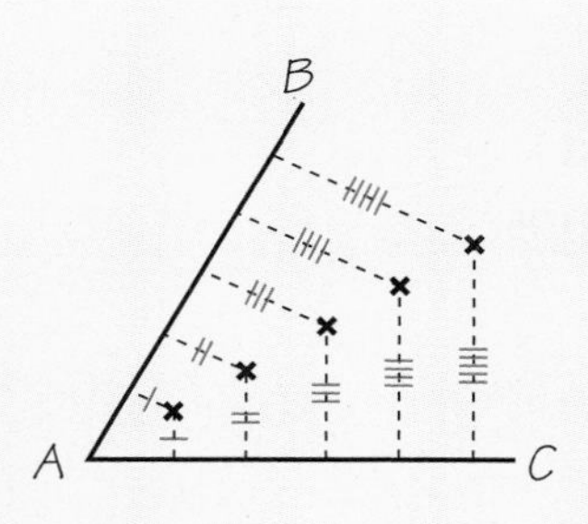

To construct the locus follow the steps in Example 3.

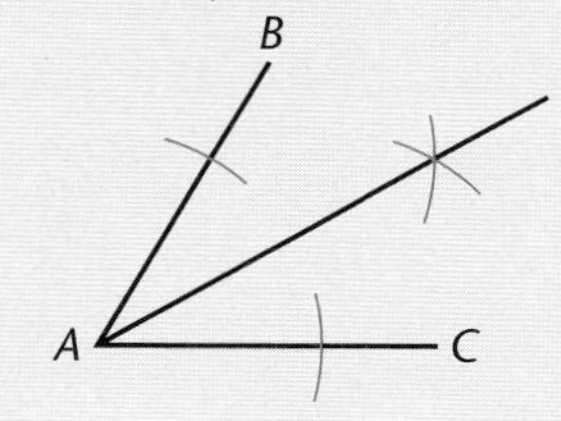

EXAMPLE 7

Construct the locus of points which are exactly 2 cm from the fixed line segment AB.

A ——————— B

Using a ruler, draw 2 lines, each 3 cm long, 2 cm from AB. Make sure these are accurate.

A ——————— B

Now join the two lines with semi-circles.

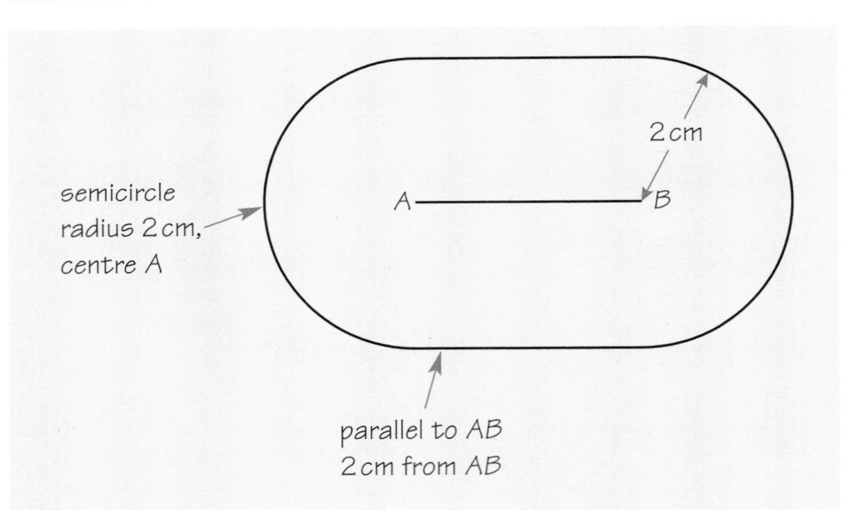

This shows the locus you need to construct. The locus will be an athletics track shape around the outside of the line segment AB.

The locus of points which are a fixed distance from a line segment is an athletics track shape. The shape has two lines parallel to AB and two semicircular ends.

For locus questions
- think about which constructions you will need
- construct the locus using standard constructions.

EXAMPLE 8

In $\triangle ABC$, $AB = 4$ cm, $AC = 6$ cm and $BC = 5$ cm. Shade the region inside the triangle where the points are

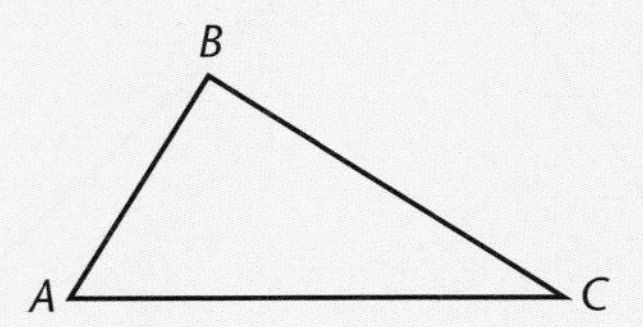

Construct the diagram accurately.

(a) less than 3 cm from B **(b)** nearer to C than A.

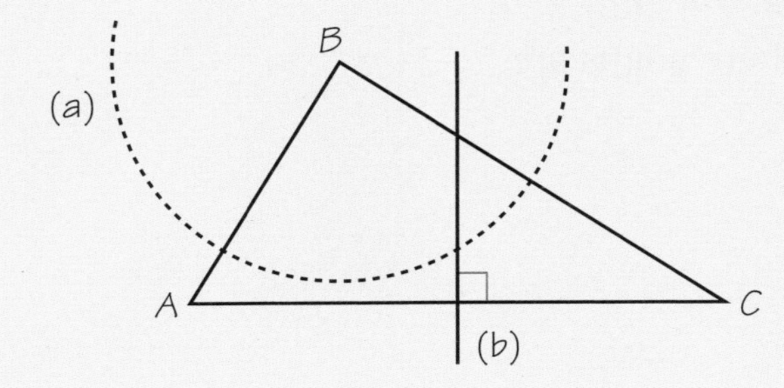

(a) Points less than 3 cm from *B* are in a circle.
(b) Points nearer to *C* than *A* are to the right of the perpendicular bisector of *AC*.

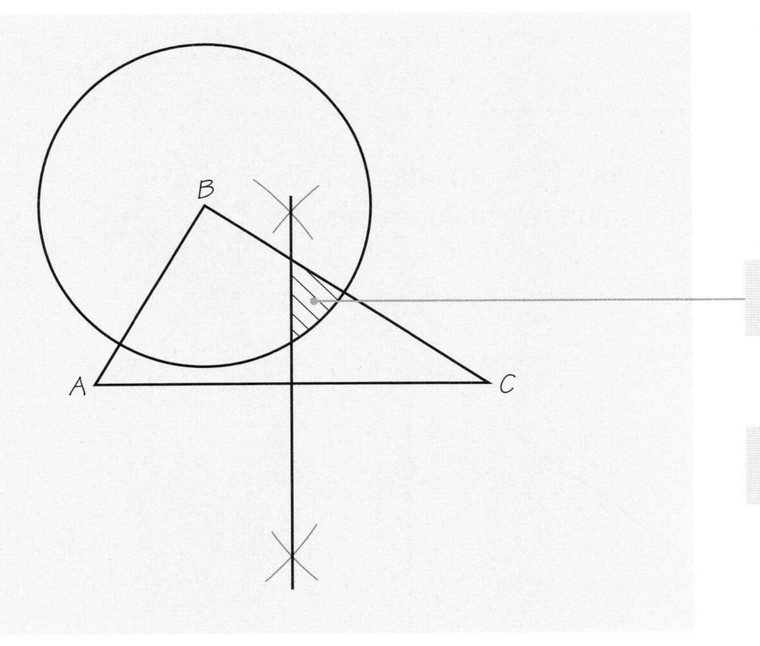

The points that satisfy **(a)** and **(b)** are in this region.

For constructing a perpendicular bisector see Example 2.

EXERCISE 16E

Use accurate constructions to answer these questions.

B

A C

1 In $\triangle ABC$, $AB = 5$ cm, $AC = 7$ cm and $BC = 6$ cm.
Copy the triangle and shade the region where points are

(a) less than 4 cm from C

(b) nearer to A than B.

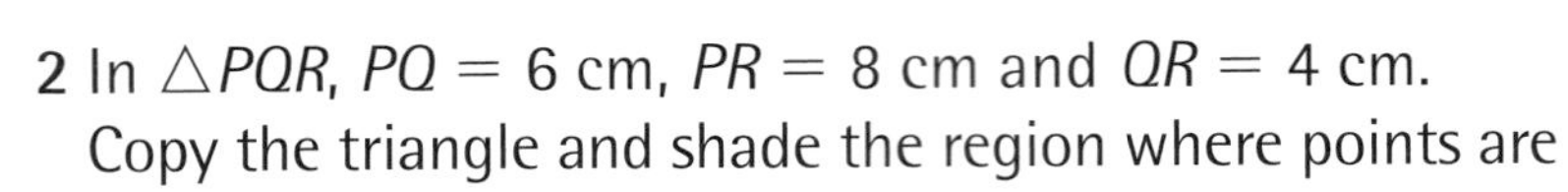

2 In $\triangle PQR$, $PQ = 6$ cm, $PR = 8$ cm and $QR = 4$ cm.
Copy the triangle and shade the region where points are

(a) nearer to R than P

(b) closer to PR than QR.

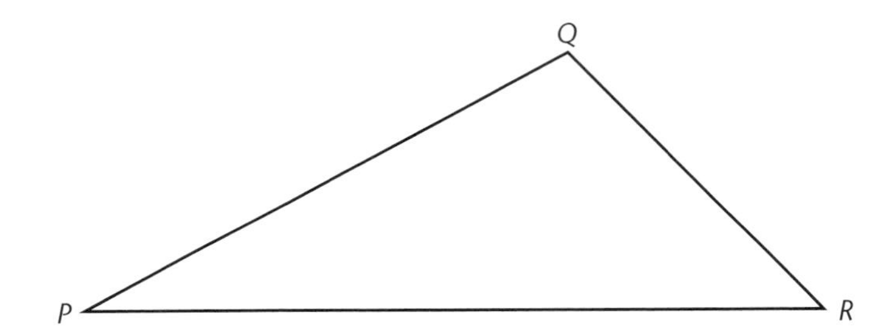

3 In $\triangle XYZ$, $XY = 6$ cm, $XZ = 8$ cm and $YZ = 10$ cm.
Copy the triangle and shade the region where points are

(a) closer to YX than YZ

(b) more than 7 cm from Z.

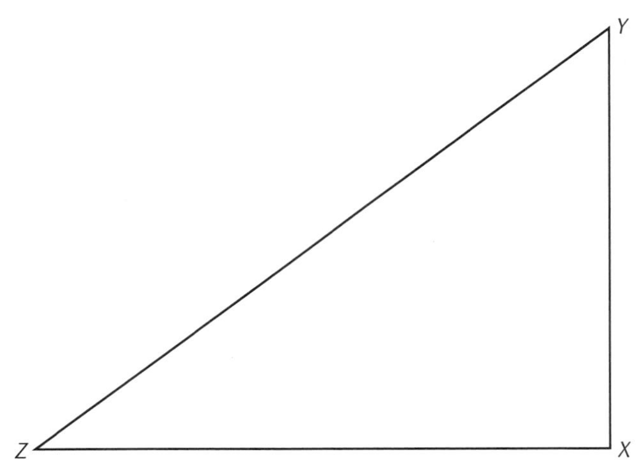

16.6 Using bearings and scale

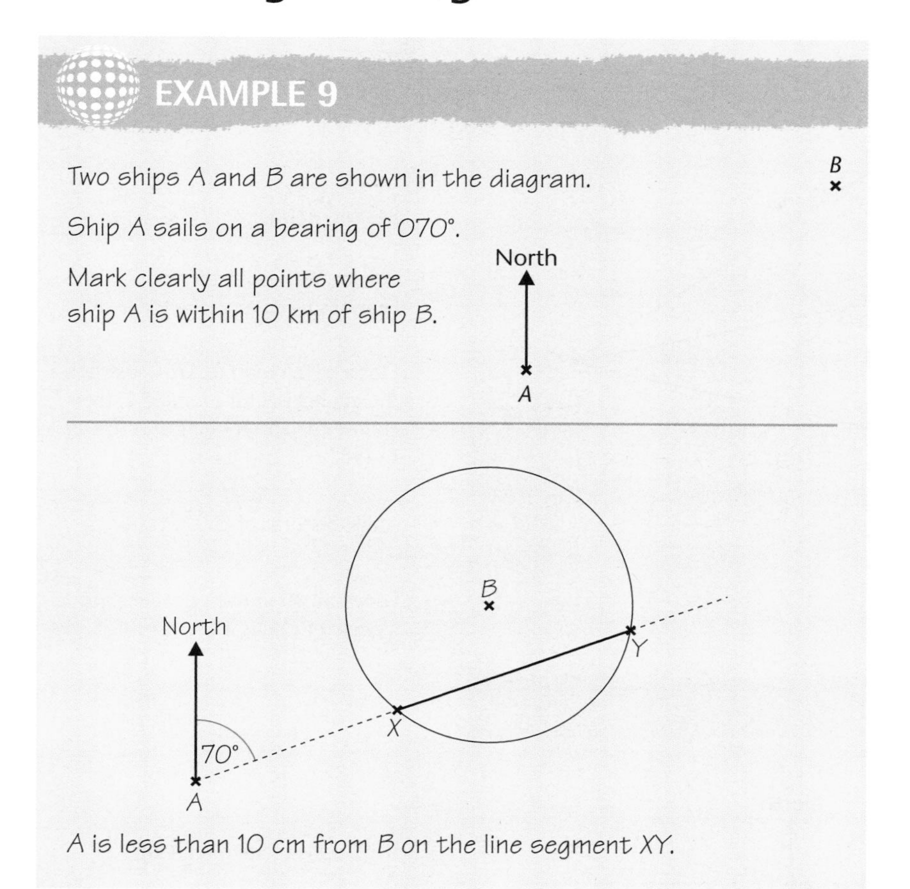

EXAMPLE 9

Two ships *A* and *B* are shown in the diagram.

Ship *A* sails on a bearing of 070°.

Mark clearly all points where ship *A* is within 10 km of ship *B*.

> Look back at bearings in Section 6.2 if you need some help.

> Use a scale of 1 cm for 5 km.

> Draw in the line that *A* sails on, using a protractor for the angle.

> The locus of points 10 km from *B* is a circle, centre *B*. On the diagram using the scale 1 cm for 5 km, this is a circle with radius 2 cm.

A is less than 10 cm from *B* on the line segment *XY*.

EXAMPLE 10

The diagram shows a triangular field *ABC* where *AB* = 60 m, *AC* = 80 m and *BC* = 70 m. Some treasure is buried in the field.

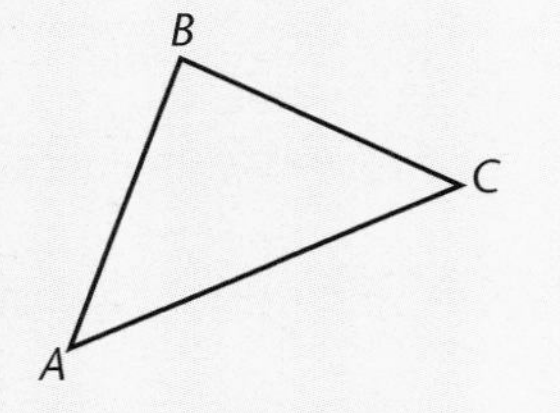

(a) It is between bearings of 040° and 060° from *A*.

(b) It is closer to *CA* than to *CB*.

(c) It is less than 40 m from *B*.

Using a scale of 1 cm for 10 m, make accurate constructions to find the area in which the treasure lies. Shade the region.

Continued ▼

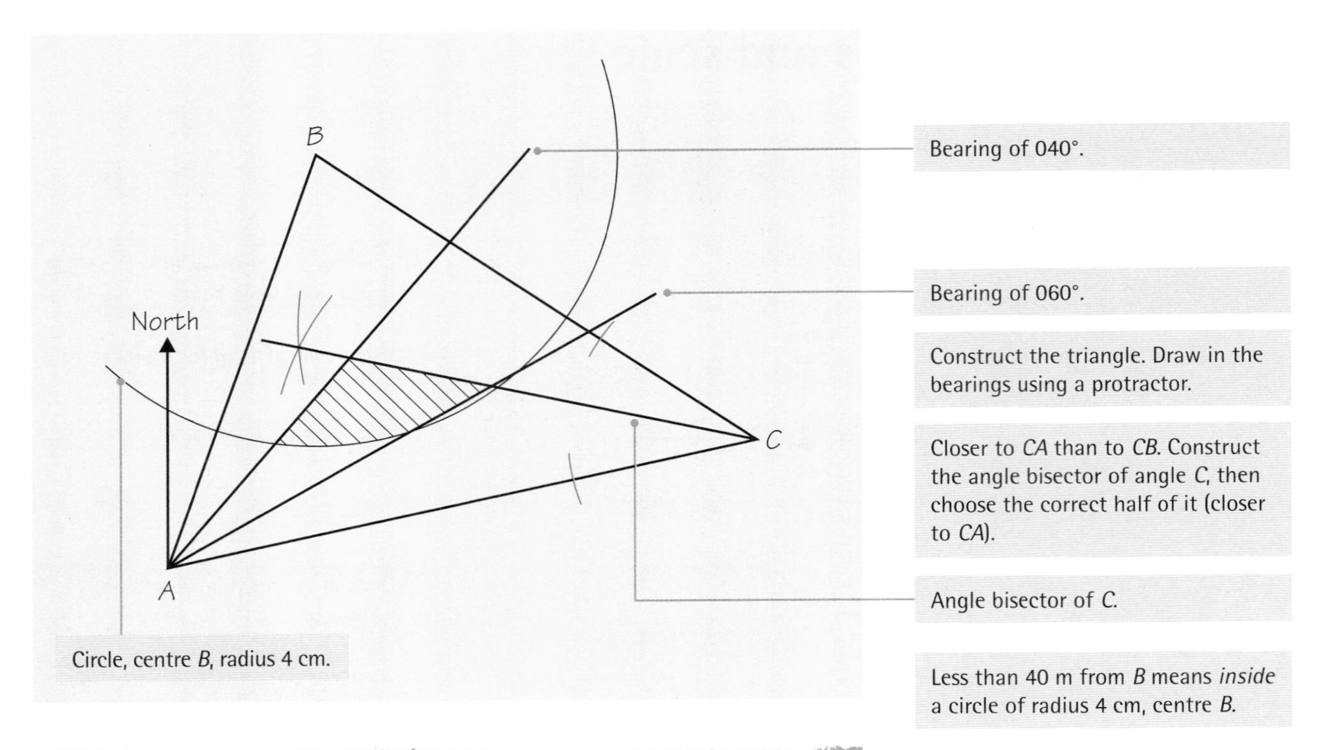

Less than 40 m from *B* means *inside* a circle of radius 4 cm, centre *B*.

EXERCISE 16F

1 Two ships *A* and *B* are shown in the diagram. Ship *A* sails on a bearing of 110° while ship B stays stationary. Copy the diagram and draw the locus of all points where ship *A* is within 12 km of ship *B*.

B

Use a scale of 1 cm for 4 km.

2 Main roads AB, AC and BC connect towns A, B and C.
$AB = 16$ km, $AC = 14$km and $BC = 12$km.
A new leisure centre is to be built so that it is,

(a) closer to road AC than it is to road AB

(b) between 8km and 10km from B.

Use a scale of 1 cm for 2km.

Copy the diagram and construct the region where the leisure centre will be built.

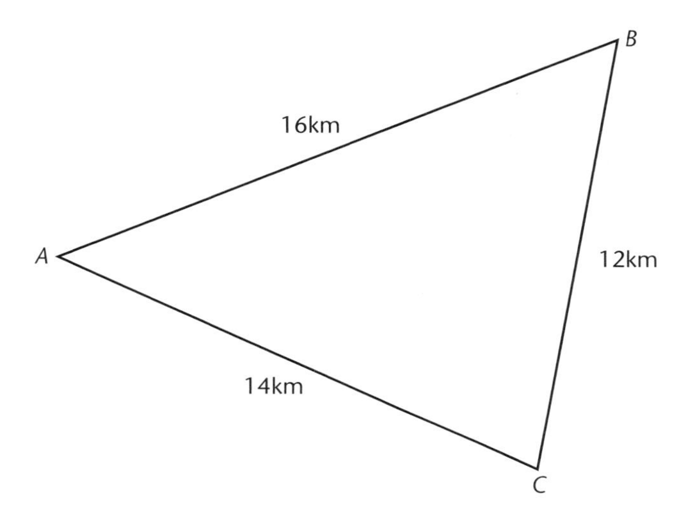

3 A, B and C are places on Treasure Island.
$AB = 200$ m, $AC = 160$ m and $BC = 120$ m.
The hidden treasure is,

(a) nearer to C than B

(b) closer to AB than to AC

(c) less than 70 m from C.

Using a scale of 1 cm for 20 m, copy the diagram and construct the region where the treasure is hidden.

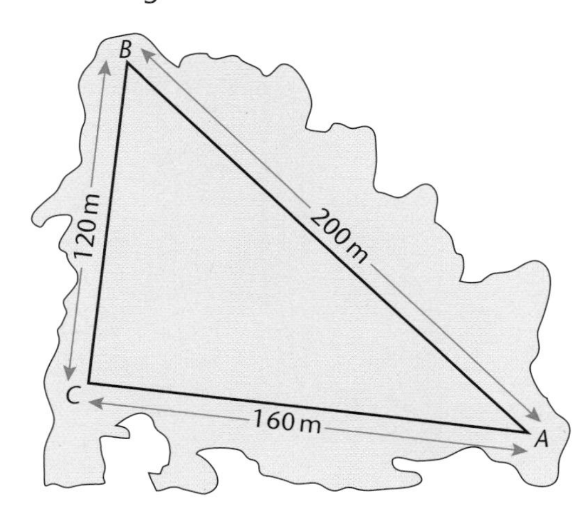

EXAMINATION QUESTIONS

1 **(a)** Construct **accurately** a parallelogram $ABCD$ which has sides $AB = 9$ cm and $BC = 7$ cm, and the diagonal $AC = 13$ cm. [3]

(b) (i) Using a straight edge and compasses only, construct the locus of points which are equidistant from A and B. [2]

(ii) By taking a suitable measurement, calculate the area of the parallelogram, to the nearest square centimetre. [2]

(c) (i) Using a straight edge and compasses only, construct the bisector of angle BAD. [2]

(ii) Copy and complete the following sentence.
The bisector of angle BAD is the locus of points which are … [2]

(d) Shade the region inside the parallelogram which contains all points which are nearer to A and more than 7 cm from B. [2]

(CIE Paper 3, Jun 2000)

---

2 Using a straight edge and compasses only, copy and construct accurately the locus of the points which are equidistant from the two points A and B. [2]

B
×

×
A

(CIE Paper 1, Jun 2001)

3

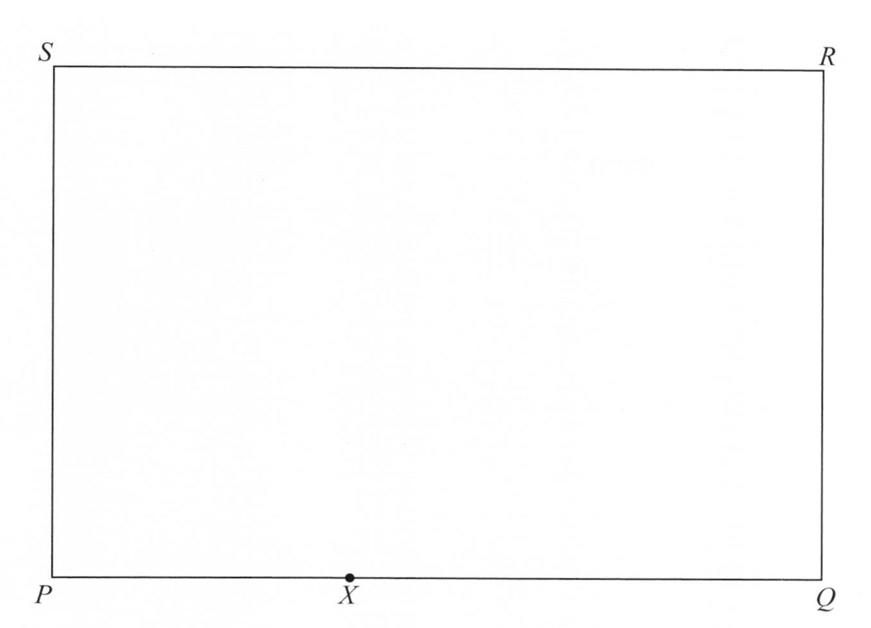

Copy the diagram accurately. Construct, using a straight edge and compasses only, the locus of points **inside** the rectangle which are

(i) 3cm from P, [1]
(ii) equidistant from X and S, [3]
(iii) equidistant from PQ and QR. [3]

(CIE Paper 3, Nov 2001)

4

Copy the diagram accurately.

(a) On your diagram draw accurately the locus of points inside the rectangle which are
 (i) 6 cm from D, [1]
 (ii) equidistant from AB and BC. [2]

(b) Shade the region inside the rectangle containing the points which are more than 6 cm from D and nearer to AB than to BC. [1]

(CIE Paper 1, Jun 2002)

5

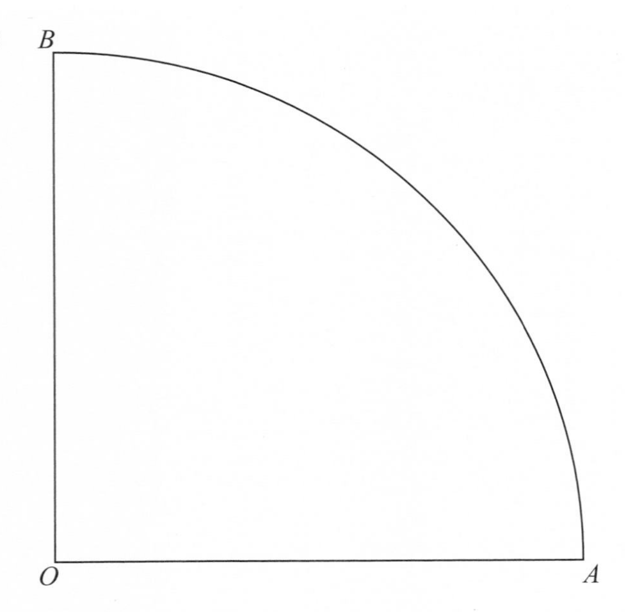

The quarter-circle above has centre O and radius 7 cm.
Copy the diagram accurately.

(a) Using a straight edge and compasses only, construct

(i) the perpendicular bisector of AO, [2]

(ii) the locus of points inside the quarter-circle which are 5 cm from O. [2]

(b) Shade the region, inside the quarter-circle, containing the points which are more than 5 cm from O and nearer to A than O. [1]

(c) The line OX bisects angle AOB and is 12 cm long.
Draw OX accurately. [2]

(CIE Paper 3, Jun 2003)

Chapter 17

Co-ordinates and graphs

This chapter will show you how to

- ✔ write co-ordinates in all four quadrants
- ✔ plot and draw straight-line graphs and draw quadratic curves
- ✔ work out co-ordinates of points of intersection when two graphs intersect
- ✔ plot conversion graphs
- ✔ interpret and use distance–time graphs

17.1 Co-ordinates and line segments

Co-ordinates in all four quadrants

Co-ordinates describe the position of a point on a grid. In the diagram point A has co-ordinates (3, 4). The first value (3) gives the number of units left or right from the **origin** (O). The second value (4) gives the number of units up or down from the origin.

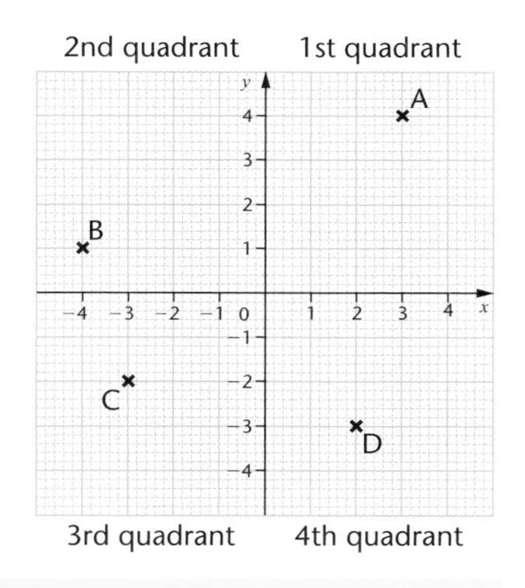

Values to the right and up from the origin are positive. Values to the left and down are negative.

Co-ordinates are always given as (x, y).
The grid is called a **co-ordinate grid**.

The four quarters of the grid are called **quadrants**.

$A(3, 4)$ is in the 1st quadrant.
$B(-4, 1)$ is in the 2nd quadrant.
$C(-3, -2)$ is in the 3rd quadrant.
$D(2, -3)$ is in the 4th quadrant.

Left/right first ... x
Up/down second ... y

x comes before y in the alphabet.

'Quad' means four. Think of quad bikes, quadrilateral.

EXAMPLE 1

Write down the co-ordinates of the points marked *A*, *B*, *C*, *D* and *E* on the x-y grid.

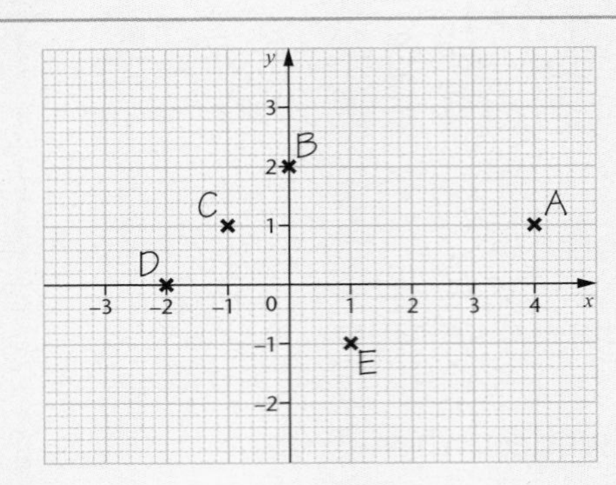

A(4, 1) B(0, 2) C(−1, 1) D(−2, 0) E(1, −1)

Write the x-value then the y-value in brackets, with a comma between.

EXAMPLE 2

(a) Draw a co-ordinate grid with a horizontal axis (x-axis) from −4 to 4 and a vertical axis (y-axis) from −6 to 6.

(b) Plot these points on the grid.
A(4,−3), B(−3,−3), C(−3,1) and D(4, 6).

(c) Join the points *ABCD* in order with straight lines.
What shape have you drawn?

Use a ruler to join the points.

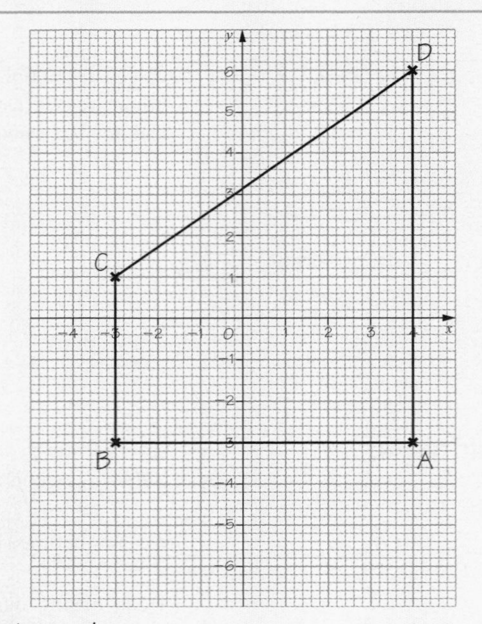

A trapezium is a four-sided shape with one pair of parallel sides.

The shape is a trapezium.

EXERCISE 17A

1 Write down the co-ordinates of the points P, Q, R, S and T.

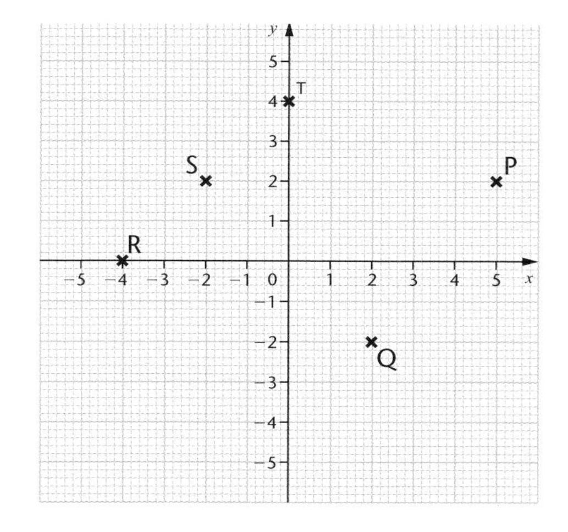

2 (a) On squared paper draw a co-ordinate grid with an x-axis from 0 to 10 and a y-axis from 0 to 10.

(b) Plot these points on the grid
$A(1, 4)$, $B(4, 9)$, $C(8, 9)$, $D(8, 3)$ and $E(5, 0)$.
Join them up in order.
What shape have you drawn?

3 For each line segment write down the co-ordinates of the end points.

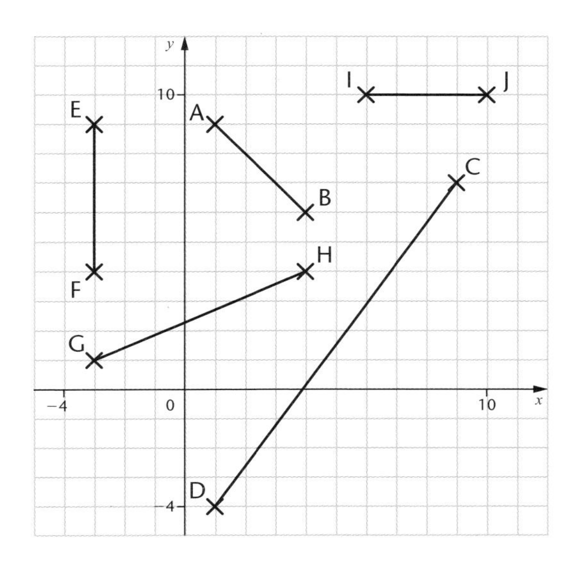

17.2 Plotting straight-line graphs

When points lie in a straight line on a grid, it means that there is a connection between the x-value and the y-value of their co-ordinates.

In a straight-line graph there is a linear relationship between x and y.

Plotting the points may help you to find the relationship between x and y.

You can describe this relationship using a **linear equation**.

Lines parallel to the x-axis or y-axis

The graph shows four points in a straight line A, B, C and D.

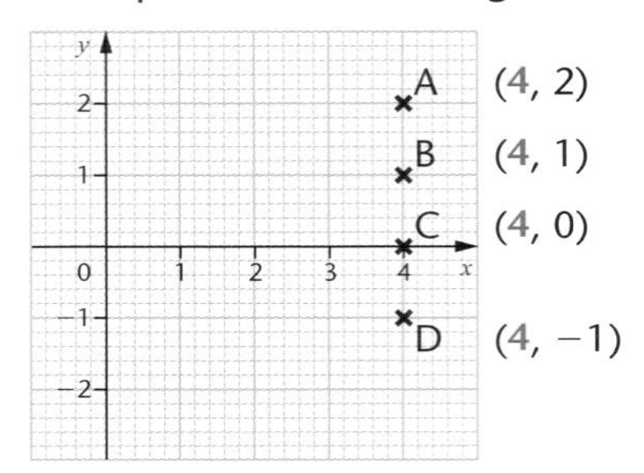

The points (4, 100) and (4, −100) also lie on this line. Any point with x-value 4 lies on the line.

Every point has exactly the same x-value. The relationship between this set of points is the equation of the line $x = 4$.

A line parallel to the y-axis has equation $x = a$, where a is a number.

The four points E(2, −2), F(1, −2), G(0, −2), H(−1, −2) all have the same y-value. The relationship is the equation of the line, $y = -2$.

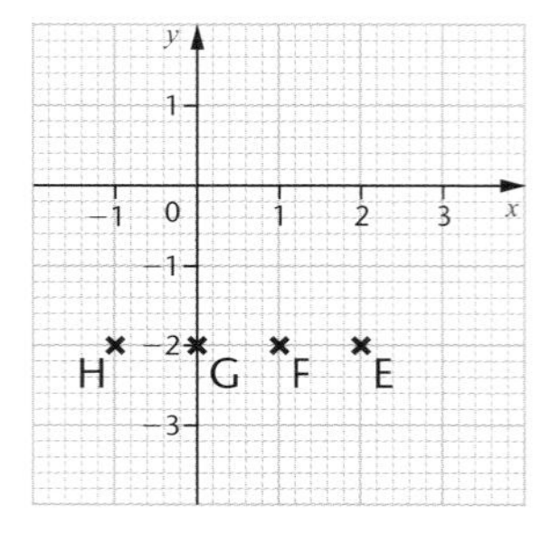

The equation of the x-axis is $y = 0$.
The equation of the y-axis is $x = 0$.

A line parallel to the x-axis has equation $y = b$, where b is a number.

EXAMPLE 3

Write down the equations of these lines.

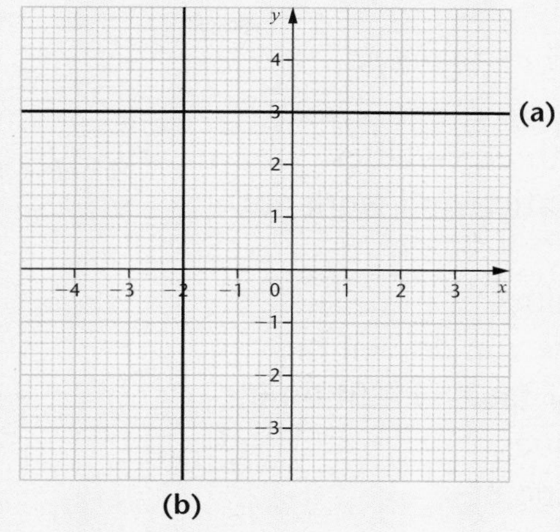

(a) All the points on the line have y–co-ordinate 3, so $y = 3$.

(b) All the points on the line have x–co-ordinate -2, so $x = -2$.

(0, 3), (−2, 3), (7, 3),
(−2, 4), (−2, 0), (−2, −1)

EXERCISE 17B

1 Write down the equations of the lines

(i) parallel to the x-axis

(ii) parallel to the y-axis.

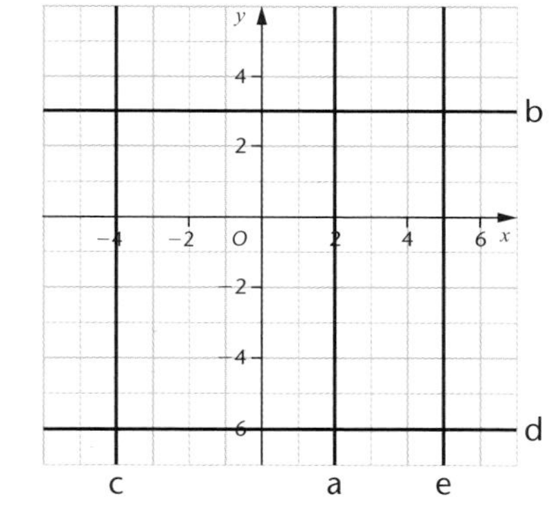

2 Draw a co-ordinate grid with both the x-axis and y-axis from -5 to 5.
On your grid, draw and label the lines

(a) $x = 4$ (b) $y = -1$ (c) $x = -5$ (d) $y = 4$.

More straight-line graphs

For lines that are not parallel to the x-axis or the y-axis, follow these steps to draw the graph.

1. Choose a minimum of three values for x (always include 0).
2. Draw a table of values and write in three x-values.
3. Substitute these values into the equation and work out the corresponding values for y.
4. Write the y-values in the table of values.
5. Draw a co-ordinate grid, making sure you draw it big enough so that all the points in your table will fit on it.
6. Plot the points from the table of values.
7. Draw a straight line through all the points.
8. Label the line with its equation.

Pair up each x-value with its y-value to give co-ordinates (x, y).

EXAMPLE 4

On a co-ordinate grid, using the same scale on both axes, draw x- and y-axes between -5 and $+5$.

Draw the graph of $y = x + 1$ for values of x from -3 to $+3$.

x	-3	0	3
y	-2	1	4

Substitute $x = 3$ into
$y = x + 1$
$y = 3 + 1 = 4$.

Draw a table of values, choosing values of x from -3 to $+3$.

Use the equation of the line to work out the values of y. Plot the x- and y-values as co-ordinate pairs on the grid. Join them with a straight line.

Always extend your line beyond the end points.

Label your axes and label the line with the equation.

Plot the points $(-3, -2)$, $(0, 1)$ and $(3, 4)$.

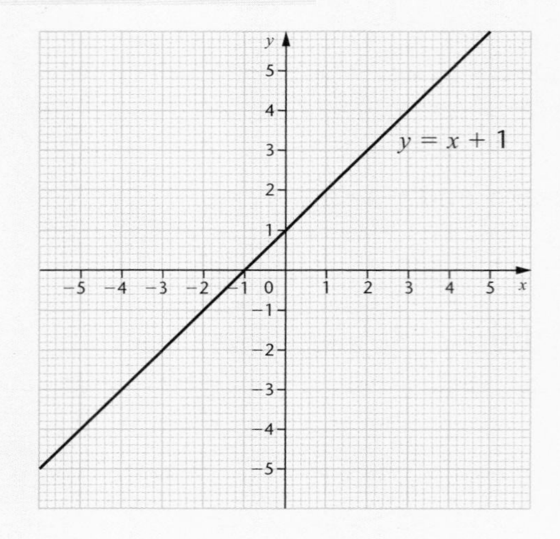

You need a minimum of three points to draw a straight-line graph.

Two points give the line – the third point is a 'check'.

EXERCISE 17C

1 (a) Copy and complete the table of values for $y = x + 3$.

x	−3	0	3
y		3	

(b) Using the same scale on both axes, draw a co-ordinate grid with the x-axis labelled from -3 to $+3$ and the y-axis labelled from -6 to $+6$.

(c) On your co-ordinate grid, draw the graph of $y = x + 3$.

2 (a) Copy and complete the table of values for $y = 2x + 1$.

x	−3	0	3
y		1	

(b) Using the same scale on both axes, draw a co-ordinate grid with the x-axis labelled from -3 to $+3$ and the y-axis labelled from -5 to $+7$.

(c) On your co-ordinate grid, draw the graph of $y = 2x + 1$.

3 In parts **(a)** to **(f)**, use the same scale on both axes and draw x-axis and y-axis between -5 and $+5$.

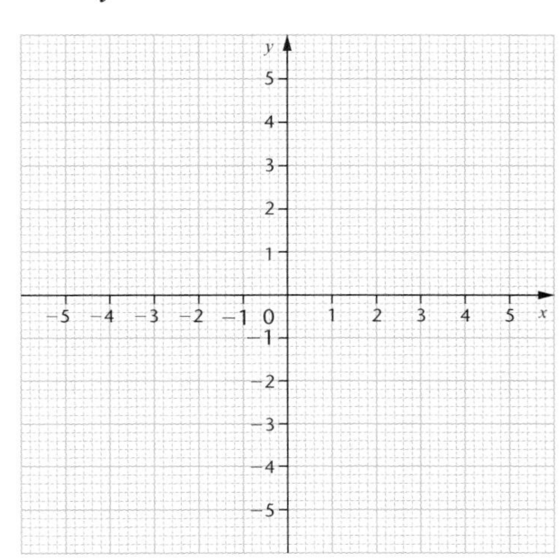

Draw these straight-line graphs. For each one

- Make a table of values for at least three values of x (including 0).
- Substitute the x-values into the equation of the line to work out the corresponding y-values.
- Plot the points and draw a straight line through them.

(a) $y = x - 3$ (b) $y = 2x + 3$

(c) $y = 4 - x$ (d) $y = 3x - 1$

(e) $y = 2 - 2x$ (f) $y = 1 - \frac{1}{2}x$

4 Using the same scale on both axes, draw x-axis and y-axis from -4 to $+4$.

(a) Draw the graphs of $y = \frac{1}{2}x + 1$ and $y = -2x + 1$ on the same co-ordinate grid.

(b) Write down the co-ordinates of the point where each of these lines crosses the

(i) y-axis (ii) x-axis.

5 (a) Copy and complete the table of values for $y = 5 - 2x$.

x	−2	0	4
y		5	

(b) On a co-ordinate grid with the x-axis labelled from -2 to $+4$ and the y-axis labelled from -3 to $+9$, draw the graph of $y = 5 - 2x$.

(c) Use your graph to find the value of x when $y = 6$.

6 Using the same scale on both axes, draw the graphs of $y = 2x$ and $y = 2x - 1$ on the same co-ordinate axes for values of x from -4 to $+4$.

What do you notice about these graphs?

7 Using the same scale on both axes, draw x-axis and y-axis from -3 to $+3$.
Draw the graphs of $y = x$ and $y = -x$ on the same co-ordinate grid.
What do you notice about these graphs?

8 Using the same scale on both axes, draw x-axis and y-axis from -5 to $+5$.
Draw the graphs of $y = 2x - 1$ and $y = \frac{1}{2}x + 2$ on the same co-ordinate grid.
Write down the co-ordinates of the point where these graphs cross each other.

17.3 Equations of straight-line graphs

Calculating the gradient

In Example 4 the **slope** of the line is from left to right in an upward direction. For every 1 unit moved to the right, the line rises by 1 unit.

The fraction $\dfrac{\text{vertical distance}}{\text{horizontal distance}}$

gives a measure of the **steepness** of the slope of the line and is called the **gradient** of the line.

Positive gradient

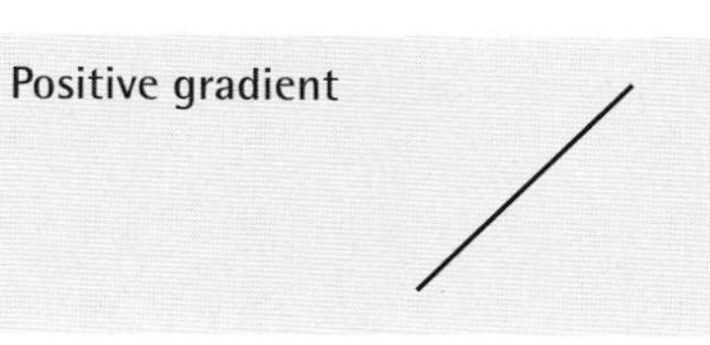

Negative gradient

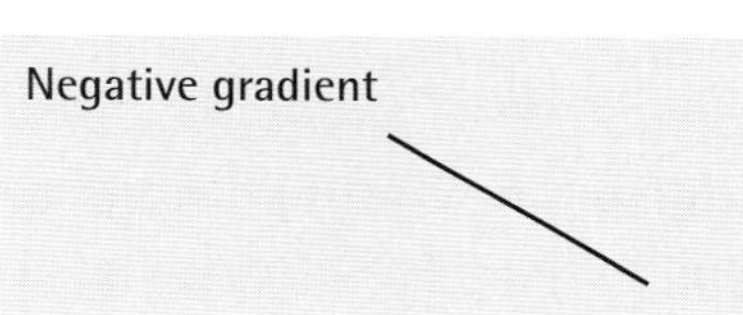

$$\text{Gradient} = \frac{\text{vertical distance}}{\text{horizontal distance}}$$

If the line slopes downward, the gradient is negative.

EXAMPLE 5

Find the gradient of the line joining the points

(a) (1, 2) and (5, 10) **(b)** (21, 6) and (2, 23)

(c) (24, 21) and (0, 22).

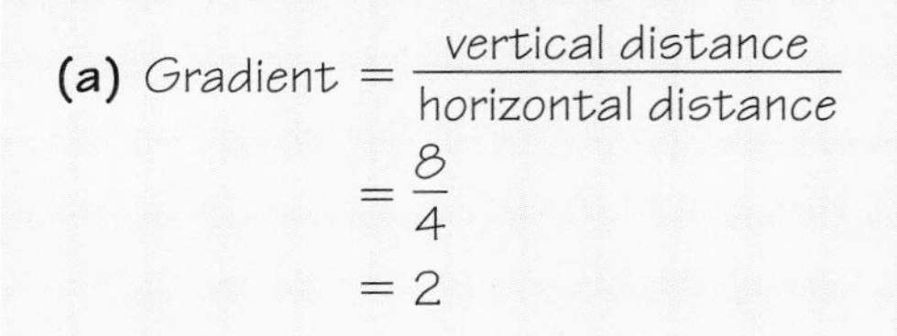

(a) $\text{Gradient} = \dfrac{\text{vertical distance}}{\text{horizontal distance}}$

$= \dfrac{8}{4}$

$= 2$

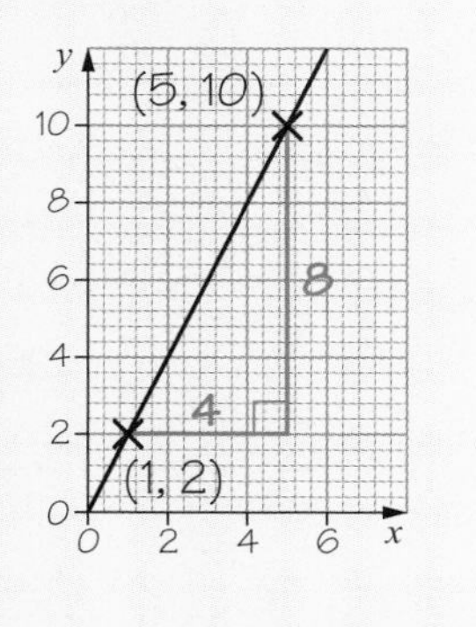

Plot the points and join them with a line.

Draw a right-angled triangle as shown.

Use gradient $= \dfrac{\text{vertical distance}}{\text{horizontal distance}}$

Gradient 2 means for every 1 you go across, the line goes 2 up.

continued ▼

(b) Gradient $= \frac{\text{vertical distance}}{\text{horizontal distance}}$

$= -\frac{9}{3}$

$= -3$

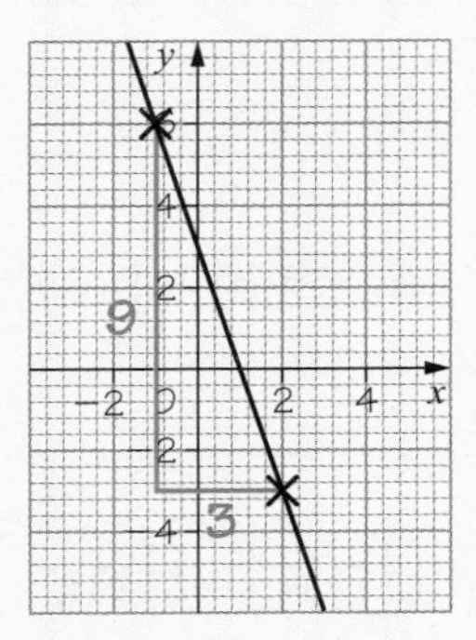

The line slopes downward, so the gradient is negative.

For every 1 you go across, the line goes down 3.

(c) Gradient $= \frac{\text{vertical distance}}{\text{horizontal distance}}$

$= -\frac{1}{4}$

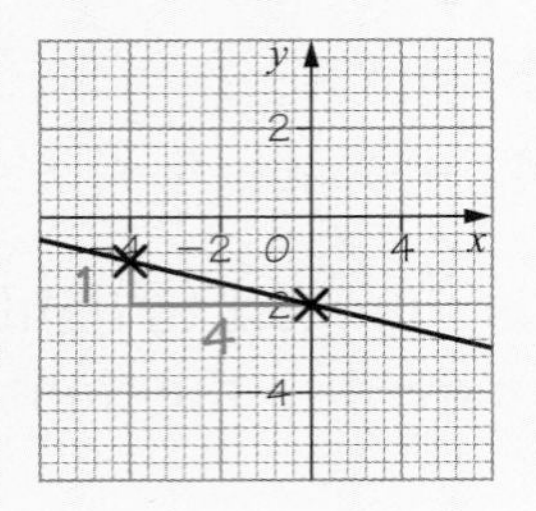

For every 1 you go across, the line goes down $\frac{1}{4}$.
Or, for every 4 you go across, the line goes 1 down.

You can work out the gradient of a line from its graph.

EXAMPLE 6

Find the gradient of these lines

(a) $y = 3x - 4$ (b) $y = -\frac{1}{2}x + 2$

(a) $y = 3x - 4$

x	0	1	2
y	−4	−1	2

Gradient is positive.

Gradient $= \frac{6}{2}$

$= 3$

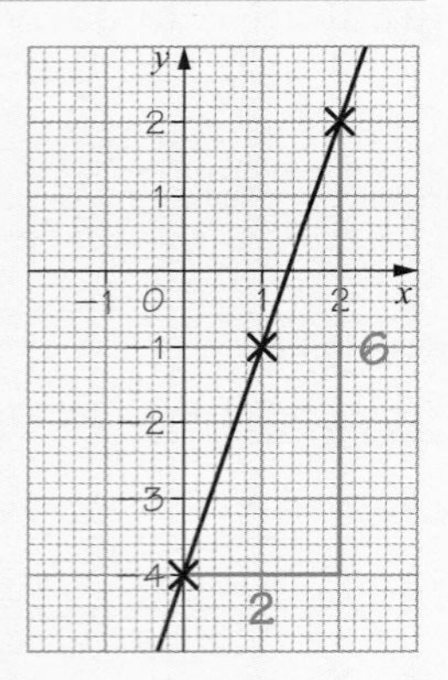

First draw the graph. Calculate and plot at least three (x, y) values.

To find the gradient, choose *any* two points on the line and draw a right-angled triangle (as shown).

Remember, $3 = \frac{3}{1}$. For every 1 across, the line goes 3 up.

(b) $y = -\frac{1}{2}x + 2$

x	−2	0	2
y	3	2	1

Gradient is negative.

Gradient $= -\frac{2}{4}$

$= -\frac{1}{2}$

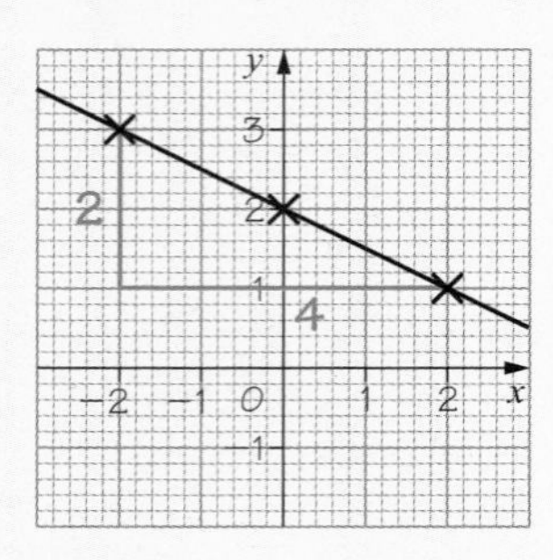

EXERCISE 17D

Example 5 will help you.

1 In each part

- plot the points
- join them with a straight line
- find the gradient of the line.

(a) (2, 8) and (5, 14) (b) (1, 1) and (5, 6)
(c) (2, 2) and (3, 5) (d) (6, 0) and (7, 9)
(e) (5, 5) and (8, 8) (f) (2, 0) and (0, 8)
(g) (3, 1) and (0, 7) (h) (−4, −1) and (6, 4)
(i) (−1, 5) and (−4, 5) (j) (−1, 3) and (3, −1)

Example 6 will help you.

2 In each part

- make a table of values
- plot the points
- draw the line
- find the gradient of the line.

(a) $y = 2x - 1$ (b) $y = x + 3$
(c) $y = 3x + 1$ (d) $y = \frac{1}{2}x - 4$
(e) $y = 2 - x$ (f) $y = 6 - 2x$

Finding the gradient from the equation

It is a very slow process to draw the graph and then find the gradient.

If we look at the gradients for some lines, we can see a pattern.

Equation	Gradient
$y = 4x + 6$	4
$y = 3x + 5$	3
$y = 2x$	2
$y = x - 4$	1
$y = -x + 9$	−1
$y = -2x - 5$	−2
$y = -3x + 6$	−3

This is the same as $y = 1x - 4$.

The number in front of x is called the **coefficient** of x.

The gradient is the same as the number in front of the x term. Notice that all the equations begin with $y =$.

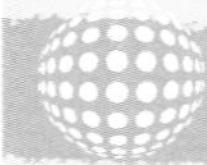

EXAMPLE 7

Write down the gradients of these lines.

(a) $y = x + 8$ (b) $y = -3x + 3$

(c) $y = -\frac{1}{2}x - 5$ (d) $y = 2.6x + 0.4$.

(a) gradient = 1 (b) gradient = −3

(c) gradient = $-\frac{1}{2}$ (d) gradient = 2.6

Gradients can also be fractions or decimals.

When the equation does not start with $y =$ it must be rearranged.

EXAMPLE 8

Find the gradients of these lines.

(a) $y - 3x = 1$

(b) $2x + y = 9$

(c) $2x + 4y - 5 = 0$

(a)
$$y - 3x = 1$$
$$y - 3x + 3x = 1 + 3x$$
$$y = 1 + 3x$$
gradient = 3

(b)
$$2x + y = 9$$
$$2x - 2x + y = 9 - 2x$$
$$y = 9 - 2x$$
gradient = −2

(c)
$$2x + 4y - 5 = 0$$
$$2x + 4y - 5 + 5 = 0 + 5$$
$$2x + 4y = 5$$
$$2x - 2x + 4y = 5 - 2x$$
$$4y = 5 - 2x$$
$$y = \frac{5}{4} - \frac{2}{4}x$$
$$y = \frac{5}{4} - \frac{1}{2}x$$
gradient = $-\frac{1}{2}$

Gradient of parallel lines

The graph shows the lines $y = 2x$, $y = 2x + 2$ and $y = 2x + 4$.

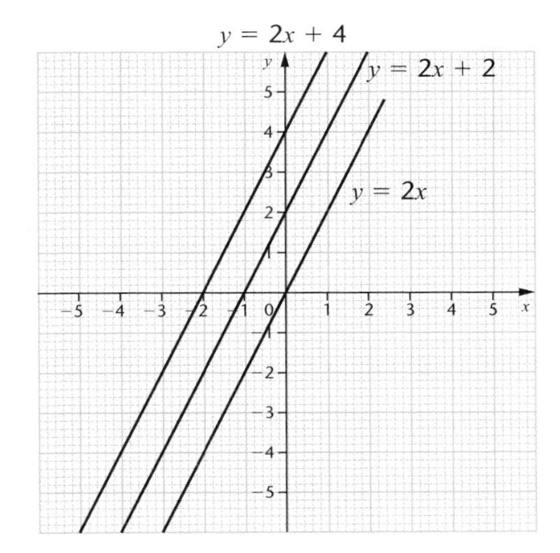

The lines are parallel.
You can find the gradients from the equations.

Equation	Gradient
$y = 2x$	2
$y = 2x + 2$	2
$y = 2x + 4$	2

Straight lines that are parallel have the same gradient.

When the equations are in the form $y =$ _____, lines are parallel if they have the same coefficient of x.

EXERCISE 17E

1 Write down the gradient of each of these lines.

(a) $y = 3x + 2$ (b) $y = 4x + 3$ (c) $y = 7x - 4$
(d) $y = -3x + 2$ (e) $y = -2x + 8$ (f) $y = -x + 4$
(g) $y = \frac{1}{2}x - 7$ (h) $y = \frac{1}{4}x + 3$ (i) $y = -\frac{1}{10}x$
(j) $y = 1.6x - 4$ (k) $y = 0.1x - 5$ (l) $y = 3.1x - 2$
(m) $y = x$ (n) $y = 4$ (o) $y = -5$.

2 Find the gradient of these lines.

(a) $x + y = 9$ (b) $2x + y = 7$ (c) $y - 3x = 1$
(d) $\frac{1}{2}x + y = 1$ (e) $x - y = 2$ (f) $x + 2y = 6$
(g) $x + y + 1 = 0$
(h) $x + 2y - 5 = 0$
(i) $3x + 4y - 8 = 0$

3 Which of the following lines are parallel to $y = x - 1$?

(a) $y = x + 1$ (b) $y = x - 7$

(c) $y = 2x - 1$ (d) $2y = x - 1$

(e) $y = x + \frac{1}{2}$ (f) $2y - 3 = 2x$

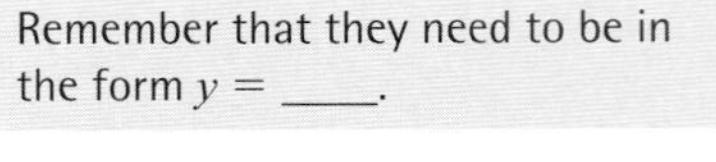

Remember that they need to be in the form $y =$ ____.

4 Which of the following lines are parallel to $y = -3x - 2$?

(a) $y = 3x - 2$ (b) $y = -3x + 2$

(c) $y = -3x + 2$ (d) $y = -2 - 3x$

(e) $y = 2 + 3x$ (f) $x = -3y - 2$

The y-intercept

Another important feature of a straight line is where it crosses the y-axis. This is called the **y-intercept**.

The lines in the diagram opposite all have gradient 2 but they cross the y-axis at different points.

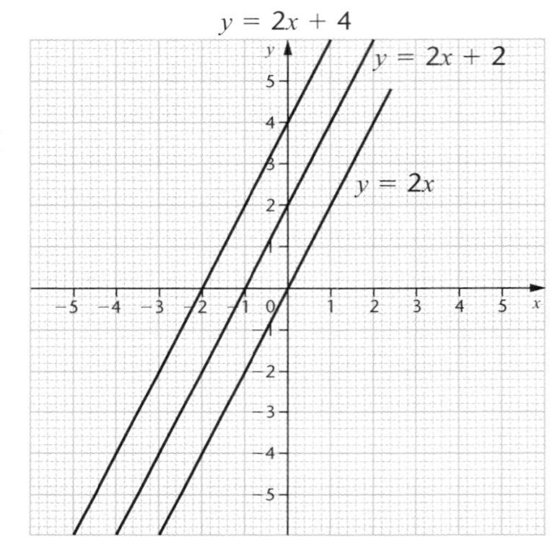

Equation	y-intercept
$y = 2x$	0
$y = 2x + 2$	2
$y = 2x + 4$	4

When the equation of the line is $y =$ ____ the y-intercept is the number term in the equation.

$y = 2x + 0 \rightarrow y = 2x$
$y = 2x$ is a straight line with gradient 2 passing through the origin.

In the equation $y = x + 1$ the y-intercept is +1.

Straight lines that pass through the origin (0, 0) have y-intercept of 0 so their equations do not have a number term in them. For example, $y = 2x$.

The graph crosses the y-axis at (0, 1). The y-intercept is 1.

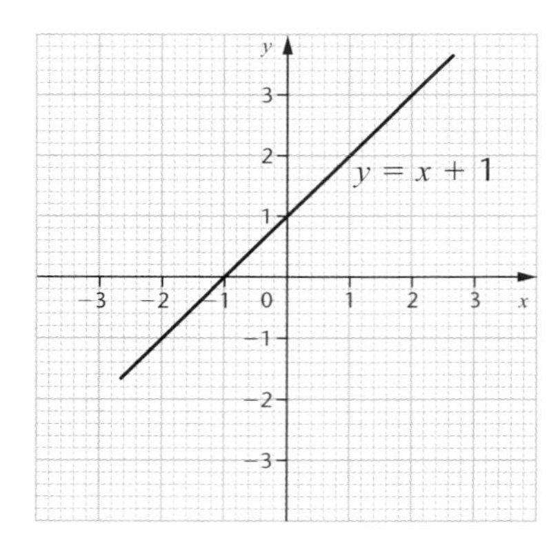

EXAMPLE 9

What are the y-intercept values of the lines with these equations?

(a) $y = x + 8$ (b) $y = -3x + 3$

(c) $y = -\frac{1}{2}x - 5$ (d) $y = 2.6x + 0.4$

(e) $y = 7x$ (f) $y = 6 - 2x$

(a) y-intercept $= 8$ (b) y-intercept $= 3$

(c) y-intercept $= -5$ (d) y-intercept $= 0.4$

(e) y-intercept $= 0$ (f) y-intercept $= 6$

Intercepts can be fractions or decimals.

$y = 0 - 2x$
The ↑ intercept is the number term.

If the equation of a line is *not* in the form $y =$ ____ you can use algebra skills to rearrange the equation.

See Chapter 7.

EXAMPLE 10

Find the y-intercepts of these straight lines.

(a) $2y = x + 6$ (b) $3x - y = 5$ (c) $2x - 5y + 20 = 0$

All the equations need to be rearranged to the form $y =$ ____.

(a) $2y = x + 6$

$$\frac{2y}{2} = \frac{x}{2} + \frac{6}{2}$$

$$y = \frac{1}{2}x + 3$$

The y-intercept is $+3$

The y can be on the left of the equation or on the right.
The y-intercept is the number term in the final equation.

(b) $3x - y = 5$

$$3x - y + y = 5 + y$$

$$3x = 5 + y$$

$$3x - 5 = 5 - 5 + y$$

$$3x - 5 = y$$

The y-intercept is -5

Remember the y-intercept is when $x = 0$.

(c) $2x - 5y + 20 = 0$

$$2x - 5y + 5y + 20 = 0 + 5y$$

$$2x + 20 = 5y$$

$$\frac{2x}{5} + \frac{20}{5} = \frac{5y}{5}$$

$$\frac{2x}{5} + 4 = y$$

The y-intercept is $+4$

or put $x = 0$

$$2 \times 0 - 5y + 20 = 0$$

$$-5y + 5y + 20 = 0 + 5y$$

$$20 = 5y$$

$$\frac{20}{5} = \frac{5y}{5}$$

$$4 = y$$

EXERCISE 17F

1 Find the y-intercepts of these lines.

(a) $y = 3x + 2$ (b) $y = -x + 3$

(c) $y = 4x - 7$ (d) $y = -3x - 4$

(e) $y = \frac{2}{3}x + 8$ (f) $y = 0.8x - 0.3$

(g) $y = x + 3$ (h) $y = 2x + 1$

(i) $y = 5 - 2x$ (j) $y = 4x$

2 *Without plotting* the graphs of these lines, find their y-intercepts.

(a) $y = x + 1$ (b) $y + 7 = x$

(c) $y = x + \frac{1}{2}$ (d) $2y = x - 1$

(e) $3y = 2x - 1$ (f) $2y - 3 = 2x$

(g) $y + 2 = 3x$ (h) $3x + y = 2$

(i) $3y = 2 + 3x$ (j) $x = -3y - 2$

(k) $2y + 3x = 2$ (l) $4y + 2 + 3x = 0$

You will need to rearrange some of the equations into the form $y =$ ____.

The general equation of a straight line

Equations of straight lines can always be written in the form

$$y = mx + c$$

where m is the gradient and c is the y-intercept.

This is an important result – you must remember it.

When the equation of a straight line is in the form $y =$ ____

- The gradient is the number in front of the x in the equation (the coefficient of x).
- The y-intercept is the number term in the equation.

It has to be $y =$ ____ not $2y$ or $3y$ or $-y$ or any other number.

EXAMPLE 11

Write the following equations in the form $y = mx + c$.

(a) $2y = 6x - 12$ (b) $5y = -10x + 20$ (c) $3x = y - 5$

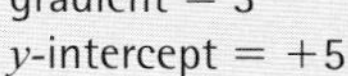

Rearrange each equation using algebra.

(a) $2y = 6x - 12$

$y = \frac{6x}{2} - \frac{12}{2}$

$y = 3x - 6$

Divide both sides by 2.

gradient = 3
y-intercept = −6

(b) $5y = -10x + 20$

$y = -\frac{10x}{5} + \frac{20}{5}$

$y = -2x + 4$

Divide both sides by 5.

gradient = −2
y-intercept = +4

(c) $3x = y - 5$

$3x + 5 = y$

$y = 3x + 5$

Add 5 to both sides and rearrange the equation.

gradient = 3
y-intercept = +5

EXERCISE 17G

1 Write the following equations in the form $y = mx + c$.
State the gradient and the y-intercept of each line.

Example 11 will help you.

(a) $3y = 9x + 18$ (b) $2y = -8x - 4$

(c) $8y = -24x + 8$ (d) $y - 7 = -2x$

(e) $2x + 6 = 2y$ (f) $-3x = 4 + y$

(g) $3x - y = 7$ (h) $2y - 10 = x$

(i) $15 - 5y = 10x$ (j) $6x + 2y - 7 = 0$

17.4 Curved graphs

Graphs that have x^2 in their equation are curves.

They are called **quadratic graphs.**

All quadratic graphs are U-shaped. The U shape can be the right way up or upside down.

EXAMPLE 12

Draw the graph of $y = x^2$ for values of x from −3 to +3.

Draw a table of values. Include all whole-number values of x in the range.

Write the terms in the equation in the rows, so you can calculate one at a time.

x	−3	−2	−1	0	1	2	3
x^2	9	4	1	0	1	4	9
$y = x^2$	9	4	1	0	1	4	9

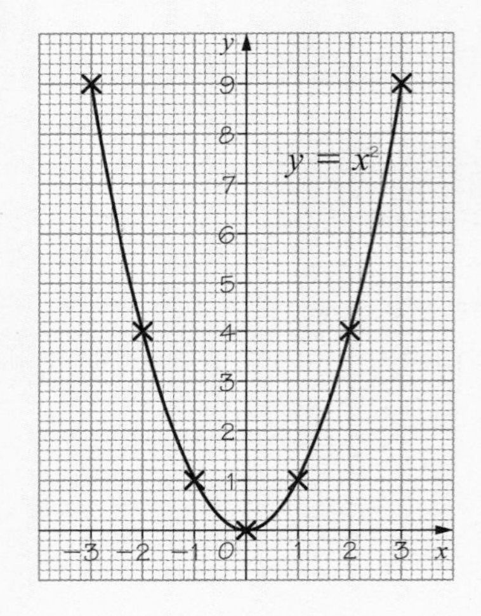

Plot the (x, y) points from your table of values and join them up with a *smooth* curve.

You will lose marks if you join the points with straight lines.

EXAMPLE 13

Make a table of values and draw the graph of $y = x^2 - 4$ for values of x from −3 to +3.

x-values from −3 to +3.

Draw the table with a row for each term in the equation.
Add the rows to get the y-values.

x	−3	−2	−1	0	1	2	3
x^2	9	4	1	0	1	4	9
−4	−4	−4	−4	−4	−4	−4	−4
$y = x^2 - 4$	5	0	−3	−4	−3	0	5

Continued ▼

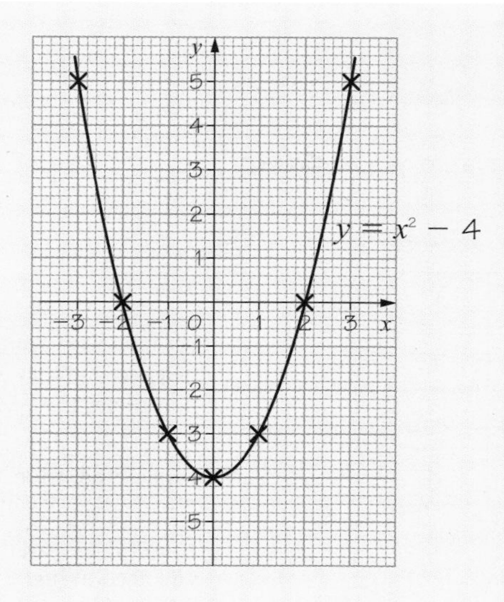

Look at the y-values in your table to see how far to draw the y-axis. y-values are from −4 to +5.

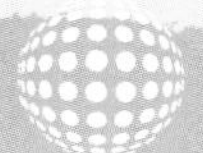

EXAMPLE 14

Complete the table of values and plot the graph of $y = x^2 + 3$ for values of x from −2 to +2.

x	−3	−2	−1	0	1	2	3
x^2	9	4	1	0	1	4	9
+3	3	3	3	3	3	3	3
$y = x^2 + 3$	12	7	4	3	4	7	12

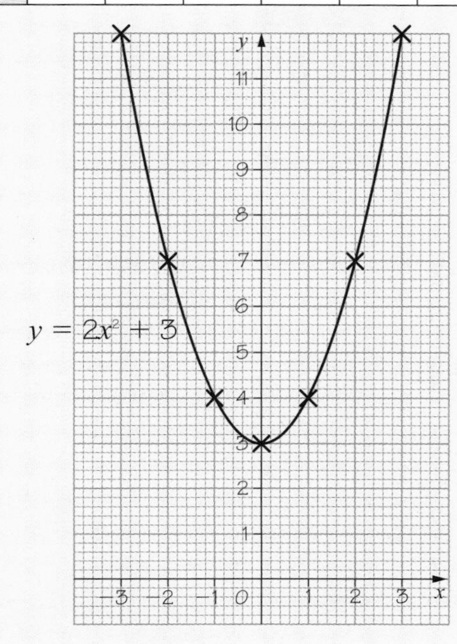

EXAMPLE 15

Complete the table of values and plot the graph of $y = x^2 - 3x$ for values of x from -1 to $+4$.

x	-1	0	1	2	3	4
x^2	1	0	1	4	9	16
$-3x$	3	0	-3	-6	-9	-12
$y = x^2 - 3x$	4	0	-2	-2	0	4

For help with multiplying by a negative number, see Chapter 1. Add to get the y-values.

y-values are from -2 to $+4$.

To work out how low the graph goes you need to find y when $x = 1.5$. When $x = 1.5$, $y = 2.25$.

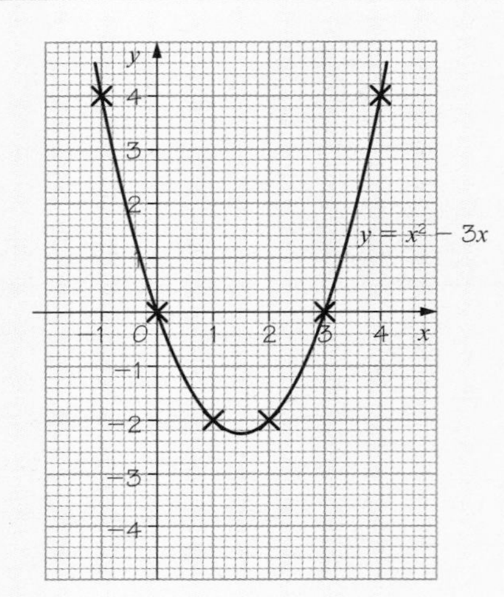

EXERCISE 17H

For each of these graphs

- make a table of values
- draw the graph.

Use a scale of 1 cm for 1 unit on each axis.

1 $y = x^2 - 5$ for values of x from -3 to $+3$

Use each value from -3 to $+3$ in your table.

2 $y = x^2 + 1$ for values of x from -2 to $+2$

3 $y = x^2 + 2$ for values of x from -3 to $+3$

4 $y = x^2 - 5x$ for values of x from 0 to $+5$

5 $y = x^2 + 3x$ for values of x from -4 to $+1$

6 $y = x^2 + 4x$ for values of x from -6 to $+2$

7 $y = x^2 - 6x$ for values of x from -1 to $+7$

Symmetry in quadratic graphs

The graphs in Examples 12, 13 and 14 are all symmetrical about the y-axis. The graph in Example 15 is symmetrical about the line $x = 1.5$

All quadratic graphs are symmetrical about a line parallel to the y-axis.

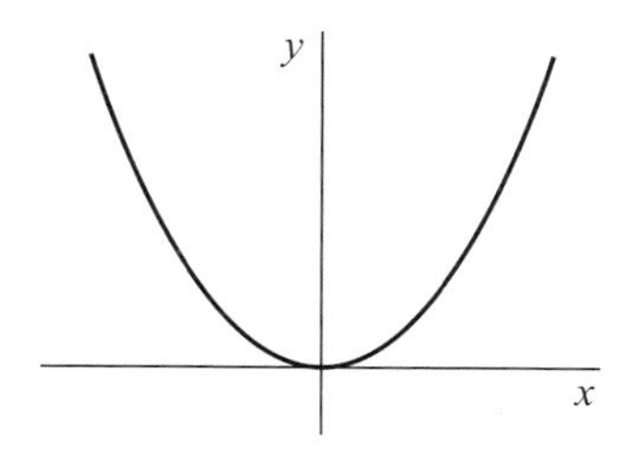

Quadratic graphs have x^2 in their equation.

A line parallel to the y-axis is a vertical line.

EXAMPLE 16

Write a table of values and plot the graph of $y = x^2 + 2x - 3$ for values of x from -4 to $+2$.
State its line of symmetry.

x	-4	-3	-2	-1	0	1	2
x^2	16	9	4	1	0	1	4
$+2x$	-8	-6	-4	-2	0	2	4
-3	-3	-3	-3	-3	-3	-3	-3
$y = x^2 + 2x - 3$	5	0	-3	-4	-3	0	5

There are three terms in the equation. Give each a line of its own.

Add to find y for each x-value.

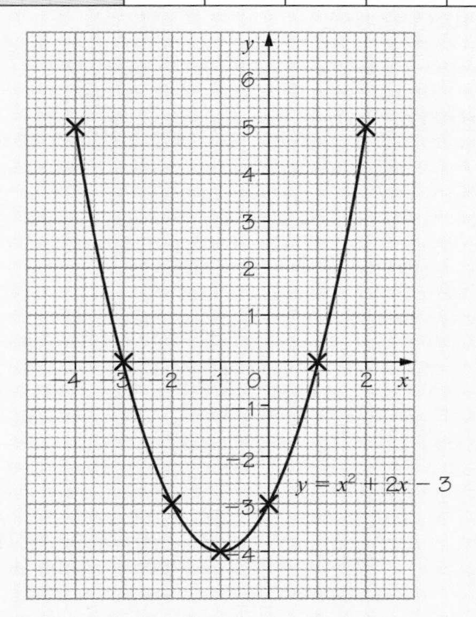

The line of symmetry is $x = -1$.

The line of symmetry passes through the lowest point on the curve $(-1, -4)$.
All points on this line have x-co-ordinate -1.

EXERCISE 17I

1 State the line of symmetry for each of the graphs you drew for Exercise 17H.

2 For each of these quadratic graphs
- make a table of values
- draw the graph for the given range of values
- state the line of symmetry.

Choose a sensible scale for each axis.

(a) $y = x^2 + x - 2$ for values of x from -3 to $+2$

(b) $y = x^2 + 4x - 1$ for values of x from -5 to $+1$

(c) $y = x^2 - 2x - 5$ for values of x from -2 to $+4$

(d) $y = x^2 + 3x + 1$ for values of x from -5 to $+2$

(e) $y = x^2 - 5x - 4$ for values of x from -1 to $+6$

All the points on the graph must fit your squared paper.

Keep these graphs because you will need them in Excercise 17K.

17.5 Graphs that intersect

Where do straight line graphs meet?

You can use graphs to help you find solutions to equations.

For example, to solve the equation $x + 1 = 0$ you can use the graph of $y = x + 1$.

$x + 1 = 0$ means that $y = 0$, so look to see where the line crosses the x-axis.

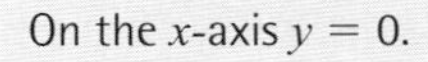

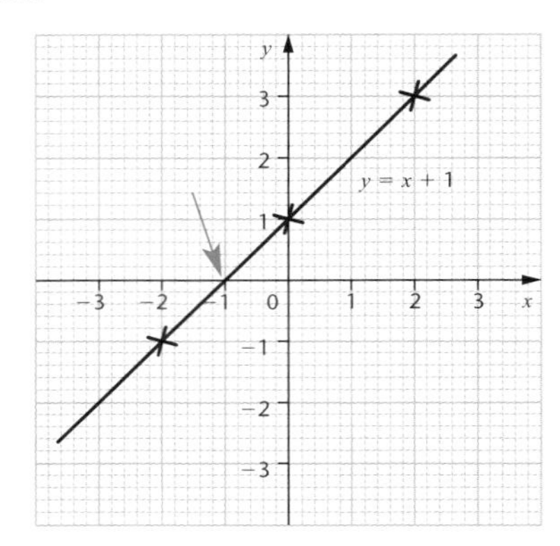

The line crosses the x-axis at $(-1, 0)$.

You could also find this by rearranging using algebra.

The solution to the equation $x + 1 = 0$ is $x = -1$.

You can use this graph to solve any equation of the form $x + 1 = \ldots$

For example, to solve $x + 1 = 3$, find the **point of intersection** of the graph $y = x + 1$ with the line $y = 3$.

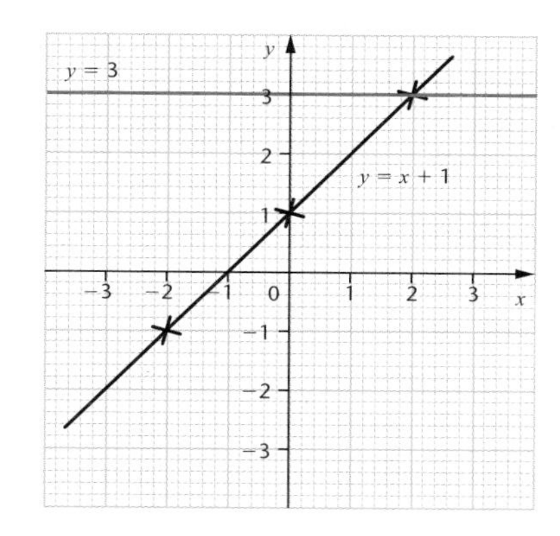

The point of intersection is the point where two graphs cross.

The solution to $x + 1 = 3$ is $x = 2$.

Check $2 + 1 = 3$ ✓

EXAMPLE 17

(a) Plot the graph of $y = 3x - 2$.

(b) It intersects the line $y = 3$ at a point P. What are the co-ordinates of P?

(c) What is the solution of the equation $3x - 2 = 3$?

(a)

x	-2	0	2
$3x$	-6	0	6
-2	-2	-2	-2
y	-8	-2	4

Make a table of values. Use x-values of -2, 0 and 2.

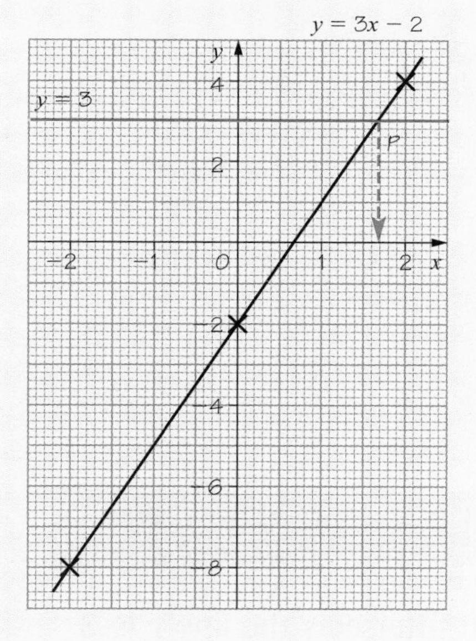

Plot the graph.

Draw in the line $y = 3$.

Draw your graph as accurately as possible.

In this example the point of intersection is not a whole number.

(b) The two straight lines intersect where $x = 1.7$ and $y = 3$, i.e. at the point (1.7, 3).

(c) Using the graph, the solution of the equation $3x - 2 = 3$ is $x = 1.7$.

The exact co-ordinates are $(1\frac{2}{3}, 3)$ but you are only expected to give your answer to an accuracy of 'half a small square'.

You can find the point of intersection of any two straight lines.

EXAMPLE 18

Draw the graphs of the straight lines $y = 3x - 2$ and $y = x + 4$ on the same co-ordinate grid.

What are the co-ordinates of the point where these lines intersect?

Draw a table of values for each graph.

$y = 3x - 2$

x	0	1	2
$3x$	0	3	6
-2	-2	-2	-2
y	-2	1	4

$y = x + 4$

x	0	1	2
$+4$	4	4	4
y	4	5	6

Draw the graphs so that the lines cross.
Make sure the y-axis extends far enough.

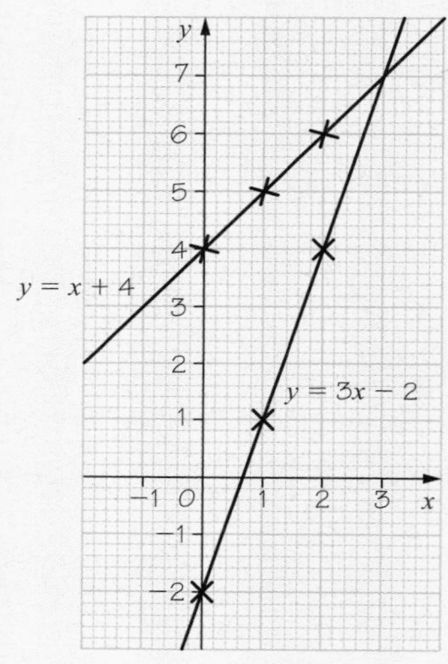

The graphs intersect at the point (3, 7).

EXERCISE 17J

1 Draw the graph $y = 2x + 5$. It intersects the line $y = 3$ at a point P.
What are the co-ordinates of P?

2 The line $y = 7$ intersects another line $y = 4x + 9$ at a point Q.
What are the co-ordinates of Q?
Where does the line $y = 4x + 9$ cross the x-axis?

Draw the graph.

3 For each part **(a)**, **(b)** and **(c)**, draw the two straight line graphs on the same co-ordinate grid.
Write down the co-ordinates of the point of intersection of the lines.

(a) $y = 2x + 1$ and $y = 10 - x$

(b) $y = 9 - 3x$ and $y = 2x - 1$

(c) $y = x - 2$ and $y = 3x + 1$

Solving quadratic equations graphically

A **quadratic equation** has an x^2 term. You can solve quadratic equations using quadratic graphs.

When a quadratic graph crosses the x-axis it does so in two places.

For the curve $y = x^2 - 4$ the graph crosses the x-axis at $x = -2$ and $x = 2$.

$x^2 - 4 = 0$ is a quadratic equation because it has an x^2 term.

This is where $y = 0$.

So $x = -2$ and $x = 2$ are the solutions to the equation $x^2 - 4 = 0$.

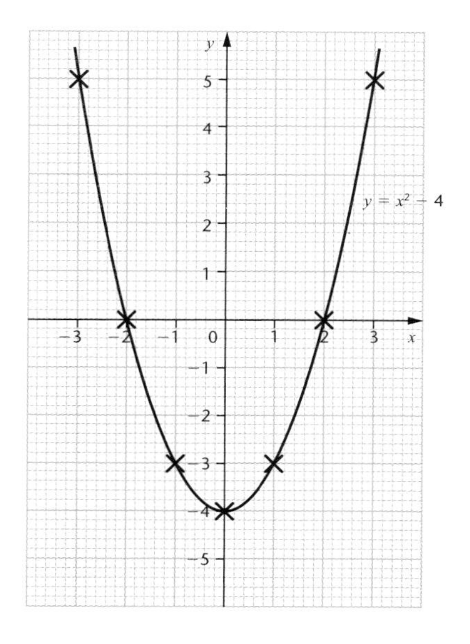

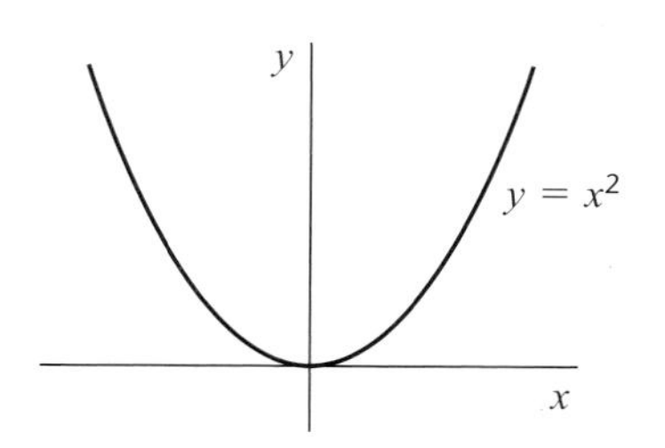

The graph of $y = x^2$ just touches the x-axis at the point (0, 0), so the equation $x^2 = 0$ has only one solution, $x = 0$.

See Example 12.

Some quadratic graphs do not cross the x-axis at all.

$y = x^2 + 3$ is one of these. This means there are no solutions to the equation $x^2 + 3 = 0$.

A quadratic graph can intersect a line parallel to the x-axis.

This graph just touches the line $y = 3$ at the point (0, 3). So the equation $x^2 + 3 = 3$ has one solution, $x = 0$.

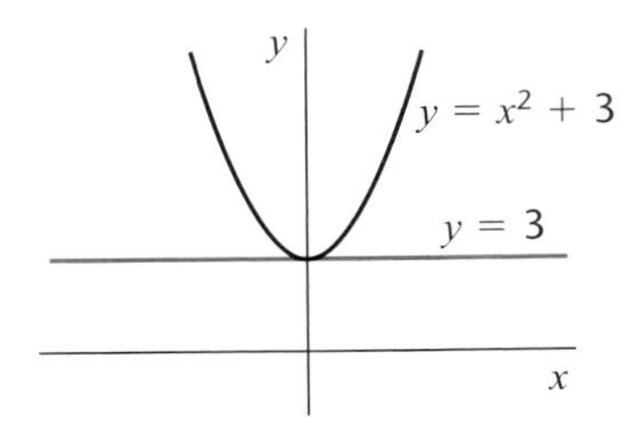

See Example 14.

EXAMPLE 19

(a) Draw a pair of axes with the x-axis from −3 to +3 and the y-axis from −5 to +10. Draw the graph of $y = x^2 - 2$.

(b) What are the solutions to the equation $x^2 - 2 = 0$?

(c) Draw the straight line $y = 3$. Write down the co-ordinates of both points where the curve and line intersect.

(d) Write down the equation that is solved by the x-values in part **(c)**.

Draw a table of values for the graph.

(a)

x	−3	−2	−1	0	1	2	3
x^2	9	4	1	0	1	4	9
−2	−2	−2	−2	−2	−2	−2	−2
$y = x^2 - 2$	7	2	−1	−2	−1	2	7

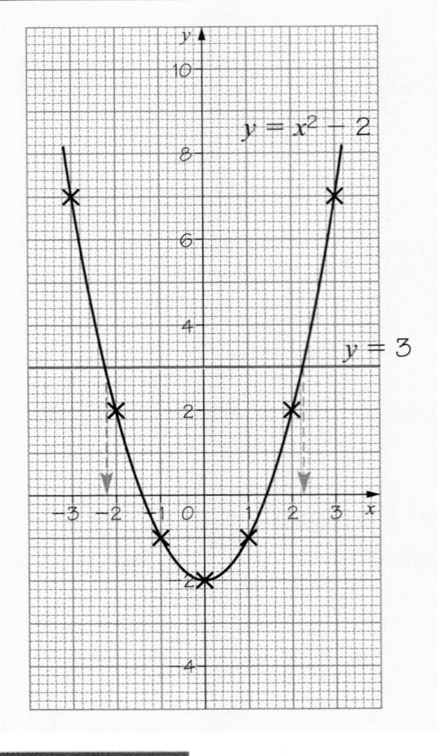

The solutions to $x^2 - 2 = 0$ are where the curve crosses the line $y = 0$ (x-axis).

Draw the line $y = 3$ on the graph. The two dashed lines show the x-values of the points of intersection.

(b) The curve crosses the x-axis ($y = 0$) at points $x = -1.4$ and $x = 1.4$.
The solutions to $x^2 - 2 = 0$ are $x = -1.4$ and $x = 1.4$.

(c) The co-ordinates of the points of intersection are $(-2.3, 3)$ and $(2.3, 3)$.

(d) The equation is $x^2 - 2 = 3$ which can be simplified to $x^2 = 5$.

You are solving
'curve' = 'line'
in other words, $x^2 - 2 = 3$.

EXAMPLE 20

(a) Using the graph of $y = x^2 + 2x - 3$, draw the graph of the straight line $y = 1$ on the same axes.

(b) Write down the x-values of the points where these graphs intersect.

(c) What equation is solved by these x-values?

(d) Use your graph to solve the equation $x^2 + 2x - 3 = 0$.

$y = x^2 + 2x - 3$ was plotted in Example 16.

(a)

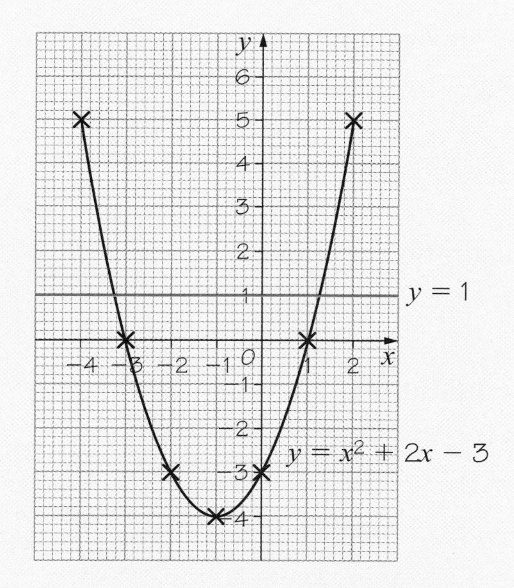

This is the graph from Example 16 with the line $y = 1$ added.

You need to draw the graph accurately to get accurate answers.

(b) The x-values of the points of intersection are $x = -3.2$ and $x = 1.2$.

(c) The equation solved is $x^2 + 2x - 3 = 1$ which simplifies to $x^2 + 2x - 4 = 0$.

You are solving 'curve' = 'line'
in other words, $x^2 + 2x - 3 = 1$.

(d) The solutions are $x = -3$ and $x = 1$.

To solve $x^2 + 2x - 3 = 0$ look to see where the curve crosses the line $y = 0$ (the x-axis).

You can solve a quadratic equation by seeing where its graph intersects a horizontal line.

Using $y = x^2 + 2x - 3$

- solving $x^2 + 2x - 3 = 0$ means looking to see where the curve crosses the x-axis ($y = 0$).
  The solutions are $x = -3$ and $x = 1$.
- Solving $x^2 + 2x - 3 = 1$ means looking to see where the curve crosses the line $y = 1$.
  The solutions are $x = -3.2$ and $x = 1.2$.

For the graph of $x^2 + 2x - 3$ see Example 19.

EXERCISE 17K

1 (a) Draw the graph of $y = x^2 + 4x + 2$ for values of x from -5 to $+1$.
(b) On the same axes draw the graph of the line $y = 5$.
(c) Write down the x-values of the points of intersection of the curve and the line $y = 5$.
(d) What equation is solved by these x-values?
(e) Use your graph to solve the equation $x^2 + 4x + 2 = 0$.

2 The graphs you drew in Exercise 17I question 2 are listed below.
For each, use the graph to solve the quadratic equations written alongside them.
Give your answers correct to 1 d.p.

	Graph	Solve these equations	
(a)	$y = x^2 + x - 2$	$x^2 + x - 2 = 0$	$x^2 + x - 2 = 2$
(b)	$y = x^2 + 4x - 1$	$x^2 + 4x - 1 = 0$	$x^2 + 4x - 1 = -3$
(c)	$y = x^2 - 2x - 5$	$x^2 - 2x - 5 = 0$	$x^2 - 2x - 5 = -4$
(d)	$y = x^2 + 3x + 1$	$x^2 + 3x + 1 = 0$	$x^2 + 3x + 1 = 8$
(e)	$y = x^2 - 5x - 4$	$x^2 - 5x - 4 = 0$	$x^2 - 5x - 4 = -6$

EXAMPLE 21

Make a table of values for $y = \frac{1}{x}$ and draw the graph.

x	−4	−3	−2	−1	−0.5	−0.25
y	−0.25	−0.33	−0.5	−1	−2	−4

x	0.25	0.5	1	2	3	4
y	4	2	1	0.5	0.33	0.25

$\frac{1}{x}$ means $1 \div x$.
So, for example, $\frac{1}{0.25} = 1 \div 0.25$
$= 4$

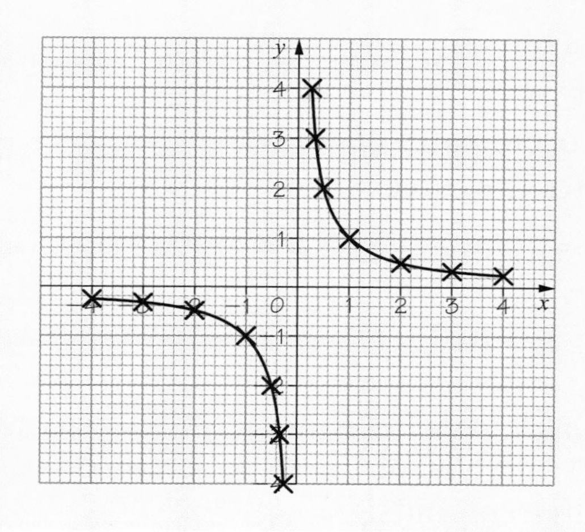

Note that you cannot find a value for $\frac{1}{x}$ at $x = 0$, as dividing 1 by 0 is undefined.

EXERCISE 17L

1 (a) Copy and complete the table of values for $y = \frac{2}{x}$.

x	−3	−2	−1	−0.5	−0.25	0.25	0.5	1	2	3
y	−0.67			−4		8		2		0.67

(b) Draw the graph of $y = \frac{2}{x}$ for values of x from −3 to +3.

2 (a) Copy and complete the table of values for $y = -\frac{3}{x}$.

x	−3	−2	−1	−0.5	−0.25	0.25	0.5	1	2	3
y	1	1.5		6		−12	−6			−1

(b) Draw the graph of $y = -\frac{3}{x}$ for values of x from −3 to +3.

3 Make a table of values for $y = -\frac{10}{x}$ for values of x from 1 to 10 and draw the graph.

4 Make a table of values for $y = -\frac{1}{x}$ for values of x from −4 to +4 and draw the graph.

5 (a) Using the same axes, draw the graphs of $y = \frac{1}{x}$ and $y = x$ for values of x from −3 to +3.

(b) Use your graphs to write down the co-ordinates of the points where the line $y = x$ crosses the curve $y = \frac{1}{x}$.

17.6 Conversion graphs

You can use a **conversion graph** to convert one type of measurement into another, usually with different units.

Conversion graphs can be linear or curved.

EXAMPLE 22

The table below shows the approximate conversion between degrees Celsius (°C) and degrees Fahrenheit (°F) using the equation $F = 2C + 30$.

This is only an approximation to make it easy to draw a graph. The actual formula is $F = 1.8C + 32$.

C	−40	−20	0	20	40	60	80	100
2C	−80		0		80		160	
+30	30		30		30		30	
F	−50		30		110		190	

(a) Complete the table.

(b) Draw a co-ordinate grid with C-values (x-axis) from −40 to 100 and F-values (y-axis) from −50 to 250. Plot the points in the table and draw a straight line.

Continued ▼

(c) What is the temperature in °F when it is 10°C?

(d) If the temperature is 160°F, what is this in °C?

(a)

C	−40	−20	0	20	40	60	80	100
2C	−80	−40	0	40	80	120	160	200
+30	30	30	30	30	30	30	30	30
F	−50	−10	30	70	110	150	190	230

(b)

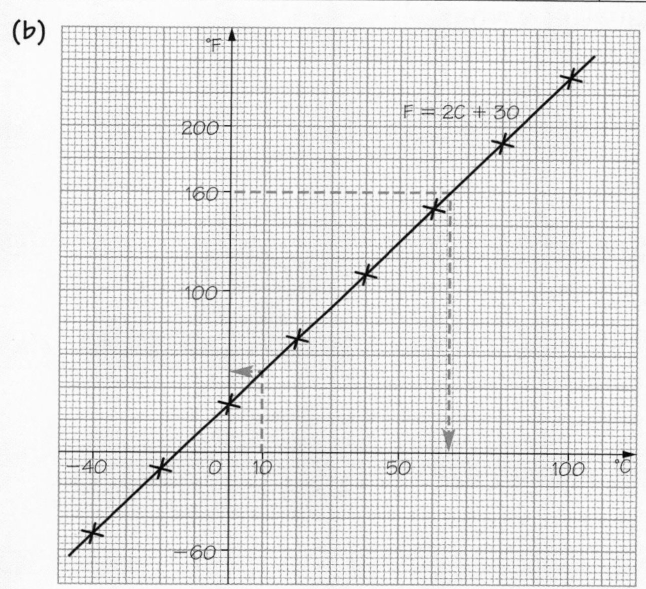

(c) 10°C = 50°F

(d) 160°F = 65°C

Draw dashed lines on the graph at 10°C and 160°F to meet the conversion line. From these points draw another dashed line to meet the axes and read off the answers.

EXERCISE 17M

1 (a) Copy and complete this conversion table between kilometres and miles using the conversion 5 miles = 8 km.

miles (x)	0	5	50	100
km (y)		8		160

(b) Draw a conversion graph with x-values from 0 to 100 miles and y-values from 0 to 200 km.

(c) How many miles are equivalent to 75 km?

(d) How many km are equivalent to 75 miles?

2 (a) Copy and complete this conversion table between centimetres (to 1 d.p.) and inches.

cm	0	2.5						30
inches	0	1	2	3	5	6	10	12

This is an approximation.

(b) Draw a conversion graph for cm and inches.

(c) How many cm are equivalent to 4 inches?

(d) How many inches are equivalent to 20 cm.

3 A stone is dropped down a well. The following table gives its distance from the top of the well after each second, using the approximate formula $d = 5t^2$.

$d = 5t^2$ means
$d = 5 \times t^2$

Time in seconds (t)	0	1	2	3
Distance in m (d)	0	5	20	45

(a) Draw a co-ordinate grid with x-values between 0 and 4 and y-values between 0 and 50. Plot the points and join them up with a smooth curve.

(b) How far did the stone drop in 1.5 seconds?

(c) How long did it take the stone to fall 35 m?

(d) If the well is 60 m deep, for how many seconds did the stone fall?

17.7 Distance–time graphs

A **distance–time graph** shows information about a journey. You always plot 'time' on the horizontal axis (x-axis) and 'distance' on the vertical axis (y-axis).

You can use a distance–time graph to work out the **speed** for part of a journey using the relationship

$$\text{speed} = \frac{\text{distance}}{\text{time}}$$

If distance is measured in metres (m) and time in seconds then speed will be in m/s. Another unit for speed is km/h.

You can use a distance–time graph to help you work out an **average speed** for a journey using the relationship

$$\text{average speed} = \frac{\text{total distance}}{\text{total time}}$$

Total distance includes the return journey.

EXAMPLE 23

The distance–time graph shows a railway journey from A to D.

(a) For how long did the train stop at station B?

(b) For how long did it stop at station C?

(c) What was the speed of the train between stations A and B?

(d) What was the average speed of the train over the whole journey from A to D?

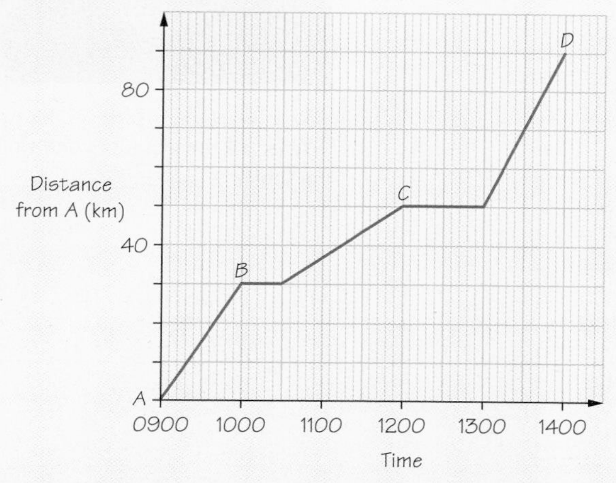

From the graph you can work out the times at each station and the distance travelled between each of the stations.

Read the times off the graph.

(a) The train arrived at station B at 1000 and departed at 1030. It stopped at B for 30 minutes.

(b) The train arrived at station C at 1200 and departed at 1300. It stopped at C for 1 hour.

(c) From A to B

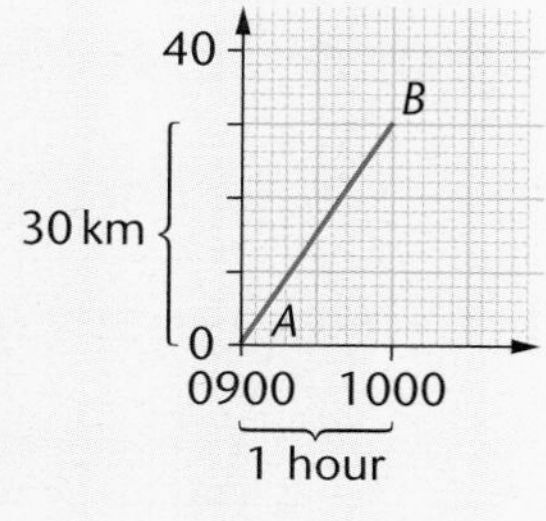

Distance in km
Time in hours
Speed in km/h

distance $= 30$ km

time $= 1$ hr

$$\text{speed} = \frac{\text{distance}}{\text{time}} = \frac{30}{1} = 30 \text{ km/h}$$

Total distance and total time for the journey (include the stops).

(d) Total distance from A to D = 90 km

Total time from A to D = 5 hours

$$\text{Average speed} = \frac{\text{total distance}}{\text{total time}} = \frac{90}{5} = 18 \text{ km/h}$$

In Example 23 the train was travelling *away* from A. On the return journey from D to A the distance of the train from A is decreasing. The graph looks like this

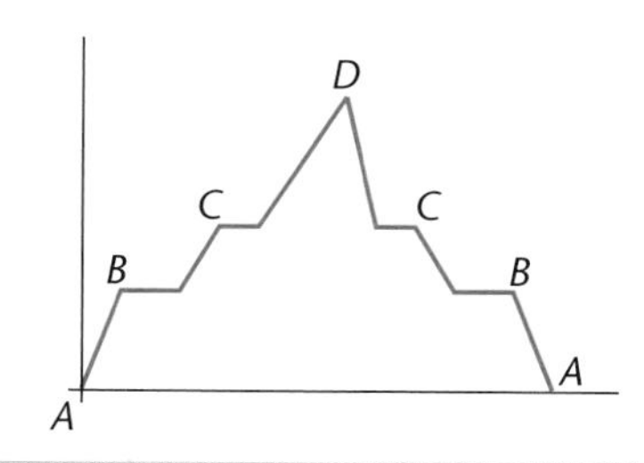

Graphs that represent an outward journey and a return journey always look like this.

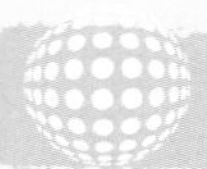

EXAMPLE 24

Céline drove to town in her car, a distance of 10 km in 15 minutes. She spent 45 minutes in town and then set off home again.

On the way home she stopped to get some fuel. The garage was 6 km from town and it took her 10 minutes to drive there.

Her stop for fuel took 5 minutes. She drove the rest of her journey home in 10 minutes.

(a) Draw a distance–time graph for Céline's journey.

(b) What was her average speed on her journey into town?

Write all the distances and times in a table.

Céline's journey		
	Distance	**Time**
To town	10 km	15 mins
In town	0 km	45 mins
To garage	6 km	10 mins
At garage	0 km	5 mins
To home	4 km	10 mins

Total time
$= 15 + 45 + 10 + 5 + 10$
$= 85$ mins

Largest distance from home = 10 km

(a)

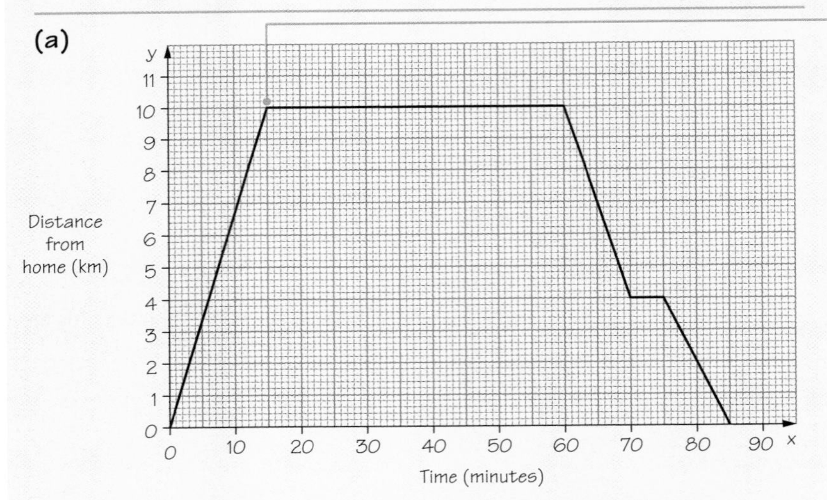

15 minutes, 10 km

The long horizontal line is when she was in town. The short horizontal line is when she stopped for fuel.

The garage is 6 km *from town* which means 4 km from home.

(b) Average speed for journey into town $= \dfrac{\text{total distance}}{\text{total time}}$

$= \dfrac{10}{15}$ km per min

$= \dfrac{10}{15} \times 60$ km/h

$= 40$ km/h

Distance in km and time in minutes will give speed in km per minute.

There are 60 minutes in 1 hour.
To convert km per min to km/h you multiply by 60.

EXERCISE 17N

1 Each morning Chloe walks to school and stops at the shop on the way. The distance–time graph shows her journey to school.

(a) What time does she get to the shop?

(b) How long is she in the shop?

(c) How long does it take Chloe to walk to school?

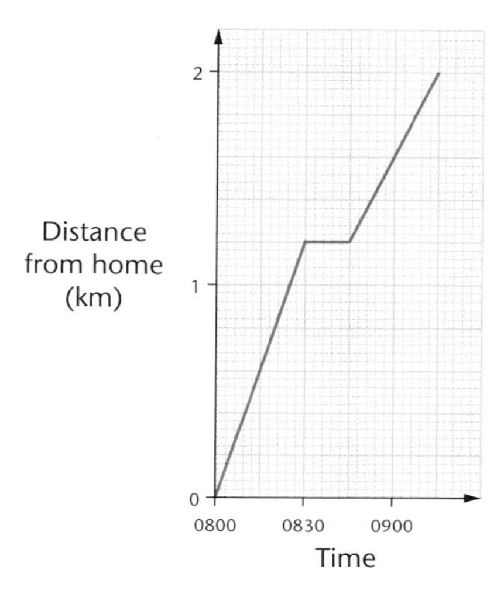

2 The distance–time graph shows Amina's cycle journey from home.
She stopped twice for breaks before returning home.

(a) How long did Amina's cycle ride take?

(b) On which part of her journey was she travelling at the greatest speed? How can you tell?

(c) Calculate her average speed for the whole cycle ride.

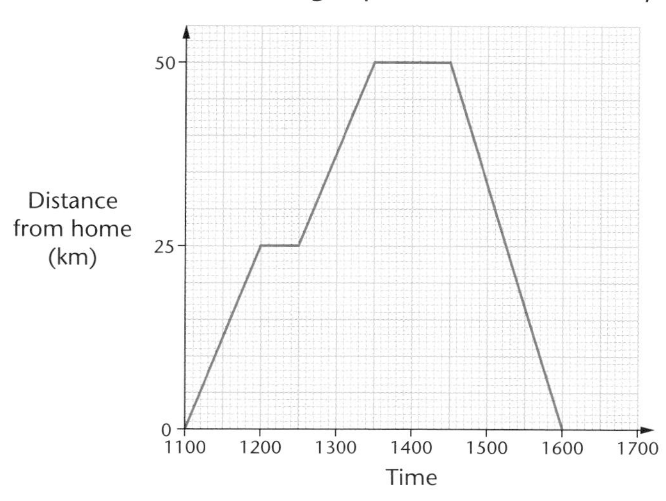

3 Jose walked to the village shop, a distance of 800 m. It took him 10 minutes to get there and he was in the shop for 5 minutes.
He then walked to the Post Office, another 200 m down the street. This took 3 minutes and he was in the Post Office for 6 minutes.
He then walked back home which took 16 minutes.
Draw a distance–time graph for his journey.

(a) How fast did he walk to the shop?

(b) What was his average speed for the whole journey?

Before you start to draw the distance–time graph, write the times and the distances for each part, including the stops, in a table. Then you can see the total time and the total distance involved.

Look at Example 23 if you need help.

This will help you decide what scales to use for the axes.

4 Henri is going to school. He leaves home at 0815 and walks 400 m to the bus stop in 6 minutes. He waits 5 minutes for the bus to arrive.
The bus journey is 4000 m and the bus arrives at the bus stop near school at 0838.
Henri then walks another 100 m to school in 2 minutes.

(a) Draw a distance–time graph for Henri's journey.

(b) What was the average speed of the bus journey? Give your answer in metres per minute then convert it to km/h.

Use a sensible scale for the axes.

Not the *whole* journey.

5 Rob went to town in his car. He drove the 12 kilometres into town in 25 minutes. He was in town for $1\frac{1}{4}$ hours. Then he left to drive home.
After 20 minutes of his journey home, 9 kilometres from town, he dropped off some library books at his aunt's house and stayed for a cup of tea for $\frac{1}{2}$ hour.
He then continued his journey home, taking another 10 minutes to get there.

(a) Draw a distance–time graph for his journey.

(b) What was his average speed on the way to town? Give your answer in km/h.

EXAMINATION QUESTIONS

1 **(a)** Copy and complete the table of values for $y = \frac{6}{x}$. [3]

x	-6	-5	-4	-3	-2	-1	1	2	3	4	5	6
y		-1.5	-2	-3				3	2	1.5		

(b) On a grid, draw the graph of $y = \frac{6}{x}$ for $-6 \leqslant x \leqslant -1$ and $1 \leqslant x \leqslant 6$. [4]

(c) Copy and complete the table of values for $y = \frac{x}{2}$. [2]

x	-6	0	6
y			

(d) On the same grid, draw the graph of $y = \frac{x}{2}$ for $-6 \leqslant x \leqslant 6$. [1]

(e) Write down the co-ordinates of the points of intersection of the two graphs, giving your answer to 1 decimal place. [2]

(CIE Paper 3, Jun 2000)

---

2 The graph shows the hire charges when a truck travels different distances.

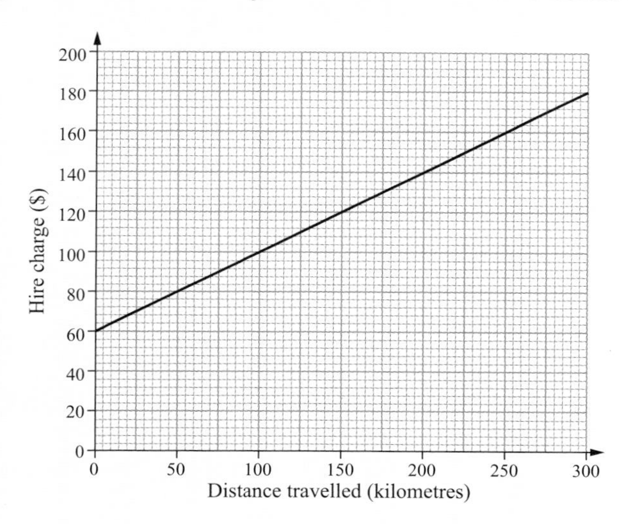

(a) What is the hire charge when the truck travels 150 kilometres? [1]
(b) What is the distance travelled by a truck when the hire charge is $130? [1]
(c) What is the increase in the hire charge, in cents, for each extra kilometre the truck travels? [$1 = 100 cents] [2]

(CIE Paper 1, Nov 2000)

3 **(a)** Copy and complete the table of values for $y = x^2 - 4x$. [3]

x	-2	-1	0	1	2	3	4	5	6
y		5	0		-4	-3	0		

(b) On a grid, plot these points and then draw the graph of $y = x^2 - 4x$ for $-2 \leqslant x \leqslant 6$. [4]

(c) Use your graph to find two values of x when $x^2 - 4x = 2$. [2]

(d) Copy and complete the table of values for $y = 3 - x$. [2]

x	-2	2	6
y			

(e) On the same grid, draw the graph of $y = 3 - x$ for $-2 \leqslant x \leqslant 6$. [2]

(f) Write down the co-ordinates of the two points where the graphs intersect. [2]

(CIE Paper 3, Nov 2000)

---

4 **(a)** Copy and complete the table of values for $y = x^2 - x - 3$. [3]

x	-3	-2	-1	0	1	2	3	4
y		3	-1	-3	-3	-1		

(b) On a grid, draw the graph of $y = x^2 - 4x - 3$ for $-3 \leqslant x \leqslant 4$. [4]

(c) Use your graph to

(i) find x when $y = -2$, [2]

(ii) solve $x^2 - x - 3 = 0$. [2]

(d) Copy and complete the table of values for $y = x + 2$. [1]

x	-3	0	2	4
y	-1	2		

(e) On the same grid, draw the graph of $y = x + 2$ for $-3 \leqslant x \leqslant 4$. [2]

(f) Write down the co-ordinates of the points of intersection of the graphs $y = x^2 - 4x - 3$ and $y = x + 2$. [2]

(CIE Paper 3, Nov 2001)

5

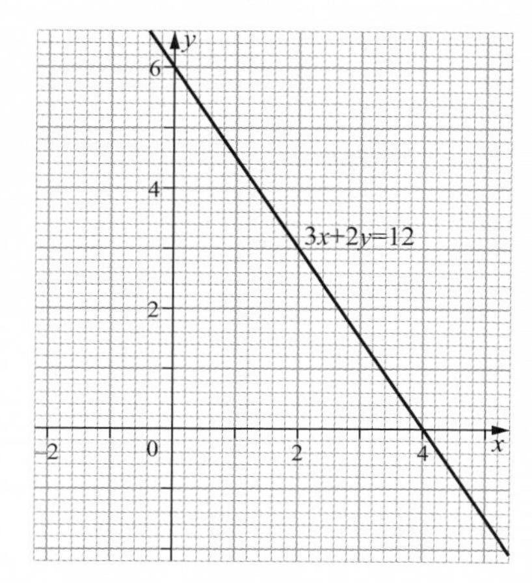

The graph of $3x + 2y = 12$ is drawn on the grid above.

(a) Copy and complete the table of value for $y = 3x - 1$. [2]

x		0	1	2
y			2	

(b) On a copy of the grid above, draw the graph of $y = 3x - 1$ for $0 \leqslant x \leqslant 2$. [1]

(c) Use the graph to find the solution of the simultaneous equations

$$3x + 2y = 12,$$
$$y = 3x - 1.$$ [2]

(CIE Paper 3, Nov 2002)

---

6 Students try to find the best price at which to sell their school newspaper.
When the price was 10 cents, they sold 200 newspapers.
When the price was 60 cents, they sold only 75 newspapers.
They drew the graph below using this information.

Use the graph to answer these questions.

(a) At what price will no-one buy the newspapers? [1]

(b) 150 newspapers are sold. What was the price? [1]

(CIE Paper 3, Nov 2002)

Chapter 18

Transformations

This chapter will show you how to

- ✔ recognise and use the four types of transformation – reflection, rotation, translation and enlargement
- ✔ find the centre and scale factor of an enlargement
- ✔ identify congruent shapes
- ✔ identify similar shapes
- ✔ use the congruence properties to identify congruent triangles

18.1 Types of transformation

A **transformation** changes the position and/or the size of an object. The transformed shape is called the **image**.

You need to know about four types of transformation.

- **reflection**
- **rotation**
- **translation**
- **enlargement**

Reflection, rotation and translation only change the position of an object. The size and shape of the image and the object are identical – they are **congruent**.

'Congruent' means identical.

Enlargement changes the position of an object and its size. The object and image shapes are **similar**.

They are the same shape, but different sizes.

18.2 Reflection

For a **reflection** you need a **mirror line**. You reflect the object in the mirror line to produce an image. This image is exactly the same size and shape as the object.

When you reflect a shape in a mirror line, the object and image are congruent.

Points on the images are the same distance behind the mirror line as the corresponding points on the object are in front of the mirror line.

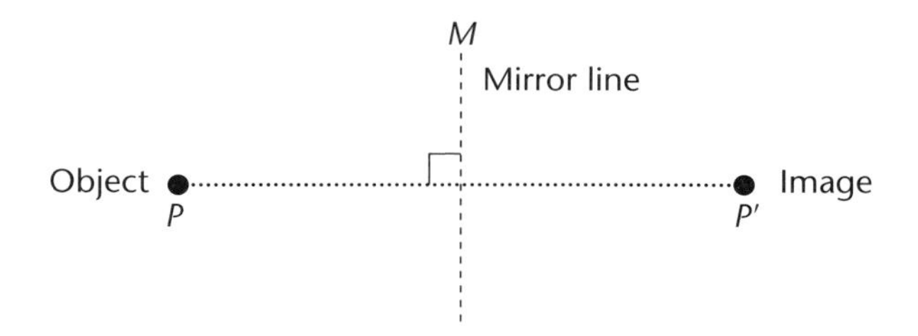

If you join a point P on the object to its corresponding point P' on the image, the line PP' crosses the mirror line M at right angles and

$PM = P'M$

P' is the reflection of P. You say 'P dash'.

Every point on an object reflects to an image point behind the mirror line.

EXAMPLE 1

The object $ABCD$ is reflected in the mirror line shown. Draw the image of the object and label it $A'B'C'D'$.

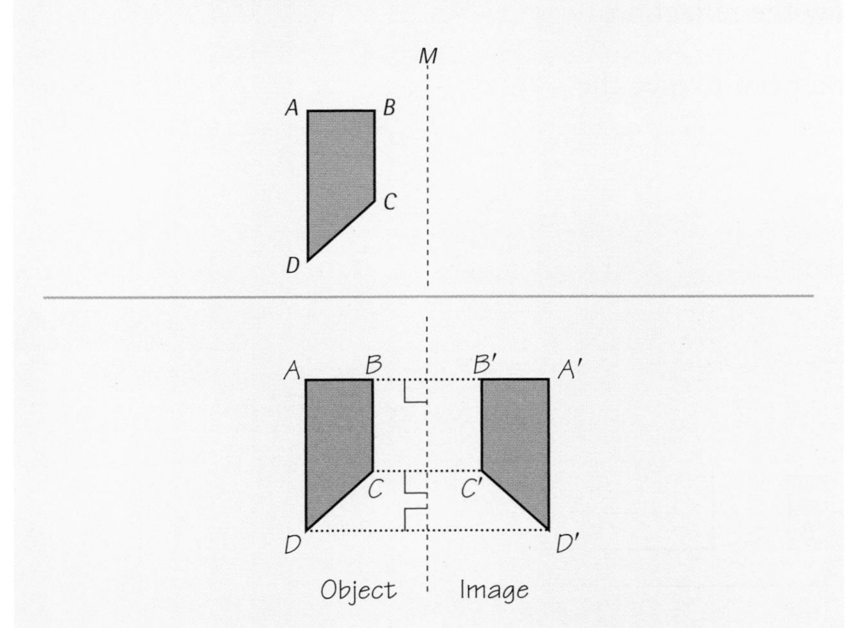

You could trace the object and the mirror line. Turn over the tracing paper and line up the mirror lines to see the reflection.

Reflect each corner (vertex) in the mirror line, then join them up. Draw a line from C at right angles to the mirror line. Continue this line to a point the same distance the other side and label it C'.
Repeat this for points A, B and D.
Join the points in order with a straight line.

The object shape is labelled A, B, C and D in a clockwise direction. After reflection the image shape is labelled A', B', C' and D' in an anticlockwise direction.

Sometimes the mirror line passes *through* the object. The reflections go in both directions.

EXAMPLE 2

The triangle ABC is reflected in the mirror line shown.
Draw the reflection of this triangle and label it $A'B'C'$.

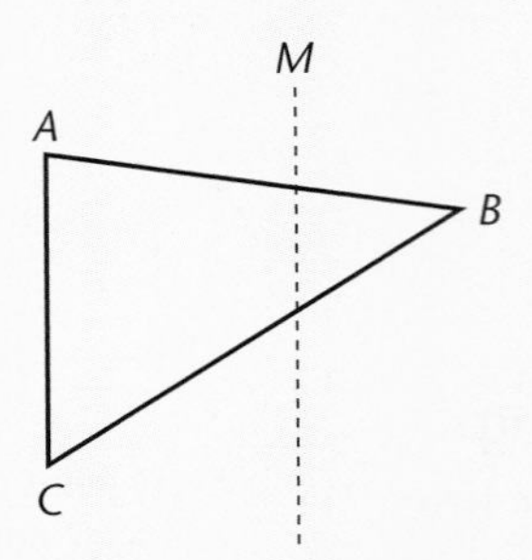

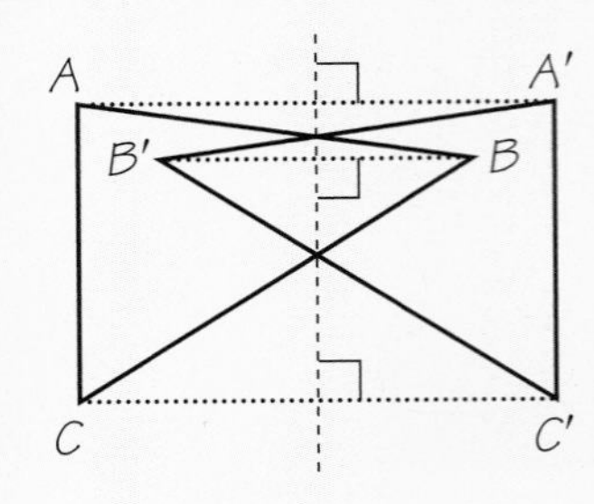

Start by finding the image of C and A. For B the reflection is from the right side to the left. Lines AB and $A'B'$ cross exactly on the mirror line. So do BC and $B'C'$. This is always true for lines that cross the mirror line.

Mirror lines can be vertical or horizontal.
You use the same method to draw the reflection.

To describe a reflection fully you need to give the equation of the mirror line.

EXAMPLE 3

The diagram shows an object P and its image Q after reflection.

(a) Draw in the mirror line as a dashed line.

(b) What is the equation of this mirror line?

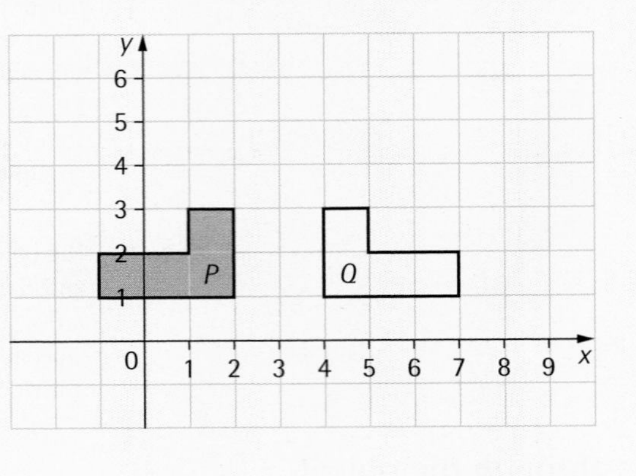

Continued ▼

(a)

(b) The equation of the mirror line is $x = 3$.

The mirror line must be the same distance from *P* as from *Q*.

The mirror line is parallel to the *y*-axis. The *x*-values of all points on it are 3. So the equation is $x = 3$.

EXERCISE 18A

You will need squared paper for each question.

1 Each diagram shows an object with its coloured image. Copy these diagrams onto squared paper and draw in the mirror line in each case.

(a)

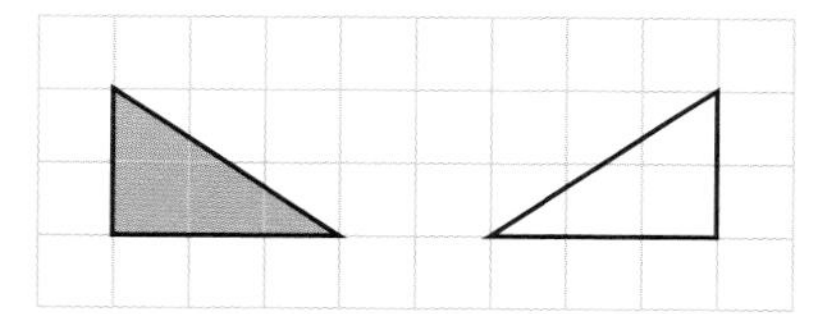

(b)

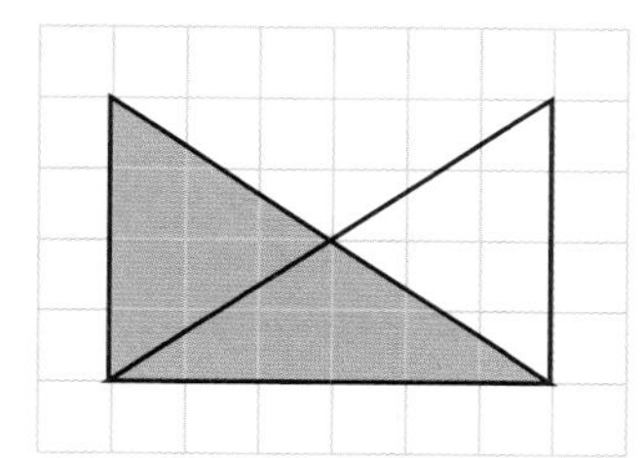

(c)

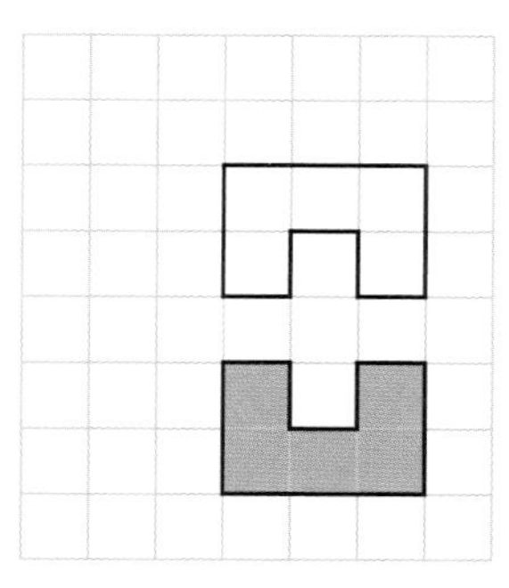

(d)

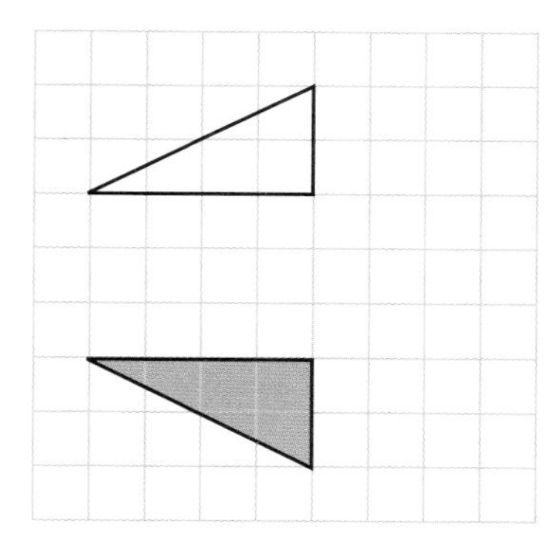

2 Copy these shapes and the dashed mirror line onto squared paper. Draw the reflected image in each case.

(a)

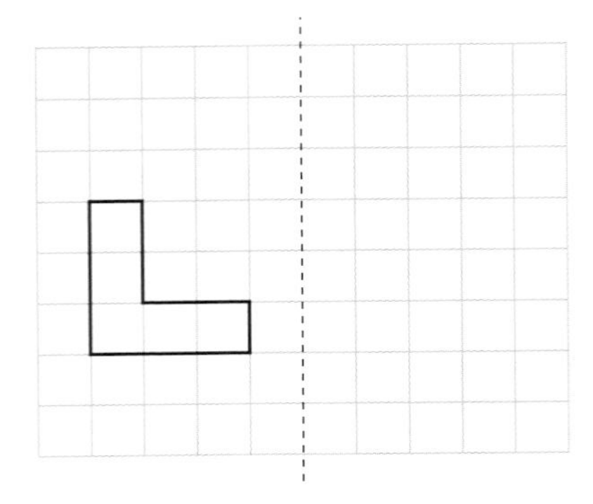

(b)

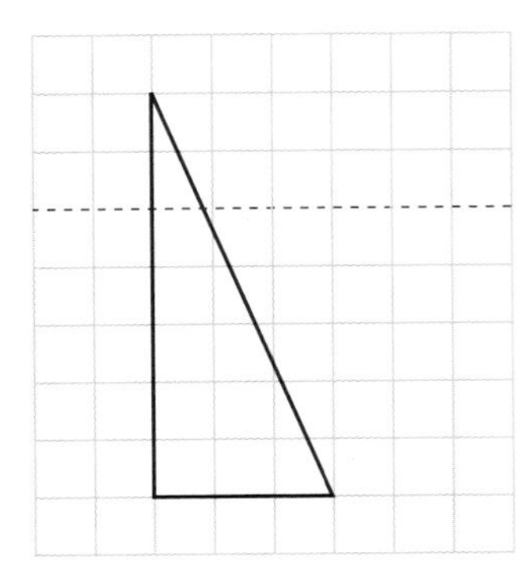

(c)

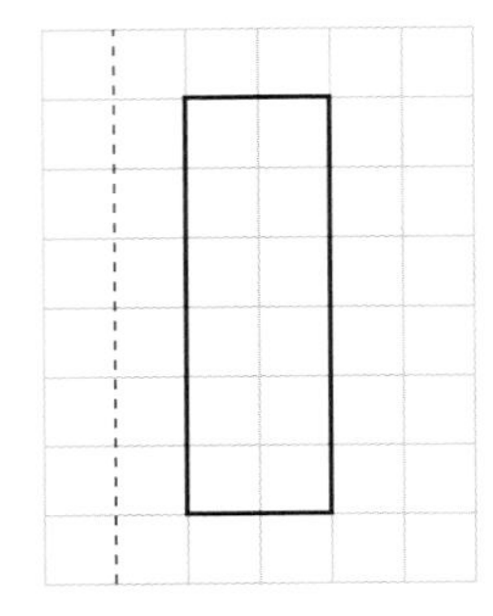

3 Copy the axes and triangle PQR onto squared paper.

(a) Reflect the triangle PQR in the y-axis and label the image $P'Q'R'$.

(b) Reflect the triangle PQR in the x-axis and label it $P''Q''R''$.

Use the y-axis as the mirror line.

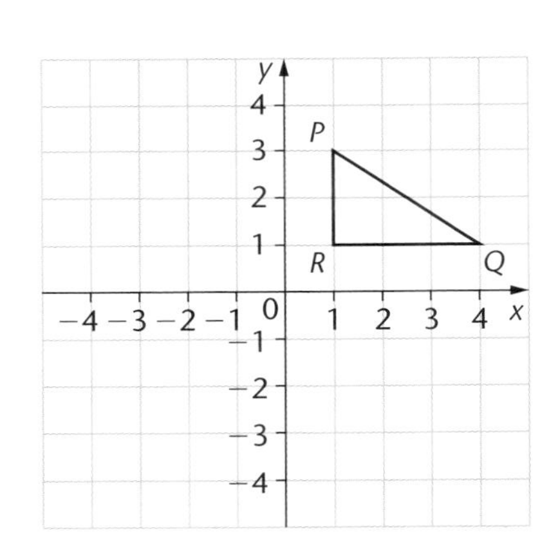

4 Copy the axes and the shape onto squared paper. Reflect the shape in the mirror line given by the equation $x = 1$.

The line $x = 1$ is shown by a dashed line.

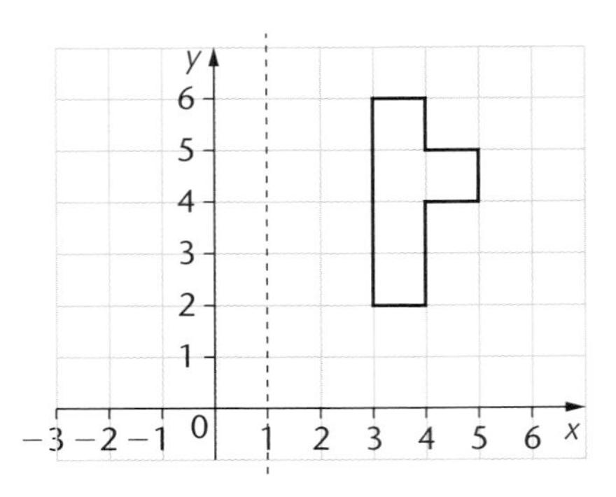

5 This five-sided polygon is to be reflected in the mirror line with equation $y = 1$.

(a) Copy the axes and the polygon onto squared paper.

(b) Draw in the mirror line as a dashed line.

(c) Draw the image after the reflection.

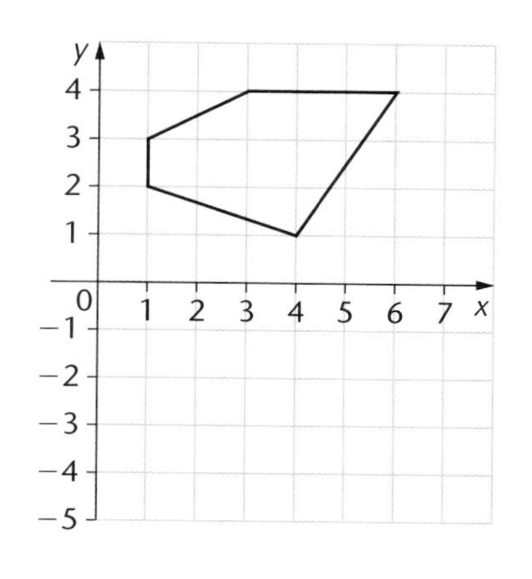

6 (a) Reflect triangle P in the line $x = -1$. Call this P'.

(b) Reflect triangle P' in the line $y = 1$. Call this P''.

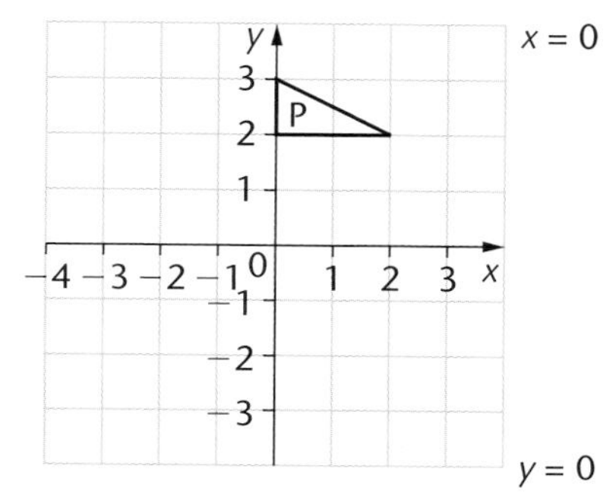

18.3 Rotation

A **rotation** turns an object either clockwise or anticlockwise through a given angle about a fixed point.

Always look carefully to see whether the rotation is clockwise or anti-clockwise.

The fixed point is called the **centre of rotation**. It can be either on the object or at the origin.

If it is on the object, it will be at a vertex or at the mid-point of one of the sides.

An object can be rotated through **quarter turn** (90°), **half turn** (180°) or **three-quarter turn** (270°).

EXAMPLE 4

Draw the image of this shape after it has been rotated through 90° anti-clockwise about the centre of rotation at

(a) *A* (b) *O* (c) *O*.

Tracing paper is very useful in all questions on rotation.

The image after each rotation is shown shaded.

Rotation about A

(a)

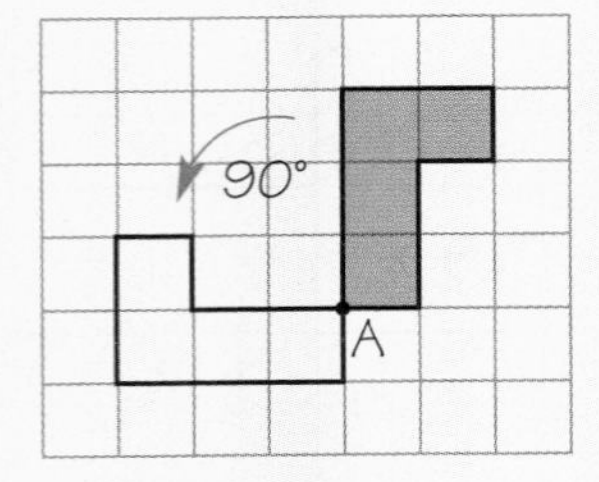

Trace the shape and the centre of rotation on to tracing paper. Hold the centre of rotation fixed with your pencil point. Turn the tracing paper through the required angle.

Draw the image you see on your diagram.

The image formed after rotation is the same size and shape as the original object.

Rotation about O

(b)

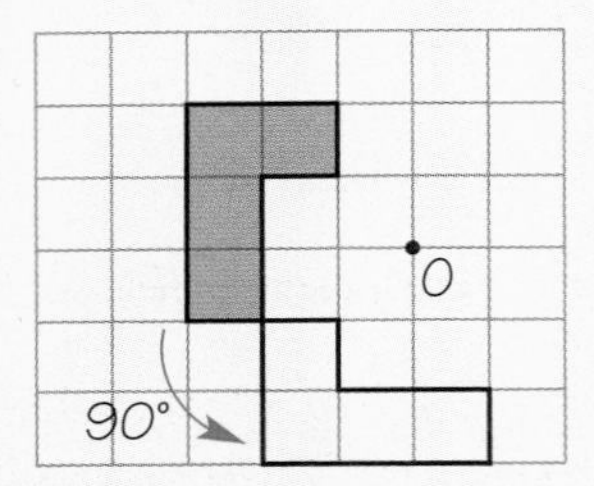

Rotation about O

(c)

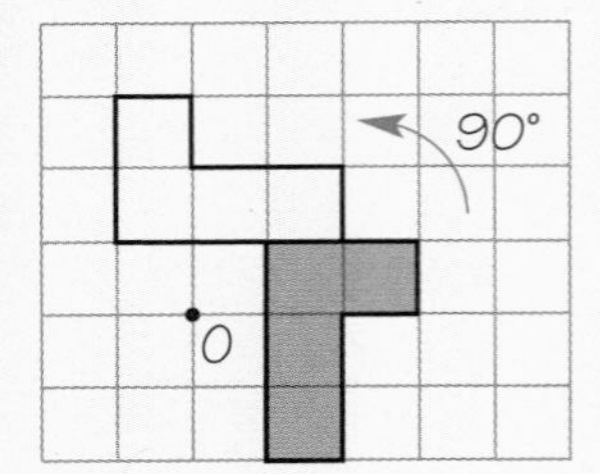

An object and its image after rotation are congruent.

EXAMPLE 5

Draw the image of the shape after rotation about O after

(a) a quarter of a turn anti-clockwise

(b) a rotation through 180° clockwise.

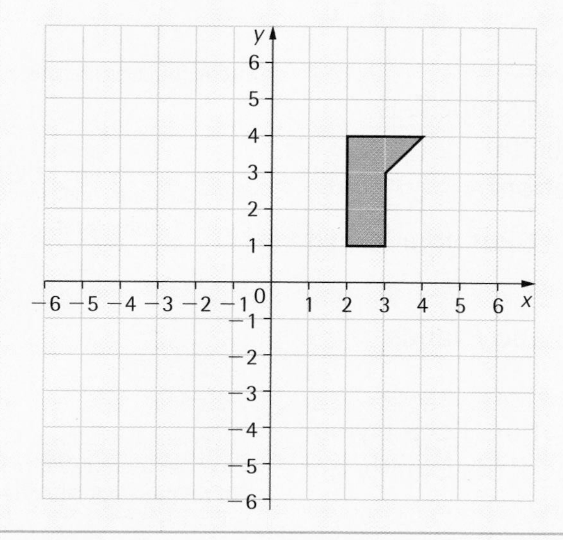

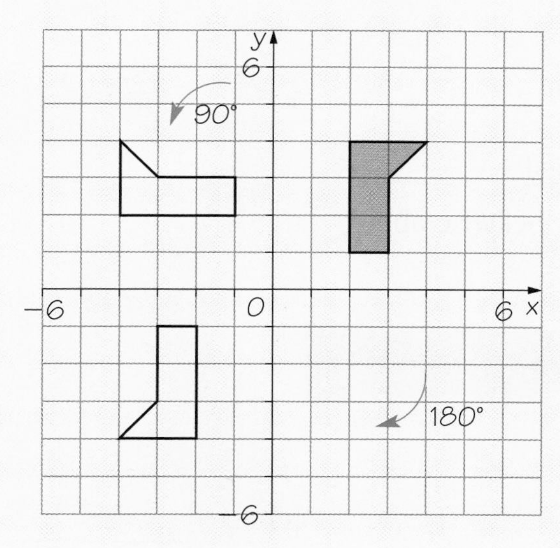

Use tracing paper to help you.

An image formed by a rotation through 180° clockwise is in the same position as an image formed by a rotation through 180° anti-clockwise.
They are both 'half turns' and it doesn't matter which direction you turn.

To describe a rotation fully you need to give

- the centre of rotation
- the angle of turn
- the direction of turn

Usually 90° or 180°.

Clockwise or anti-clockwise.

If you want to describe fully the rotation that takes shape P on to shape Q, the easiest way is to use tracing paper.

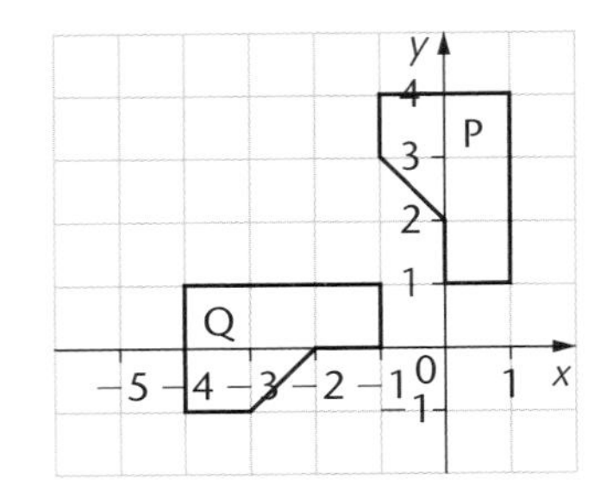

Remember to ask for some tracing paper when you attempt any transformation question on the examination paper. It is particularly useful for rotation questions.

Place the tracing paper over the whole of the diagram and draw shape P on it.

The longest side of the diagram in shape P is vertical. The corresponding side in shape Q is horizontal. So the angle of rotation must be 90° (or 270° if you rotate in the opposite direction).

Look for clues like this.

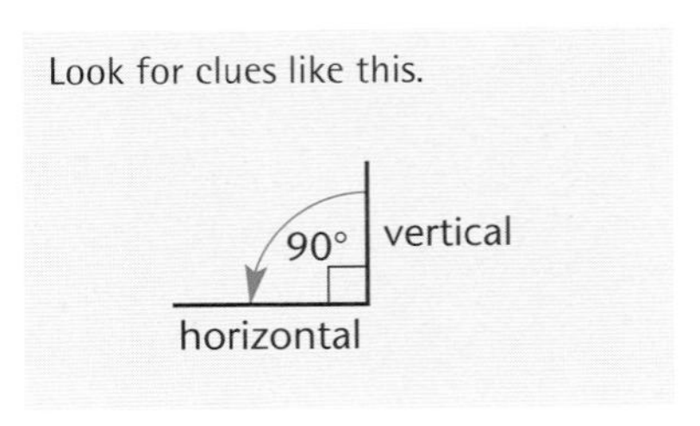

1 Try points on the diagram or the point O. Put your pencil on the point and hold it fixed.

2 Turn through 90° (or 270°) and see if shape P lands exactly on top of shape Q.

3 If it does not, try another centre of rotation. Keep trying different centres of rotation until shape P fits exactly on top of shape Q.

Once you have drawn the diagram, you can try several different centres of rotation quite quickly.

4 Describe the rotation giving
- the angle
- the direction
- the centre of rotation (as co-ordinates).

The rotation that takes P to Q is 90° anti-clockwise about point (0, 0).

EXERCISE 18B

1 Copy these shapes onto squared paper.

(a)

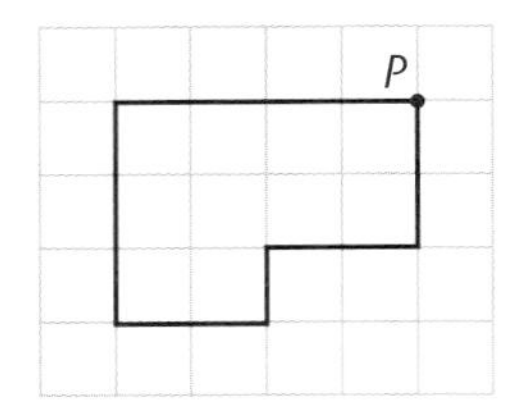

(b)

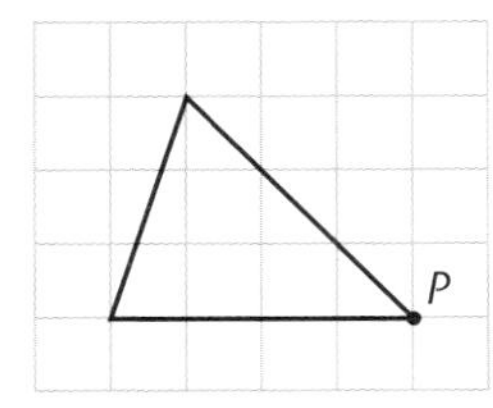

(c)

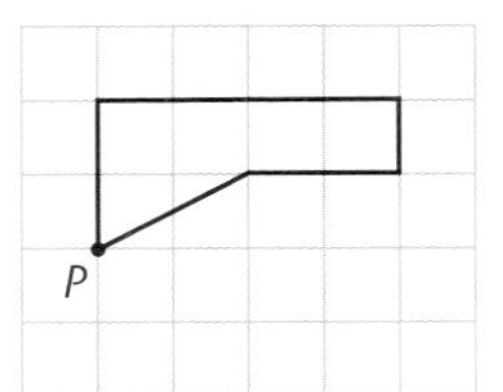

Rotate each shape
(a) a half turn clockwise
(b) a quarter turn anti-clockwise about the point P.

2 Copy this shape onto squared paper.

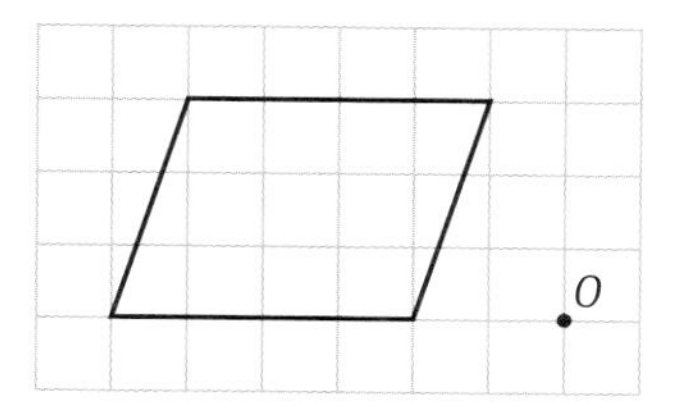

Draw the image of the shape after it has been rotated about the point O by

(a) 90° clockwise

(b) 180° anti-clockwise

(c) three-quarters of a turn clockwise.

3 On squared paper, draw x-axis and y-axis going from -6 to $+6$. Copy this shape onto your axes.

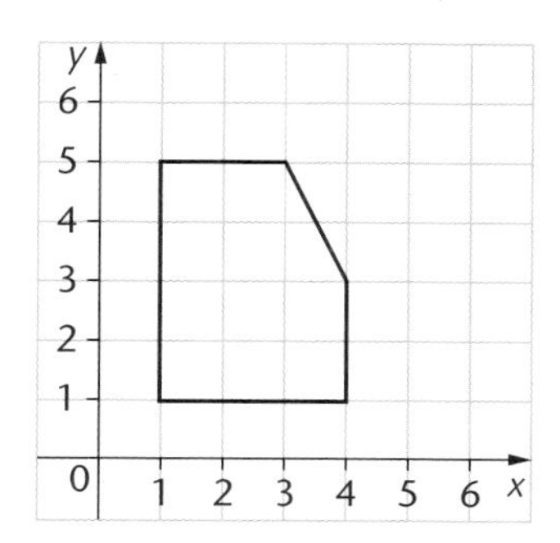

Draw the image of the shape after

(a) a quarter turn clockwise about the origin (0, 0). Label the image A.

(b) a quarter turn anti-clockwise about the origin (0, 0). Label the image B.

(c) 180° rotation anti-clockwise about the origin (0, 0). Label this image C.

4 Describe fully the single transformation which maps shape P onto shape Q.

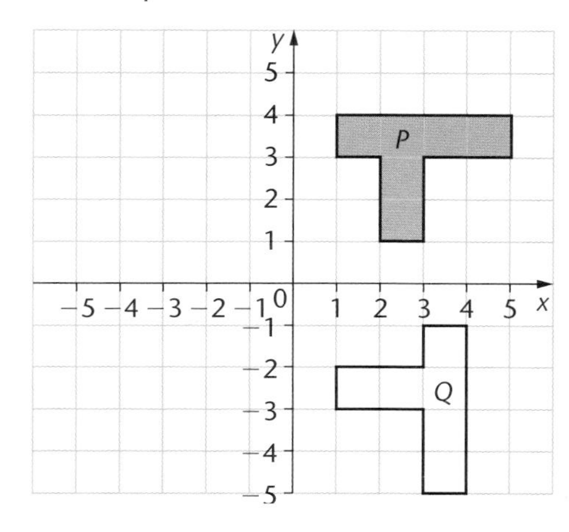

You need to give four pieces of information

- rotation
- the amount of turn
- the direction of turn
- the centre of rotation.

Use tracing paper to help you.

18.4 Translation

A **translation** slides a shape from one position to another.

In a translation every point on the shape moves the same distance in the same direction.

To describe a translation you need to give the distance and the direction of the movement.

EXAMPLE 6

Translate this shape 4 squares to the right and 2 squares up.

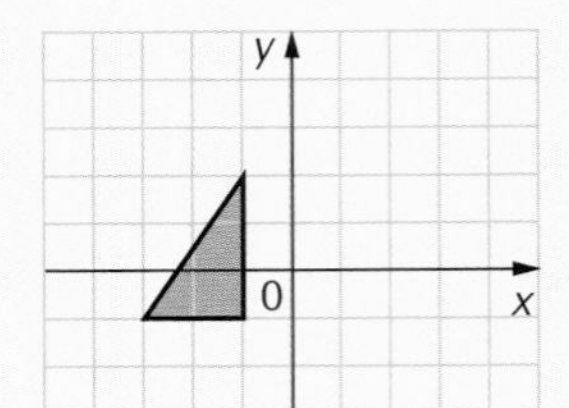

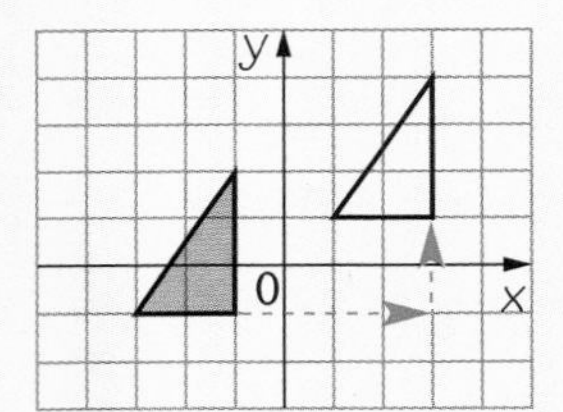

Choose any corner (vertex) of the shape and move this point 4 squares to the right and 2 squares up. Repeat this for the other vertices. Join up the vertices with straight lines.

An object and its image after a translation are congruent.

You can describe a translation by a **column vector**.

Column vectors always have tall brackets round them.

The translation in Example 6 has column vector $\begin{pmatrix} 4 \\ 2 \end{pmatrix}$.

The top number represents the movement in the x-direction and the bottom number represents the movement in the y-direction.

They are not fractions so do not draw a line between the numbers.

Notice that the translation to take the triangle back to its original position is $\begin{pmatrix} -4 \\ -2 \end{pmatrix}$.

Movements right and up are positive but left and down are negative.

EXAMPLE 7

Translate this shape by the vector $\begin{pmatrix} -3 \\ -3 \end{pmatrix}$.

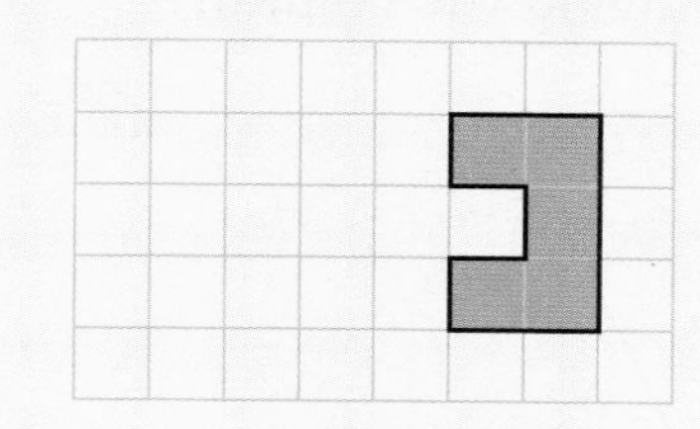

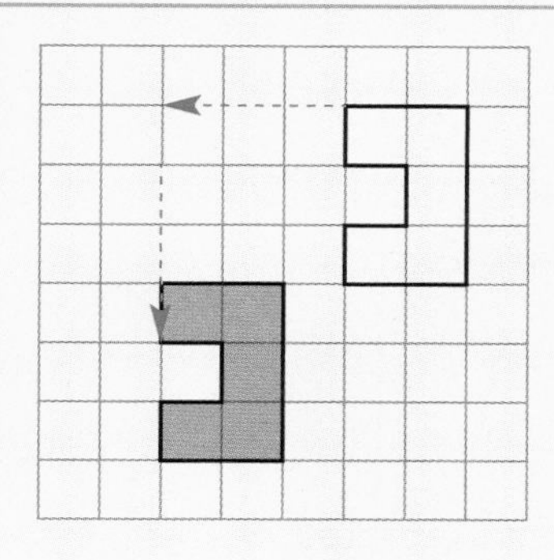

$\begin{pmatrix} -3 \\ -3 \end{pmatrix}$ means 3 squares *left* and 3 squares *down*.

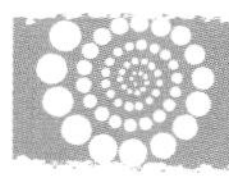

EXERCISE 18C

1 Copy these shapes onto squared paper and translate them by the amounts shown.

(i)

(ii)

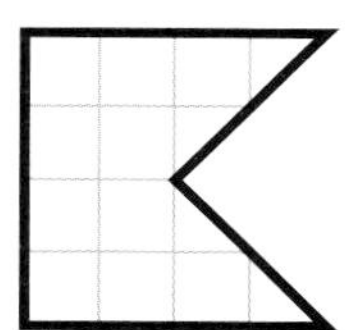

(a) $\begin{pmatrix} 3 \\ -2 \end{pmatrix}$ (b) $\begin{pmatrix} -4 \\ 2 \end{pmatrix}$ (c) $\begin{pmatrix} -5 \\ 0 \end{pmatrix}$ (d) $\begin{pmatrix} -1 \\ -6 \end{pmatrix}$ (e) $\begin{pmatrix} 0 \\ 3 \end{pmatrix}$

2 Copy this shape onto squared paper.
A translation of the shape moves the point P to the point P' on the image. Draw the complete image. What is the column vector that describes the translation?

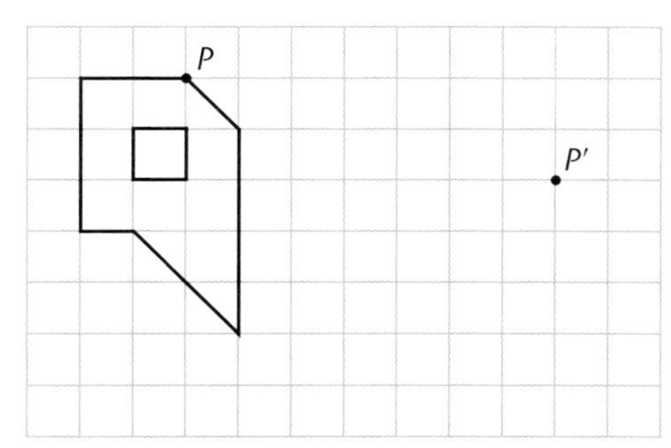

3 The triangle A is translated to new positions at B, C, D and E.
Describe each transformation by giving the column vector.

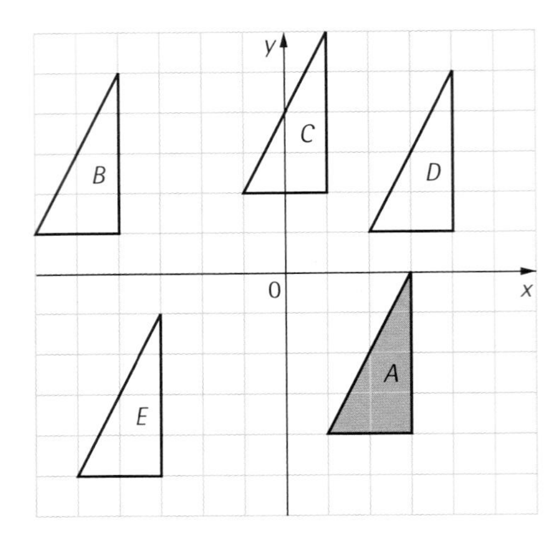

18.5 Enlargement

An **enlargement** changes the size of an object but not its shape.

The number of times the shape is enlarged is called the **scale factor**. This can be a positive whole number or a fraction.

In an enlargement, all the angles stay the same but all the lengths are changed in the same **proportion**. The image is **similar** to the object.

For help with proportionality see Chapter 12 .

Similar shapes
- have equal angles
- have lengths in the same proportion
- have perimeters in the same proportion.

EXAMPLE 8

Enlarge the rectangle ABCD by a scale factor 2.

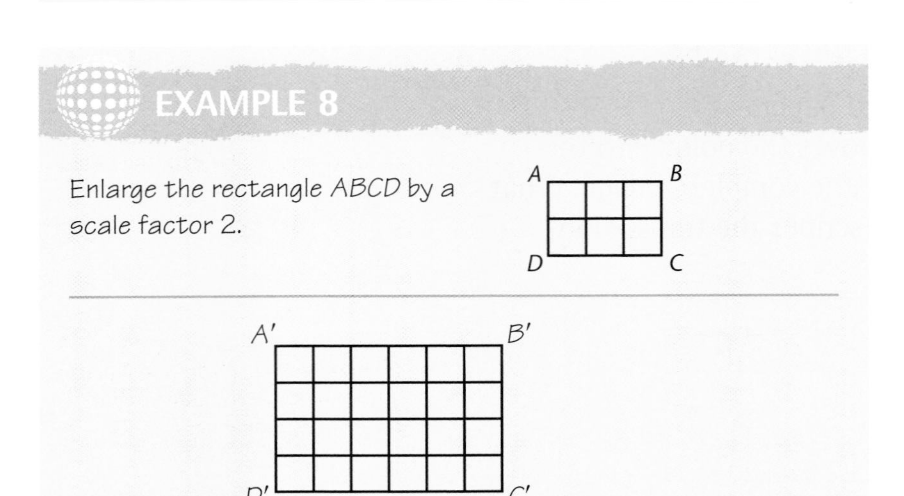

Every length on the object is multiplied by 2 on the image.

$AB = 3$, so
$A'B' = 2 \times 3 = 6$
$BC = 2$, so
$B'C' = 2 \times 2 = 4$

EXAMPLE 9

What is the scale factor of the enlargement that takes shape A to shape B?

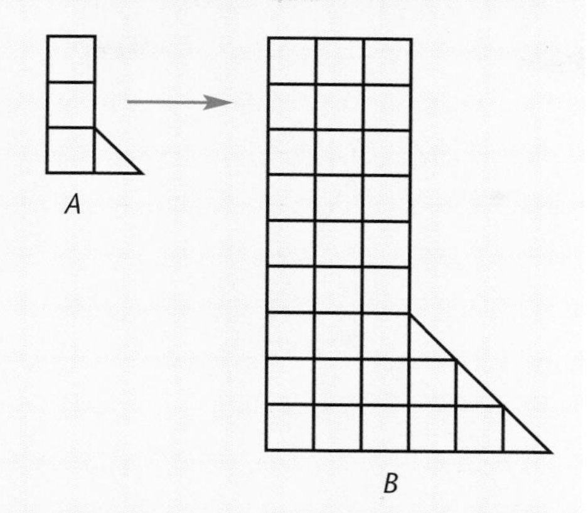

Compare the lengths of corresponding sides.

If no centre of enlargement is given you can then draw it close to the original shape.

Shape A is only 1 square wide. Shape B is 3 squares wide.
Shape A is 3 squares long. Shape B is 9 squares long.

Shape B is 3 times longer and 3 times wider than shape A.

The scale factor is 3.

The final position of an enlargement is determined by the position of the centre of enlargement. In Examples 8 and 9 there is no centre of enlargement so the image can be drawn anywhere.

When you enlarge from a centre of enlargement, the distances from the centre to each point are multiplied by the scale factor.

EXAMPLE 10

Copy the triangle ABC. Enlarge the triangle by scale factor 2 using the point O as the centre of enlargement.

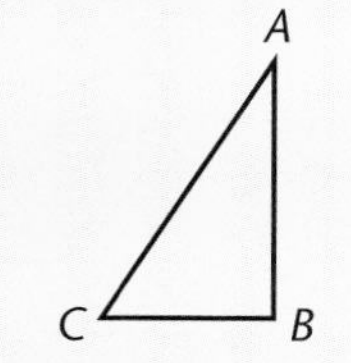

Multiply the distance OA by 2 to get OA', and similarly for the other vertices.
$OA' = 2 \times OA$
$OB' = 2 \times OB$
$OC' = 2 \times OC$

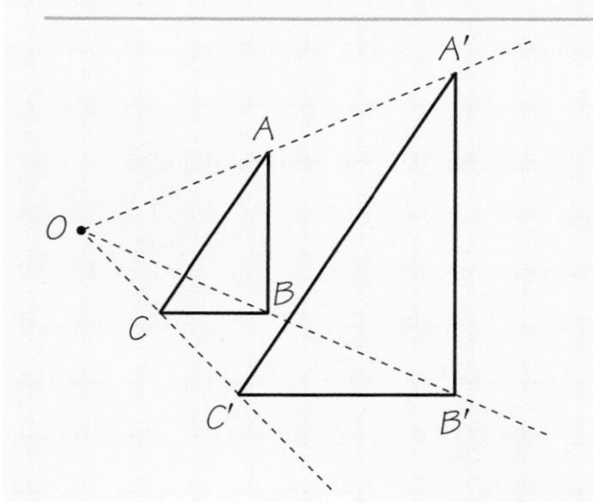

Always draw your diagram as accurately as possible using a pencil and ruler. Leave the construction lines on your diagram.

The centre of enlargement can be a point on the shape.

EXAMPLE 11

Enlarge the triangle by scale factor 3 using point P as the centre of enlargement.

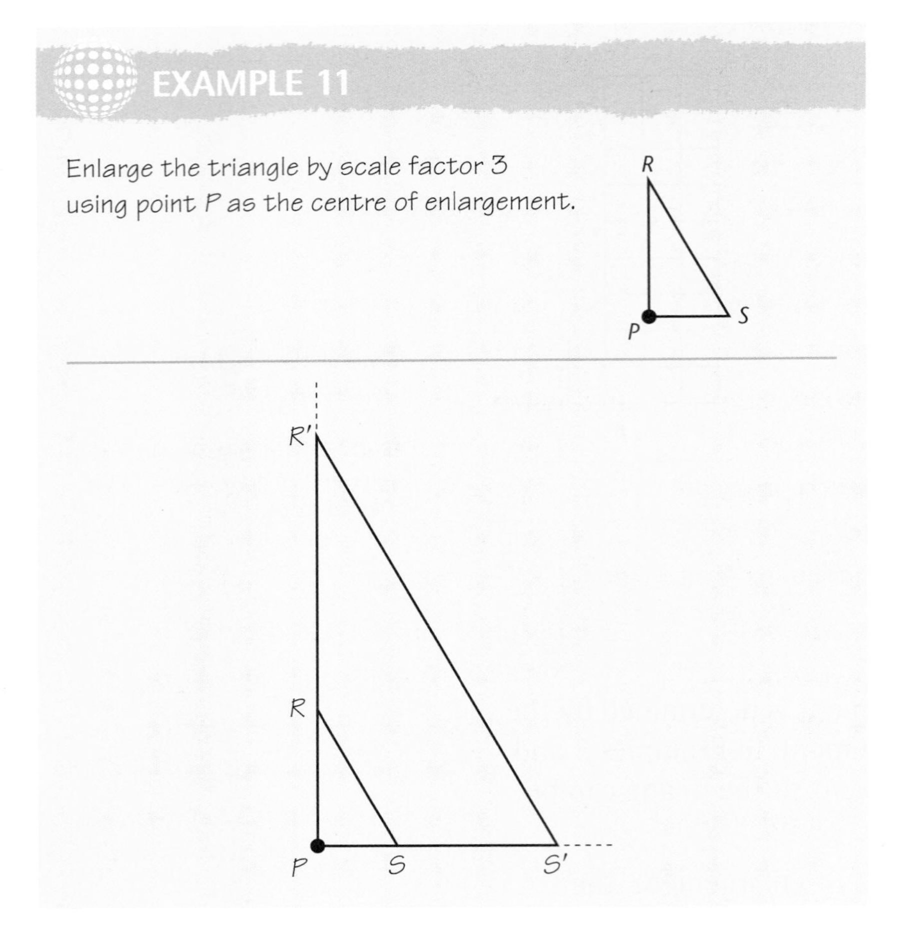

For an enlargement by a scale factor 3
$PR' = 3 \times PR$
$PS' = 3 \times PS$
The enlargement overlaps the original shape.

Sometimes the centre of enlargement is inside the shape.

To describe an enlargement fully you need to give the scale factor and the centre of enlargement.

EXERCISE 18D

1 For each of the following shapes work out the scale factor of the enlargement.

(a)

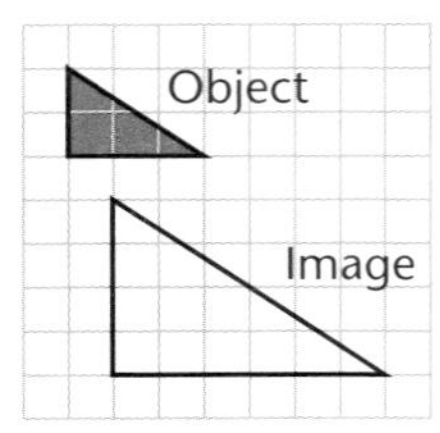

(b)

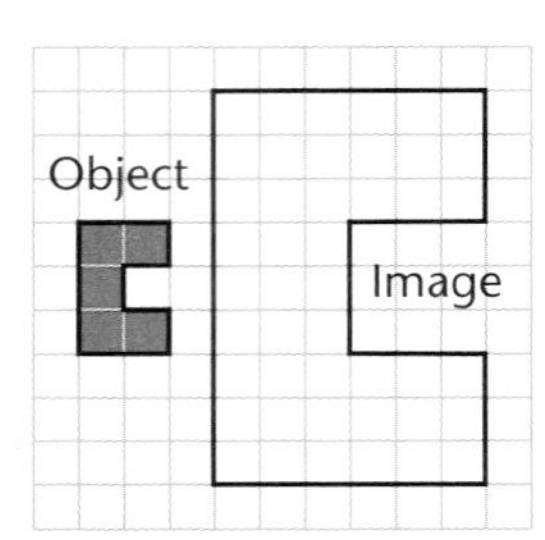

Use Example 9 to help you.

(c)

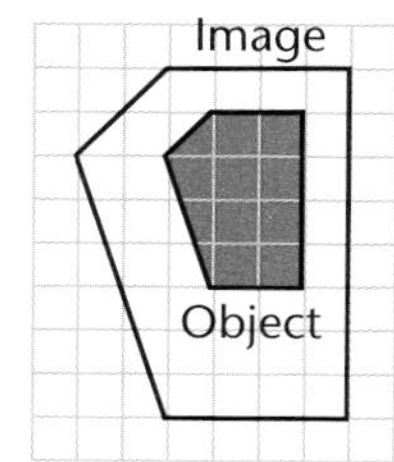

(d)

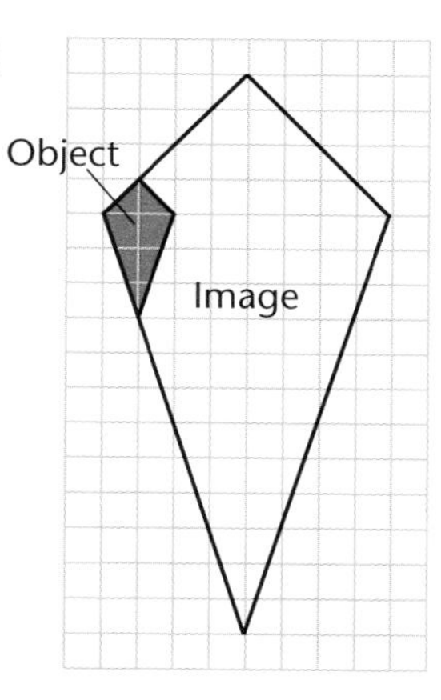

2 Copy each of the following shapes onto squared paper.
Enlarge each one by scale factor 2.

(a)

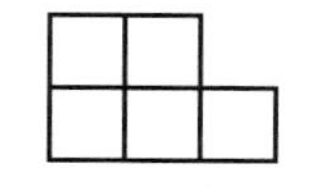

(b)

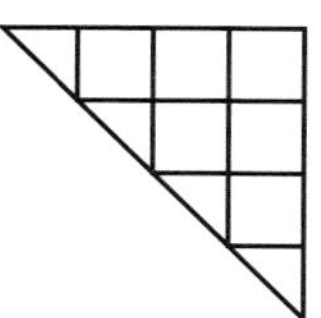

(c)

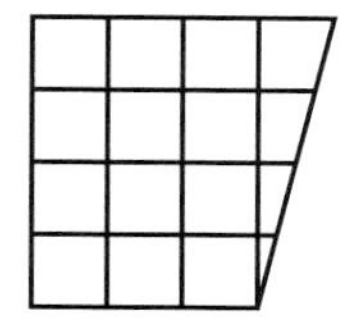

No centre of enlargement is given, so draw the enlargement close to the original shape.

3 Copy the following shapes onto squared paper.
Enlarge each one by scale factor 2 from the centre of enlargement *C*.

(a)

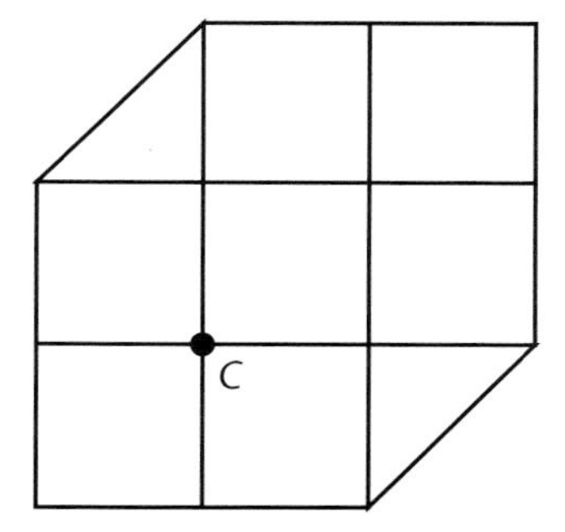

(b)

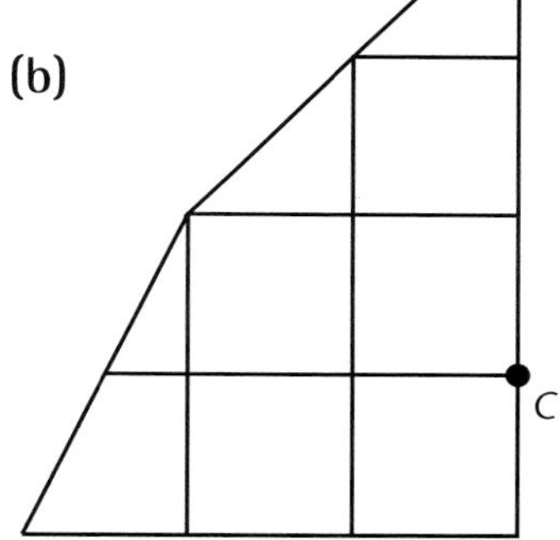

4 The vertices of triangle *K* are (1, 1), (1, 2) and (4, 1).

Enlarge the triangle *K* by scale factor 3 with (0, 0) as the centre of enlargement.

What are the co-ordinates of the image triangle *K*′?

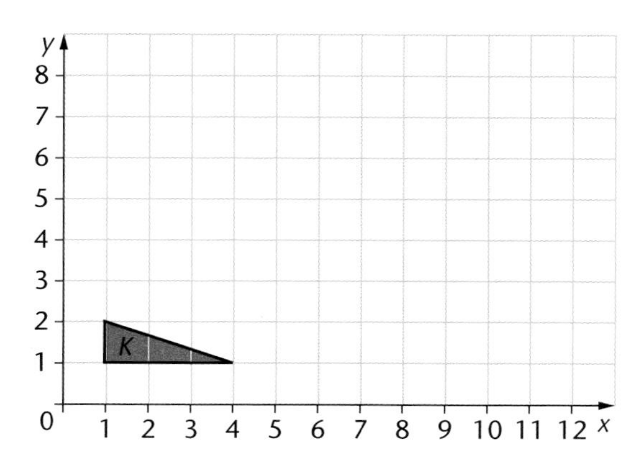

5 This right-angled triangle is to be enlarged by scale factor 4.
The centre of enlargement is at a point $P(1, 2)$.
Copy the diagram and draw the enlargement.

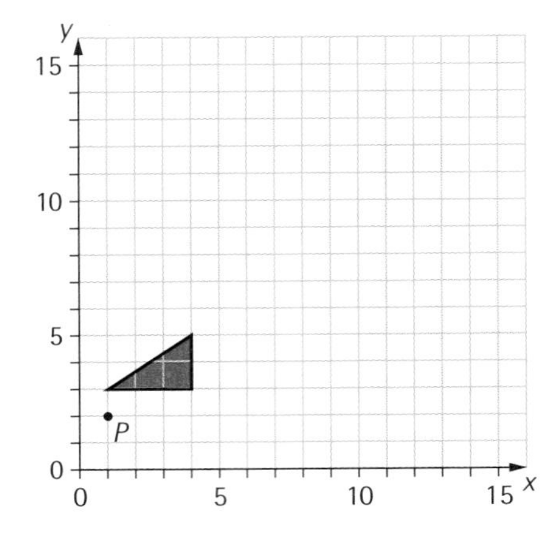

Use Example 10 to help you.

Fractional scale factors

A fractional scale factor of enlargement (<1) makes the image smaller.

Here, *ABCD* has been enlarged about centre *O*, by scale factor $\frac{1}{3}$.

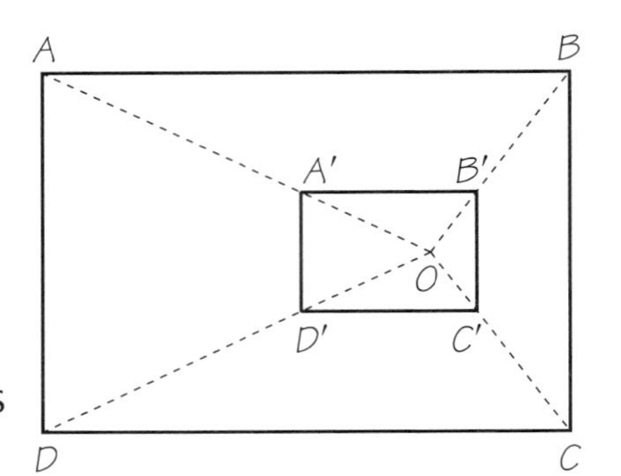

$OA' = \frac{1}{3}OA$
$OB' = \frac{1}{3}OB$
$OC' = \frac{1}{3}OC$
$OD' = \frac{1}{3}OD$

To find the position of the centre of enlargement, join vertices in the enlargement to the corresponding vertices in the original and continue these lines until they meet at a point.

The enlargement is smaller than the original.

This point is the centre of enlargement.

You draw in the construction lines for the enlargement.

EXAMPLE 12

Triangle *ABC* has been enlarged to produce the shaded triangle *A′B′C′*.

(a) What is the scale factor of the enlargement from *ABC* to *A′B′C′*?

(b) Mark on the grid the position of the centre of enlargement *P*, and give its co-ordinates.

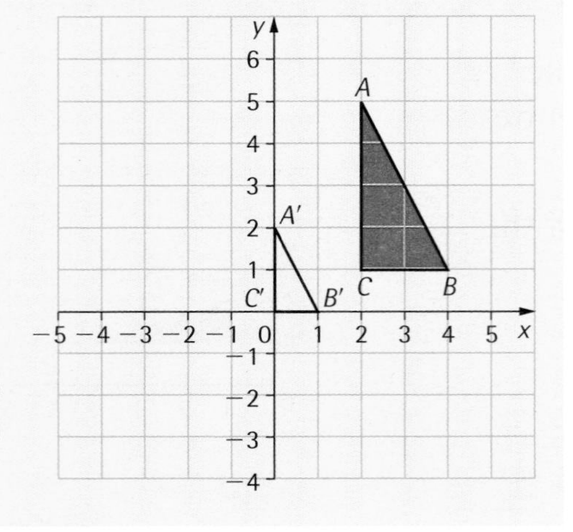

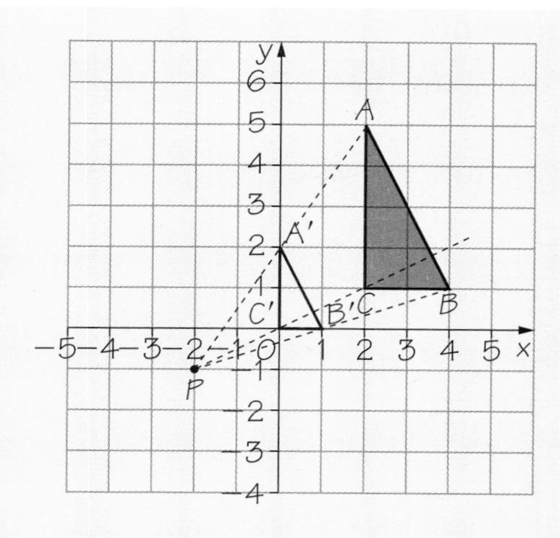

This enlargement has made a smaller image so the scale factor must be less than 1.

All lengths on triangle $A'B'C'$ are half the lengths on triangle ABC. So the scale factor is $\frac{1}{2}$.

Draw in the construction lines joining A to A', B to B' etc. These lines meet at point $P(-2, -1)$, the centre of enlargement.

(a) The scale factor is $\frac{1}{2}$.

(b) The centre of enlargement is at $P(-2, -1)$.

- A scale factor of 1 produces an image the same size as the object. The object and image are congruent.
- A scale factor > 1 produces an image larger than the object and the two shapes are similar.
- A scale factor < 1 produces an image that is smaller than the object and the two shapes are similar.

EXERCISE 18E

1 The rectangle X is enlarged to produce the image Y.

Use Example 12 to help you.

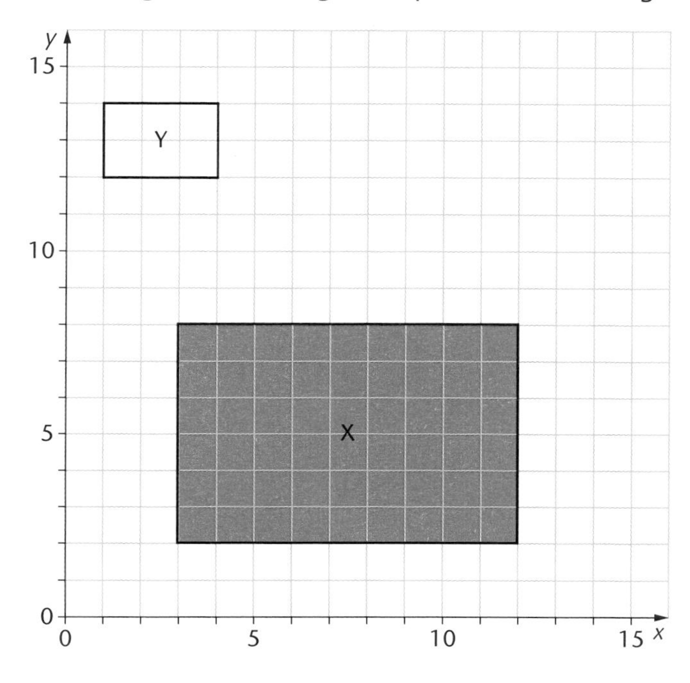

(a) What is the scale factor of this enlargement?

(b) Copy the diagram and construct lines to show the position of the centre of enlargement.

(c) What are the co-ordinates of the centre of enlargement?

2

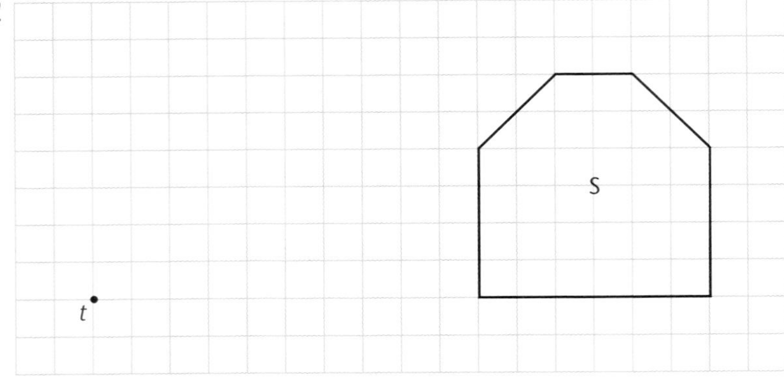

Copy the diagram and draw the image of shape S after an enlargement with scale factor $\frac{1}{2}$ and centre of enlargement at point *t*.

3

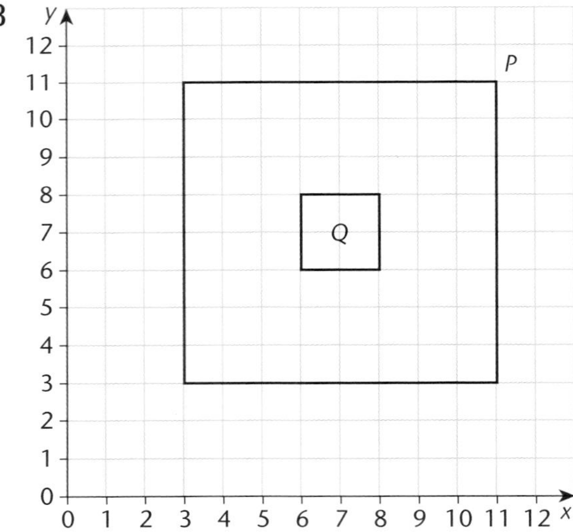

(a) Find the co-ordinates of the centre of enlargement when shape *P* has been enlarged to shape *Q*.

(b) What is the scale factor of this enlargement?

18.6 Congruence and similarity

Congruent shapes are exactly the same shape and exactly the same size.

You may need to turn shapes over before they fit exactly. Reflections in a mirror line are congruent.

All corresponding lengths and angles in the object and image are equal.

Similar shapes have exactly the same shape but are not the same size.

All corresponding angles in the object and image are equal. All lengths are in the same ratio or proportion. The ratio is the same as the scale factor of the enlargement.

All circles are mathematically similar.
All squares are mathematically similar.

EXAMPLE 13

(a) Which shapes are congruent to shape *A*?

(b) Which shapes are similar to shape *A*?

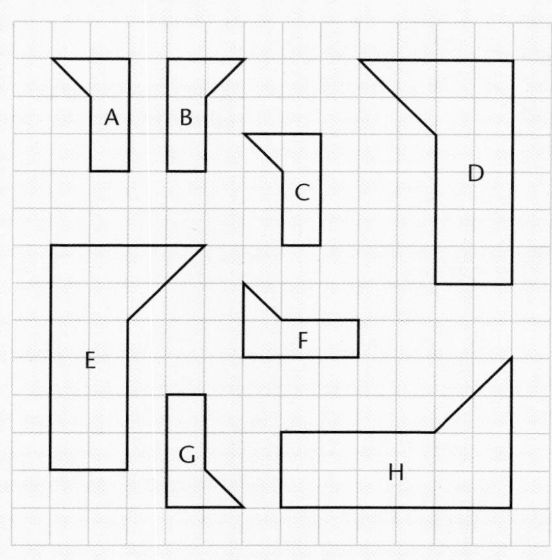

(a) *B*, *C*, *F* and *G* are all congruent to *A*.

(b) *D*, *E* and *H* are all similar to *A*.

D, *E* and *H* are all enlargements, scale factor 2, of *A*.

EXAMPLE 14

Which of these shapes are similar to B?
Explain your answer.

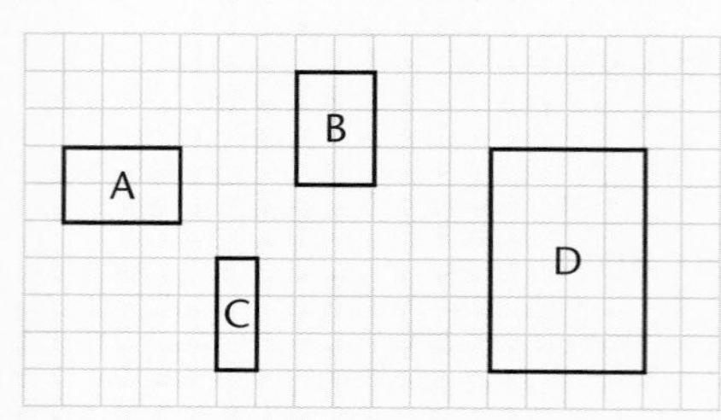

Shape A	shapes A and B are both of length 3 units and width 2 units so they are identical.
Shape C	length the same as B width not the same as B C and B are not similar.
Shape D	length of D = 2 × length of B width of D = 2 × width of B D and B are similar.

A and B are not similar but they are congruent.

D is an enlargement of *B*, scale factor 2.

There are four different rules you can use to prove that two triangles are congruent.

You need to remember these rules.

- Three sides equal (Side, Side, Side – SSS)

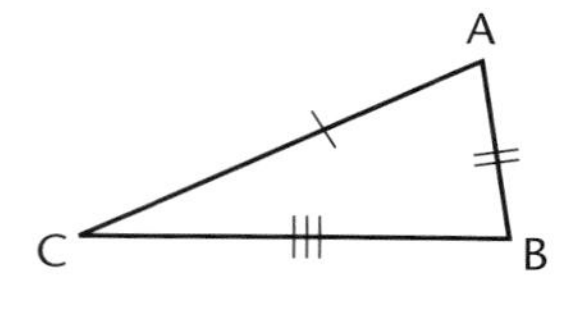

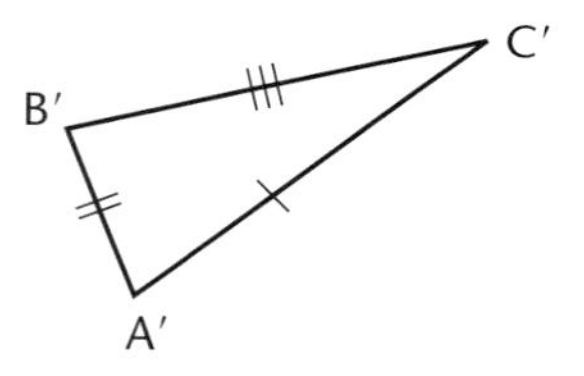

When two triangles have 3 equal sides they *must* be congruent.

- Two sides equal and the included angle the same (Side, Angle, Side – SAS)

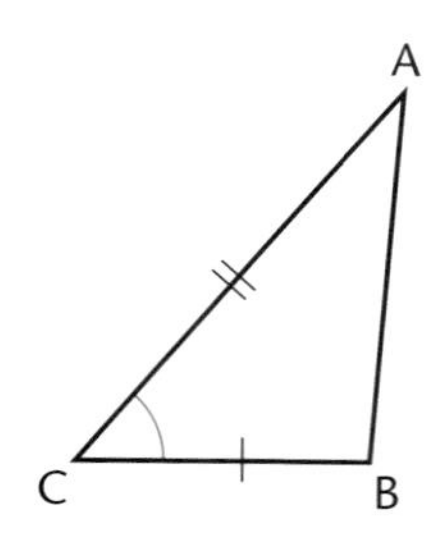

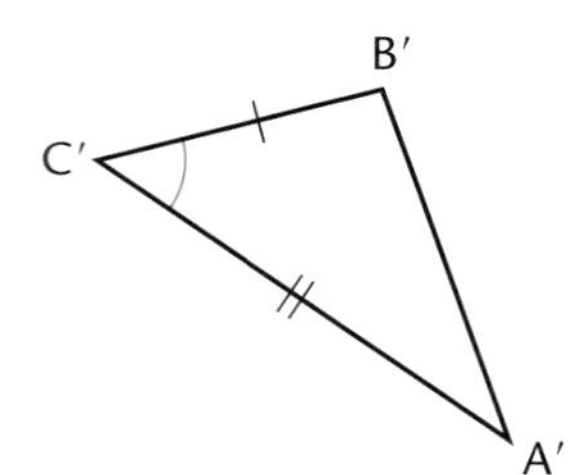

This case is easy to spot.

- Two angles the same and a corresponding side is equal (Angle, Side, Angle – ASA or Side, Angle, Angle – SAA)

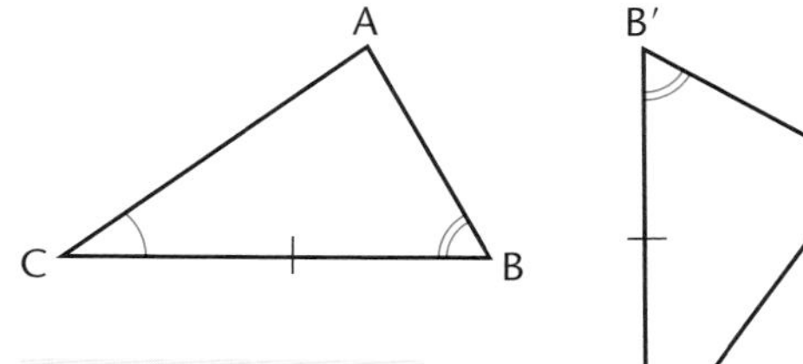

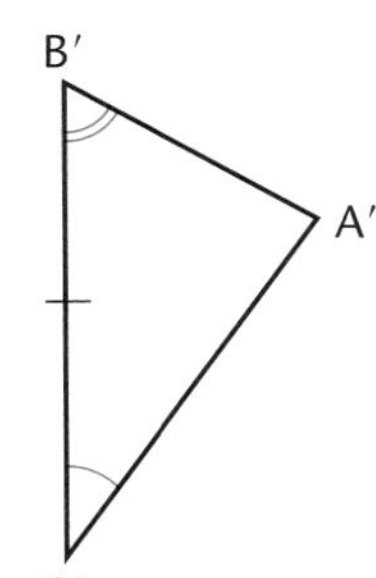

BC and *B′C′* are the corresponding sides.

or

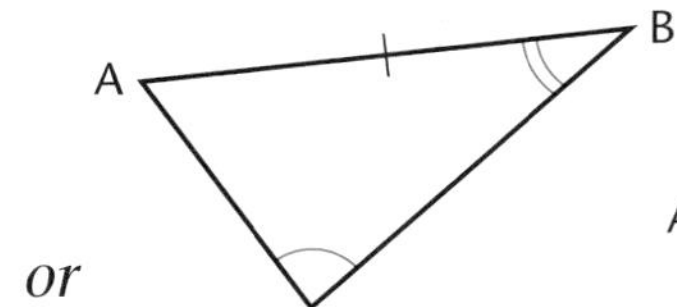

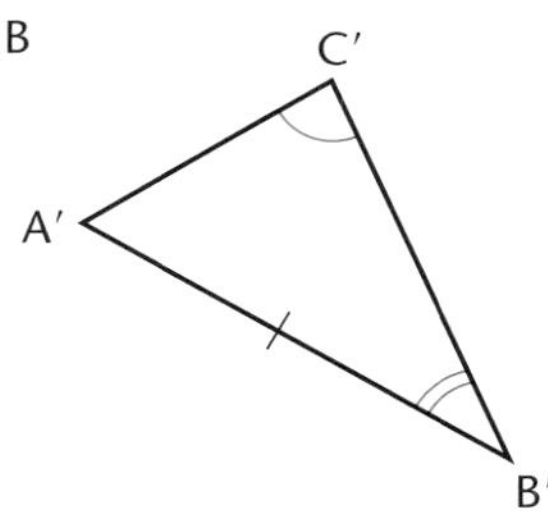

AB and *A′B′* are the corresponding sides.

Corresponding side means 'the side in the same position in both triangles'.

- A right angle, a hypotenuse and a corresponding side are equal (Right angle, Hypotenuse, Side – RHS)

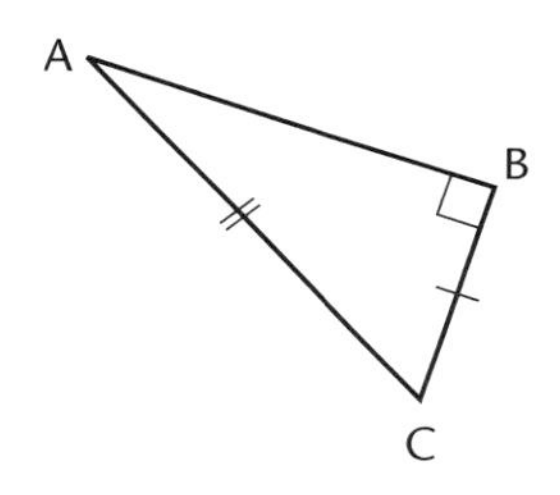

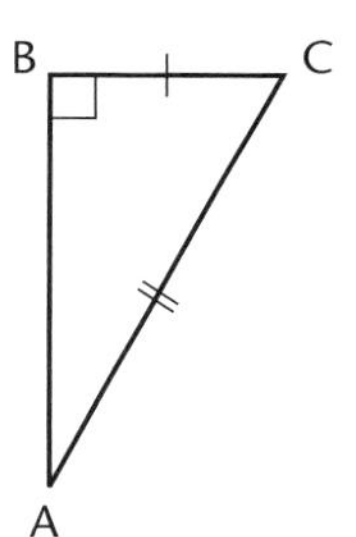

The hypotenuse is the side opposite the right angle.

BC and *B′C′* are the corresponding sides. They are the shortest sides in each triangle.

EXAMPLE 15

State whether any two of these triangles are congruent and give your reasons.

A

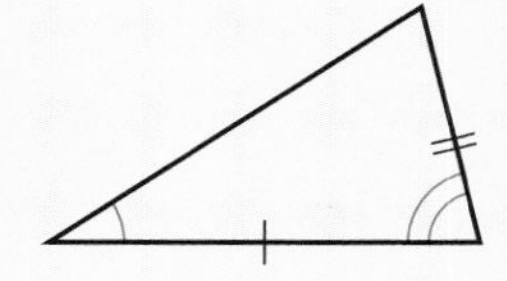

B

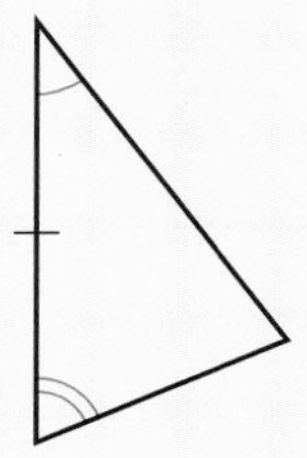

C

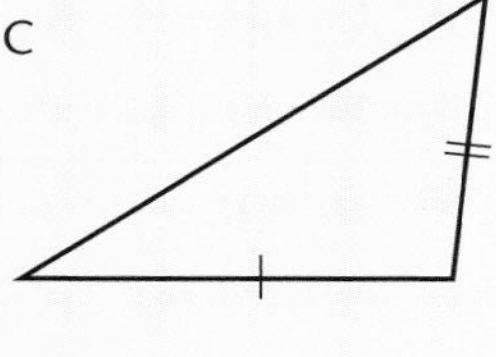

A and B are congruent, as they have 2 pairs of equal angles, and corresponding sides (the side between the angles) equal.

Triangle C cannot be congruent with either A or B, as C has an obtuse angle and A and B each have 3 acute angles.

There is one case where triangles are sometimes thought to be congruent, but they are not. Look at these diagrams

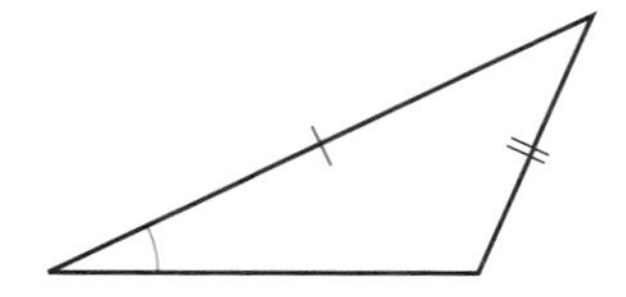

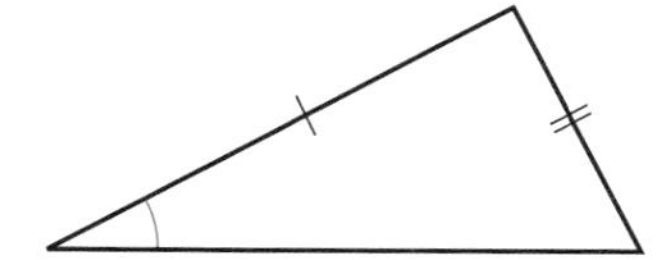

These triangles have two sides which are the same length and an angle which is the same size.

You might think that this is a reason for congruence but one look at these triangles should convince you that it is not!

EXERCISE 18F

1 Look at the shapes in this diagram.

(a) Which shapes are congruent to shape *A*?

(b) Which shapes are similar to shape *A*?

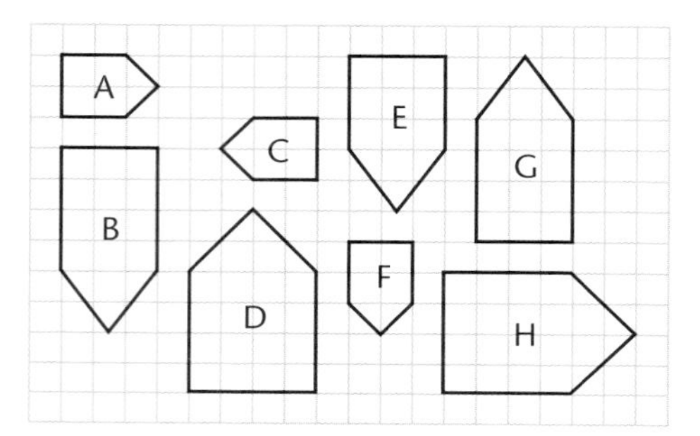

2 Copy this shape onto squared paper. On the same squared paper draw one shape that is similar and one shape that is congruent to this shape.

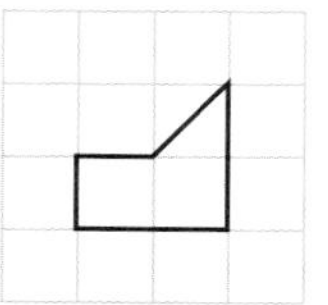

In similar shapes, all sides are enlarged by the same scale factor.

3 For each set of shapes, write down the letters of the shapes that are similar to each other.

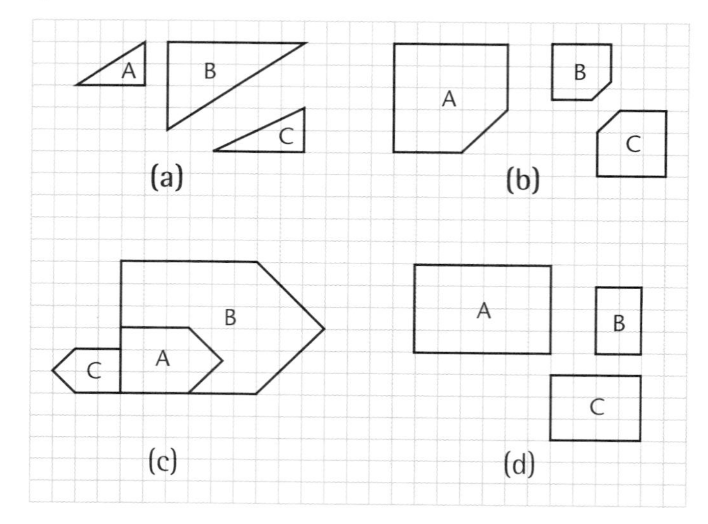

4 Look at the following pairs of triangles. State which pairs are congruent and give the appropriate reason.

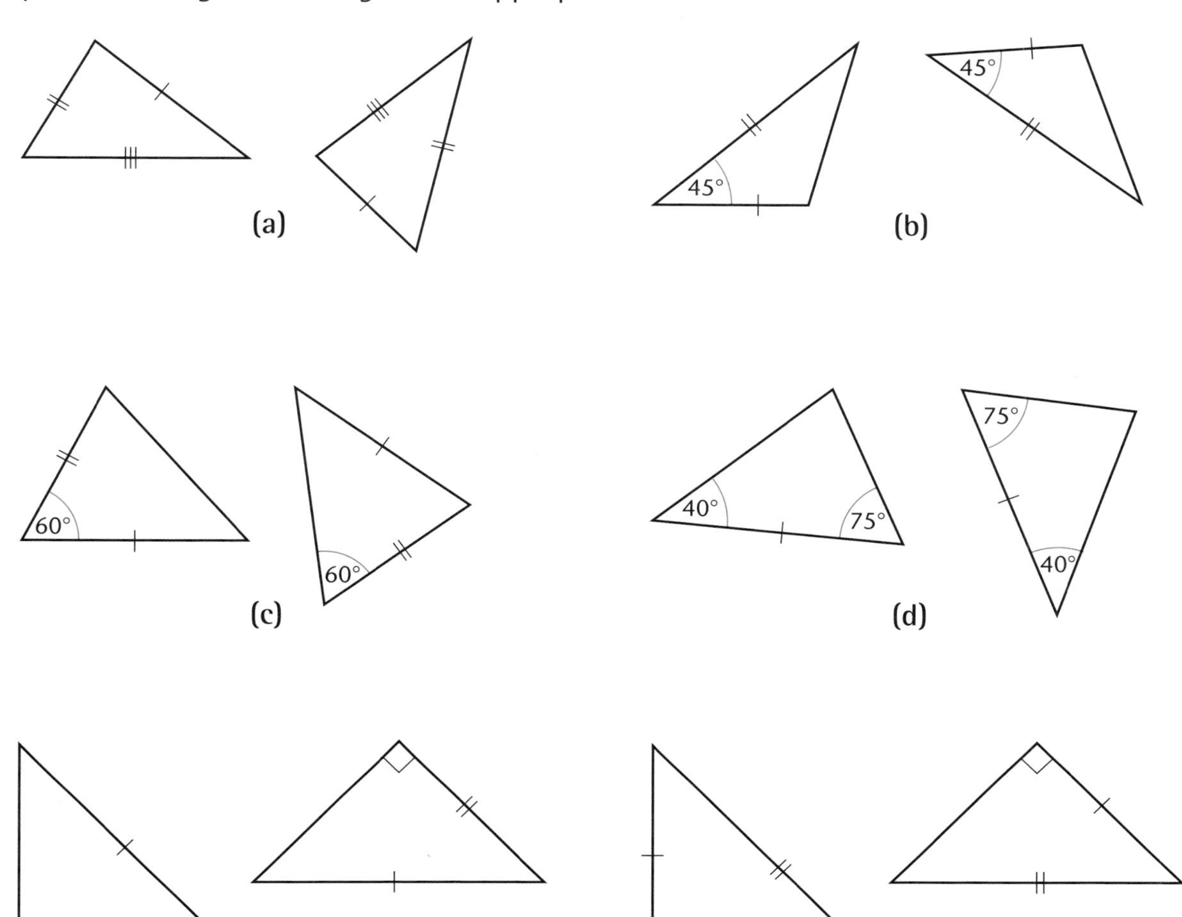

EXAMINATION QUESTIONS

1

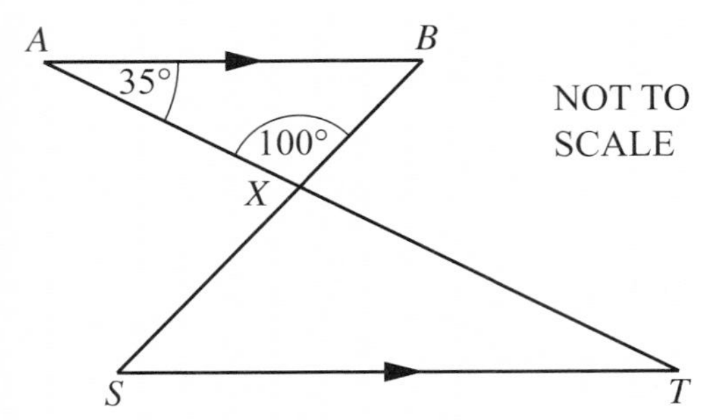

In the diagram AB and ST are parallel. AT and BS meet at X.
Angle $AXB = 100°$ and angle $BAX = 35°$.

(a) Write down the value of angle XTS. [1]

(b) Write down the correct mathematical word to complete the following sentence.
"Triangles ABX and TSX are ...." [1]

(c) $AB = 3.0$ cm, $ST = 6.0$ cm, $XS = 3.5$ cm and $XT = 4.3$ cm.
Calculate the length of AX. [2]

(CIE Paper 1, Nov 2001)

2

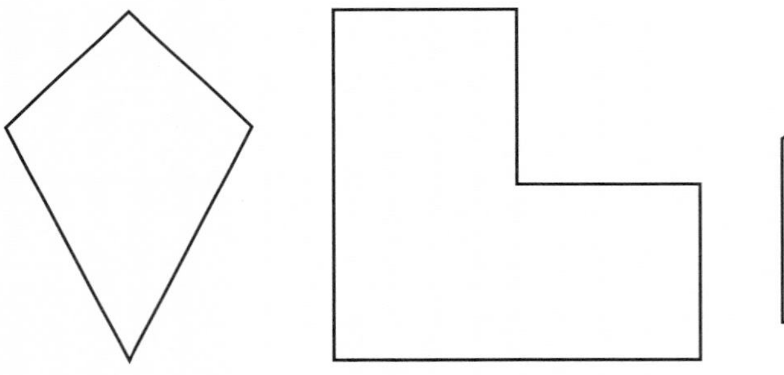

Copy each of the shapes above and draw **one** line which divides each one into two congruent shapes. [3]

(CIE Paper 1, Nov 2003)

3

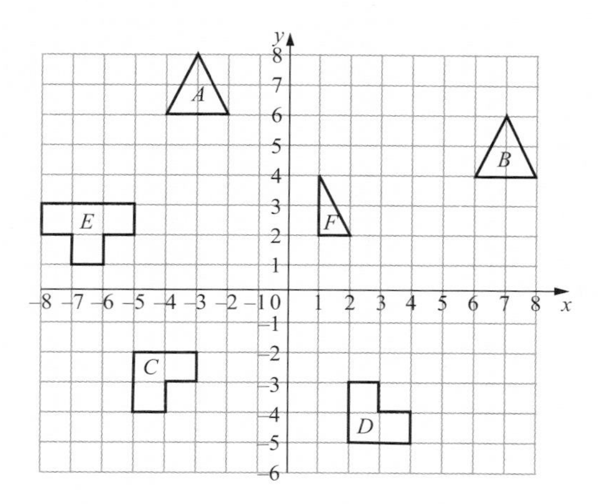

(a) Describe fully the single transformation that maps
 (i) shape A onto shape B, [2]
 (ii) shape C onto shape D.

(b) On a copy of the grid above, draw
 (i) the reflection of shape E in the y-axis, [2]
 (ii) the enlargement of shape F, with scale factor 2 and centre (0, 0).

(CIE Paper 3, Jun 2003)

4

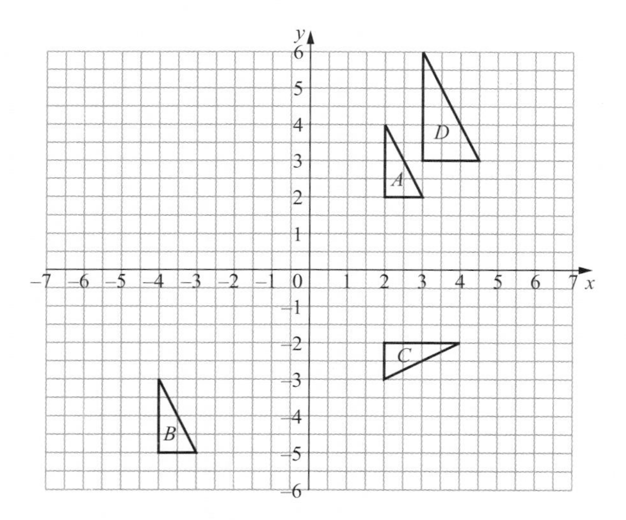

(a) Describe fully the single transformation that maps triangle A onto triangle B. [3]

(b) Describe fully the single transformation that maps triangle A onto triangle C. [3]

(c) Find the centre and the scale factor of the enlargement that maps triangle A onto triangle D. [2]

(d) On a copy of the grid
 (i) draw the image of triangle A under a reflection in the line $x = -1$, [2]
 (ii) draw the image of triangle B under a rotation of 180° about $(-4, -3)$. [2]

(CIE Paper 3, Nov 2004)

5

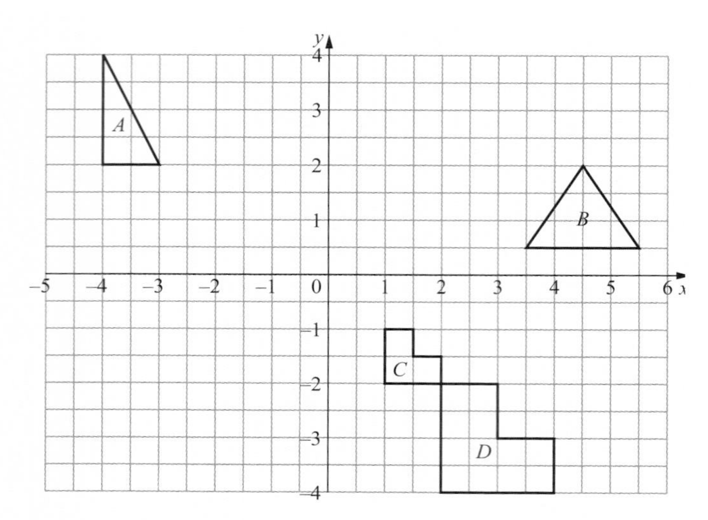

(a) A translation is given by $\begin{pmatrix} 6 \\ 3 \end{pmatrix} + \begin{pmatrix} -3 \\ -4 \end{pmatrix}$.

(i) Write this translation as a single column vector. [2]

(ii) On the grid, draw the translation of triangle A using this vector. [2]

(b) Another translation is given by $-2\begin{pmatrix} 1 \\ -1 \end{pmatrix}$.

(i) Write this translation as a single column vector. [2]

(ii) On a copy of the grid, draw the translation of triangle B using this vector. [2]

(c) Describe fully the single transformation that maps shape C onto shape D. [3]

(d)

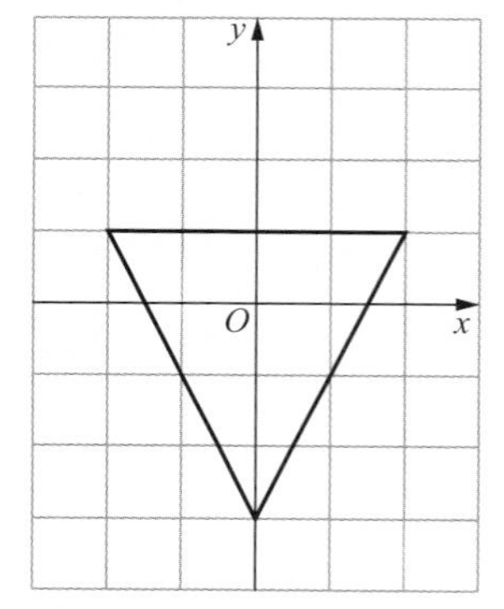

The triangle in the diagram above is isosceles.

(i) How many lines of symmetry does this triangle have? [1]

(ii) Write down the order of rotational symmetry of this triangle. [1]

(iii) On the grid above, draw the rotation of this triangle about O through 180°. [2]

(iv) Describe fully another single transformation that maps this triangle onto your answer for part (d)(iii). [2]

(CIE Paper 3, June 2005)

Chapter 19

Vectors

This chapter will show you how to

- ✔ define a vector quantity
- ✔ use the correct notation
- ✔ multiply by scalars
- ✔ add and subtract vectors

19.1 Definition of a vector

You have encountered **vectors** before in bearings and also in translation.

Velocity is a vector quantity.

Examples of vectors are

(a) 20km on a bearing 045°

(b) translate a shape by the column vector $\begin{pmatrix} 6 \\ -3 \end{pmatrix}$

See p 511.

A vector is a quantity which has both magnitude (size) and direction.

Quantities which have magnitude but no direction are called **scalar** quantities.

Mass is a scalar.
Speed is a scalar.

A vector can be shown by a **directed line segment** which is just a line with an arrow on it.

Length of line = magnitude of the vector.
Direction of arrow = direction of vector.

19.2 Vector notation

You have already met the column vector which gives the movement in the x and y directions.

There are other ways of writing vectors and you will need to know both of them.

1. Using capital letters.

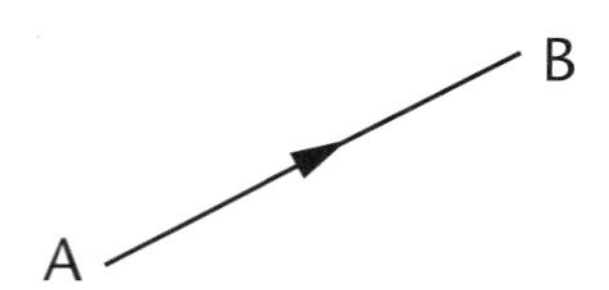

The vector shown is written $\overrightarrow{AB}$

2. Using single small letters.

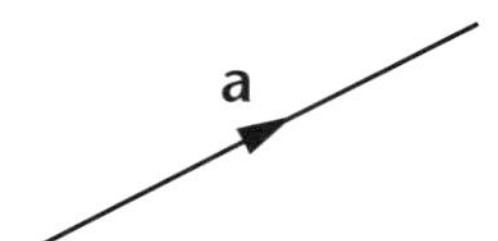

Textbooks will use **bold** letters like **a** but as you can't do this very easily it is normal to underline the vector and write $\underline{a}$.

EXAMPLE 1

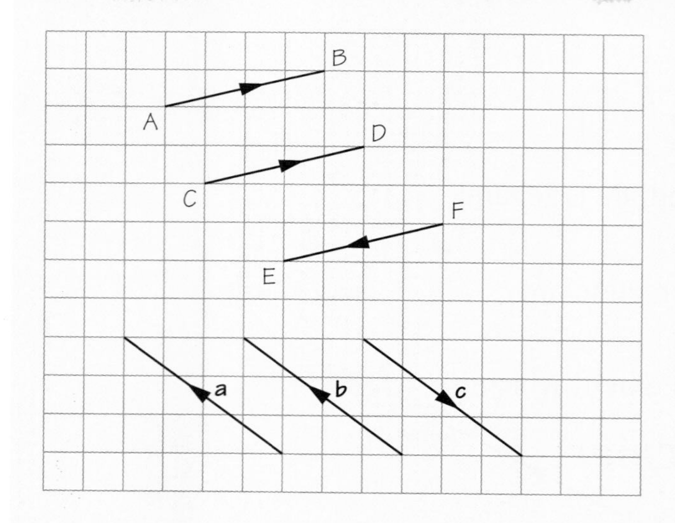

Use the grid to write down the column vectors for each of the following vectors.

(a) $\overrightarrow{AB}$ (b) $\overrightarrow{CD}$ (c) $\overrightarrow{EF}$

(d) **a** (e) **b** (f) **c**

(a) $\overrightarrow{AB} = \begin{pmatrix} 4 \\ 1 \end{pmatrix}$ (b) $\overrightarrow{CD} = \begin{pmatrix} 4 \\ 1 \end{pmatrix}$ (c) $\overrightarrow{EF} = \begin{pmatrix} -4 \\ -1 \end{pmatrix}$

(d) $\mathbf{a} = \begin{pmatrix} -4 \\ 3 \end{pmatrix}$ (e) $\mathbf{b} = \begin{pmatrix} -4 \\ 3 \end{pmatrix}$ (f) $\mathbf{c} = \begin{pmatrix} 4 \\ -3 \end{pmatrix}$

Did you notice that $\overrightarrow{AB}$ and $\overrightarrow{CD}$ are equal?
Look at the diagram, they have the same magnitude **and** direction.
Did you notice that $\overrightarrow{AB} \neq \overrightarrow{EF}$?
They have the same magnitude but not the same direction.

You can see that the vector **c** is in exactly the opposite direction to **a** and **b**.
In the column vector both signs have changed.
A negative sign in front of a vector reverses its direction.

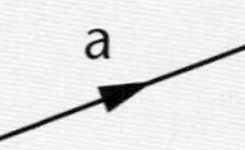

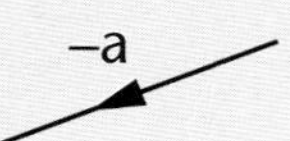

EXERCISE 19A

1 Use the grid below to write down the column vectors for each of the following vectors.

(a) $\overrightarrow{AB}$ (b) $\overrightarrow{CD}$ (c) $\overrightarrow{EF}$

(d) a (e) b (f) c

(g) d (h) e (i) f

(j) g

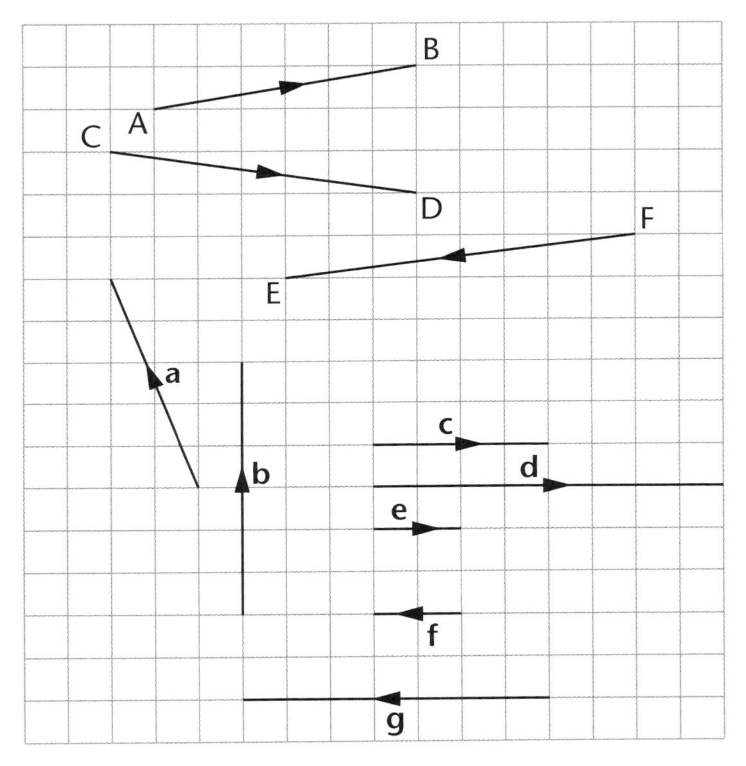

19.3 Multiplication by a scalar

Vectors are equal if they have the same magnitude and direction.

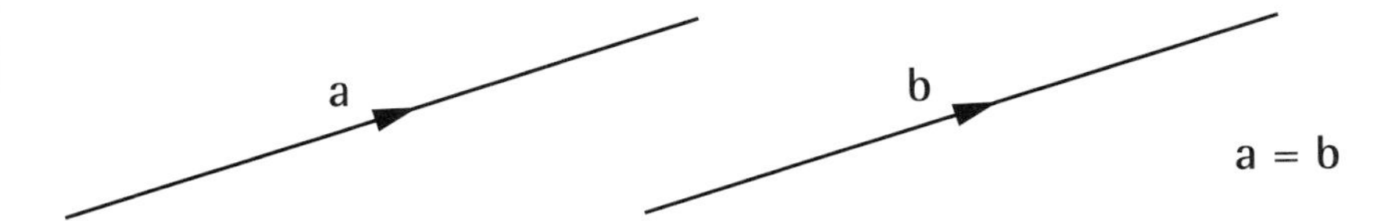

Vectors can be multiplied by scalars (numbers).

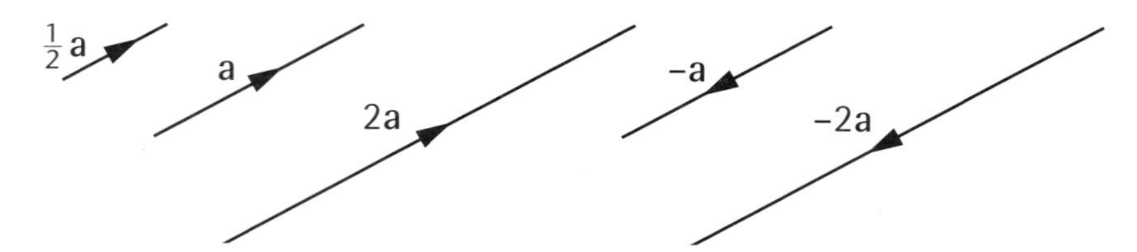

All the vectors will be parallel but they have different lengths and direction depending on the multiplier.

EXAMPLE 2

$\mathbf{a} = \begin{pmatrix} 3 \\ -1 \end{pmatrix}$ $\mathbf{b} = \begin{pmatrix} -4 \\ 2 \end{pmatrix}$ $\mathbf{c} = \begin{pmatrix} 0 \\ 4 \end{pmatrix}$

(a) Draw these vectors on a square grid.

(i) **a** (ii) **b** (iii) **c** (iv) −**a** (v) 2**b** (vi) $\frac{1}{2}$**c**
(vii) $-1\frac{1}{2}$ (viii) −2**c**

(b) What is the column vector notation for

(i) 2**a** (ii) $\frac{1}{2}$**b** (iii) 2**c**?

(a)

$a = \begin{pmatrix} 3 \\ -1 \end{pmatrix}$ which is 3 squares right and 1 square down.

$b = \begin{pmatrix} -4 \\ 2 \end{pmatrix}$ which is 4 squares left and 2 squares up.

$c = \begin{pmatrix} 0 \\ 4 \end{pmatrix}$ which is 4 squares up.

(b) (i) $2\mathbf{a} = 2\begin{pmatrix} 3 \\ -1 \end{pmatrix} = \begin{pmatrix} 6 \\ -2 \end{pmatrix}$

(ii) $\frac{1}{2}\mathbf{a} = \frac{1}{2}\begin{pmatrix} -4 \\ 2 \end{pmatrix} = \begin{pmatrix} -2 \\ 1 \end{pmatrix}$

(iii) $-\mathbf{c} = -\begin{pmatrix} 0 \\ 4 \end{pmatrix} = \begin{pmatrix} 0 \\ -4 \end{pmatrix}$

EXERCISE 19B

1 $\mathbf{a} = \begin{pmatrix} -2 \\ 5 \end{pmatrix}$ $\mathbf{b} = \begin{pmatrix} 0 \\ -3 \end{pmatrix}$ $\mathbf{c} = \begin{pmatrix} 4 \\ 2 \end{pmatrix}$

(a) Draw these vectors on a square grid.
(i) $\mathbf{a}$ (ii) $\mathbf{b}$ (iii) $\mathbf{c}$ (iv) $-\mathbf{a}$ (v) $\frac{1}{2}\mathbf{c}$ (vi) $-2\mathbf{b}$

(b) Write these as column vectors.
(i) $3\mathbf{a}$ (ii) $-4\mathbf{b}$ (iii) $-1\frac{1}{2}\mathbf{c}$

2 $\mathbf{p} = \begin{pmatrix} 2 \\ -4 \end{pmatrix}$ $\mathbf{q} = \begin{pmatrix} 1 \\ 3 \end{pmatrix}$ $\mathbf{r} = \begin{pmatrix} -3 \\ -2 \end{pmatrix}$

(a) Draw these vectors on a square grid.
(i) $\mathbf{p}$ (ii) $\mathbf{q}$ (iii) $\mathbf{r}$ (iv) $\frac{1}{2}\mathbf{p}$ (v) $2\mathbf{q}$ (vi) $-\mathbf{r}$

(b) Write these as column vectors.
(i) $5\mathbf{r}$ (ii) $-2\mathbf{q}$ (iii) $1\frac{1}{2}\mathbf{p}$

19.4 Adding vectors

EXERCISE 19C

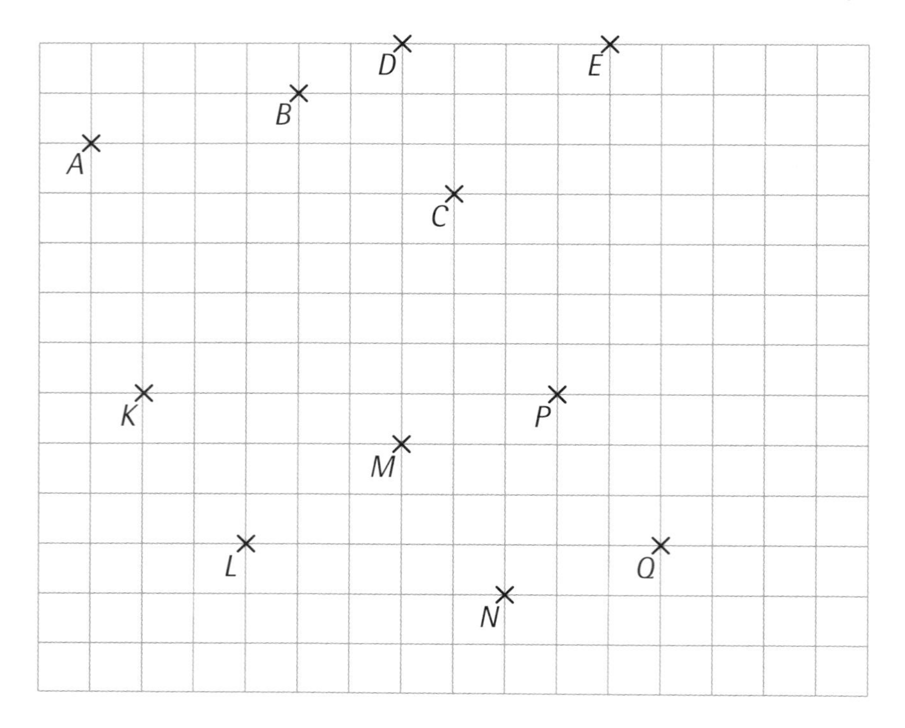

1 Write down as column vectors $\overrightarrow{AB}$, $\overrightarrow{BC}$, $\overrightarrow{AC}$.
Work out $\overrightarrow{AB} + \overrightarrow{BC}$. What do you notice?

2 Write down as column vectors $\overrightarrow{BD}$, $\overrightarrow{DE}$ and $\overrightarrow{BE}$.
Work out $\overrightarrow{BD} + \overrightarrow{DE}$. What can you write down about these three vectors?

3 Write down a vector statement about $\overrightarrow{AB}$, $\overrightarrow{BP}$ and $\overrightarrow{AP}$.

4 Write down as column vectors $\overrightarrow{BD}$, $\overrightarrow{DC}$, $\overrightarrow{CA}$ and $\overrightarrow{AB}$.
Work out $\overrightarrow{BD} + \overrightarrow{DC} + \overrightarrow{CA} + \overrightarrow{AB}$. What do you notice?

5 Write down a statement about $\overrightarrow{KM}$, $\overrightarrow{MP}$, $\overrightarrow{PQ}$, $\overrightarrow{QN}$, $\overrightarrow{NL}$.

6 Write down a vector statement about $\overrightarrow{KL}$, $\overrightarrow{LM}$, $\overrightarrow{MP}$ and $\overrightarrow{PQ}$.

You should now have the idea that adding vectors can be written as

$\overrightarrow{AB} + \overrightarrow{BC} = \overrightarrow{AC}$

or drawn as

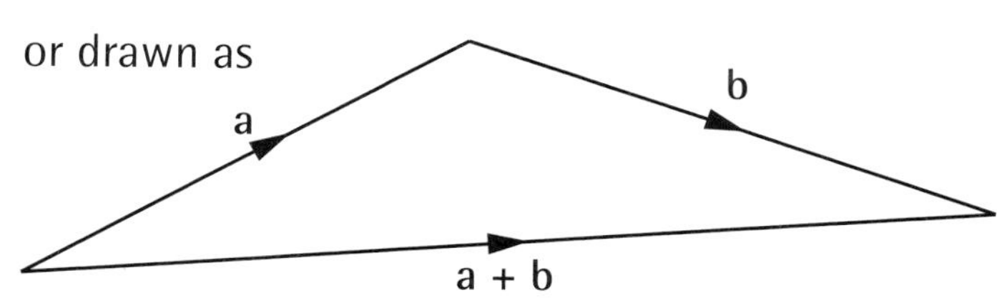

19.5 Subtracting vectors

We will need to make use of the fact that a negative sign in front of a vector reverses its direction.

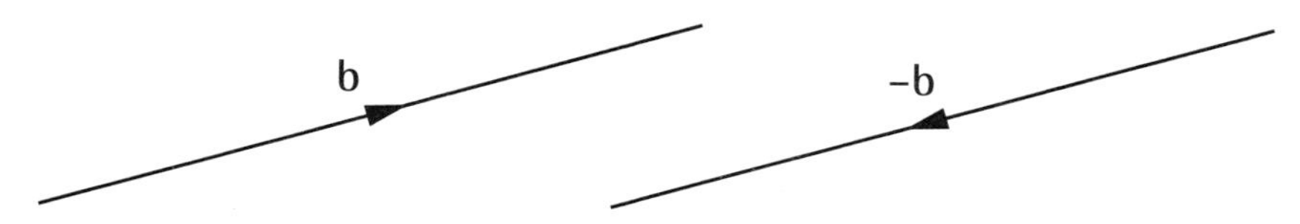

and also that **a** − **b** can be written as **a** + −**b**.

By adding **a** and −**b** we will have a picture that looks like the following.

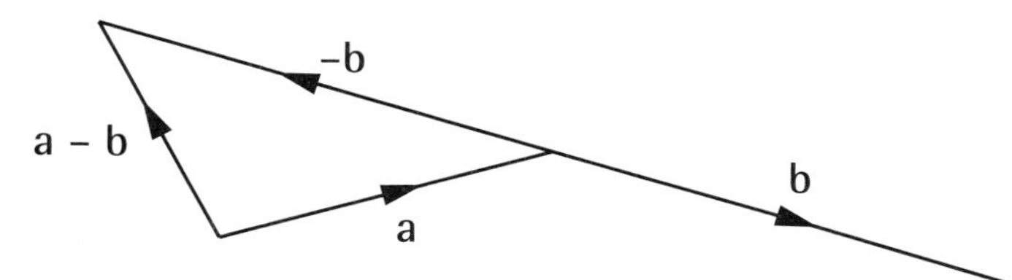

EXAMPLE 3

$\mathbf{a} = \begin{pmatrix} 3 \\ -2 \end{pmatrix}$ $\mathbf{b} = \begin{pmatrix} -4 \\ 5 \end{pmatrix}$ $\mathbf{c} = \begin{pmatrix} 0 \\ 6 \end{pmatrix}$

Write these as column vectors.

(a) **a** + **b** (b) **b** − **a** (c) 2**b** + 3**c** (d) **c** − 2**a**

(a) $\mathbf{a} + \mathbf{b} = \begin{pmatrix} 3 \\ -2 \end{pmatrix} + \begin{pmatrix} -4 \\ 5 \end{pmatrix} = \begin{pmatrix} -1 \\ 3 \end{pmatrix}$

(b) $\mathbf{b} - \mathbf{a} = \begin{pmatrix} -4 \\ 5 \end{pmatrix} - \begin{pmatrix} 3 \\ -2 \end{pmatrix} = \begin{pmatrix} -7 \\ 7 \end{pmatrix}$

(c) $2\mathbf{b} + 3\mathbf{c} = \begin{pmatrix} -8 \\ 10 \end{pmatrix} + \begin{pmatrix} 0 \\ 18 \end{pmatrix} = \begin{pmatrix} -8 \\ 28 \end{pmatrix}$

(d) $\mathbf{c} - 2\mathbf{a} = \begin{pmatrix} 0 \\ 6 \end{pmatrix} - \begin{pmatrix} 6 \\ -4 \end{pmatrix} = \begin{pmatrix} -6 \\ 10 \end{pmatrix}$

EXAMPLE 4

$\mathbf{a} = \begin{pmatrix} 2 \\ -1 \end{pmatrix}$ $\mathbf{b} = \begin{pmatrix} 4 \\ 3 \end{pmatrix}$

On a square grid draw diagrams to illustrate these vectors.

$\mathbf{a} + \mathbf{b}$ $\mathbf{b} - \mathbf{a}$ $2\mathbf{a} - \mathbf{b}$

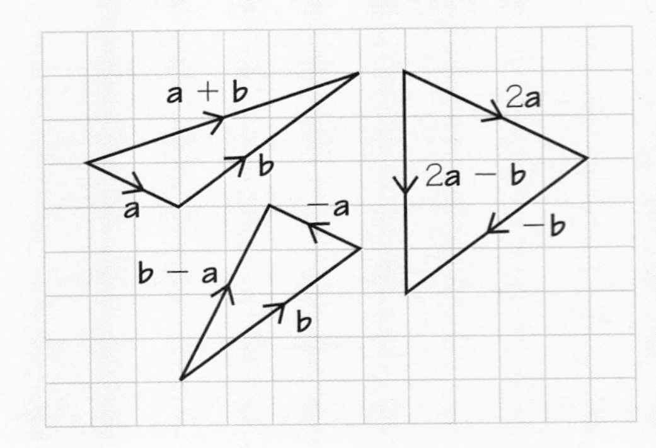

EXAMPLE 5

The diagram shows a grid of congruent parallelograms.

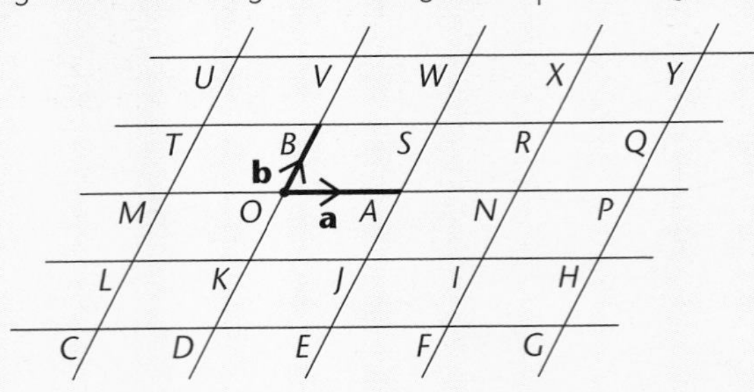

Write, in terms of **a** and **b**,

(a) $\overrightarrow{OP}$ (b) $\overrightarrow{OV}$ (c) $\overrightarrow{OR}$ (d) $\overrightarrow{OG}$

(e) $\overrightarrow{AB}$ (f) $\overrightarrow{NE}$ (g) $\overrightarrow{FB}$ (h) $\overrightarrow{UG}$

(a) $\overrightarrow{OP} = 3\mathbf{a}$

(b) $\overrightarrow{OV} = 2\mathbf{b}$

(c) $\overrightarrow{OR} = \overrightarrow{ON} + \overrightarrow{NR}$
$= 2\mathbf{a} + \mathbf{b}$

(d) $\overrightarrow{OG} = \overrightarrow{OP} + \overrightarrow{PG}$
$= 3\mathbf{a} - 2\mathbf{b}$

(e) $\overrightarrow{AB} = \overrightarrow{AO} + \overrightarrow{OB}$
$= -\mathbf{a} + \mathbf{b}$

(f) $\overrightarrow{NE} = \overrightarrow{NA} + \overrightarrow{AE}$
$= -\mathbf{a} - 2\mathbf{b}$

(g) $\overrightarrow{FB} = \overrightarrow{FD} + \overrightarrow{DB}$
$= -2\mathbf{a} + 3\mathbf{b}$

(h) $\overrightarrow{UG} = \overrightarrow{UY} + \overrightarrow{YG}$
$= 4\mathbf{a} - 4\mathbf{b}$

$\overrightarrow{OP}$, $\overrightarrow{OV}$, $\overrightarrow{OR}$, $\overrightarrow{OG}$, are all position vectors because they start at the origin *O*.
The others are free vectors.

There are alternative solutions. For example,
(g) $\overrightarrow{FB} = \overrightarrow{FR} + \overrightarrow{RB}$
$= 3\mathbf{b} - 2\mathbf{a}$

EXAMPLE 6

Using the parallelogram grid for Example 5, answer the following.

(a) Write down a vector equal to $3\mathbf{a} - \mathbf{b}$.

(b) Write down three other vectors equal to $3\mathbf{a} - \mathbf{b}$.

(c) Write down a vector parallel to $\overrightarrow{CP}$.

(d) Write down three vectors that are half the size of CP and in the opposite direction.

(e) Write down all the vectors that are equal to $3(\mathbf{a} + \mathbf{b})$.

(f) Write down, in terms of **a** and **b**, the vectors

$\overrightarrow{CK}$ $\overrightarrow{CA}$ $\overrightarrow{CR}$ $\overrightarrow{CY}$.

What is the connection between these answers and the positions of C, K, A, R and Y on the grid?

(a) $\overrightarrow{OH}$

(b) $\overrightarrow{VQ}$ $\overrightarrow{TN}$ $\overrightarrow{MI}$

(c) $\overrightarrow{CP} = 2\mathbf{b} + 4\mathbf{a}$

A parallel vector is $\overrightarrow{LQ}$

d) New vector $= -\mathbf{b} - 2\mathbf{a}$

$\overrightarrow{QA}$ $\overrightarrow{PJ}$ $\overrightarrow{HE}$

e) $\overrightarrow{CR}$ $\overrightarrow{LX}$ $\overrightarrow{DQ}$ $\overrightarrow{KY}$

f) $\overrightarrow{CK} = \mathbf{a} + \mathbf{b}$

$\overrightarrow{CA} = 2\mathbf{a} + 2\mathbf{b}$

$\overrightarrow{CR} = 3\mathbf{a} + 3\mathbf{b}$

$\overrightarrow{CY} = 4\mathbf{a} + 4\mathbf{b}$

C, K, A, R and Y are all points on a straight line.

You can choose *any* vectors from the grid that involve 3 steps in the same direction as **a** and one in the opposite direction to **b**.

Again, there is more than one choice. An alternative is $\overrightarrow{MY}$.

Multiply $\overrightarrow{CD}$ by $-\frac{1}{2}$.
There are many examples of $-\mathbf{b} - 2\mathbf{a}$ that you could choose here.

EXERCISE 19D

1 $\mathbf{d} = \begin{pmatrix} -4 \\ 1 \end{pmatrix}$ $\mathbf{e} = \begin{pmatrix} 2 \\ -2 \end{pmatrix}$ $\mathbf{f} = \begin{pmatrix} -1 \\ -5 \end{pmatrix}$

Write these as column vectors

(a) $\mathbf{d} + \mathbf{e}$ (b) $\mathbf{f} - \mathbf{d}$ (c) $2\mathbf{f} + \mathbf{e}$

(d) $\mathbf{d} - 3\mathbf{e}$ (e) $\mathbf{d} + \mathbf{e} + \mathbf{f}$ (f) $3\mathbf{f} + 2\mathbf{d}$

(g) $4\mathbf{e} - 3\mathbf{d}$ (h) $2\mathbf{d} - \mathbf{e} - 2\mathbf{f}$

2 $\mathbf{u} = \begin{pmatrix} -2 \\ 1 \end{pmatrix}$ $\mathbf{v} = \begin{pmatrix} 3 \\ 4 \end{pmatrix}$ $\mathbf{w} = \begin{pmatrix} -4 \\ 0 \end{pmatrix}$

On a square grid draw diagrams to illustrate these vectors

(a) $\mathbf{u} + \mathbf{v}$ (b) $\mathbf{v} - \mathbf{u}$ (c) $\mathbf{w} - 2\mathbf{u}$

(d) $2\mathbf{v} + \mathbf{w}$ (e) $\mathbf{u} - \mathbf{v} - \mathbf{w}$ (f) $\mathbf{v} + \mathbf{w} - 3\mathbf{u}$

3 The diagram shows a grid of congruent parallelograms. The origin is labelled O and the position vectors of points A and B are given by $\overrightarrow{OA} = \mathbf{a}$ and $\overrightarrow{OB} = \mathbf{b}$.

Write, in terms of $\mathbf{a}$ and $\mathbf{b}$,

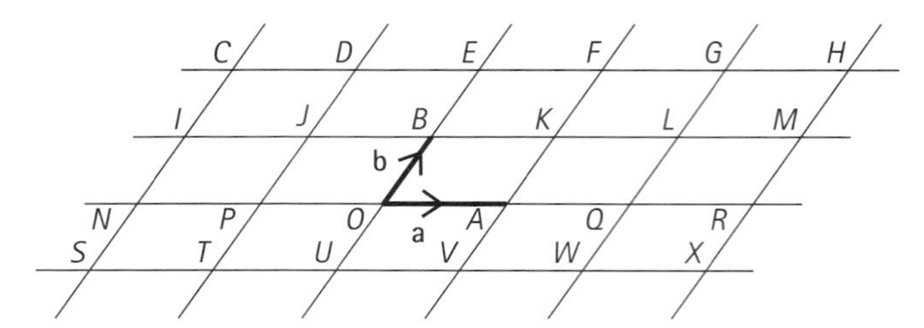

(a) $\overrightarrow{OR}$ (b) $\overrightarrow{OU}$ (c) $\overrightarrow{OC}$ (d) $\overrightarrow{OX}$ (e) $\overrightarrow{GP}$

(f) $\overrightarrow{IW}$ (g) $\overrightarrow{HR}$ (h) $\overrightarrow{MJ}$ (i) $\overrightarrow{UF}$ (j) $\overrightarrow{ET}$

4 Here is another grid of congruent parallelograms. The origin is labelled O and the position vectors of points A and B are given by $\overrightarrow{OA} = \mathbf{a}$ and $\overrightarrow{OB} = \mathbf{b}$.

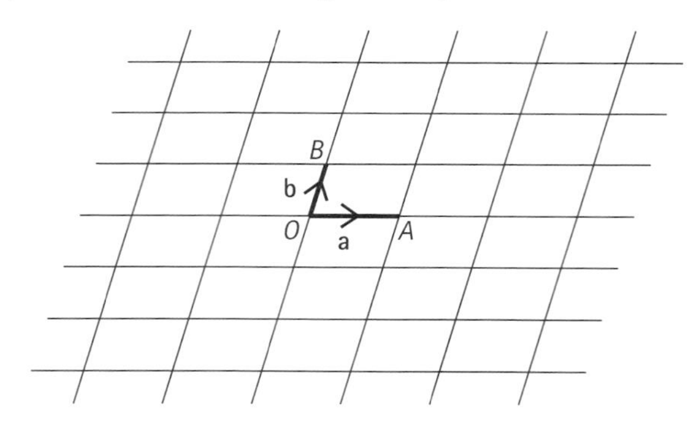

On a copy of the grid, mark the points C to L where

(a) $\overrightarrow{OC} = 2\mathbf{a} + 2\mathbf{b}$ (b) $\overrightarrow{OD} = 3\mathbf{a} + \mathbf{b}$

(c) $\overrightarrow{OE} = 2\mathbf{a} - \mathbf{b}$ (d) $\overrightarrow{OF} = \mathbf{a} - 3\mathbf{b}$

(e) $\overrightarrow{OG} = 2\mathbf{a} - 2\mathbf{b}$ (f) $\overrightarrow{OH} = 3\mathbf{b}$

(g) $\overrightarrow{OI} = \mathbf{a} + \frac{3}{2}\mathbf{b}$ (h) $\overrightarrow{OJ} = 2\frac{3}{2}\mathbf{a} - 2\mathbf{b}$

(i) $\overrightarrow{OK} = 2\mathbf{a} + \frac{1}{2}\mathbf{b}$ (j) $\overrightarrow{OL} = \frac{5}{2}\mathbf{a} - \frac{3}{2}\mathbf{b}$

EXAMINATION QUESTIONS

1

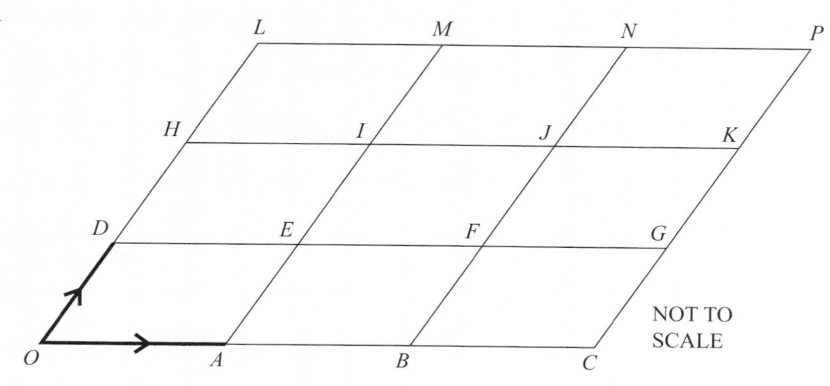

The diagram consists of nine equal parallelograms.

$\overrightarrow{OA} = \mathbf{s}$ and $\overrightarrow{OD} = \mathbf{t}$.

(a) Write, in terms of **s** and/or **t**, the vectors

(i) $\overrightarrow{EF}$, [1]

(ii) $\overrightarrow{BN}$, [1]

(iii) $\overrightarrow{OE}$. [1]

(b) If $\overrightarrow{OA} = \begin{pmatrix} 3 \\ 0 \end{pmatrix}$ and $\overrightarrow{OD} = \begin{pmatrix} -1 \\ -2 \end{pmatrix}$ write as a column vector,

(i) $\overrightarrow{HK}$, [1]

(ii) $\overrightarrow{OF}$, [1]

(iii) $\overrightarrow{LC}$. [1]

(CIE Paper 3, June 2001)

2

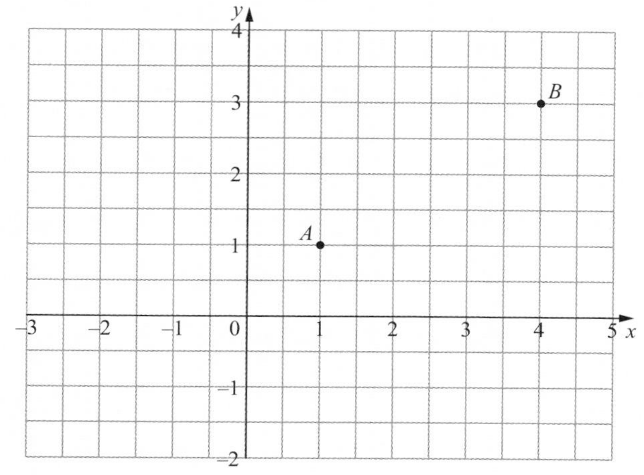

In the diagram, A is the point (1, 1) and B is the point (4, 3).

(a) Write $\overrightarrow{AB}$ as a column vector. [1]

(b) The point C is such that $\overrightarrow{BC} = 2\overrightarrow{BA}$.

(i) Copy the diagram and draw $\overrightarrow{BC}$ on the diagram. [1]

(ii) Write down the co-ordinates of C. [1]

(CIE Paper 1, June 2002)

---

3 $\mathbf{a} = \begin{pmatrix} 3 \\ -4 \end{pmatrix}$ and $\mathbf{b} = \begin{pmatrix} -1 \\ 2 \end{pmatrix}$

Work out $\mathbf{a} - 2\mathbf{b}$. [2]

(CIE Paper 1, Nov 2004)

---

4 $\mathbf{p} = \begin{pmatrix} 2 \\ -3 \end{pmatrix}$ and $\mathbf{q} = \begin{pmatrix} 3 \\ 1 \end{pmatrix}$

(a) Write $\mathbf{p} + \mathbf{q}$ as a column vector [2]

(b) The point O is marked on the grid below.

Draw the vector $\overrightarrow{OP}$ where $\overrightarrow{OP} = \mathbf{p}$. [1]

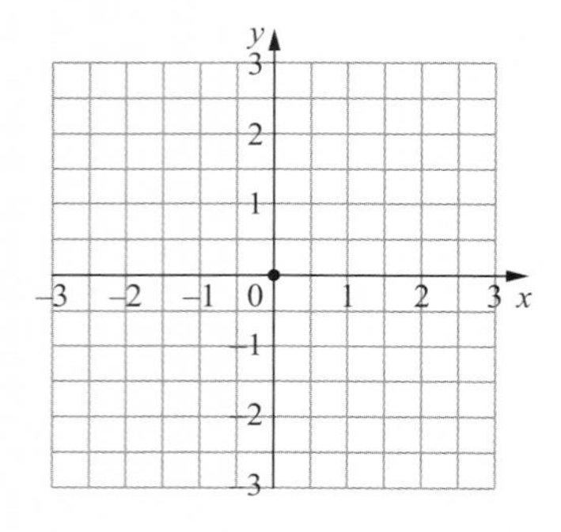

(CIE Paper 1, Nov 2005)

Chapter 20

Averages

This chapter will show you how to

- ✔ find the 'averages' – mean, median and mode for discrete data
- ✔ find the range and understand what is meant by the spread of the data
- ✔ find the mean, median and mode from frequency distributions for discrete data
- ✔ compare data and offer sensible interpretations

20.1 Averages and spread

You will need to know

- **that discrete data is data that can be counted**
- **how to order numbers**

'On average girls are better at mathematics than boys.'
'The average cost of a car, new or used, is $8500.'
'Jasmine is of average height.'
'Boris and Ali were pleased when they saw their examination results because they thought that they had both done better than average.'

In these statements 'average' describes something that typically happens.

In mathematics, an **average** is a value that represents a set of data.

For a set of discrete data you can find three 'averages' – the **mean**, the **mode** and the **median**. The **range** describes how the data values are **spread**.

The mean, mode, median and range describe the main features of the data. You can use these measures to compare data sets. For some data sets, one of these measures is more suitable than the others.

The mean

This list shows the amounts of pocket money received each week by nine friends.

Juana	$12	Beth	$7.20	Amina	$8.60
Maria	$7.50	Emmie	$10	Antonio	$7.50
Siobhan	$13.40	Hannah	$10	Imogen	$7.50

The mean of a set of data is the sum of all the values divided by the number of values.

Add up all the values. Divide by the number of values.

Mean =

$$\frac{12 + 7.20 + 8.60 + 7.50 + 10 + 7.50 + 13.40 + 10 + 7.50}{9}$$

$$= \frac{\$83.70}{9}$$

$$= \$9.30$$

This is the amount each person would have if the total amount was shared equally between them.

The mode

The mode of a set of data is the value that occurs most often.

Mode = $7.50

The median

The median of a set of data is the middle value when the data is arranged in order of size.

These are the amounts in ascending order.

$7.20, $7.50, $7.50, $7.50, $8.60, $10, $10, $12, $13.40

First write the data in order.

The median is $8.60.

$8.60 is the middle value in the list.

There is an *odd* number of data values, so there is *only one middle value*. Notice that there are the same number of items either side of the median.

If you have an *even* number of data values then there are *two middle values* and the median is the number mid-way between them.

The data set
1 5 7 9 10 12 14
15 15 17
has 10 values.
The two middle values are 10 and 12.
The median is

$$\frac{10 + 12}{2} = \frac{22}{2} = 11$$

The range

The range or spread of a set of data is the difference between the highest and the lowest values.

Range = \$13.40 − \$7.20 = \$6.20

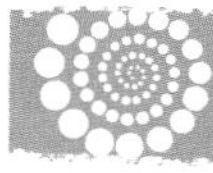

EXERCISE 20A

1 Find the mean, mode, median and range for these sets of data.

(a) 4 4 7 10 3 7 6 6 3 1 4

(b) 34 62 24 80 76 61 80 54 51

(c) 27 21 15 16 31 19 32 23

(d) 0.5 0.25 0.75 0.25 1.25 1.5 2.5 0.25

EXAMPLE 1

Find the mean, median, mode and range for this set of values.

17 6 11 12 12 16 2 2 7 12 8 9

$$\text{mean} = \frac{17 + 6 + 11 + 12 + 12 + 16 + 2 + 2 + 7 + 12 + 8 + 9}{12}$$

$$= \frac{114}{12} = 9.5$$

17 + 6 + 11 + ...
(the sum of all the data values)
12 is the number of pieces of data.

Writing the data in ascending order

2, 2, 6, 7, 8, (9), (11), 12, 12, 12, 16, 17

$$\text{Median} = \frac{9 + 11}{2} = 10$$

12 is even so there are two middle values. The median is halfway between them.

Mode = 12

The value 12 appears most often in the list.

Range = 17 − 2 = 15

The difference between the largest and the smallest.

EXAMPLE 2

(a) Work out the mean of these amounts of money.

\$10 \$15 \$12 \$9 \$14 \$12

(b) When one more amount is added the new mean value is \$11.

What amount of money is added?

(a) Mean $= \dfrac{10 + 15 + 12 + 9 + 14 + 12}{6} = \dfrac{72}{6}$

mean = \$12

(b) New mean value = \$11

There are now 7 pieces of data.

So total $= 7 \times 11$

Amount added = sum of 7 values − sum of 6 values

$= 77 - 72$

Amount added = \$5

From **(a)**, the sum of 6 values is 72.

EXERCISE 20B

1 Find the mean, median, mode and range of these sets of data.

(a) 6 3 9 12 1 9 8 8 6 1 3 6

(b) 20.1 20.7 21.4 22.7 29.6 22.6

(c) $\frac{1}{4}$ $\frac{1}{2}$ $\frac{3}{4}$ $\frac{1}{4}$ $\frac{5}{4}$ $\frac{3}{2}$ $\frac{5}{2}$ $\frac{1}{4}$

(d) 151 154 161 179 180 124 162 180 134

2 In a test these percentage marks were recorded.

75 61 52 82 64 71 90 46 55 57 64 63 67

Find the mean, median and mode for the marks.

3 In a ten-pin bowling game these scores were recorded.

7 8 4 1 10 8 3 6 5 9

Find the mean score (to the nearest whole number) and the mode.

4 These are the prices of a small sports car in different garages.

\$9900 \$10 200 \$9625 \$9865 \$10 150 \$9950

What is the mean price for these sports cars?

5 (a) Work out the mean of these amounts of money

$150 $75 $62 $87 $46 $102

When one more amount is added the new mean value is $83.

(b) What amount of money is added?

6 The mean height of 13 men in a theatre company is 179 cm. The mean height of 12 women in the company is 166 cm. What is the mean height for the whole company? Give your answer to the nearest cm.

7 In a mental test these results were recorded.

21 25 18 27 22 23 19 16 21 27
24 24 16 18 23 24 25 20 28

Find **(a)** the mean score **(b)** the median score **(c)** the mode and **(d)** the range of scores.

Which is the best average to use?

Sometimes an 'average' can give the wrong impression of a data set.

You need to decide which average is the most sensible one to use.

EXAMPLE 3

A factory making components for a mobile phone has a managing director, a works' supervisor and 12 employees.

Their weekly wages (in $) are

135 135 135 135 150 150 150
164 164 178 193 193 276 957

Calculate the mean, mode and median wage.

$$\text{Mean} = \frac{135 + 135 + \ldots + 276 + 957}{14}$$

$$= \frac{3115}{14}$$

$$= \$222.50$$

Mode = $135

$$\text{Median} = \frac{150 + 164}{2} = \$157$$

Continued ▼

Which average makes the most sense?

'The average wage is $222.50'

This is the mean value.

All of the 12 employees earn less than this. Some of them earn nearly $90 less! The only reason the **mean** is so high is because the managing director's $957 is included in the total of $3115.

If you re-calculate the mean **without** the $957 you get

$$\text{Mean} = \frac{2158}{13}$$
$$= \$166$$

which is much more **representative** of the data.

The value 957 is an **extreme value.**

An extreme value is either a lot larger or a lot smaller than the rest of the data.
It is also called an **outlier**.

'The average wage is $135'

This is the mode.

This is not representative of the data. No-one earns less than this and eight of the staff earn quite a lot more.

'The average wage is $157'

This is the median.

All of the 12 employees wages are fairly close to this.
As it is the **median**, it is in the middle of the list and is the most sensible answer to use in this case.

This value gives a fairer picture of people's earnings.

You need to use your common sense to decide which is the best average to use in a situation.

EXERCISE 20C

1 The wickets taken by members of a village cricket team one season were as follows

39 38 38 35 34 33 29 28 26 9

(a) Find the mode, median and mean and the range for this set of data.

(b) Which 'average' would best represent the data? Explain your reasons.

2 The numbers of cars parked during the day in a city centre car park during a two-week period were

52 45 61 67 48 70 12 56 41 57 53 62 70 9

(a) Find the mode, median and mean number of cars. Comment on your results.

(b) Which would be the best 'average' to use?

20.2 Frequency distributions

You will need to know

- **how to tally results and record the frequency**

When you have a large amount of data, you can tally it into a **frequency table**.

The **frequency** is the number of times an answer or result occurs in the data.

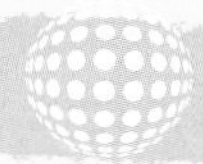

EXAMPLE 4

In a local soccer tournament the numbers of goals scored in each game were

0 2 5 4 3 2 0 0 3 5 3 4 2 2 2 3 1 1 0 1
6 1 0 0 3 1 0 1 2 2 1 4 3 0 0 1 2 1 2 1
5 0 1 3 2 0 6 3 1 1 1 0 2 0 2 1 0 2 2 0
1 2 3 0 4 2 3 1 5 1 6 1 1 3 2 2 0 0 3 0

Put this data into a frequency table.

Find **(a)** the mean **(b)** the median **(c)** the mode
(d) the range of goals scored in the tournament.

Remember we are finding a representative number for the number of goals scored in each game.

Number of goals scored (x)	Number of games (f)	Total number of goals (fx)
0	19	0
1	20	20
2	18	36
3	12	36
4	4	16
5	4	20
6	3	18
Totals	80	146

80: the total number of games played.

146: the total number of goals scored.

Use x for the data. Use f for the frequency (the number of times each value occurs).

Tally the data into the correct box.

The column fx records the total number of goals.
For example, 12 teams scored 3 goals, which is $12 \times 3 = 36$ goals in total.

(a) Mean $= \dfrac{\text{total goals scored}}{\text{total games played}}$

$= \dfrac{146}{80}$ $x = 1.83$ goals per game (3 s.f.)

This is the same as the number of items of data.

Continued ▼

There are 80 items of data. When they are arranged in order of size, the median will be midway between the 40th and 41st values.

In 80 pieces of data, the 40th and 41st pieces are the two middle values.

Number of goals scored (x)	Number of games (f)	Total number of goals (fx)
0	19	0
1	20	20
2	18	36

First 19 pieces of data.

20 + 19 = 39 pieces of data in first 2 rows.

40th and 41st pieces must be in here.

(b) The 40th and 41st values are both 2.
Median = 2 goals per game

(c) Mode = 1 goal per game

The mode is the number of goals with the highest frequency.

(d) Range = 6 − 0 = 6 goals

The range is the difference between the largest and the smallest number of goals.

How to find the median

- Odd number of data items.
  Middle of 5 and 1 is $\frac{5+1}{2} = 3$
  so median value is 3rd item.

1st 2nd 3rd 4th 5th
median value

- Even number of data items.
  Middle of 4 and 1 is $\frac{4+1}{2} = 2.5$
  so the median is the $\frac{\text{2nd} + \text{3rd}}{2}$th item.

1st 2nd 3rd 4th
median value

So we can generalise by saying for n items in a data set

1st 2nd 3rd 4th … nth

the median value is the $\left(\frac{n+1}{2}\right)$th item.

EXERCISE 20D

1 A fair, six-sided dice is rolled 50 times. The scores are recorded in a frequency table.

Number on dice (x)	Frequency (f)	Total score (xf)
1	7	
2	9	
3	6	
4	11	
5	10	
6	7	
Totals		

$xf = f \times x$

(a) Copy and complete the table.

(b) Work out the mean, median and mode of the scores for this dice.

2 The police recorded the speeds of cars (to the nearest 10 km/h) on a road in a built-up area.

Speed km/h (x)	Frequency (f)	Total score (xf)
10	0	
20	23	
30	48	
40	16	
50	2	
60	1	
Totals		

(a) Copy and complete the table.

(b) Find the mean, median and mode speeds for this road.

(c) What do you think is the likely speed limit for this road?

3 In an ice hockey tournament the following number of goals were scored in different games.

Goals scored	x	0	1	2	3	4	5	6	7
Frequency	f	12	14	12	9	4	6	0	3

Find the mean, median and mode of the goals scored in the tournament.

4 75 job applicants were given a mental arithmetic test. These are their scores out of 10.

5 7 8 2 1 9 3 8 7 4 2 8 4 9 8
2 4 9 5 5 6 6 5 8 9 10 6 10 7 3
8 3 4 2 5 7 3 7 7 8 9 7 8 4 10
6 4 3 8 2 9 9 2 3 8 5 6 6 8 10
5 6 6 2 8 10 5 5 6 7 1 8 7 4 6

(a) Put this data into a frequency table.

(b) Find the mean, median, mode and range of the scores.

EXAMINATION QUESTIONS

1 One day Abdul carried out a survey on the shoe sizes of 87 students.
He put the results in the table below.

Shoe sizes	33	34	35	36	37	38	39	40
Number of students	5	12	10	16	15	17	9	3

(a) Write down the modal shoe size. [1]
(b) Find the median shoe size. [2]
(c) Calculate the mean shoe size. [3]

(CIE Paper 1, Jun 2000)

---

2 Arantxa had the following scores on the eighteen holes of a golf course.
6, 2, 5, 4, 6, 5, 4, 6, 6, 4, 5, 6, 2, 6, 5, 4, 7, 6

(a) Copy and complete the frequency table below.

Score	2	3	4	5	6	7
Frequency						

(b) Write down
(i) the mode, [1]
(ii) the median. [2]

(CIE Paper 1, Nov 2000)

---

3 Marcos recorded the daily maximum temperature during one week.
The temperatures were 8°C, 9°C, 10°C, 11°C, 14°C, 15°C and 8°C.
(a) Write down the modal temperature. [1]
(b) Find the median temperature. [2]
(c) Calculate the mean temperature. [2]
(d) Marcos recorded the maximum temperature on the eighth day.
The mean temperature for the eight days was 11°C.
Calculate (i) the temperature on the eighth day, [2]
(ii) the median temperature for the eight days. [1]

(CIE Paper 1, Nov 2001)

---

4 Fifty students take part in a quiz.
The table shows the results.

Number of correct answers	5	6	7	8	9	10	11	12
Number of students	4	7	8	7	10	6	5	3

(a) How many students had 6 correct answers? [1]
(b) How many students had less than 11 correct answers? [1]
(c) Find
(i) the modal number of correct answers, [1]
(ii) the median number of correct answers, [1]
(iii) the mean number of correct answers. [2]

(CIE Paper 3, Jun 2003)

---

5 A dentist recorded the number of fillings that each of a group of 30 children had in their teeth. The results were

2 4 0 5 1 1 3 2 6 0
2 2 3 2 1 4 3 0 1 6
1 4 1 6 5 1 0 3 4 2

(a) Complete this frequency table. [2]

Number of fillings	0	1	2	3	4	5	6
Frequency							

(b) What is the modal number of fillings? [1]
(c) Find the median number of fillings. [2]
(d) Work out the mean number of fillings. [2]

(CIE Paper 3, Nov 2003)

---

6 The list shows marks in an examination taken by a class of 10 students.

65, 51, 35, 34, 12, 51, 50, 75, 48, 39

(a) Write down the mode. [1]
(b) Work out the median. [2]
(c) Calculate the mean. [2]

(CIE Paper 3, Jun 2004)

Chapter 21

Pythagoras' theorem and trigonometry

This chapter will show you how to
- ✔ understand Pythagoras' theorem
- ✔ calculate the longest side (the hypotenuse) of a right-angled triangle
- ✔ calculate a shorter side of a right-angled triangle
- ✔ apply Pythagoras' theorem to solving 'real' problems
- ✔ use the basic trigonometric functions sine, cosine and tangent

21.1 Pythagoras' theorem

You will need to know
- how to square numbers
- how to find square roots

Square numbers

These come from the area of a square.

3 cm

9 cm^2

3 cm

The area of the square is $3 \times 3 = 9$ cm^2.

The shorthand way of writing this is $3^2 = 9$.

This is pronounced 'three squared equals nine'.

Your calculator will have a button for working this out. 3 x^2 9.

EXERCISE 21A

Work these out.

1 4^2	**2** 7^2	**3** 5^2	**4** 9^2	**5** 12^2
6 15^2	**7** 18^2	**8** 22^2	**9** 1.6^2	**10** 3.5^2
11 0.8^2	**12** 0.55^2	**13** 1^2	**14** 1.8^2	**15** $(\frac{1}{2})^2$

Square roots

This is the inverse of squaring.
It will find the length of the side if you know the area of the square.
If the length is not exact, give your answer correct to 3 s.f.

The area of a square is 64 cm^2.
The length of a side will be 8 cm because $\sqrt{64} = 8$

Your calculator will have a button for this.

64 √ 8 or √ 64 = 8

EXERCISE 21B

Work these out.

1 $\sqrt{81}$	**2** $\sqrt{49}$	**3** $\sqrt{16}$	**4** $\sqrt{4}$	**5** $\sqrt{169}$
6 $\sqrt{289}$	**7** $\sqrt{576}$	**8** $\sqrt{1}$	**9** $\sqrt{2}$	**10** $\sqrt{19}$
11 $\sqrt{57}$	**12** $\sqrt{3}$	**13** $\sqrt{5}$	**14** $\sqrt{15.8}$	**15** $\sqrt{0.4}$

Over 2500 years ago Pythagoras made a discovery about the areas of squares drawn on right-angled triangles.

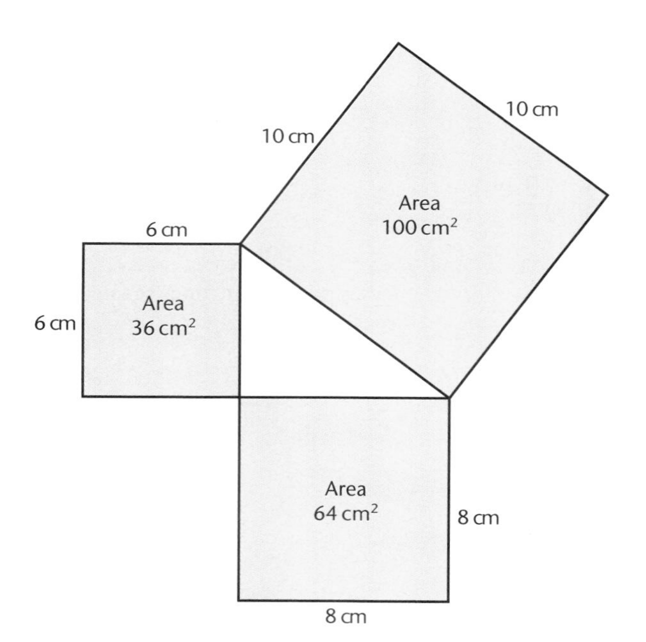

He saw that the larger area 100 was equal to the two smaller areas 36 + 64.

$$100 = 36 + 64$$

This can be written $10^2 = 6^2 + 8^2$

This idea will work for any **right-angled** triangle. It will **not** work for any other type of triangle.

For a right-angled triangle with sides of length a, b and c as shown, Pythagoras' theorem states that $c^2 = a^2 + b^2$.

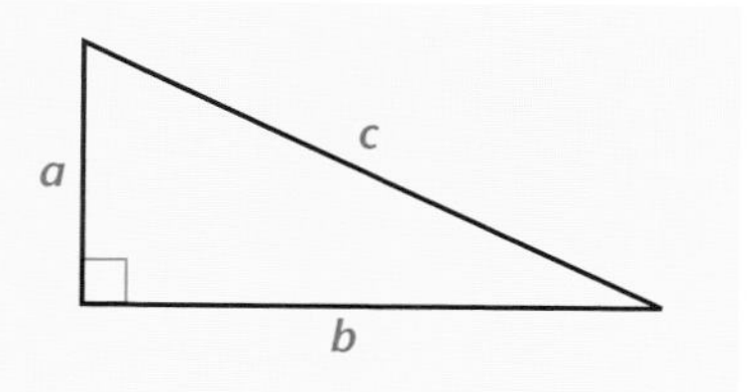

(c) is the longest side. It is called the **hypotenuse** and is opposite the right angle.

EXERCISE 21C

1 In the following triangles, write the letter that represents the hypotenuse.

(a)

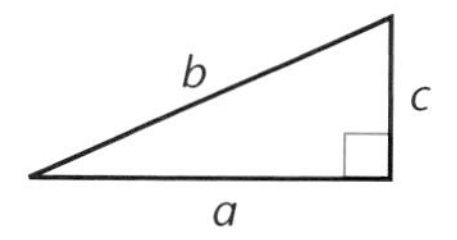

(b)

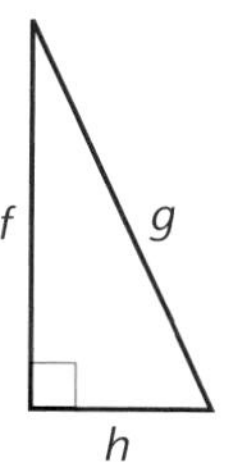

(c)

2 For the triangles above, write out the formula using Pythagoras' theorem.

21.2 Finding the hypotenuse

You can use Pythagoras' theorem to find the length of the hypotenuse.

EXAMPLE 1

Work out the length of the hypotenuse (the side marked x) in this right-angled triangle.

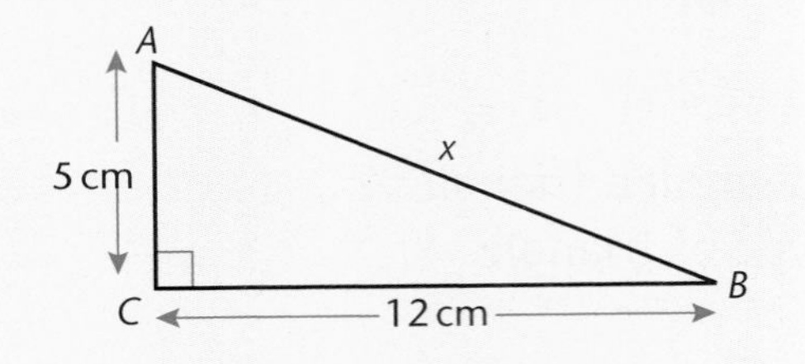

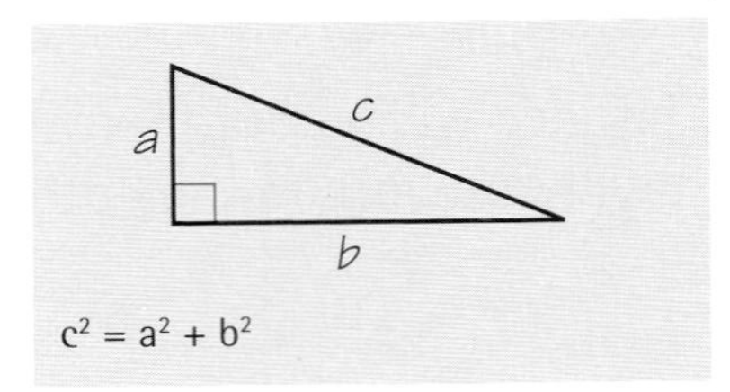

$c^2 = a^2 + b^2$

$x^2 = 5^2 + 12^2$

$x^2 = 25 + 144$

$x^2 = 169$

$x = \sqrt{169}$

$x = 13$ cm

EXAMPLE 2

In the triangle PQR, angle Q = 90°, QP = 4 cm and QR = 7 cm. Calculate y, the length of PR.

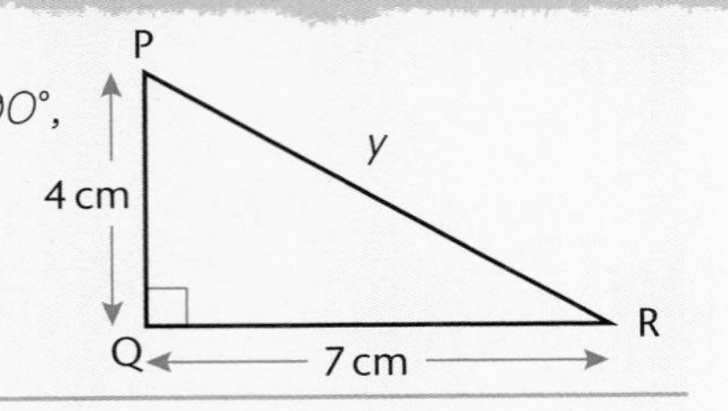

$y^2 = 4^2 + 7^2$

PR is the hypotenuse, because it is opposite the right angle.

$y^2 = 16 + 49$

$y^2 = 65$

$y = \sqrt{65}$

Use your calculator to find the square root.

$y = 8.06$ cm (to 3 s.f.)

Always put the units in your answer.

EXERCISE 21D

1 Calculate the length marked with a letter in each triangle.

(a)

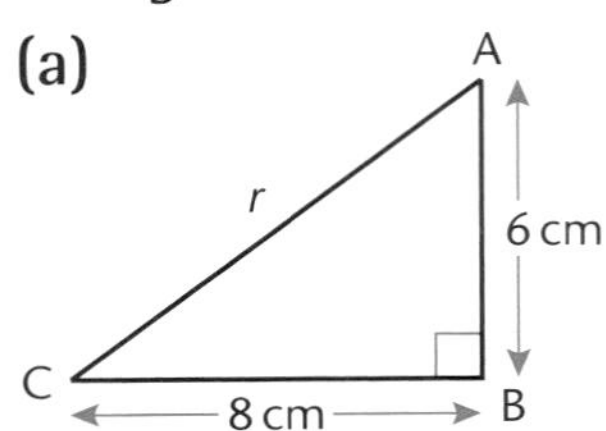

(b)

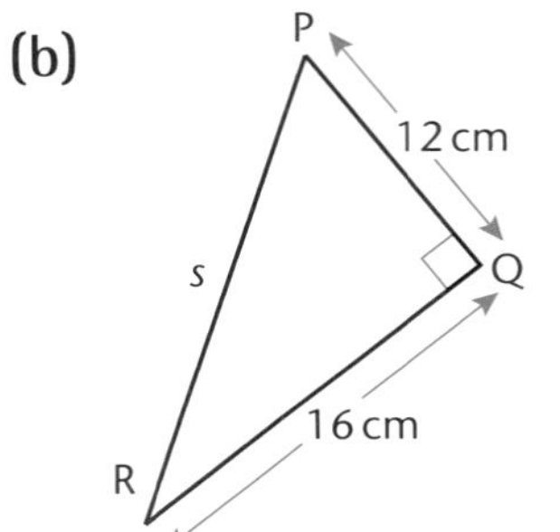

(c)

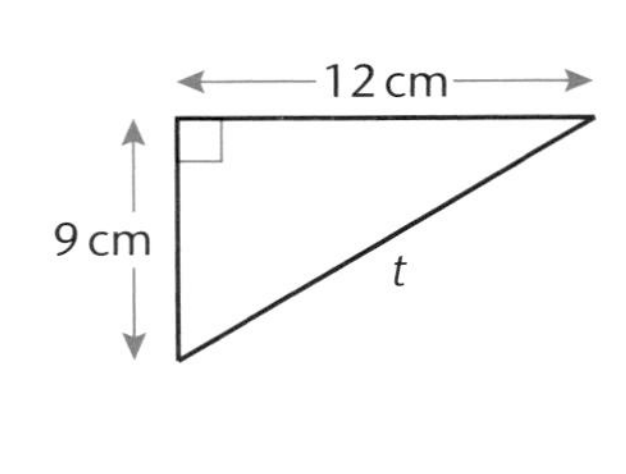

2 Calculate the lengths marked with letters to 1 d.p.

(a) (b) (c)

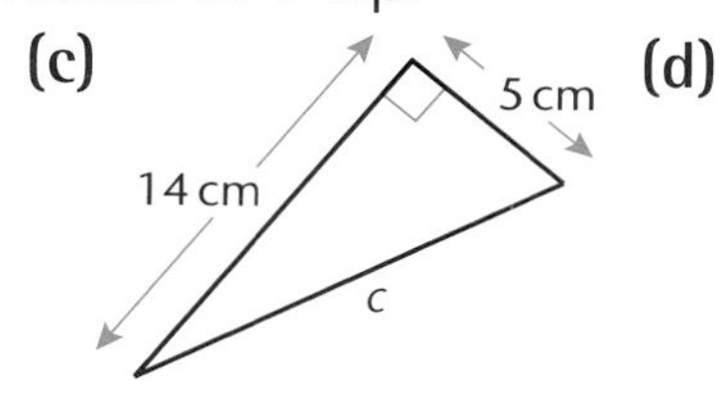

(d)

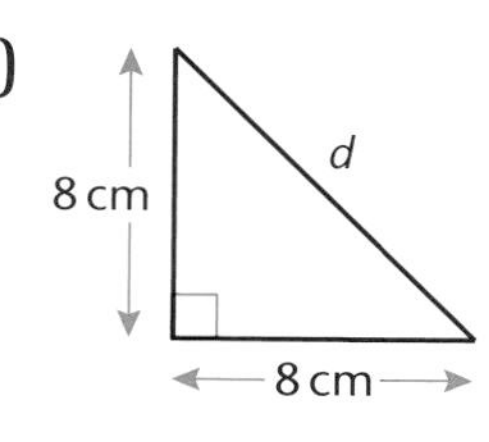

3 Calculate the lengths marked with letters in these triangles. Give your answers to 2 decimal places.

(a)

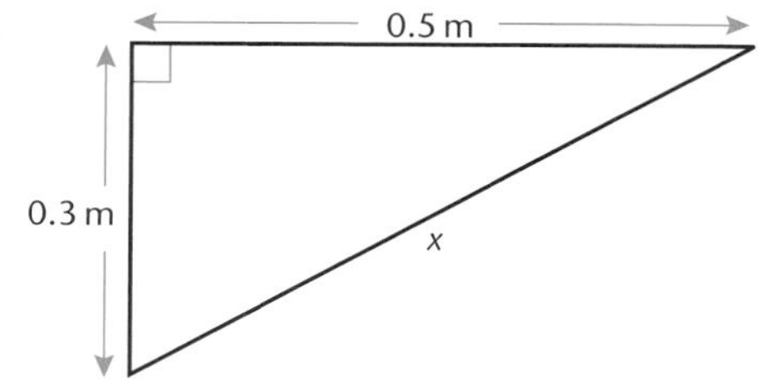

(b)

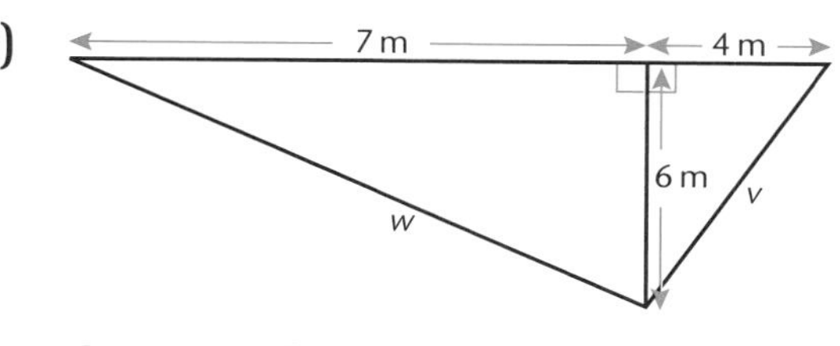

(c)

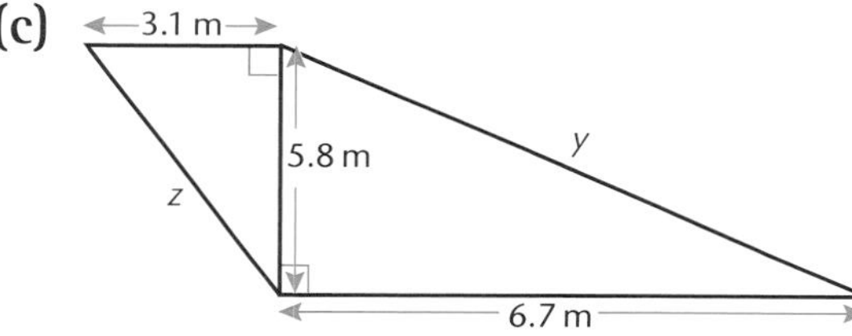

4 The diagram shows the cross-section of a roof.
The roof is symmetrical.
Work out the length marked *r*.
Give your answer to the nearest cm.

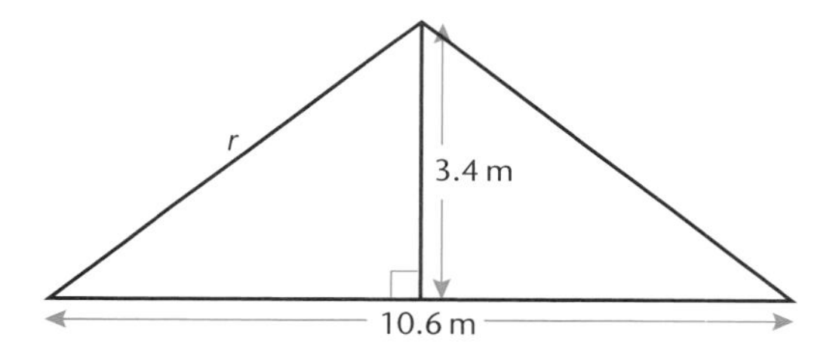

5 A rectangular gate measures 3.5 m long by 1.5 m high.
Work out the length of wood needed to make the diagonal.

6 A ladder rests against a vertical wall just below a window.
PR = 2.6 m and the foot of the ladder is 1.2 m away from the wall (RQ).
How long is the ladder?

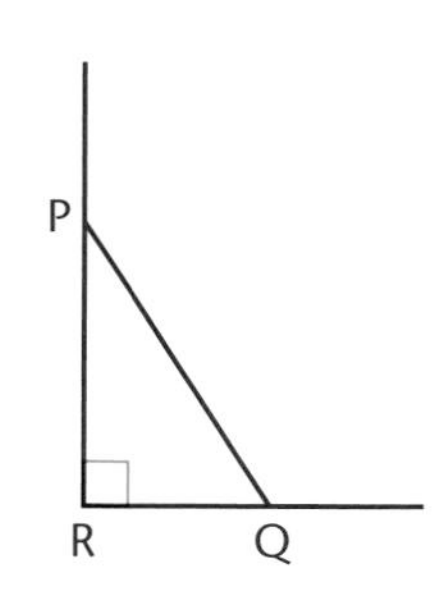

7 A rectangular room has length 5.3 m and width 3.9 m. A spider walks diagonally across the room. Work out, to the nearest cm, the distance the spider walks.

8 Calculate the length of a diagonal of a square of side 30 cm.

9 In the triangle ABC, angle ABC = 90°, AB = 7.5 cm and BC = 5.9 cm.
Calculate AC. Give your answer to 2 decimal places.

For questions 10 and 11, sketch a diagram and label it.

10 Fluella is flying her kite in a strong breeze. The kite is flying at a height of 18 m and is 20 m away horizontally from Fluella. How long is the kite string?

11 A boat sails due east for 24 km then due south for 10 km. How far is the boat from its starting point?

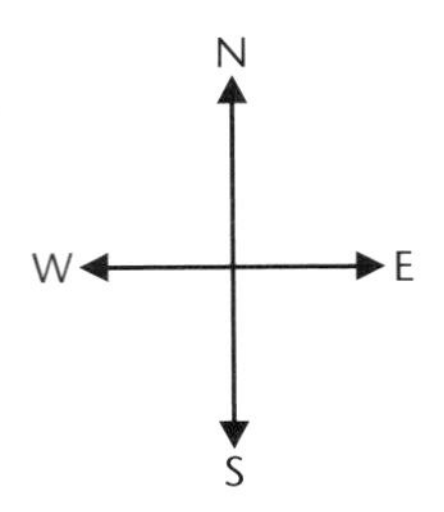

21.3 Finding the length of a shorter side

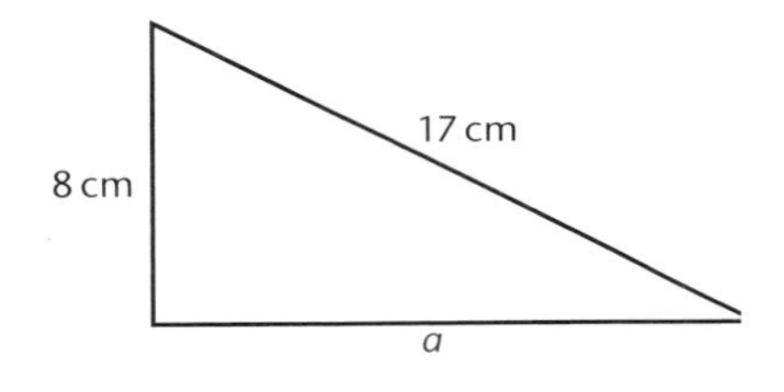

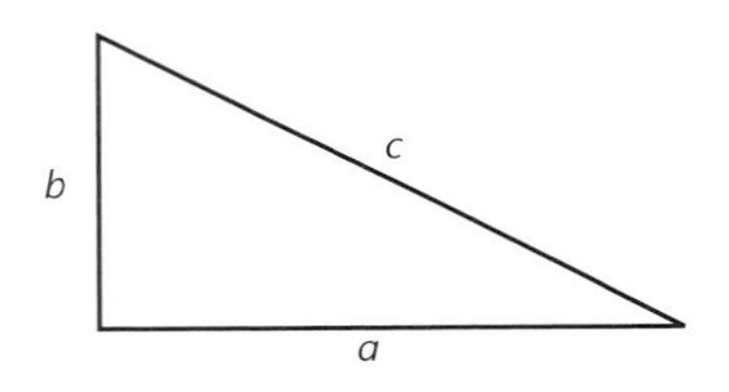

$a^2 + b^2 = c^2$

Using the theorem will give the equation

$$a^2 + 8^2 = 17^2$$
$$a^2 + 64 = 289$$
$$a^2 + 64 - 64 = 289 - 64$$
$$a^2 = 225$$
$$a = 15 \text{ cm}$$

17 is the hypotenuse.

$8^2 = 64$; $17^2 = 289$

$\sqrt{225} = 15$

EXERCISE 21E

1 Calculate the lengths marked with letters in these triangles.

(a)

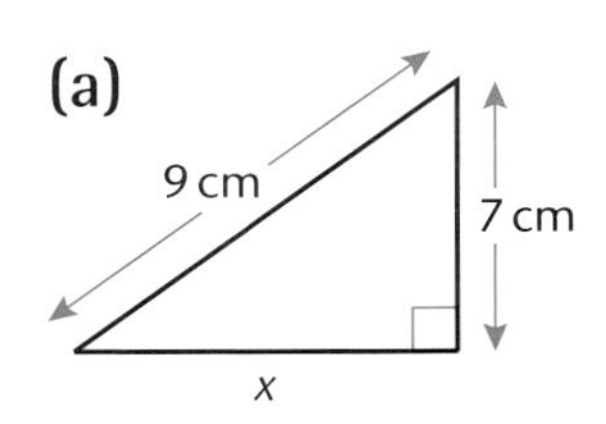

(b)

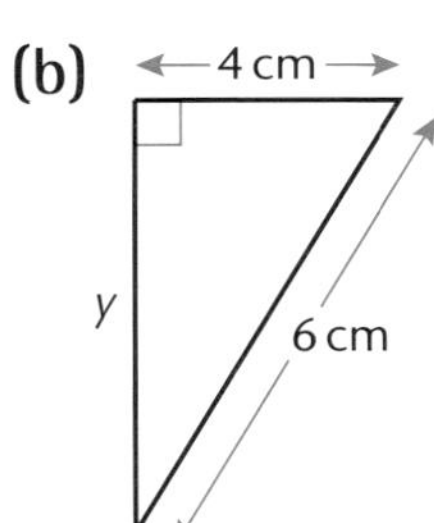

(c)

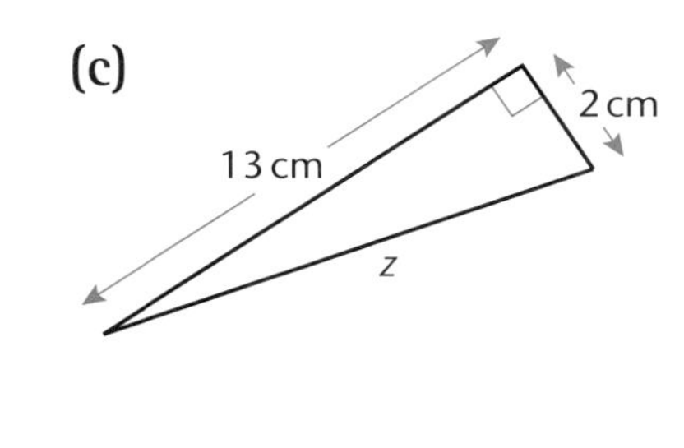

(d)

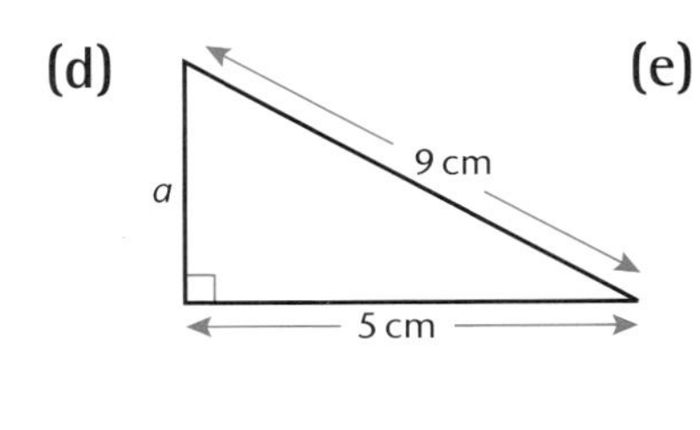

(e)

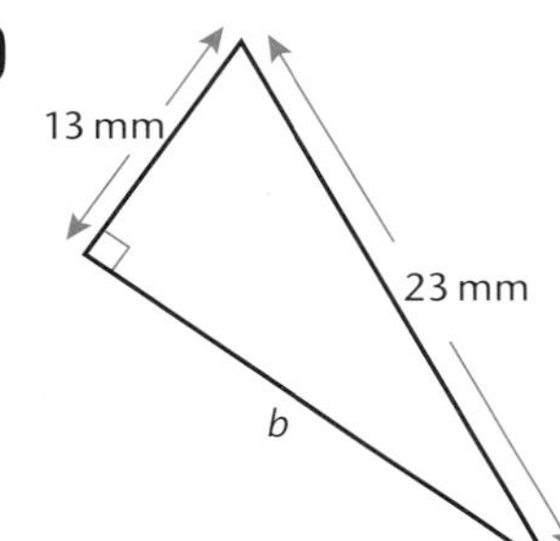

(f)

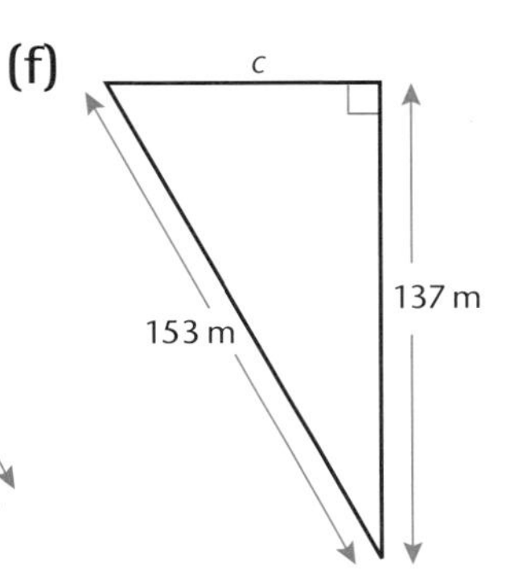

2 Calculate the lengths (a) XY (b) QR (c) BC (d) NM.

(a)

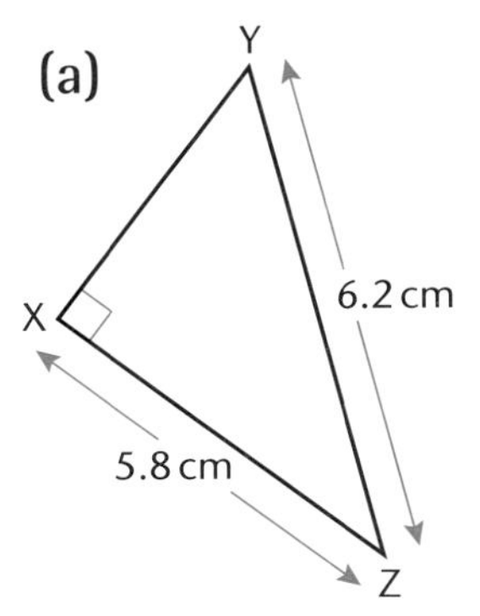

(b)

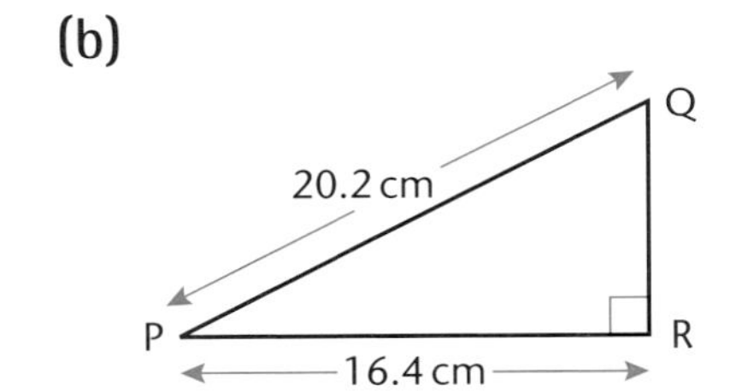

(c)

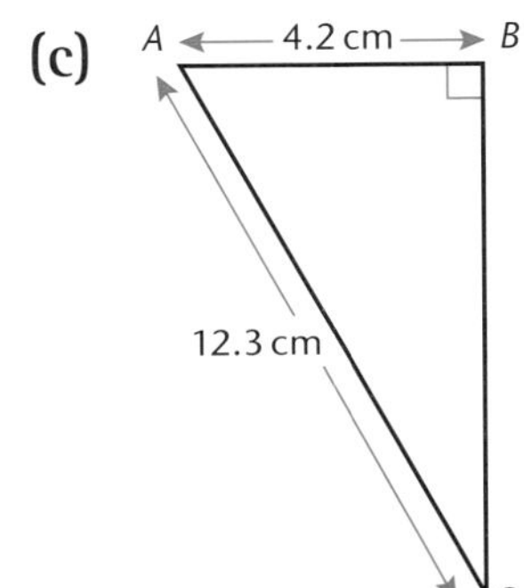

(d)

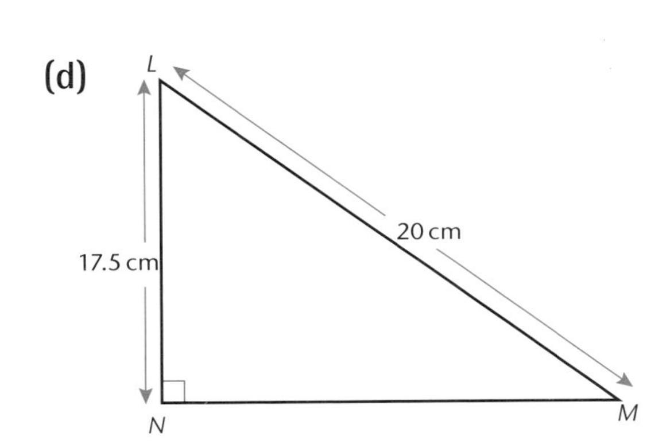

3 In the diagram,

(a) calculate the length of the side marked a. Give your answer to 2 decimal places.

(b) using your answer to part (a), calculate the length of the side marked b.

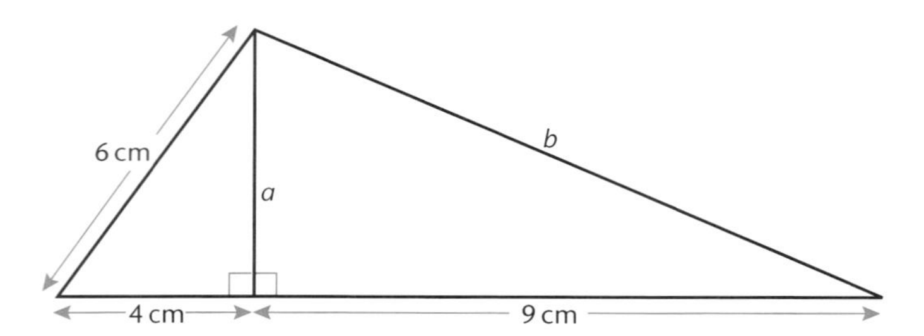

21.4 Problem solving

When you have a diagram you must identify a right angled triangle.

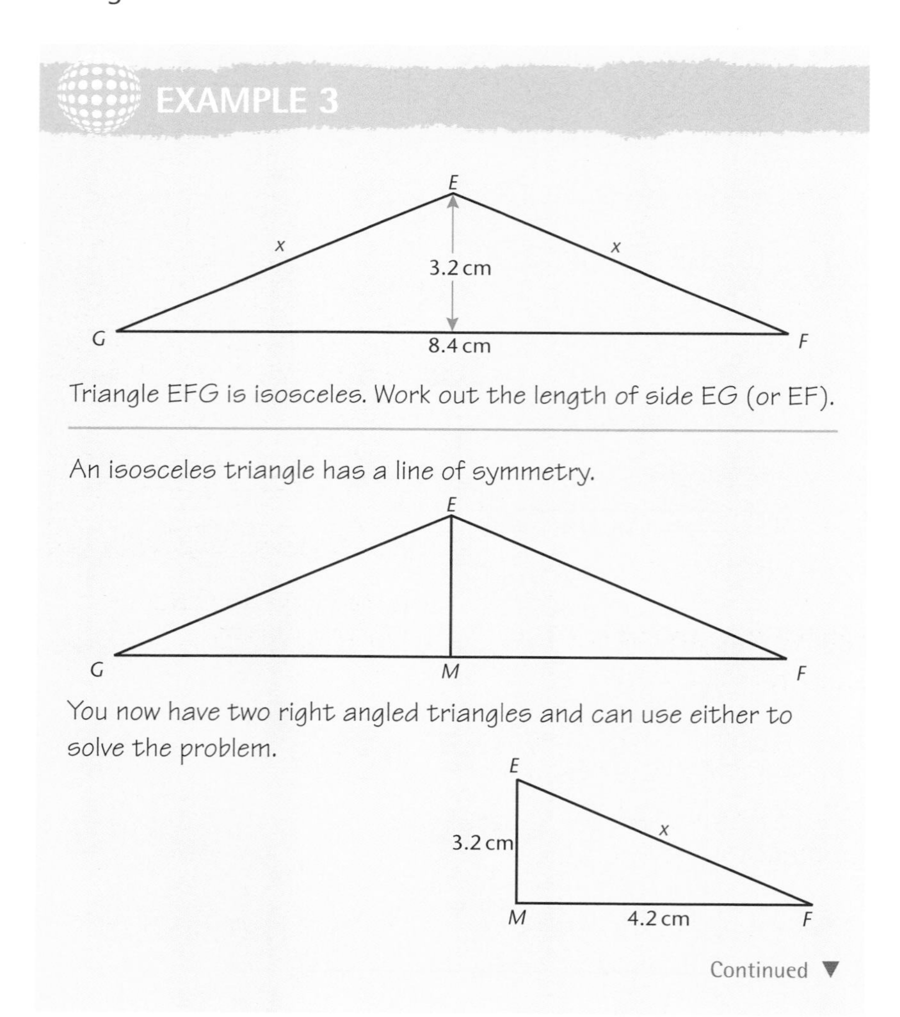

M is the mid-point of GF.

$x^2 = 4.22 + 3.22$
$x^2 = 17.64 + 10.24$
$x^2 = 27.88$
$x = \sqrt{27.88}$
$x = 5.28$ (3 s.f.) cm.

If the question is written in words only then you must draw a diagram first.

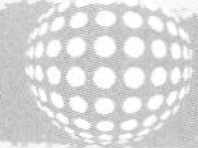

EXAMPLE 4

A ship sails 9km due north and then 17km due east.
How far is the ship from its starting point?

First draw a diagram.

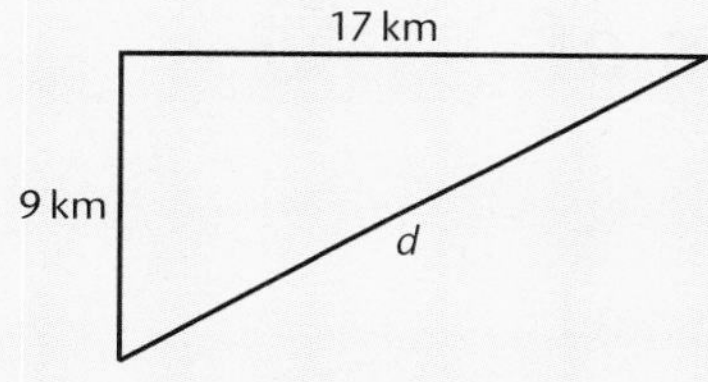

Now you can see the right-angled triangle.

$d^2 = 9^2 + 17^2$
$d^2 = 81 + 289$
$d^2 = 370$
$d = \sqrt{370}$
$d = 19.2$ km (3 s.f.)

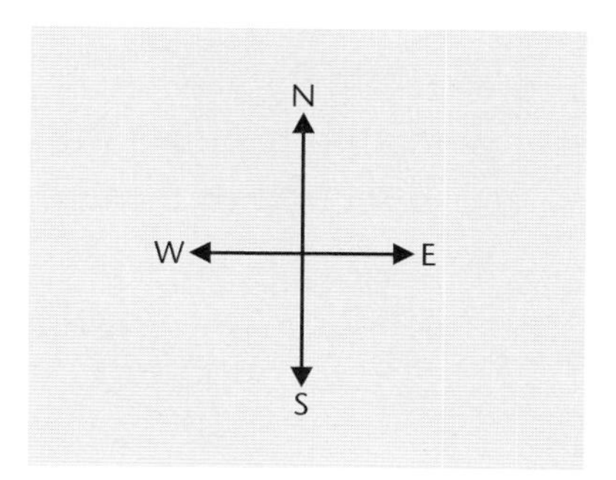

For questions 3 to 5, sketch a diagram and label it.

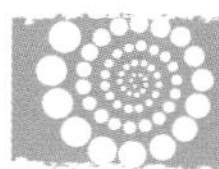

EXERCISE 21F

1 A children's slide is 3.6 m long.
The vertical height of the slide above the ground is 2.1 m.
Work out the horizontal distance between each end of the slide.

2 Jotinder climbs 120 m up a steep slope. From the map he sees he has walked a horizontal distance of only 95 m.
Work out the height he has climbed.

3 A ladder 5.2 metres long is leaning against a wall. The top of the ladder is 4.6 metres up the wall. How far from the bottom of the wall is the foot of the ladder?

4 The diagram shows the cross-section of a shed. Work out the width of the shed, w.

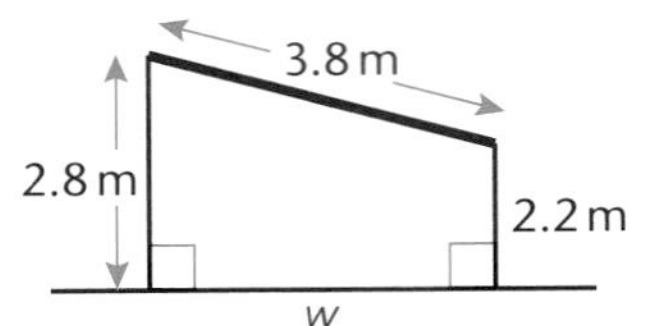

21.5 Trigonometry – the ratios of sine, cosine and tangent

Trigonometry is concerned with calculating sides and angles in triangles and involves three ratios called **sine, cosine** and **tangent**.

Using a calculator for trigonometry

When using a calculator in trigonometry **make sure** it is working **in degrees**.

You will be able to check this because at the top of your screen there will be either DEG or just D in the display.

If it is not in this mode , press the MODE key until

you see

DEG	RAD	GRA
1	2	3

appear

then press 1.

Remember that different makes of calculator have different keys. You may have to read the instruction booklet to find out how to do this on your calculator.

Look at the right-angled triangle *ABC* shown below.

The side opposite the angle *x* is called the **opposite**.

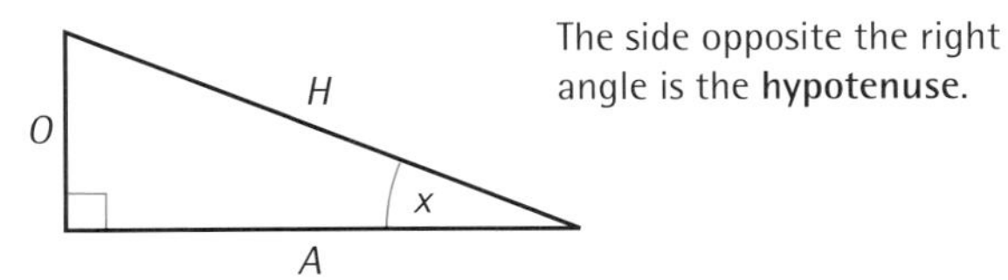

The side opposite the right angle is the **hypotenuse**.

The side next to the angle *x* is called the **adjacent**.

Questions on trigonometry are always on the examination paper.

The hypotenuse is always the same side but the adjacent and opposite sides switch depending on which angle you are using.

EXERCISE 21G

Copy these triangles and label the sides *A*, *O* and *H*, where *x* is the angle to be used.

1

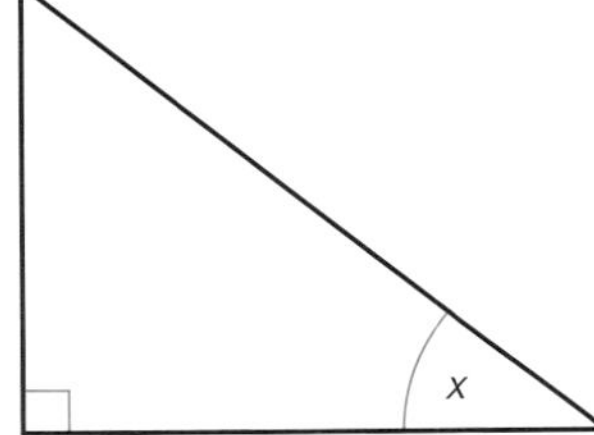

2

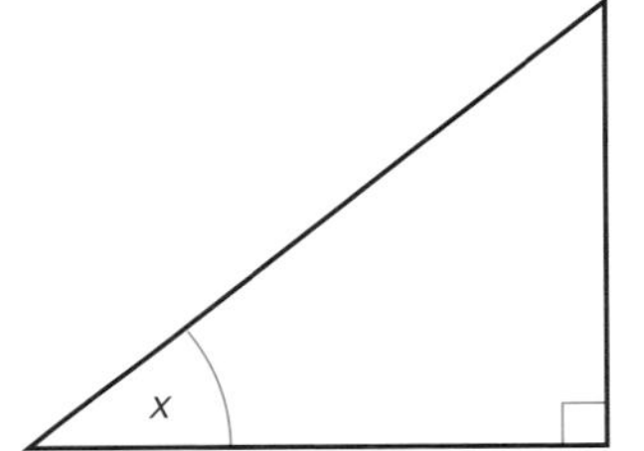

3

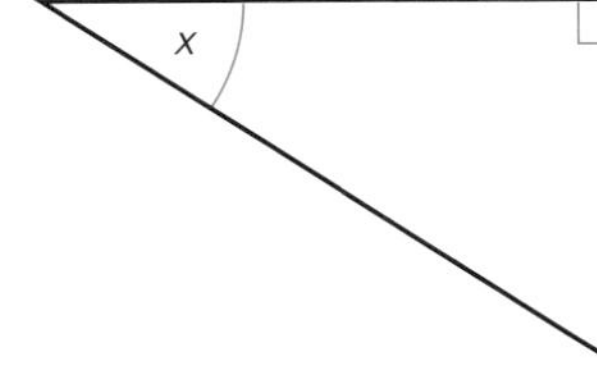

4

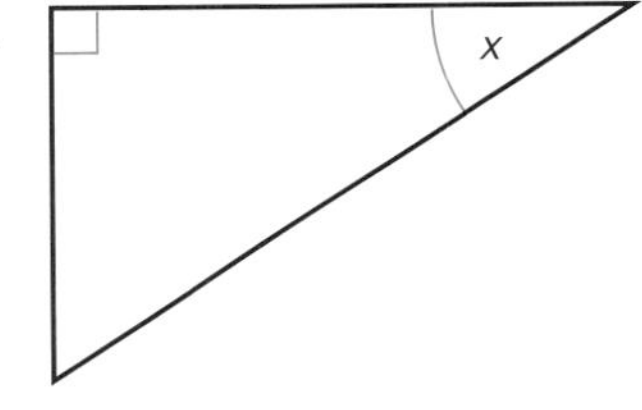

5

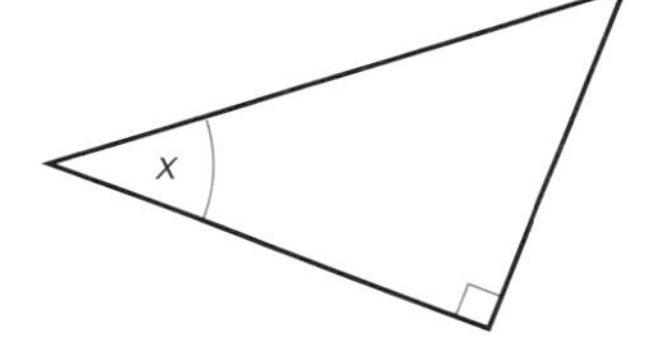

6

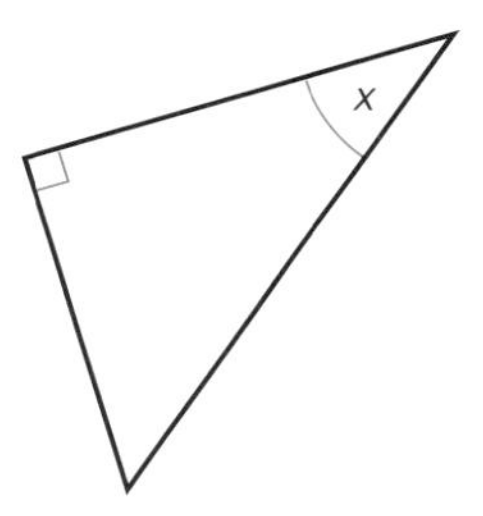

Tangent

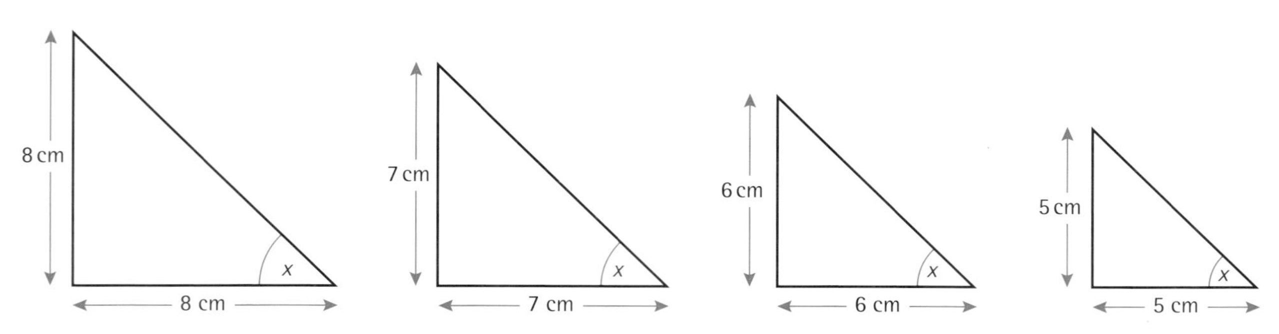

The opposite and adjacent sides are shown for four right angled triangles.

If you calculate the value of $\frac{\text{opposite side}}{\text{adjacent side}}$ for each triangle you will find that it is 1.

If you draw the triangle accurately and measure the angle x you will find that it is 45°.

When you enter **tan** 45 **=** into your calculator you will find that it also gives the answer of 1.

Change the angle to 30°, draw some right-angled triangles and calulate the ratio – you will get a different value (close to 0.6). Your calculator will give a much more accurate value of 0.577350...

This method that the tangent of an angle is the value of the ratio $\frac{\text{opposite side}}{\text{adjacent side}}$. This is usually shortened to tan $x = \frac{\text{opposite side}}{\text{adjacent side}}$.

This idea can be used to find an angle or a missing length.

You will sometimes have to work backwards.

For example if tan $x = 0.6$ and you want to find the angle, then you must use the inverse function. This may be labelled **SHIFT** or **2nd F** or **INV**

So to find the angle when tan $x = 0.6$ you press

You will see that it gives 30.96375.... = 31.0° (1 d.p.)

This calculation is sometimes written $\tan^{-1} 0.6 = 31.0°$

EXERCISE 21H

1 Use a calculator to work out the value of the following tangents correct to 4 s.f.

(a) tan 36° (b) tan 4° (c) tan 76° (d) tan 67.8°
(e) tan 29.5° (f) tan 1° (g) tan 89.5° (h) tan 90°

The answer to part (h) is written as ∞.

2 Use your calculator to find the value of each angle correct to 1 d.p.

(a) tan $a = 0.8$ (b) tan $b = 2$ (c) tan $c = 0.5$
(d) tan $d = 4.8$ (e) tan $e = 200$ (f) tan $f = 1$

We are now going to use this idea to find unknown angles and lengths.

EXAMPLE 5

In the triangle work out the size of angle x.

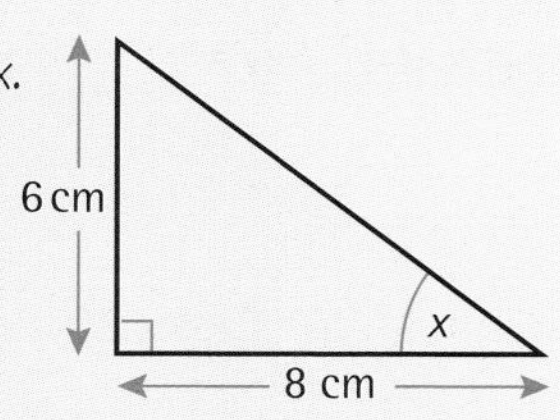

$$\tan x = \frac{\text{opposite}}{\text{adjacent}} = \frac{6}{8} = 0.75$$

$$x = \tan^{-1} 0.75$$
$$= 36.9° \text{ (1 d.p.)}$$

Give angle answers to 1 d.p.

EXERCISE 21I

Find the angles marked with a letter.

1

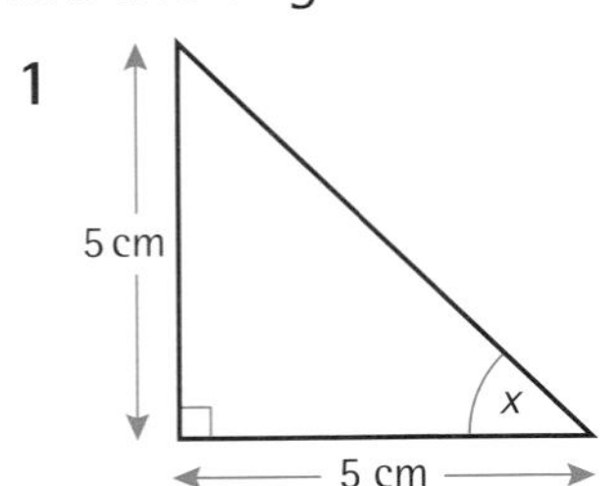

2

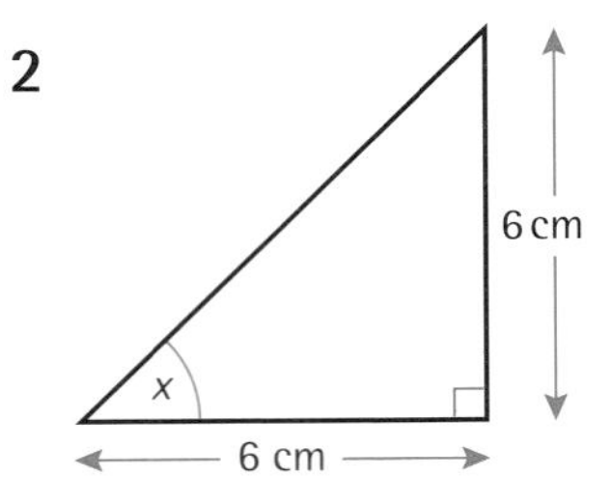

3

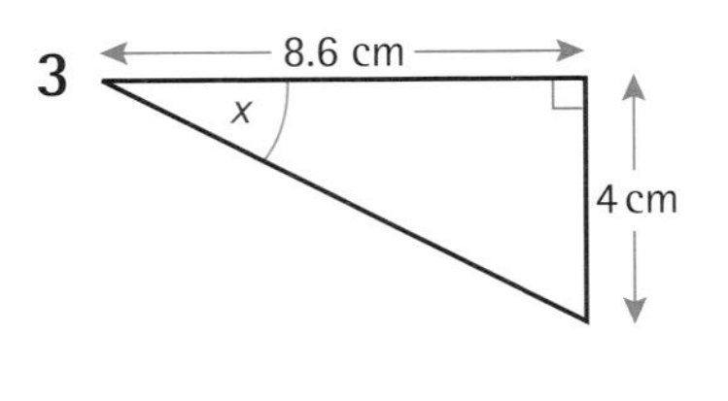

4

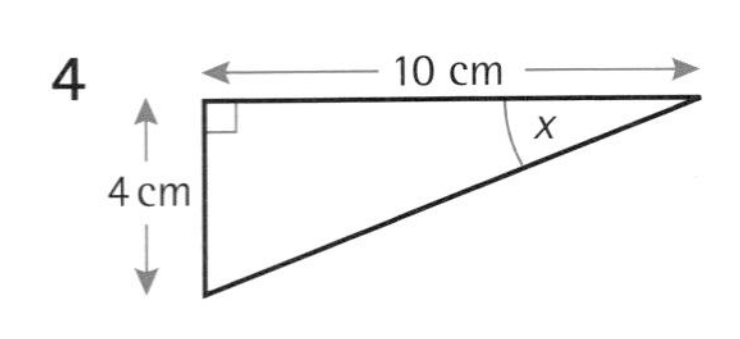

5

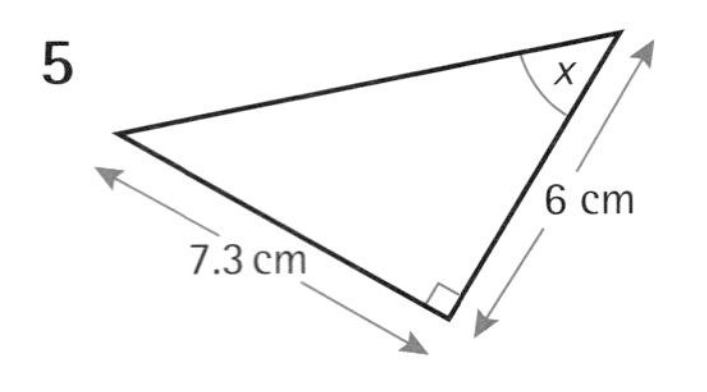

6

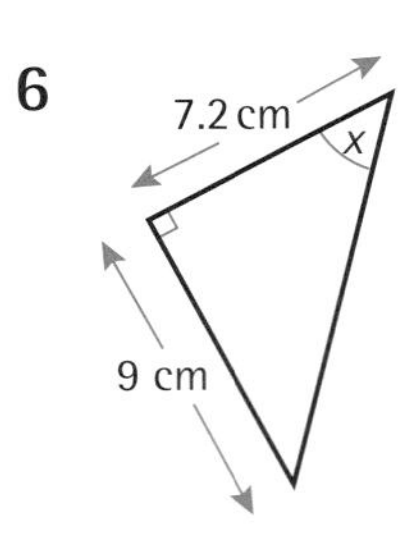

7

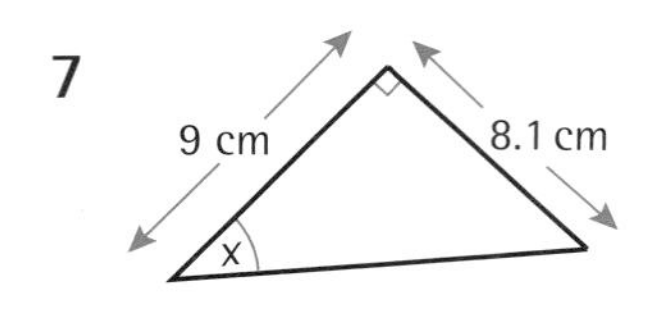

8

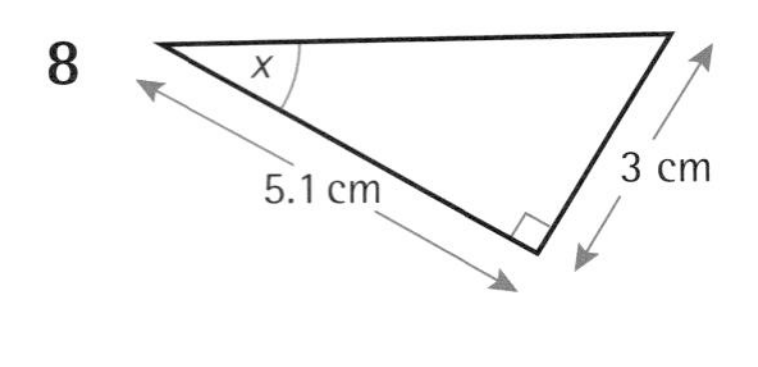

9

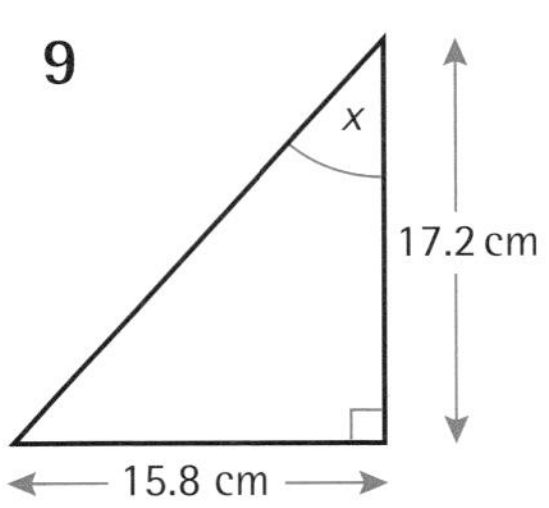

We are now going to use this idea in reverse to find the length of a side.

EXAMPLE 6

In the triangle work out the length of the side labelled y.

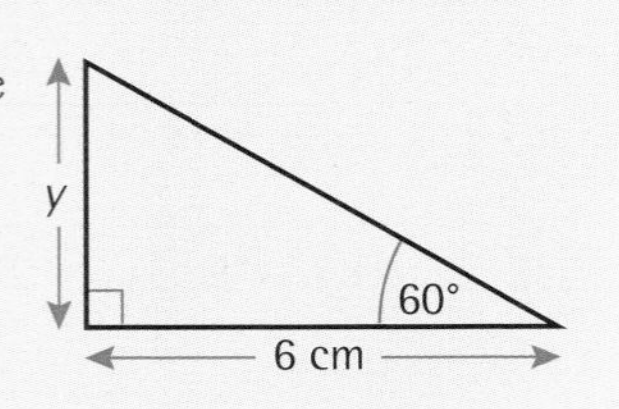

$$\tan 60 = \frac{\text{opposite}}{\text{adjacent}} = \frac{y}{6}$$

$$y = 6 \times 1.732$$

$$= 10.392\ldots = 10.4 \text{ cm (3 s.f.)}$$

Use your calculator to find tan 60°.
Always work to 4 s.f. or use your calculator value.
Give the final answer to 3 s.f.

EXERCISE 21J

Find the lengths marked with a letter.

1

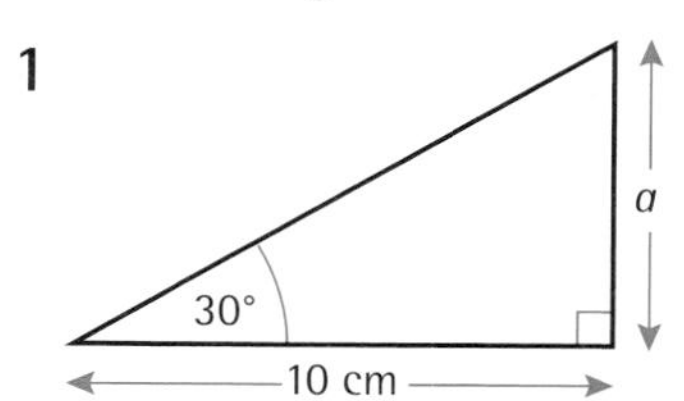

2

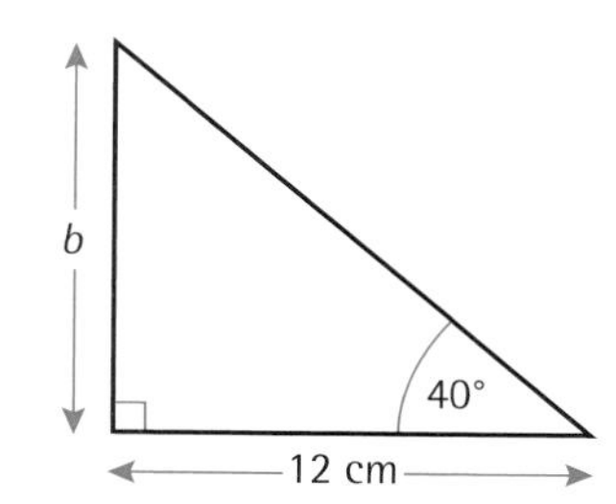

3

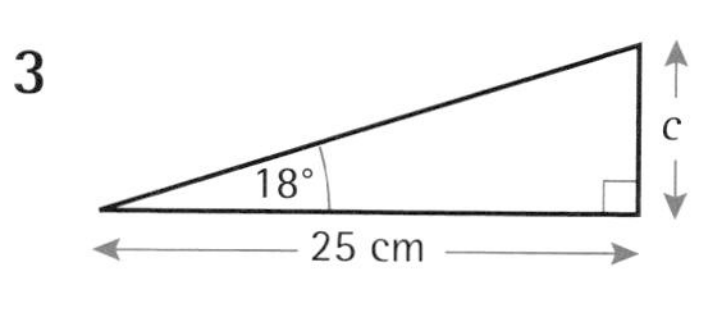

4

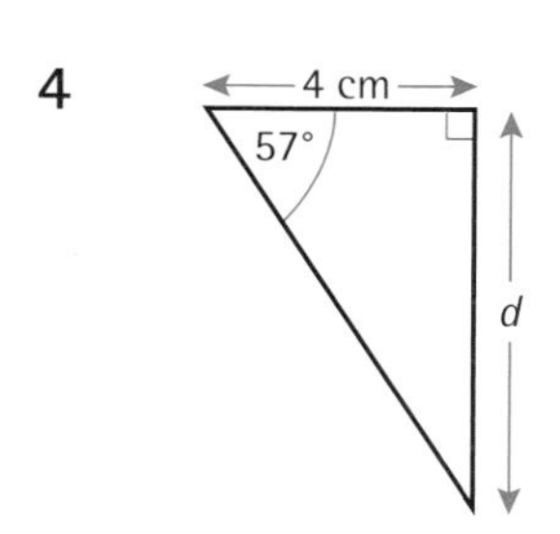

5

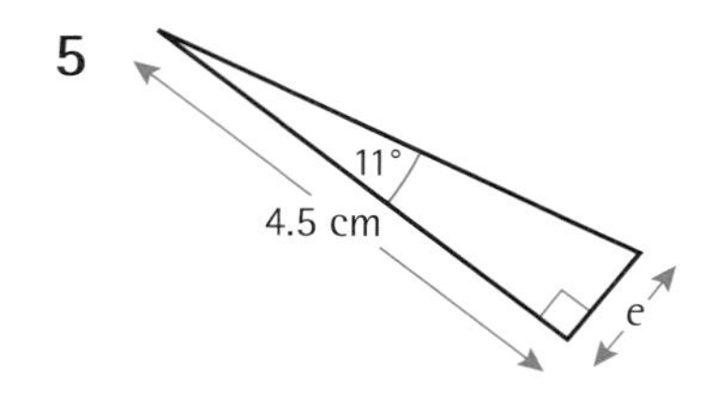

6

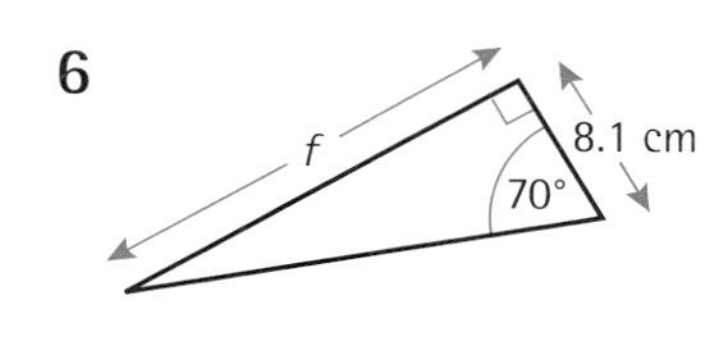

7

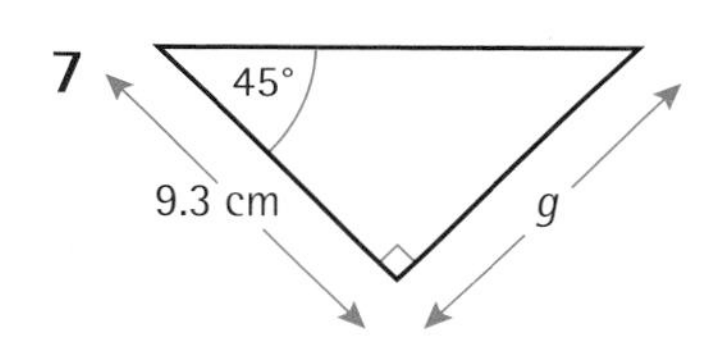

8

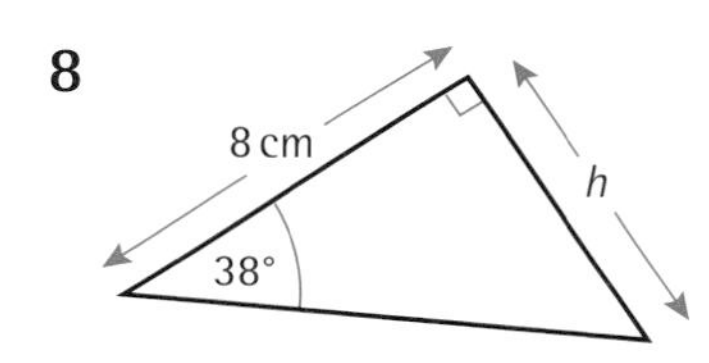

9

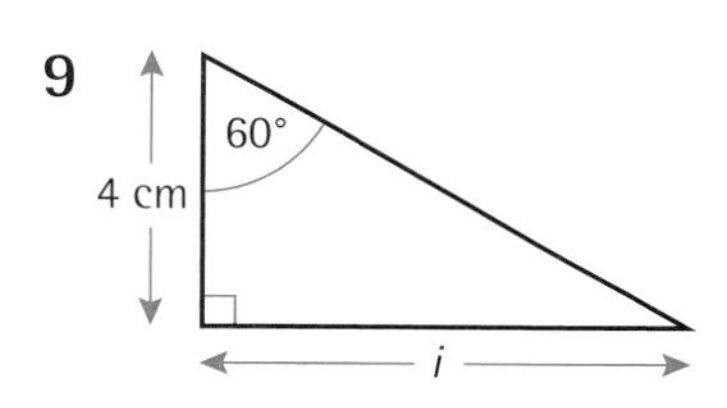

Sine

This ratio uses the opposite side and the hypotenuse but otherwise it works in the same way as the tangent did.

You would find that if you drew a right-angled triangle with an angle of 30° that the opposite side divided by the hypotenuse would give an answer of 0.5.

Putting 30 into your calculator and pressing 

would also give 0.5.

Your calculator may work differently to this.

So we can write

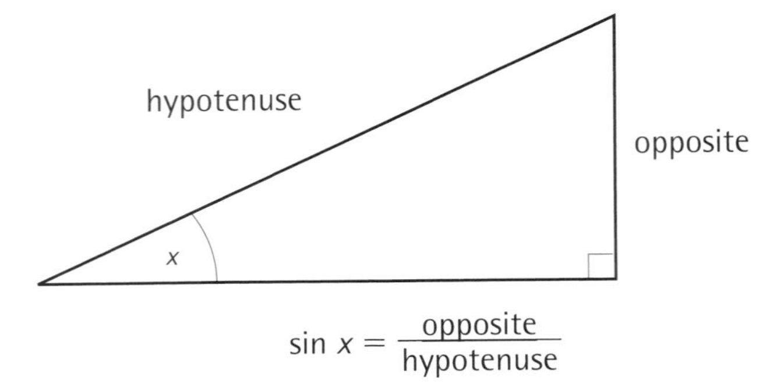

$$\sin x = \frac{\text{opposite}}{\text{hypotenuse}}$$

EXAMPLE 7

In the triangle work out the size of angle x.

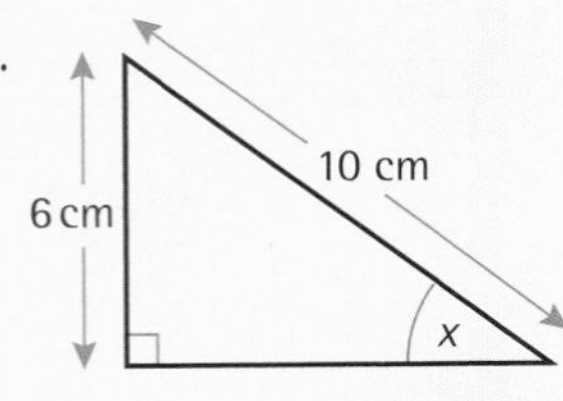

$\sin x = \frac{\text{opposite}}{\text{hypotenuse}} = \frac{6}{10} = 0.6$

$x = \sin^{-1} 0.6$

$= 36.9°$ (1 d.p.)

EXAMPLE 8

In the triangle work out the length of the side labelled y.

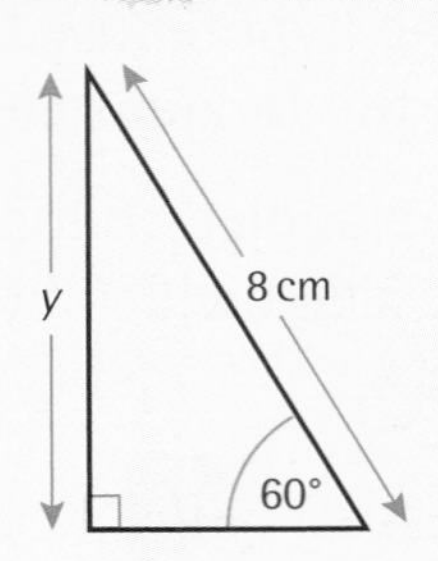

$\sin 60 = \frac{\text{opposite}}{\text{hypotenuse}} = \frac{y}{8}$

$0.8660 = \frac{y}{8}$

$y = 8 \times 0.8660$

$= 6.928\ldots = 6.93$ cm (3 s.f.)

Use your calculator to find sin 60°. Always work to 4 s.f. or use your calculator value.

EXERCISE 21K

1 Use a calculator to work out the value of the following sines correct to 4 s.f.

(a) sin 16° (b) sin 44° (c) sin 76° (d) sin 27.4°
(e) sin 79.5° (f) sin 2° (g) sin 89° (h) sin 90°

2 Use your calculator to find the value of each angle correct to 1 d.p.

(a) sin $a = 0.5$ (b) sin $b = 0.2$ (c) sin $c = 0.1617$
(d) sin $d = 0.8660$ (e) sin $e = 1$ (f) sin $f = 2$

You can't find an angle where the sine x is larger than one as the hypotenuse is the longest side.

3 Find the angles marked with a letter.

(a)

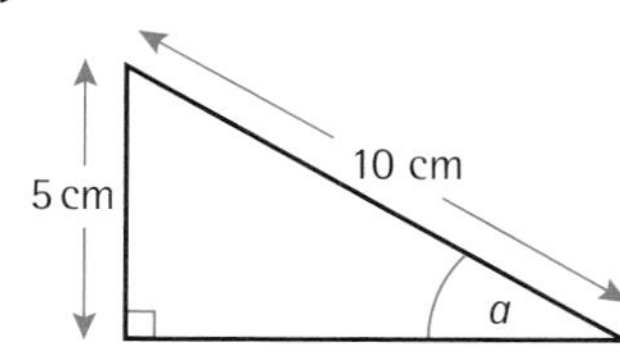

(b)

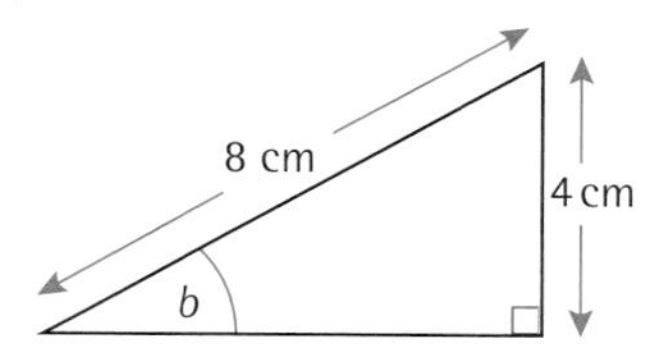

(c)

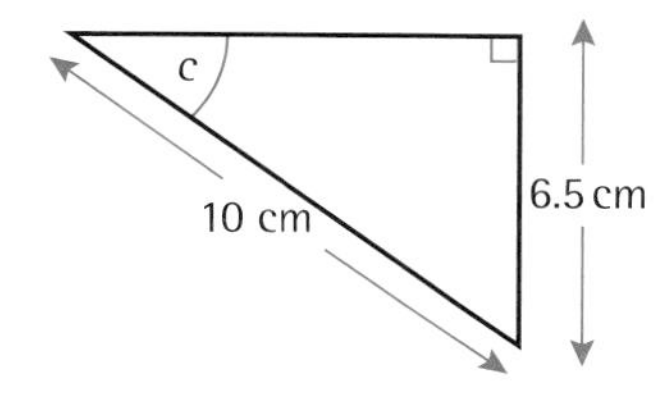

(d)

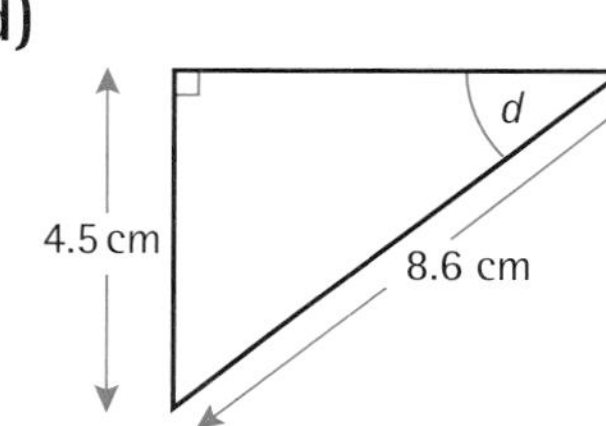

(e)

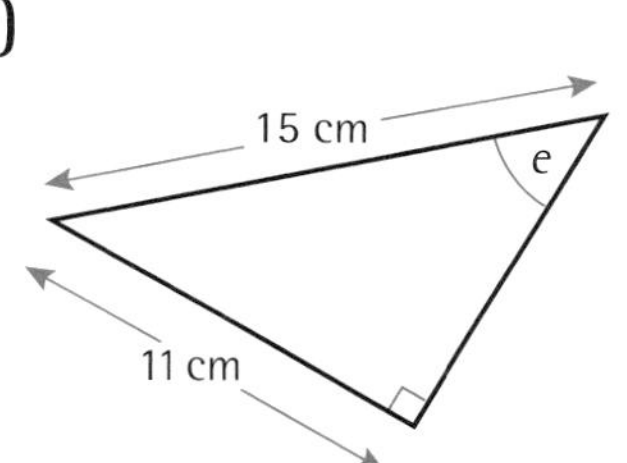

(f)

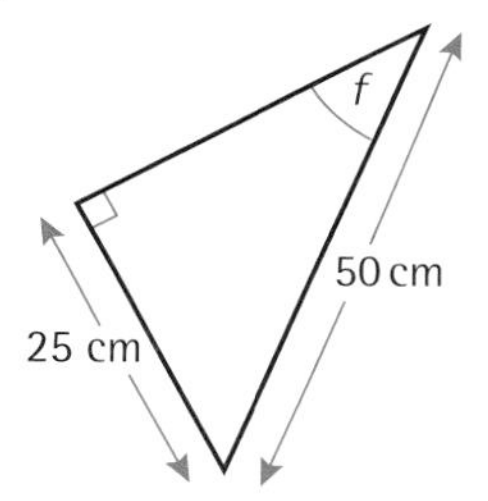

4 Find the lengths marked with a letter.

(a)

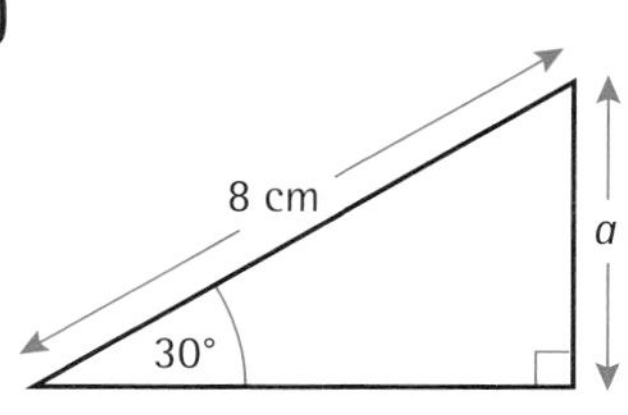

(b)

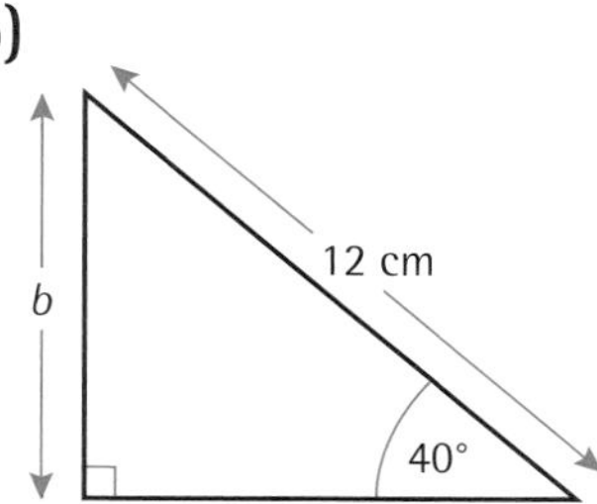

(c)

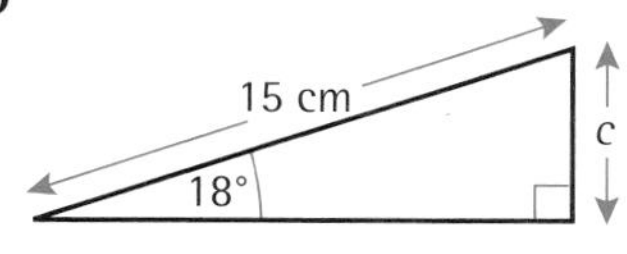

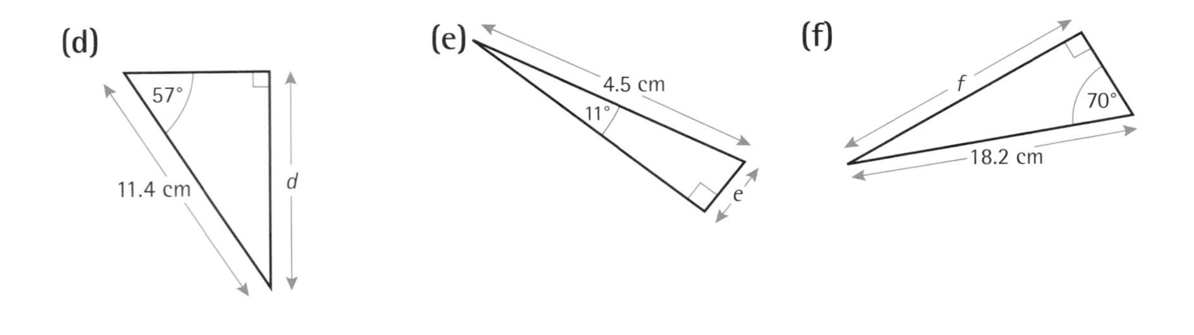

Cosine

Finally we have the cosine of x which uses the sides adjacent and hypotenuse.

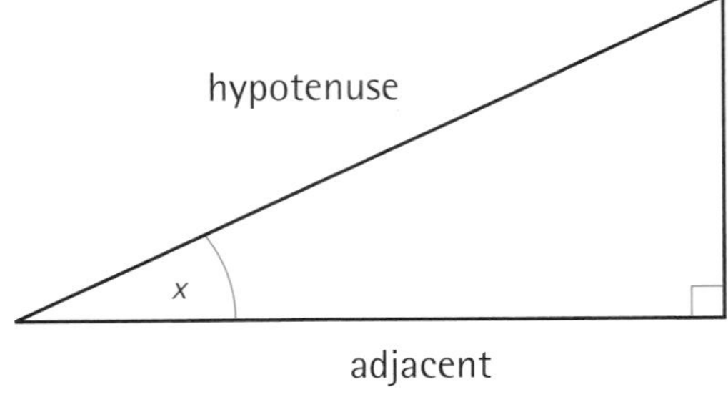

$$\cos x = \frac{\text{adjacent}}{\text{hypotenuse}}$$

EXAMPLE 9

In the triangle work out the size of angle x.

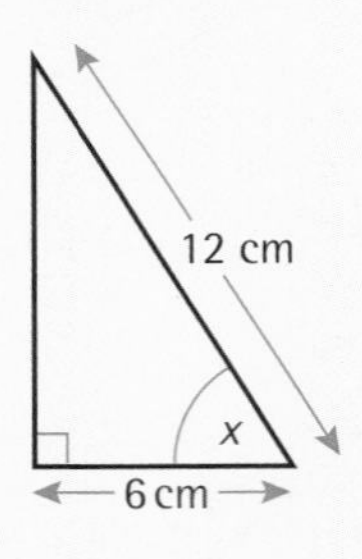

$$\cos x = \frac{\text{adjacent}}{\text{hypotenuse}} = \frac{6}{12} = 0.5$$

$$x = \cos^{-1} 0.5$$

$$= 60°$$

EXAMPLE 10

In the triangle work out the length of the side labelled *a*.

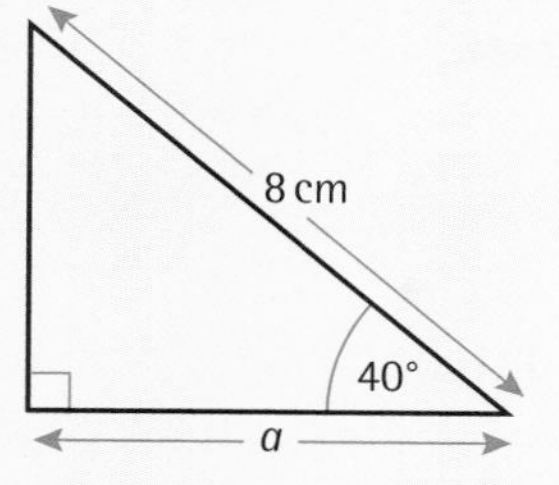

$$\cos 40 = \frac{\text{adjacent}}{\text{hypotenuse}} = \frac{a}{8}$$

$$0.7660 = \frac{a}{8}$$

$$a = 8 \times 0.7660$$

$$= 6.128\ldots = 6.13 \text{ cm (3 s.f.)}$$

Use your calculator to find cos 40°. Always work to 4 s.f. or use your calculator value.

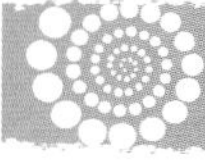

EXERCISE 21L

1 Use a calculator to work out the value of the following cosines correct to 4 s.f.

(a) cos 36° (b) cos 81° (c) cos 25° (d) cos 15.4°
(e) cos 0° (f) cos 66.2° (g) cos 89° (h) cos 90°

2 Use your calculator to find the value of each angle correct to 1 d.p.

(a) $\cos a = 0.5$ (b) $\cos b = 0.3$ (c) $\cos c = 0.5517$
(d) $\cos d = 0.8660$ (e) $\cos e = 1$ (f) $\cos f = 2$

You can't find an angle where cos x is larger than one as the hypotenuse is the longest side.

3 Find the angles marked with a letter.

(a)

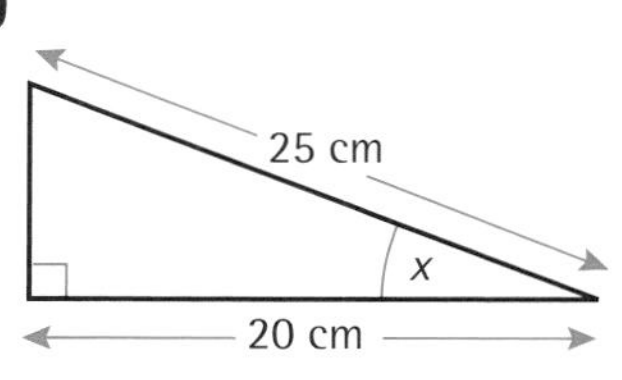

(b)

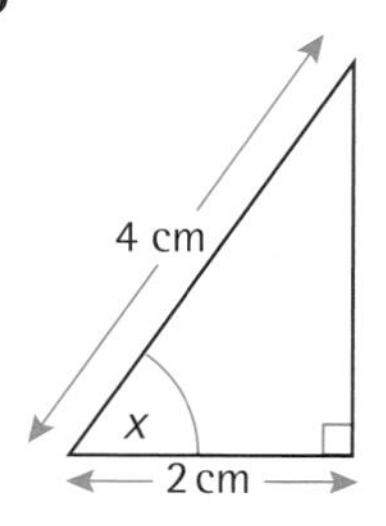

(c)

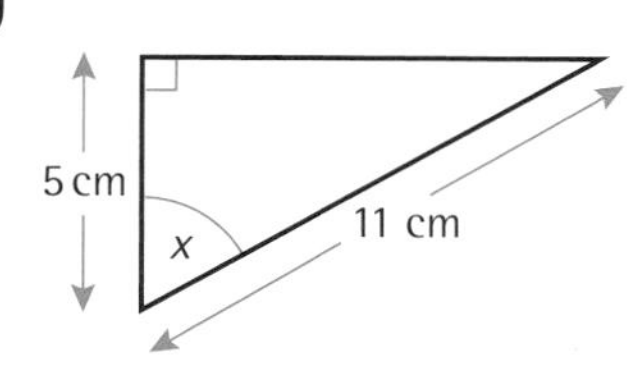

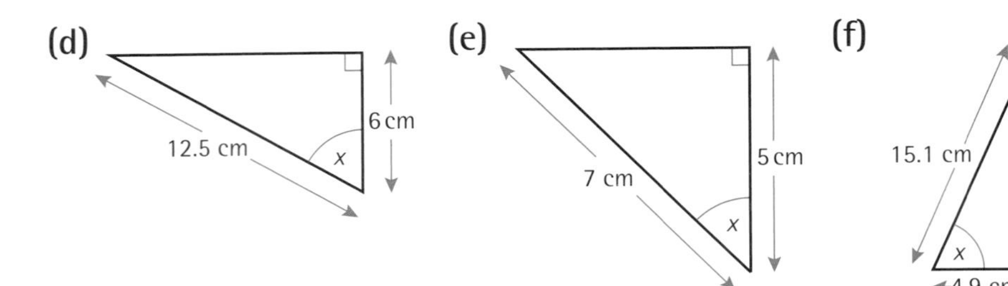

4 Find the lengths marked with a letter.

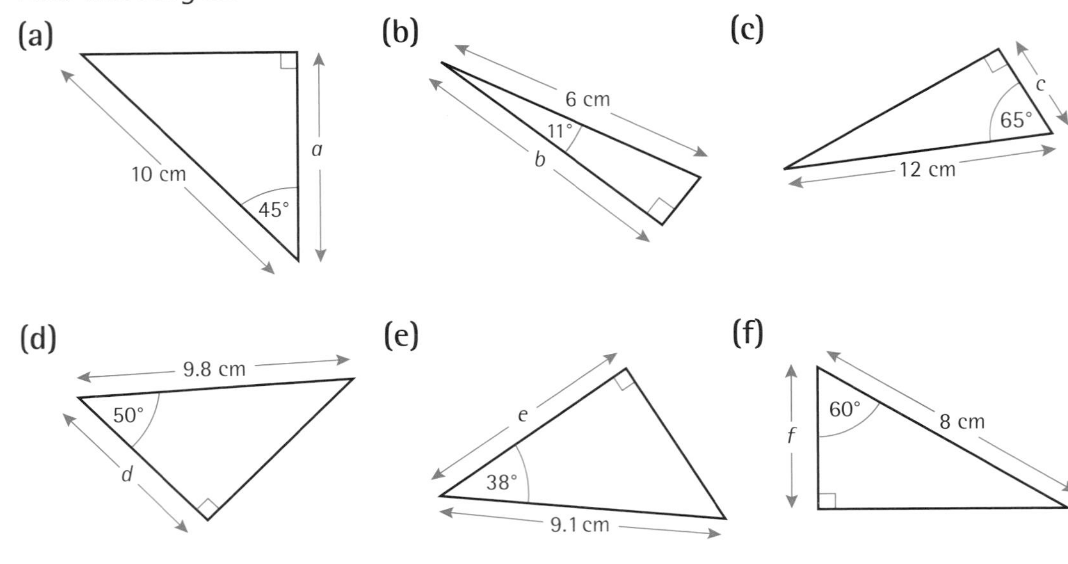

To summarise, we now have three trigonometric ratios.

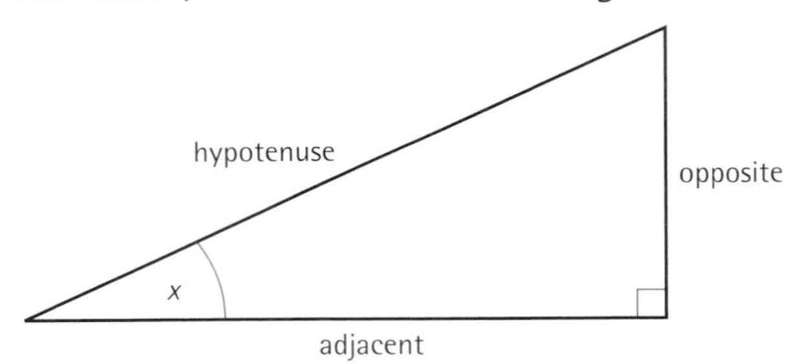

$\sin x = \dfrac{\text{opposite}}{\text{hypotenuse}}$ $\quad \cos x = \dfrac{\text{adjacent}}{\text{hypotenuse}}$ $\quad \tan x = \dfrac{\text{opposite}}{\text{adjacent}}$

This is sometimes written as

$\text{Sin } x = \dfrac{\text{O}}{\text{H}}$ $\qquad \text{Cos } x = \dfrac{\text{A}}{\text{H}}$ $\qquad \text{Tan } x = \dfrac{\text{O}}{\text{A}}$

to make the memory aid SOHCAHTOA.

Problem solving

We can now begin to use trigonometry to solve problems.

EXAMPLE 11

In the triangle PQR, $QR = 15$ cm, angle $RPQ = 90°$ and angle $PRQ = 50°$. Calculate the length of PR.

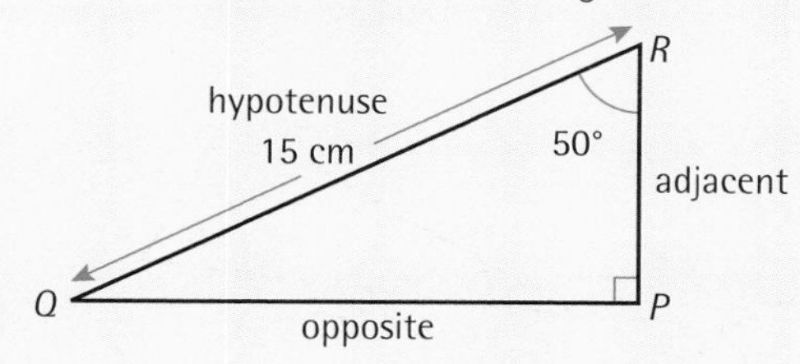

Start by labelling the sides in order of hypotenuse, opposite and then adjacent.

You can then see that the two sides in this problem are the hypotenuse (15 cm) and adjacent (PR).

This means that the ratio required is cosine.

$$\cos 50° = \frac{PR}{15}$$

$$PR = 15 \times \cos 50°$$

$$PR = 9.6418\ldots \text{ cm}$$

$$= 9.64 \text{ cm (3s.f.)}$$

EXERCISE 21M

1 Calculate the unknown length marked on each diagram to 1 d.p. (all lengths are in cm).

(a)

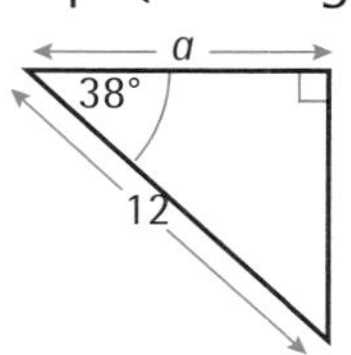

(b)

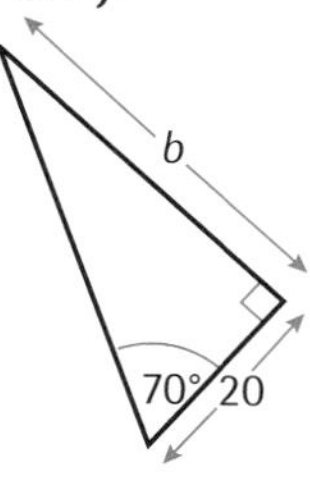

(c)

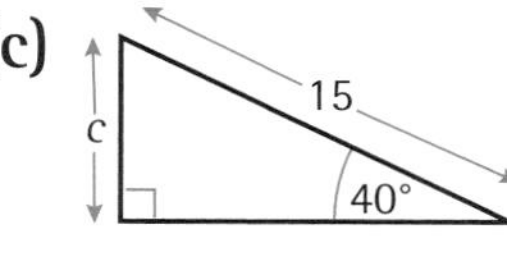

(d)

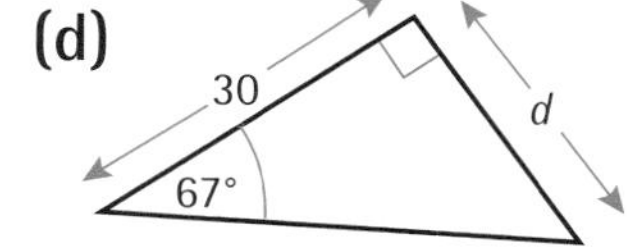

(e)

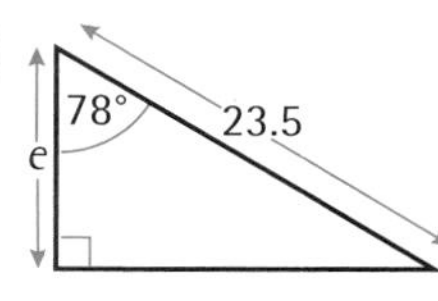

(f)

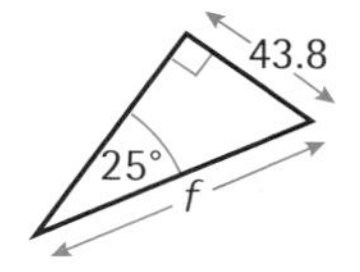

(g)

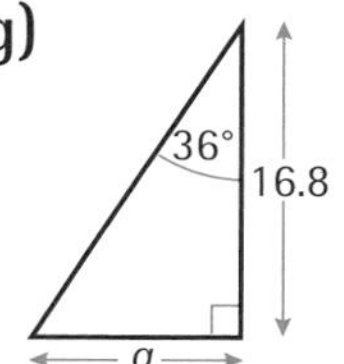

(h)

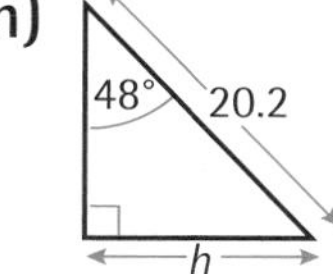

EXAMINATION QUESTIONS

1

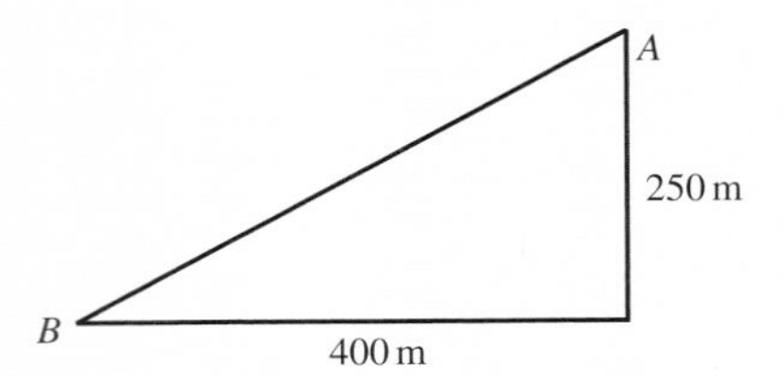

NOT TO SCALE

Renata rows her boat directly from A to B.
The point B is 250 m south and 400 m west of the point A.
Calculate

(a) how far Renata rows, [2]

(b) the bearing on which she rows. [3]

(CIE Paper 1, Jun 2000)

2

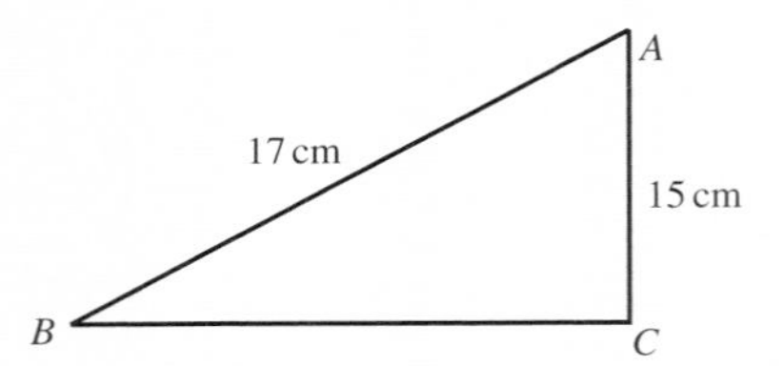

NOT TO SCALE

In triangle ABC, $AB = 17$ cm, $AC = 15$ cm and angle $ACB = 90°$.
Calculate the length of BC.

(CIE Paper 1, Nov 2000)

3

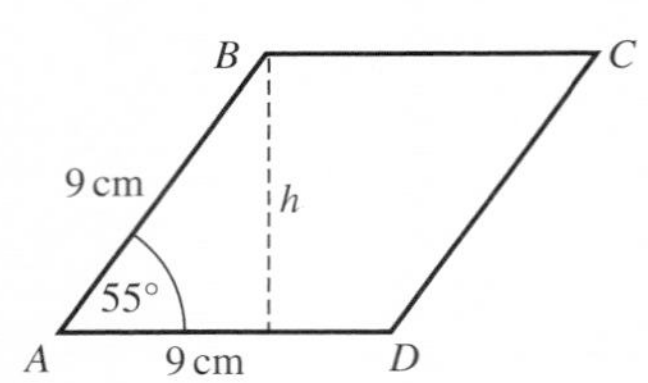

NOT TO SCALE

The diagram shows a rhombus with sides of length 9 cm. Angle $BAD = 55°$.
Calculate

(a) h, the height of the rhombus, [2]

(b) the area of the rhombus. [1]

(CIE Paper 3, Nov 2000)

4

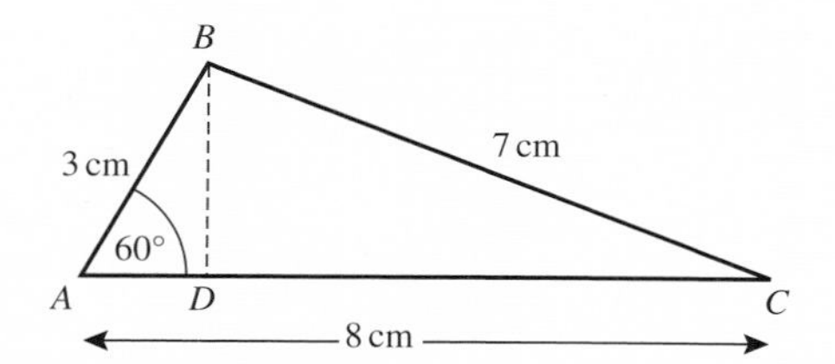

NOT TO SCALE

In triangle ABC, $AB = 3$cm, $BC = 7$cm, $AC = 6$cm and angle $A = 60°$.
BD is perpendicular to to AC.
Calculate

(a) the length of AD, [2]
(b) the length of DC, [1]
(c) the size of angle C. [2]

(CIE Paper 1, Nov 2000)

---

5 (a) Write down the value of tan 63.5°, correct to three decimal places. [1]
(b) Write down the angle whose cosine is 0.25, correctc to the nearest tenth of a degree. [1]

(CIE Paper 1, Jun 2001)

6

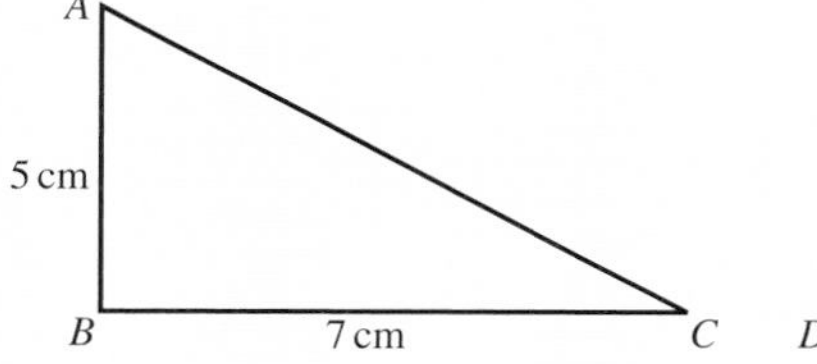

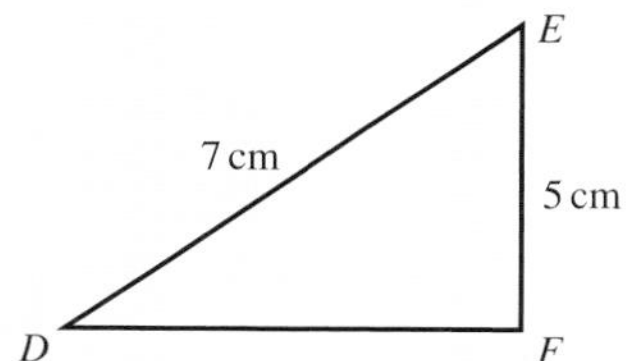

NOT TO SCALE

The diagram shows two different right-angled triangles, each with sides of 7 cm and 5 cm.

(a) Work out the length of AC. [2]
(b) Work out angle ACB. [2]
(c) Work out the length of DF. [2]
(d) Which triangle has the greater perimeter? [1]

(CIE Paper 3, Jun 2001)

7

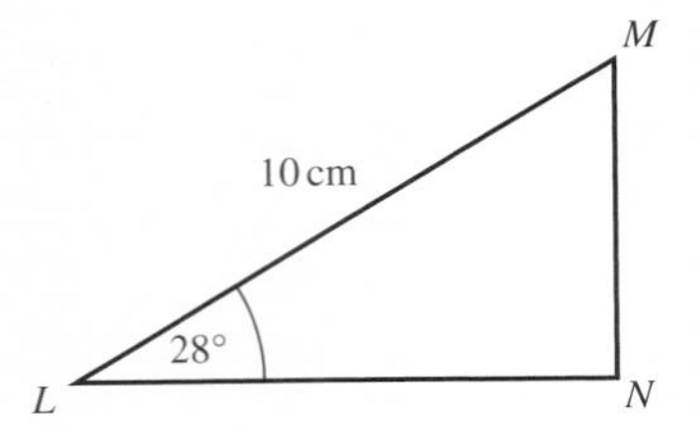

NOT TO SCALE

In triangle LMN, angle $LNM = 90°$, angle $MLN = 28°$ and $LM = 10$cm.
Calculate

(a) MN, [2]

(b) LN, [2]

(c) the area of triangle LMN. [2]

(CIE Paper 3, Nov 2002)

8

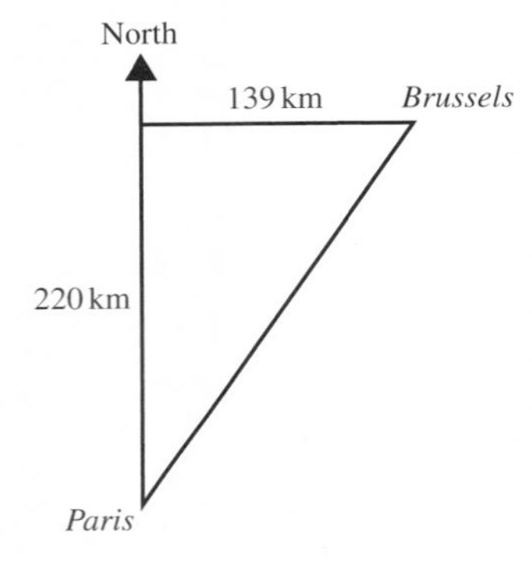

NOT TO SCALE

Brussels is 220km north and 139km east of Paris.
Calculate the bearing of Brussels from Paris, to the nearest degree. [3]

(CIE Paper 1, Nov 2002)

9

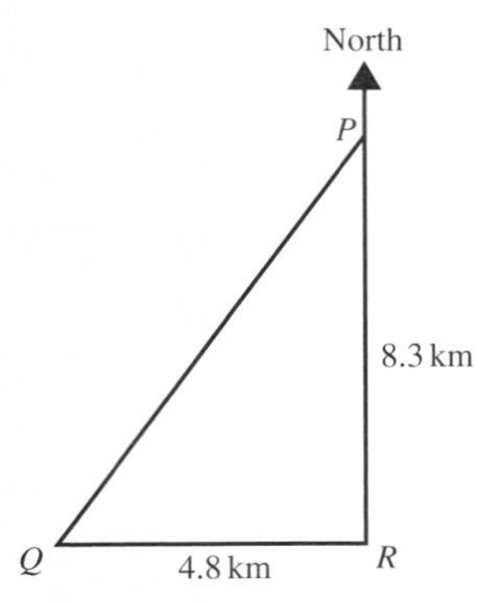

NOT TO SCALE

A straight road between P and Q is shown in the diagram.
R is the point south of P and east of Q.
$PR = 8.3$ km and $QR = 4.8$ km.

Calculate

(a) the length of the road PQ, [2]

(b) the bearing of Q from P. [3]

(CIE Paper 1, Jun 2003)

10

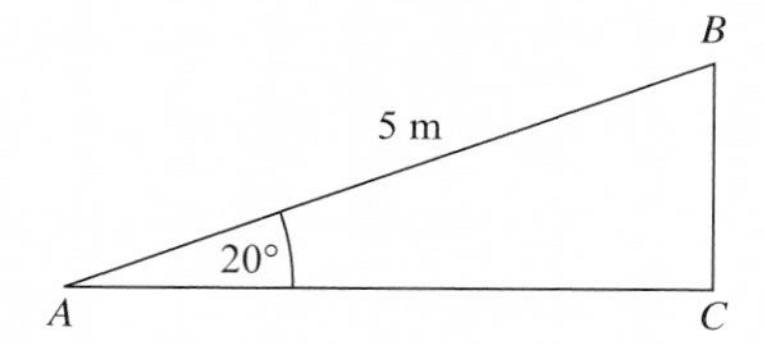

NOT TO SCALE

The diagram shows a right-angled triangle ABC with $AB = 5$m and angle $BAC = 20°$.
Calculate the length BC. [2]

(CIE Paper 1, Nov 2003)

11

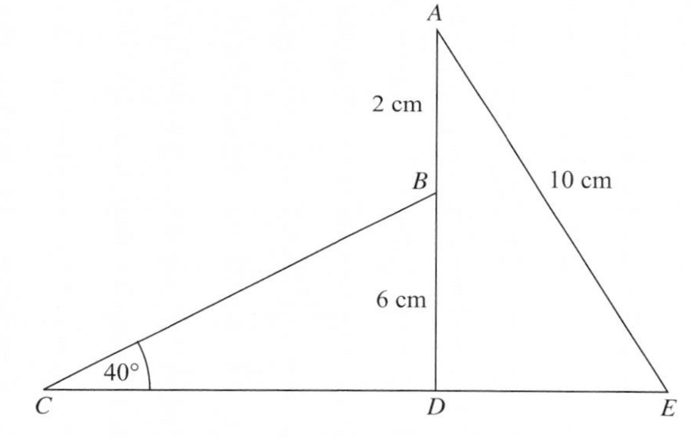

NOT TO SCALE

On the above diagram, $AB = 2$ cm, $BD = 6$ cm, $AE = 10$ cm, angle $BCD = 40°$ and angle $BDE = 90°$.

(a) Write down the length of AD. [1]

(b) Calculate the length of DE. [2]

(c) Calculate the size of angle AED. [2]

(d) Calculate the length of CD. [3]

(CIE Paper 3, Nov 2004)

Index